Myconanotechnology
Emerging Trends and Applications

Editors

Mahendra Rai
Department of Biotechnology
SGB Amravati University
Amravati, Maharashtra, India

Patrycja Golińska
Department of Microbiology
Nicolaus Copernicus University
Torun, Poland

CRC Press
Taylor & Francis Group
Boca Raton   London   New York

CRC Press is an imprint of the
Taylor & Francis Group, an **informa** business

A SCIENCE PUBLISHERS BOOK

Cover credit:
Step-wise illustration of liposome-based smart drug delivery system for cancer therapy (adapted from Hossen et al. 2019 under Creative Common Attribution (CC By-NC-ND 4.0 https://creativecommons.org/licenses/by-nc-nd/4.0/))

First edition published 2023
by CRC Press
6000 Broken Sound Parkway NW, Suite 300, Boca Raton, FL 33487-2742

and by CRC Press
4 Park Square, Milton Park, Abingdon, Oxon, OX14 4RN

© 2023 Mahendra Rai and Patrycja Golińska

CRC Press is an imprint of Taylor & Francis Group, LLC

Library of Congress Cataloging-in-Publication Data (applied for)

ISBN: 978-1-032-35540-5 (hbk)
ISBN: 978-1-032-35543-6 (pbk)
ISBN: 978-1-003-32735-6 (ebk)

DOI: 10.1201/9781003327356

Typeset in Times New Roman
by Radiant Productions

Foreword

The term myconanotechnology was first defined by Rai et al. to describe the role of fungi in the creation and exploitation of materials in the 1–100 nm size range (Rai et al. 2009). Although myconanotechnology is often perceived as a "new" development, the fact that various fungi can produce metallic microparticles has been known for many years. A frequently cited example of this was the description of the synthesis of nanometre scale crystallites of cadmium sulfide by *Candida glabrata* and *Schizosaccharomyces pombe* (Dameron et al. 1989). Since then production of metallic nanoparticles has been demonstrated in a wide range of fungi, particularly among conidial ascomycetes including species of *Penicillium*, *Aspergillus*, *Trichoderma*, *Verticillium*, *Fusarium* and *Guignardia* (Alghuthaymi et al. 2015, Ajee et al. 2021, Rai et al. 2021, Balakumaran et al. 2015). This capability does not appear to be limited to any particular fungal group and the production of metallic nanoparticles has also been reported in other fungal groups including species of *Mucor*, *Rhizoctonia*, *Trametes* and *Pleurotus* (Aziz et al. 2016, Raudabaugh et al. 2013, Gudikandula et al. 2020, Chaturvedi et al. 2020). In recent years, considerable research has been undertaken into the potential uses for biological nanoparticles as well as the optimisation of their production and delivery (Salata 2004, Rai et al. 2009, Li et al. 2011, Moghaddam et al. 2015, Guilger-Casagrande and de Lima 2019).

Looking at the breadth of work now available on fungal derived nanoparticles, it is important to consider how much progress has been made in a relatively short time. Simple searches of Google Scholar indicate that since 2009 the term "myconanotechnology" has been included in the text of around 350 publications, the terms "mycosynthesis" and "nanoparticle" appear together in over 2000 publications and the terms "fungal-derived" and "nanoparticle" appear in over 18,000 publications (searches carried out in October 2021). These figures will undoubtedly include some duplications and cross-citations and do not provide any information on context. They do however illustrate that a considerable body of work has been generated in the area in the last 12 years.

In this volume the editors have brought together a comprehensive collection of reviews that provide extensive information on the synthesis, characterisation and application of fungal nanoparticles. The chapters focussing on the use of fungi in the production of nanoparticles demonstrate the wide range of metallic and metalloid compounds that have been synthesised and separated to date. As more methods are developed it would seem likely that this list will increase. Fungi may also play an important role in providing metabolites for external coatings for nanoparticles, such as chitosan and microbial proteins. Examples of some of the characteristics and mechanisms of these have also been included.

Apart from synthesising and characterising fungal nanoparticles, a major research area in recent years has been their application in practical systems. Metallic nanoparticles produced by both inorganic and organic processes have shown to have many potential applications. Probably the most widespread application is in pharmacology and this is represented here with chapters describing various antimicrobial activities. They have also been of interest as potential agents for targeting cancerous tumours, and as other pharmaceutical products, such as drug delivery systems and vaccine adjuvants. Outside of the medical environment, nanoparticles are becoming significant in many other "delivery" systems. This is apparent from the large number of chapters included here that demonstrate such applications. These showcase a wide range of possibilities for the use of fungal derived nanoparticles in the agriculture and food environments with their potential uses as pesticides, for the delivery of micronutrients and as contributors in food preservation.

One of the features of myconanotechnology that has been raised by many authors in this volume, and elsewhere, is the potential environmental benefit in both the production and use of fungal derived nanoparticles. I believe that this will undoubtedly play an important part in their adoption in many processes in the future. The Greek orator Demosthenes said that "small opportunities are often the beginning of great enterprises", and so as new potential applications for nanotechnology continue to be described, it is hoped that fungal derived nanoparticles can continue to be utilised in this fast expanding discipline.

References

Ajee, R.S., Sharma, N., Tomar, R.S. and Kaushik, S. 2021. Myconanoparticles synthesis and their applications in the agriculture sector. *In*: Anwar Mallick, M., Manoj Kumar Solanki, Baby Kumari and Suresh Kumar Verma (eds.). Nanotechnology in Sustainable Agriculture. CRC Press, Boca Raton. eBook ISBN 9780429352003.

Alghuthaymi, M.A., Almoammar, H., Rai, M., Ernest Said-Galiev, E and Abd-Elsalam, K.A. 2015. Myconanoparticles: Synthesis and their role in phytopathogens management. Biotechnology & Biotechnological Equipment 29(2): 221–236.

Aziz, N., Pandey, R., Barman, I. and Prasad, R. 2016. Leveraging the attributes of *Mucor hiemalis*-derived silver nanoparticles for a synergistic broad-spectrum antimicrobial platform. Frontiers in Microbiology 7: 1984.

Balakumaran, M.D., Ramachandran, R. and Kalaichelvan, P.T. 2015. Exploitation of endophytic fungus *Guignardia mangiferae* for extracellular synthesis of silver nanoparticles and their *in vitro* biological activities. Microbiological Research 178: 9–17.

Chaturvedi, V.K., Yadav, N., Rai, N.K., Ellah, N.H.A., Bohara, R.A., Rehan, I.F., Marraiki, N., Batiha, G.E., Hetta, H.F. and Singh, M.P. 2020. *Pleurotus sajor-caju*-mediated synthesis of silver and gold nanoparticles active against colon cancer cell lines: A new era of herbonanoceutics. Molecules 25(13): 3091.

Dameron, C.T., Reese, R.N., Mehra, R.K., Kortan, A.R., Carroll, P.J., Steigerwald, M.L., Brus, L.E. and Winge, D.R. 1989. Biosynthesis of cadmium sulphide quantum semiconductor crystallites. Nature 338: 596–597 .

Gudikandula, K., Jaffar, S. and Charya, M.A.S. 2020. Biogenic silver nanoparticles from *Trametes ljubarskyi* (White Rot Fungus): Efficient and effective anticandidal activity. pp. 405–414. *In*:

Khasim, S.M., Long, C., Thammasiri, K. and Lutken, H. (eds.). Medicinal Plants: Biodiversity, Sustainable Utilization and Conservation. Springer, Singapore. https://doi.org/10.1007/978-981-15-1636-8_23.

Guilger-Casagrande, M. and De Lima, R. 2019. Synthesis of silver nanoparticles mediated by fungi: Areview. Frontiers in Bioengineering and Biotechnology 7: 287.

Li, X., Xu, H., Chen, Z-S. and Chen, G. 2011. Biosynthesis of nanoparticles by microorganisms and their applications. Journal of Nanomaterials 2011, Article ID 270974.

Moghaddam, A.B., Namvar, F., Moniri, M., Md. Tahir, P., Azizi, S. and Mohamad, R. 2015. Nanoparticles biosynthesized by fungi and yeast: A review of their preparation, properties, and medical applications. Molecules 20: 16540–16565.

Rai, M., Bonde, S., Golinska, P., Trzcińska-Wencel, J., Gade, A., Abd-Elsalam, K.A., Shende, S., Gaikwad, S. and Ingle, A.P. 2021. *Fusarium* as a novel fungus for the synthesis of nanoparticles: Mechanism and applications. Journal of Fungi 7(2): 139.

Rai, M., Yadav, A., Bridge, P.D. and Gade, A. 2009. Myconanotechnology: A new and emerging science. pp. 258–267. *In*: Rai, M. and Bridge, P.D. (eds.). Applied Mycology. CABI Publishing, Wallingford, UK.

Raudabaugh, D., Tzolov, M., Calabrese, J. and Overton, B. 2013. Synthesis of silver nanoparticles by a bryophilous *Rhizoctonia* species. Nanomaterials and Nanotechnology 3. Art. 2.

Salata, O. 2004. Applications of nanoparticles in biology and medicine. Journal of Nanobiotechnology 2: 3.

Paul Bridge
Former Director of Bioservices at CABI
Head of Evolutionary Biology Group at British
Antarctic Survey, and Kew Chair of Mycology at Birkbeck, London

Preface

Myconanotechnology represents the use of fungi for the synthesis of nanoparticles in the size range of 1–100 nm and the application of these nanoparticles in medicine, food, agriculture, veterinary, environment, and textiles. The term 'Myconanotechnology' was coined for the first time by Rai and his collaborators in 2009. Science is emerging at a fast pace and has garnered the attention of nanotechnologists, mycologists, biomedical experts, and agriculture scientists, among others. Over the last decade, there has been tremendous progress in this field owing to its wider and more effective applications.

The present book contains valuable chapters written by professionals and experts on the diverse aspects of myconanotechnology. The chapters have been divided into four sections, Section I introduces myconanotechnology and discusses the opportunities, emerging trends and constraints; Section II covers various biomedical applications including myconanotechnology for cancer, mycogenic nanoparticles as novel antimicrobial agents and formulations, myconanotechnology in vaccine adjuvants, challenges for crises to come, antioxidant activity of mycosynthesized nanoparticles, and fungal biofilms in nanotechnology era: diagnosis, drug delivery and treatment strategies; Section III discusses the role of myconanotechnology in agriculture, mainly strategic role in control of plant pathogens, as nanofertilizers for plant growth promotion, in food preservation and enhancement of shelf-life of Agri-food and fruits, and application for the control of insect pests; Section IV includes applications of mycogenic zinc oxide nanoparticles in medicine, agriculture, water management, textiles, etc., and a separate chapter has been dedicated to antimicrobial textiles.

This book is interdisciplinary and caters to the needs of postgraduate and research students of fungal biology, microbiology, chemistry, nanotechnology, biotechnology, pharmacology, and those who are interested in green nanotechnology. It will also enrich academicians and researchers in pharma industry.

We would like to thank the authors for their excellent work in providing cutting-edge information on the subject of their respective chapters. Their efforts will certainly enhance and update the readers' knowledge of myconanotechnology and its multiple applications in different fields. Further, we would like to express our sincere thanks to the publishers and authors of the chapters whose research work has been mentioned in the book.

Finally, we thankfully acknowledge the financial support rendered by the Polish National Agency for Academic Exchange (NAWA) (Project No. PPN/ULM/2019/1/00117/A/DRAFT/00001) Department of Microbiology, Nicolaus Copernicus University, Toruń, Poland.

Mahendra Rai, India
Patrycja Golińska, Poland

Contents

Section I
Introduction

1

Myconanotechnology:
Opportunities and Challenges

Mahendra Rai,[1,2,]* *Magdalena Wypij,*[2] *Joanna Trzcińska-Wencel,*[2]
Alka Yadav,[1] *Avinash P. Ingle,*[3] *Graciela Dolores Avila-Quezada*[4]
and *Patrycja Golińska*[2,]*

Introduction

Nanotechnology is the synthesis, manipulation, and application of materials at the nanoscale level. Nanoparticles are metal particles in the size range of 1–100 nm, that are synthesized using different methods and are employed in a wide number of applications (Rai et al. 2016). Metal nanoparticles can be synthesized using physical and chemical methods but it also involves the use of biological entities like micro-organisms, plants, enzymes, and proteins for the synthesis process (Rai et al. 2021a). The use of physical and chemical methods for the synthesis yields many toxic by-products which can be harmful to the environment and human body, so researchers have looked up to the use of biological agents for the synthesis of metal nanoparticles. Bionanotechnology can be used for the synthesis of nanoparticles through biological agents (Rai et al. 2021a). Employing the greener approach for the fabrication of metal nanoparticles yields clean, non-toxic, cost-effective, and environmentally friendly synthesis of nanoparticles (Adebayo et al. 2021). The field of bionanotechnology is rapidly enhancing due to the efficient synthesis of nanoparticles and their application in diverse fields. In recent years, the synthesis of nanoparticles using micro-organisms like bacteria, fungi, algae, yeast, and actinomycetes have gained immense attention

[1] Nanobiotechnology Laboratory, Department of Biotechnology, Sant Gadge Baba Amravati University, Amravati, Maharashtra, India.
[2] Department of Microbiology, Nicolaus Copernicus University, Torun, Poland.
[3] Biotechnology Centre, Department of Agricultural Botany, Dr. Panjabrao Deshmukh Krishi Vidyapeeth, Akola, India.
[4] Facultad de Ciencias Agrotecnologicas, Universidad Autonoma de Chihuahua, Escorza 900, Chihuahua 31000, Mexico.
* Corresponding authors: mahendrarai@sgbau.ac.in; golinska@umk.pl

(Guilger-Casagrande and de Lima 2019). The biologically synthesized nanoparticles show the presence of capping agents that are obtained from the biomolecules, produced by the biological entity used for the synthesis process. These capping agents provide stability to the biosynthesized nanoparticles (Rai et al. 2009a).

The term "Myconantechnology" deals with the synthesis of metal nanoparticles using fungi either intracellularly or extracellularly (Rai et al. 2009b, Yadav et al. 2015). Thus, myconanotechnology can be defined as an interface between nanotechnology and mycology. Fungi show the presence of reducing enzymes and proteins that help in the faster reduction of metal ions to metal nanoparticles. Also, the fungal cell wall shows active absorption of metal ions and subsequent synthesis of nanoparticles, affirming it as an efficient tool for extracellular and intracellular synthesis of nanoparticles (Garcia-Rubio et al. 2020). Consequently, the fungal system serves as a proficient mode of synthesis of nanoparticles and provides a great opportunity for researchers around the world to employ this system for the synthesis and application of metal nanoparticles (Li et al. 2021). The biosynthesized nanoparticles using fungi offer applications in biomedicine, agriculture, antimicrobial agents, textiles, etc. However, apart from the applications offered by the mycosynthesized nanoparticles, studies on the toxicity effects of nanoparticles and the accumulation of nanoparticles in the environment and their subsequent effect on the ecosystem should also be studied (Garcia-Rubio et al. 2020).

The present chapter establishes myconanotechnology as a field that utilizes fungi as a novel system for the synthesis of metal nanoparticles. The diverse applications offered by fungi-assisted-metal nanoparticles have been explained in detail, crystallizing the thought that the mycological system is one of the most efficient systems for the synthesis and application of metal nanoparticles.

Why Myconanotechnology is Important?

Synthesis of nanoparticles using micro-organisms like bacteria, fungi, algae, yeast, and actinomycetes has gained considerable attention as an environment-friendly, low cost, and uncomplicated synthesis process (Rai et al. 2021a,b). Among the different micro-organisms harnessed for the synthesis process, the fungal system has shown tremendous potential and a number of applications in different fields of science and technology. Fungi are eukaryotic microbes with simple nutrition requirements. They are easy to grow in a controlled lab environment and the handling of biomass is also much simpler (Rai et al. 2009b, Garcia-Rubio et al. 2020). Fungi significantly produce a large number of bioactive compounds which can be used in a number of applications. Due to the property of heavy metal accumulation and tolerance, they are extensively used for the synthesis and stabilization of nanoparticles (Guilger-Casagrande and de Lima 2019). Also, fungi can be used as "nanofactories" for the large-scale synthesis of nanoparticles due to the ease in the downstream processing of the biomass (Yadav et al. 2015). The fungal system has the unique quality of single step, fast and cost-effective synthesis of nanoparticles with controlled size and shape. There are several research studies based on the use of fungi for the synthesis of nanoparticles including fungal species like *Aspergillus niger*, *A. fumigatus*, *A. oryzae*, *Fusarium oxysporum*, *F. solani*, *F. semitectum*, *Phoma glomerata*, *Phytophthora* sp.,

Pleurotus florida, P. sajor-caju, Pestalotia sp., *Trichoderma versicolor, T. viridae, Verticillium* sp., etc. (Rai et al. 2009b, Yadav et al. 2015, Guilger-Casagrande and de Lima 2019). These fungal species were reported to synthesize metal nanoparticles including silver, gold, copper, zinc, platinum, palladium, iron and zinc (Khandel and Shahi 2018). The synthesis process can be achieved at a different temperature, pH, biomass, and metal ion concentration yield nanoparticles with different morphological and physicochemical characteristics (Adebayo et al. 2021). The mycosystem is an efficient system for the synthesis of metal nanoparticles due to the easy availability and diversity of the fungal species. Also, nanoparticles synthesized using fungi show different applications in agriculture, textile, cosmetics, medicine, and electronics. Thus, myconanotechnology is a distinctive branch of nanotechnology dealing with the synthesis and application of nanoparticles using different fungal species.

Biomedical Applications

Nanotechnology is currently growing at a faster pace due to its unique functionality and a wide range of applications in different fields including biomedicine (Rai et al. 2019). Biosynthesized nanoparticles in general and mycosynthesized nanoparticles in particular reported to have promising utility in biomedical sector which mainly include disease diagnosis, disease treatment, drug delivery, antimicrobial activities, development of nanosensors useful for disease diagnosis, etc. (Figure 1).

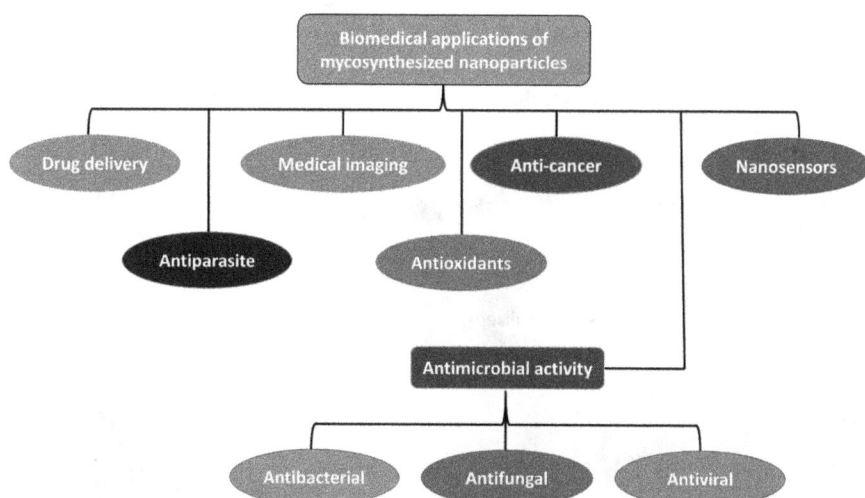

Figure 1. Biomedical applications for mycosynthesized nanoparticles.

Over the period of last few years, it has been reported that nanotechnology based immunotherapeutic agents have been used for the diagnosis and treatment of several diseases including different types of cancer. Cancer is one of the leading causes of death having a significant global health burden. According to Global cancer statistics (2018), there were around 18.1 million new cancer cases and 9.6 million cancer-related deaths recorded in 2018. However, one of the predictions suggested that by 2030, 30 million people will die from cancer each year (Lancet editorial 2018). In this context, nanotechnology has led to many promising applications in the diagnosis and treatment of cancer through drug delivery, gene therapy, drug carriage, biomarker mapping, targeted therapy, and molecular imaging (Jin et al. 2020). As far as mycosynthesized nanoparticles are concerned, metal nanoparticles like gold nanoparticles, silver nanoparticles, etc., are mostly used for cancer diagnosis at the molecular level and treatment. In addition, Rezvantalab et al. reviewed the role of Poly (lactic-co-glycolic acid) (PLGA)-based nanoparticles as drug delivery systems in the treatment of cancer (Rezvantalab et al. 2018). Similarly, Hossen et al. reviewed the role of smart nano carrier-based drug delivery systems particularly liposomes for cancer therapy and toxicity (Hossen et al. 2019). Clarance et al. demonstrated the anticancer activity of gold nanoparticles synthesized from the endophytic fungus, *Fusarium solani* (Clarance et al. 2020). The findings showed that the gold nanoparticles used in the research exhibited potential dose-dependent cytotoxicity on cervical cancer cells (HeLa) and against human breast cancer cells (MCF-7). Similarly, another study evaluated the anticancer activity of copper oxide nanoparticles synthesized from *Aspergillus terreus* against colon cancer cell lines (HT-29) (Mani et al. 2021). The results suggested that 22 μg mL^{-1} of copper oxide nanoparticles are enough to achieve 50% of cell inhibition. These findings clearly indicated that copper-based nanoparticles can play a crucial role in oncology by inhibiting tumour cell proliferation in a progressive way using their lowest doses. Figure 2, represents a step-wise illustration of a liposome-based smart drug delivery system for cancer therapy.

Similarly, a variety of nanoparticles such as silver, gold, copper, zinc oxide, etc., were effectively used in the management of human pathogens. However,

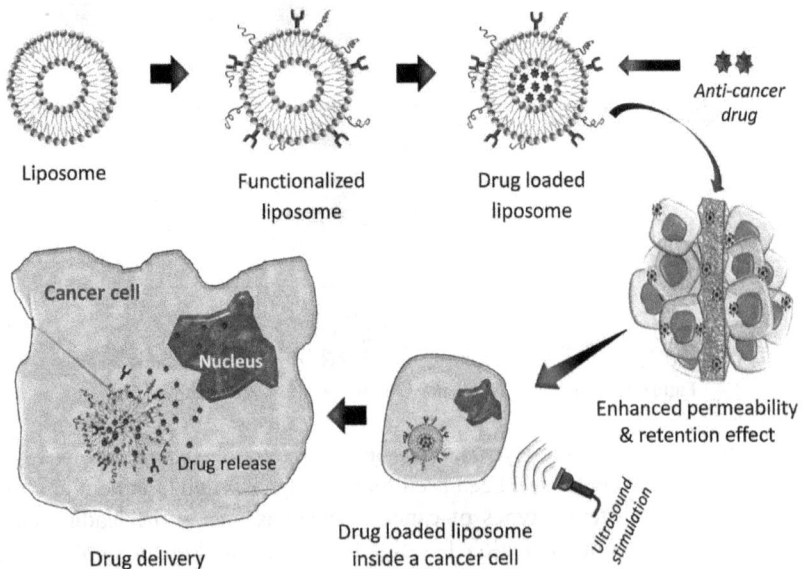

Figure 2. Step-wise illustration of liposome-based smart drug delivery system for cancer therapy reprinted from Hossen et al. 2019 under the CC BY-NC-ND 4.0.

among all these nanoparticles, silver nanoparticles were most commonly used as a new generation antimicrobial (Rai et al. 2009a, Sharma et al. 2022). In 2008, Ingle et al. coined the term "mycosynthesis" for the first time, which means the synthesis of nanoparticles using fungi (Ingle et al. 2008). In their study, mycosynthesized silver nanoparticles from *Fusarium acuminatum* were evaluated for their activity against human pathogenic and multidrug-resistant bacteria, viz., *Staphylococcus aureus, Salmonella typhi, Staphylococcus epidermidis* and *Escherichia coli*. The findings revealed the potential antibacterial activity against all tested bacteria. The maximum antibacterial activity of silver nanoparticles was shown against *S. aureus* followed by *S. epidermidis, Salmonella typhi* and the minimum by *E. coli*. Another recent study reviewed the efficacy of silver nanoparticles against various fungi such as *Aspergillus fumigates, A. niger, A. flavus, Trichophyton rubrum, Candida albicans, Penicillium* species, etc., which are responsible for a variety of infections in human beings (Mansoor et al. 2021). Recently, Mani et al., also reported antimicrobial activity of copper oxide nanoparticles synthesized from *Aspergillus terreus* against various pathogens like *Salmonella typhi, Staphylococcus aureus, Proteus mirabilis, Pseudomonas aeruginosa, Klebsiella pneuemoniae, Escherichia coli, Vibrio cholerae, Staphylococcus epidermidis, Candida albicans* and *Aspergillus niger* (Mani et al. 2021).

Moreover, mycosynthesized nanoparticles were also found to be effective against various vectors responsible for numerous diseases like dengue fever, chikungunya, zika fever, malaria, mayaro and yellow fever viruses, etc. (Sundaravadivelan and Padmanabhan 2014, Rahman et al. 2019, Rai et al. 2017). Sundaravadivelan and Padmanabhan demonstrated activity of AgNPs form *Trichoderma harzianum* against larvae and pupae of *Aedes aegypti*, a mosquito that is a vector for many diseases including dengue (Sundaravadivelan and Padmanabhan 2014). Similarly, another study evaluated the activity of *Isaria fumosorosea* mediated silver nanoparticles in the control of the mosquito species *Culex quinquefasciatus* and *A. aegypti* (Banu and Balasubramanian 2014). In both the above studies, potential concentration-dependent control of mosquitoes was reported with 100% mortality within 24 hrs in the case of 1st instar larvae.

Apart from this, these mycosynthesized nanoparticles also possess effective antioxidant activities (Mani et al. 2021, Renganathan et al. 2021) and can be used as vaccine adjuvants (Mao et al. 2021, Petkar et al. 2021).

Application in Agriculture

Agriculture is the sustenance of humanity, that produces commercial exchanges and generates peace. The world population continues to grow, the number of people in the world has more than doubled between 1961 and 2016. Therefore, there is a greater demand for food (FAO 2020). Providing food to the world's population is a challenge, especially for the estimated population growth of 6 to 9 billion by 2050 (Chen and Yada 2011). The soil provides 98.8% of food and we are currently facing a reduction in agricultural land. The global area of cropland per capita is estimated to have decreased from about 0.45 hectares per capita in 1961 to 0.21 hectares per capita in 2016 (FAO 2020). Moreover, the excessive use of agrochemicals has caused the loss of diversity in the soil and the development of resistance in pathogens and pests. In addition, in humans, many synthetic biocides show complex chronic effects (Yilmaz et al. 2020). The application of chemical biocides on crops has increased substantially, with the current annual use of 5.6 million tonnes of formulated products in 2019, with a value of $35.5 billion. Pesticide use increased by more than 50% in the 2010s compared to the 1990s. In addition, the application of pesticides per area of cultivated land increased from 1.8 to 2.7 kg ha^{-1}. However, Europe only showed 3% increase in pesticide application between the 1990s and 2010s (FAO 2021). We face great economic and environmental losses due to excessive and continuous application of pesticides. For example, the estimated losses related to pesticides in the US were $1.1 billion a year in public health, $1.5 billion for pest resistance to pesticides, $1.4 billion for production losses caused by pesticides, $2.2 billion dead birds due to pesticides, and $2.0 billion for groundwater contamination (Pimentel and Burgess 2014).

Moreover, there is excessive use of fertilizers globally. Among fertilizers, nitrogen (N) is the most important fertilizer applied to soils. The world production of N is approximately 123 million tonnes per year; a 9.5 fold increase over production in 1961 (FAO 2018). Although fertilizers are very important to produce food, their efficiency has decreased. Similarly, N use efficiency has decreased from 68% in 1961 to 47% in 2010 (Lassaletta et al. 2014). This is because less than half of the nitrogenous fertilizer is taken up by plants, and the remaining N is lost to the soil. It has been documented that only a small part of the fertilizers applied is used by the plant and a large part is lost through leaching, contaminating the soil and water (Bijay-Singh and Craswell 2021). Furthermore, the application of chemical fertilizers worldwide increased 10 fold between 1950 and 2000.

The problems facing agriculture must be solved in a sustainable way. The rapid development of technological innovations in the agricultural sector poses challenges such as sustainable production. The progress in science can offer potential alternative solutions to current production systems. Many technologies have been developed that have the potential to increase agricultural productivity besides reducing environmental costs. The use of nanotechnology is a clever way to overcome these problems (Wu et al. 2021).

The success of the application of nanotechnology in agriculture is based on the fact that the delivery of agrochemicals is directed to the site of interest to protect plant health (Worrall et al. 2018).

The development of nanosensors, nanofertilizers, nanopesticides, and nanoformulated products for pathogen control have demonstrated great potential in agriculture (Ingle et al. 2020).

Several countries are interested in nanotechnology as it provides support for epidemiological surveillance, detection, monitoring, and control of agricultural pests and diseases, such as phytosanitary epidemiological surveillance, an activity to prevent or mitigate the effects of pests (Mora-Aguilera et al. 2021).

Nanosensors are valuable devices used to detect plant pathogens. Biosensors are sensitive because they combine biology as a molecular recognition agent and nanoparticles as a physicochemical transducer (Merkoçi 2021). Biosensors detect the volatile compounds emitted by the pathogenic fungi in a short time. Enzymatic biosensors bind highly specifically to certain bio-molecules (Patel 2002). Similarly, the electronic nose (E-nose) is a device that is very precise and can differentiate between types of odors. It uses a response pattern through a series of gas sensors composed of nanoparticles such as ZnO nanowires (Sugunan et al. 2006). Moreover, nanomaterials (e.g., metallic and magnetic nanoparticles) have been reported to have effective applications in the detection and analysis of mycotoxins, produced by various pathogenic fungi that can be dangerous for plants, humans and animals (Ingle et al. 2020). Nanomaterials can be incorporated into biosensors for the purpose of achieving better analytical performance in terms of limit of detection, linear range, analytical stability, low production cost, etc. Particularly, gold nanoparticles are one of the most extensively studied and commonly used nanomaterials, which can be employed as an immobilization carrier, signal amplifier, mediator and mimic enzyme label, while developing nanobiosensors for mycotoxin detections (Wu et al. 2021). Moreover, Jogee and Rai demonstrated that various nanoparticles like gold, silver, zinc, aluminium nanoparticles, palladium, titanium, iron oxide, sulphur nanoparticles, etc., can be effectively used in the management of mycotoxin producing fungi belonging to different genera, i.e., *Aspergillus, Fusarium, Penicillium, Alternaria,* which secrete mycotoxins, viz., aflatoxins, fumonisin, patulin, ochratoxins, zearalenone, deoxynivalenol, etc. (Jogee and Rai 2020).

Wang et al. experimented with an electrical nanosensor contained in a gold electrode that was modified with copper nanoparticles, and applied for detecting the electrocatalytic oxidation of salicylic acid and the electrochemical behavior of this acid. The level of salicylic acid can be used for detection of phytopathogenic fungi. Many researchers have reported antibody-based biosensors for the detection of pathogenic fungi and virus such as *Fusarium culmorum, Phytophthora infestans, Puccinia striiformis, Aspergillus niger, Tobacco mosaic virus* and *Cowpea mosaic virus*. Another kind of nanosensors are known as DNA-based nanosensors, which can detect the disease at the beginning of the symptoms. Such sensors can also be used for bacterial pathogen detection (Figure 3).

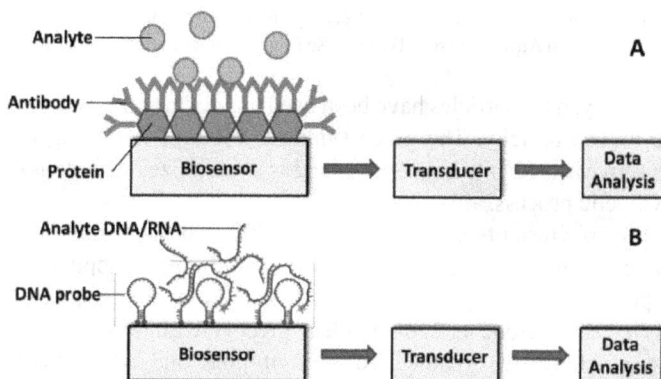

Figure 3. Schematic illustration of (A) antibody-based and (B) DNA/RNA-based biosensor for analyte detection. The specific combination of analyte and immobilized antibody (A) or DNA/RNA probe (B) produces a physicochemical change, such as mass, temperature, optical property or electrical potential. The change can be translated into a measurable signal for detection, reprinted from Fang and Ramasamy under Creative Common Rights Licence (Fang and Ramasamy 2015).

Nanocarriers play a crucial role in the efficient delivery and controlled release of fertilizers, pesticides, herbicides, plant growth regulators, etc. Indeed, mechanisms such as encapsulation and trapping, polymers and dendrimers, ionic surface, and weak bond attachments, are involved. In this way, the nanoproduct is stable in the environment, in addition to minimizing chemical runoff and reducing soil and water contamination (Worrall et al. 2018).

Nanopesticides have the advantage of controlled release of the active ingredient, which makes the product more efficient. Examples of this are clay nanotubes and porous hollow silica nanoparticles (PHSN) that are carriers of pesticides with very good contact (Liu et al. 2006). Interestingly, nanoparticles act as inhibitors of the pathogenic microbes as well as efficient carriers (Worrall et al. 2018) as shown in Figure 4.

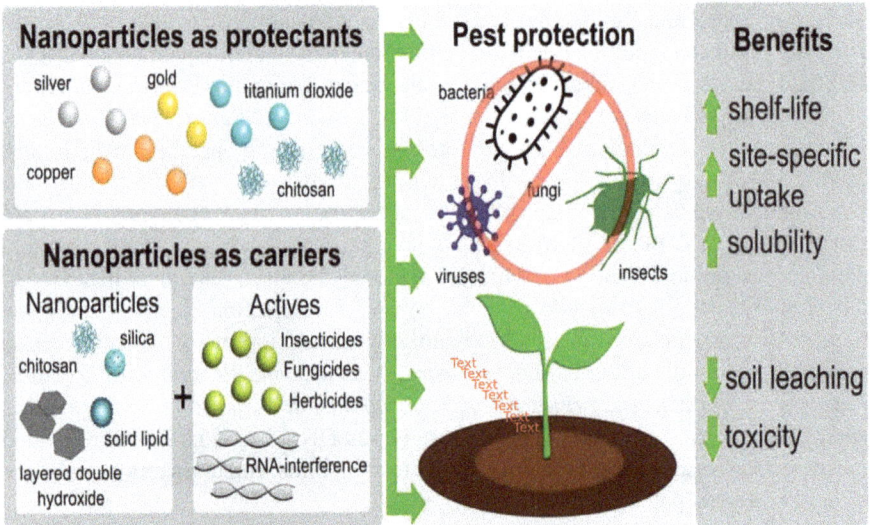

Figure 4. Nanomaterials as carriers/inhibitors of plant pathogens. Reprinted from Worrall et al. (2018) under Creative Common Attribution (CC BY) license (http://creativecommons.org/licenses/by/4.0/).

More recently, nanoparticles have been used mainly in combination with natural products to reduce toxicity. The green synthesis route of designing nanoparticles loaded with active ingredients is recommended as it minimizes hazardous components in the biosynthetic process.

The new phytosanitary products will be nanoformulations based on biodegradable polymers as they offer sustainable technology applicable to organic crops. The polymers are polysaccharides such as chitosan and starch, or polyesters such as poly-ε-caprolactone and polyethylene glycol (Alzahrani et al. 2015).

Bansod et al. (2013) synthesised NPs of gold and silver from *Fusarium oxysporum* and used in nano-PCR method for rapid identification of *Candida albicans*. The identification is more sensitive and is carried out in less time than with a traditional PCR. Nano-based PCR diagnostic methods could be applied for rapid detection of devastating pathogenic fungi such as *Pyricularia oryzae* (El-Abbasi et al. 2020), *Cephalosporium maydis* (Awad et al. 2019), and more.

More research needs to be performed to demonstrate the effectiveness of AgNPs synthesized by *F. oxysporum* to inhibit phytopathogenic fungi. Although several metal NPs have been used to control plant pathogens such as *Raffaelea* sp. (Kim et al. 2009), *Botrytis cinerea*, *Pilidium concavum*, and *Pestalotia* sp. (Bayat et al. 2021). In the last study, 100 ppm of AgNP had some effect on the pathogens, therefore a higher dose will be required for successful control.

Application in Veterinary

Silver nanoparticles prove to be beneficial in a wide range of veterinary applications. Their unique physicochemical properties may offer us a new helpful drug to treat and prevent animal diseases (Osama et al. 2020). Several studies indicated the possible use of fungal AgNPs in the improvement of wound healing as well as treatment of numerous bacterial and fungal infections of animals (Kalaiselvan and Rajasekaran 2009, Li et al. 2012, Hu et al. 2019, Feroze et al. 2020). Besides, Gherbawy et al. reported anti-fasciolasis activity of AgNPs synthesized from *Trichoderma harzianum* (TUT89) isolated from soil (Gherbawy et al. 2013). *Fasciola* species are an important parasite of sheep, cattle, rabbit, and camel (Rojas et al. 2014). Studies proved the enhanced efficiency of triclabendazole, the anti-parasitic drug, in combination with bio-AgNPs against *Fasciola* egg hatching. Results demonstrated that treatment with triclabendazole inhibited the egg hatching up to 70.6%, while the combination of drug and nanoparticles inhibited *Fasciola* egg hatching up to 94%. It was also found, that drug and bio-AgNPs caused surface distortion and perforation of eggs (Gherbawy et al. 2013). Furthermore, AgNPs synthesized by using *Duddingtonia flagrans* filtrate showed anti-nematicidal activity against infective larvae of *Ancylostoma caninum* that is pathogenic to dogs and cyathostomins, which is the primary nematode pathogens of horses (Barbosa et al. 2019, Ferraz et al. 2021). AgNPs from *D. flagrans* at concentration 43.4 μg mL^{-1} eliminated 100% of *A. caninum* by the destruction of the nematode cuticle (Barbosa et al. 2019) and 43% cyathostomin larvae reduction (Ferraz et al. 2021). Moreover, mycosynthesized AgNPs showed higher nematicidal efficacy, while compared to chemically produced AgNPs. The authors suggested that this effect was possible due to the presence of *D. flagrans* proteins on the surface of the AgNPs, which enabled the penetration of the AgNPs into the nematode. In addition, bioAgNPs showed a non-toxic effect on non-cancerous cells (L929 fibroblast) until an active concentration of 43.4 μg mL^{-1}. The use of mycosynthesized AgNPs as nematicides provides an opportunity for more effective animal parasite control to combat the already existing problem of parasite resistance. However, it requires further studies on the effects of nanoparticles on the host organisms (Barbosa et al. 2019).

Role as Catalysts

Recently, one of the major concerns for environment, plants, animals, and human health is the high levels of chemicals and other pollutants, incorporated during the industrial processes. These include a variety of non-degradable dyes that are widely used in various industrial sectors, such as food, leather, paints, paper, pharmaceutical, plastics and varnishing, etc. (Dutta et al. 2021). Myco-AgNPs showed potential as an alternative to hazardous chemical and physical routes of remediation. Therefore, silver nanoparticles, that are almost spherical, having an average size of 25 nm, synthesized by fungus *Cylindrocladium floridanum*, exhibited the homogeneous catalytic activity in the reduction of toxic and mutagenic pollutants, namely 4-nitrophenol (4-NP) to 4-aminophenol (4-AP) using sodium borohydride (Narayanan et al. 2013). Silver nanoparticles synthesized from *Penicillium oxalicum* have been developed as efficient catalysts to accelerate the reduction of methylene blue (MB). The kinetically slow process of reduction of MB was accelerated using AgNPs as an electron relay system, between the BH_4 donor and the MB acceptor (Du et al. 2015). Likewise, water dispersible and thermo-stable silver/silver (I) oxide nanoparticles (Ag/Ag$_2$O NPs) were biofabricated from endophytic fungi *Fusarium oxysporum*, which showed catalytic efficiency for degradation of MB dye with reducing agent, namely NaBH$_4$. Results indicated enhanced catalytic activity, attributed as Ag/Ag$_2$O NPs for degradation reaction, when compared with the bulk Ag$_2$O (Islam et al. 2021). Fungal xylanases from *Aspergillus niger* and *Trichoderma longibrachiatum* were used for AgNP synthesis. Biofabricated nanoparticles were spherical with sizes ranging between 15.21 and 77.49 nm and showed the ability to degrade malachite green (MG) and MB dyes. Stronger degradation effects after 24 hours of reaction were found for AgNPs from *A. niger* xylanase (nearly 80% degradation of MG and 25% of MB), when compared to AgNPs from *T. longibrachiatum* (degradation of dyes were 64.3% and 14.8% for MG and MB, respectively) (Elegbede et al. 2018). These results indicated the potential of mycosynthesized nanoparticles in biodegradation of dye as alternatives to common chemical and physical methods, mainly based on concentrating or transferring the dyes from one phase to another. In addition, due to the ability of AgNPs to accelerate the reaction at low temperatures with low catalyst loading, and any ligand source, they can be found as efficient catalysts for organic reactions (Begum et al. 2020).

Application in Textiles

The development of microorganisms on the surfaces of textile materials, damages the material, and has a negative impact on the comfort of wearing as well as human health. It is well known that metal nanoparticles have the potential for inhibiting bacterial growth and multiplication. Therefore, metallic nanoparticles such as AgNPs are increasingly embedded into textile materials (Vigneshwaran et al. 2007, Radetić et al. 2013, Deshmukh et al. 2019). Spherical and homogeneous silver nanoparticles synthesized from *Fusarium oxysporum* with a diameter of 7.30 nm were applied on cotton fabric. Textile loaded with AgNPs inhibited growth of *Staphylococcus epidemidis*, *Escherichia coli*, *Pseudomonas aeruginosa*, and *Salmonella enterica*. Neutral red uptake (NRU) and methylthiazoletetrazolium (MTT) reduction studies confirmed non-cytotoxic activity of these nanoparticles until 16 µM with IC$_{50}$ of 22 µM towards lung fibroblast cell line (V79) cells. In addition, the nanoparticles were found to be stable for several months after 30 successive washing due to the presence of biomolecules on their surface (Marcato et al. 2012). Further, antimicrobial activity of cotton fibers impregnated with nanosilver, synthesized from *Fusarium oxysporum* fungal filtrate against *Candida parapsilosis* and *Xanthomonas axonopodis*, was also reported (Ballottin et al. 2017). These activities of AgNPs were associated with strong effects on membranes of microorganisms. Approximately 10% decrease in antimicrobial activity of bioAgNPs was observed after 10 washes (Ballottin et al. 2017). Another study on cotton fabric embedded with spherical nanosilver synthesized by *Aspergillus terreus*, demonstrated that 100 ppm of AgNPs inhibited the growth of Gram-positive *Bacillus cereus* and *Staphylococcus aureus*, and Gram-negative *Pseudomonas aeruginosa*, *Escherichia coli*, and *Proteus vulgaris*. This implies that a hybrid of cotton fiber with AgNPs with an inhibitory effect on many microbes, could be helpful to combat multi-drug resistant microorganisms (Velhal et al. 2016). The PET/C blend fabric, unhydrolyzed and activated with cellulose, was placed in the solution of AgNPs that were biosynthesized from *Aspergillus fumigatus* DSM819 strain. Interestingly, it was found that the very low concentration (0.25% weight) of AgNPs in the fabrics showed antibacterial activity against *Escherichia coli* and *Bacillus mycoides* and antifungal activity against *Candida albicans*. Moreover, cells lines, namely HCT116 (human colon carcinoma), A549 (lung carcinoma cell line), MCF7 (human Caucasian breast adenocarcinoma), and PC3 (prostate cell line) exhibited a straight dose–response to concentrations of AgNPs ranging from 0.4 to 54 µg mL^{-1}. IC$_{50}$ values obtained from MMT assay ranged between 31.1 and 45.4 µg mL^{-1}. Whereas, AgNPs caused only 15.6% growth inhibition of BJ1 (normal skin fibroblast) cell line at maximum tested concentration (54 µg mL^{-1}). The above results indicated that the mycosynthesized AgNPs may have various featured applications like biomedical applications or antimicrobial surfaces, due to its unique properties and high biocompatibility (Othman et al. 2019).

Toxicity of Nanoparticles

Cytotoxicity

Nanoparticles have valuable features that enable their use in various engineering technologies and medicine (Liao et al. 2019, Santos et al. 2021). The toxicity of nanoparticles such as silver, gold, copper, zinc, magnesium, titanium oxides are a matter of concern for human health and environment (Gerloff et al. 2009, Chauhan et al. 2011, Saravanakumar 2019, Clarance et al. 2020, Ferdous et al. 2020, Santos et al. 2021). Proper understanding of the interactions between nanoparticles and mammalian cells is essential for their safe application in different fields of human life, especially in nanomedicine (Liao et al. 2019). This knowledge enables scientists to develop more functional nanoparticles with improved biocompatibility to such applications (Liao et al. 2019). The recent studies showed that biogenic nanoparticles might be effective (Xiong et al. 2018) and even better (Guilger et al. 2017) and less toxic (Datkhile et al. 2017) when compared to chemically synthesized ones.

It should be emphasized that the mechanism of nanoparticle toxicity has been most explored and described for silver nanoparticles (Ferdous and Nemmar 2020). The size of nanoparticles is a pivotal factor that determines their penetration into the cells by diffusion, endocytosis or phagocytosis (Zhang et al. 2015, Carrola et al. 2016). According to Pratsinis et al. smaller nanoparticles are much toxic compared to larger ones due to higher surface area per unit mass (Pratsinis et al. 2013). Apart from the size, factors such as shape, crystallinity, surface chemistry, including capping agents, can also effect cytotoxicity of NPs (Marambio-Jones 2010). Studies of many authors like Balakumaran et al. (2015), Hu et al. (2019), Ameen et al. (2021a) showed that AgNPs and released silver ions may cause intracellular reactive oxygen species (ROS) production leading to oxidative stress (e.g., by structural and functional disruption of mitochondrial membrane), which in turn can damage cellular components such as lipids, DNA, and proteins and processes such as ATP synthesis leading to cell death (Hussain et al. 2005, Rao et al. 2016). Moreover, the acidic lysosomal environment promotes the release of silver ions from silver nanoparticles, initiating cascades and a series of events that lead to intracellular toxicity defined as the "Trojan horse effect" (Sabella et al. 2014).

One of the most important and unanswered question is what form of metal (nanoparticles or ions) contributes more to cytotoxicity on mammalian normal and cancer cells (Jaswal and Gupta 2021). However, oxidative stress is believed to be a main mechanism of toxicity for a wide variety of nanoparticles. Silver nanoparticles synthetized from endophytic fungus *Guignardia mangiferae* showed higher cytotoxic activity against HeLa and MCF-7 cancer cells lines than normal Vero cells. In the case of tumor cells, symptoms of apoptosis, membrane damage and the presence of apoptotic bodies were observed (Balakumaran et al. 2015).

The investigation of the toxicity of nanoparticles against mammalian cells is carried out *in vitro* using well-established methodologies such as MTT Cell Viability Assay, Neutral Red Uptake (NRU), *Lactate Dehydrogenase* (LDH) and ROS level measurements. These methods estimate cell viability or cell death by determination of metabolic activity, membrane disruption, generation of reactive oxygen species, and genotoxicity tests as comet assay and Ames test (Santos et al. 2021). The results of such investigations are helpful for improving the physical and chemical properties of nanoparticles before increasing the scale of their production or carrying out more complex analyses (Santos et al. 2021).

Biodistribution of Nanoparticles

The biodistribution and transformation of nanoparticles in mammals are very important aspects for understanding their mechanisms of toxicity (Dong et al. 2020). The *in vivo* behavior of nanoparticles is determined by their physicochemical and biological properties (Arms et al. 2018). In addition, nanoparticles can combine certain properties, viz., long-term circulation, biocompatibility, bioavailability and enhanced stability (Khlebtsov and Dykman 2011). The *in vivo* studies can help in identifying the target sites of nanoparticle accumulation after exposure (Khlebtsov and Dykman 2011). Unfortunately, the results of *in vivo* tests of nanoparticles have been rarely reported and remain controversial (Khlebtsov and Dykman 2011). The nanoparticles may enter the human body by pulmonary inhalation, oral ingestion, skin penetration through lesions or scratch and injection in blood circulation.

Moreover, dose or time duration are identified as important determinants affecting biodistribution and pathogenic outcomes that cause toxic exposure (Raies and Bajic 2016). Nanoparticle administartion in mice affects their distribution and accumulation in many organs, especially in the liver, lungs and kidneys. However, transport of nanoparticles through blood vesicular system to the target organs remains unclear (Guo 2016). Kermanizadeh et al. (2014) showed *in vivo* in rodents that orally administrated metal nanoparticles accumulated in the liver, leading to tissue injury. Another study examined the mechanism of nephrotoxicity in rats exposed to copper nanoparticles by analyzing the expression profiles of renal genes (Liao and Liu 2012). Authors noted renal proximal tubule necrosis in kidneys induced with nanocopper. Guo studied *in vivo* toxicity of silver nanoparticles administarted as single or multiple intravenous injections in female Balb/c mice for period of 1, 4 and 10 days (Guo 2016). He reported the distribution and toxicity of AgNPs caused by the destruction of the endothelial barrier in the liver, lungs and kidneys. Further, rats exposed orally to nanosilver in size of 15–20 nm for 28 days were tested (van der Zande Sahu et al. 2012). Authors showed accumulation of nanosilver in stomach wall, liver, spleen, testis, kidney, lungs, heart and brain, but primarily in the liver and spleen. Hillyer and Albrecht studied size-dependent biodistribution of gold nanoparticles (4–58 nm) served to mice through the alimentary tract in the form of a supplement (200 mg mL^{-1}) to drinking water (Hillyer and Albrecht 2001). The smallest NPs were easily translocated from digestive tract and accumulated in different organs, but when the size of the nanoparticles increased to 10, 28 or 58 nm, the penetration of NPs from the gastrointestinal tract and accumulation in other organs decreased significantly. The nanogold level for 4 nm particles was highest in kidneys (approximately 0.075 µg g^{-1}) while it was much lower in other organs including small intestine, lungs, stomach, spleen, and liver (from 0.02 to 0.035 µg g^{-1}). Similarly, Dong et al. reported the presence of silver ions and nanosilver in the blood and all major organs and tissues of rats, including the liver, kidneys, testis, ovaries, olfactory bulb and brain, after administration by inhalation or injection (Dong et al. 2020). Osmond and Mccall found that the ZnO NPs inhalation showed toxic effects on human lungs (Osmond and Mccall 2010). The intensity of inflammatory disease was related to the size and surface area of zinc oxide nanoparticles (Kalpana and Devi Rajeswari 2018).

The biodistribution of nanoparticles can be analyzed and assessed by electron microscopy, liquid scintillation counting (LSC), histology, indirectly measuring drug concentrations, *in vivo* optical imaging, computed tomography (CT), magnetic resonance imaging (MRI), and nuclear medicine imaging (Arms et al. 2018).

Ecotoxicity of Nanoparticles

Ecotoxicity studies of nanoparticles are very limited and difficult to compare (Joner et al. 2008). Results from experiments show that some nanoparticles cause toxicity in many organisms, even at very low concentrations (Joner et al. 2008). Continuous development of nanotechnologies used in the industries causes increase in amount of nanoparticles in the environment, either in soil, water or air. It can directly affect human health. Therefore, to protect human health and nature from the adverse effects of nanomaterials, more comprehensive studies should be considered (Rana and Kalaichelvan 2013). The mobility of nanoparticles in soils depends on their physical–chemical properties, characteristics of the soil environment and the interaction of nanoparticles with natural colloidal material (Azeez 2021). The mammalian lung epithelium has some relevance to ecotoxicology because the lung is representative of a typical mucous epithelial tissue and is not fundamentally different in structure to other epithelia such as the gills or guts of aquatic organisms, or the body surface of earth worms (Handy et al. 2008). Fish gills are certainly sensitive to some synthesized nanoparticles such as TiO_2 NPs (Federici et al. 2007). Moreover, the allocation of nanoparticles in the water environment can be influenced by different processes, including aggregation and disaggregation, transformation, diffusion, degradation by biotic and abiotic factors, interaction with natural water components and photoreaction (Vale et al. 2015). The aggregation of nanoparticles is associated with deposition on organisms in the aquatic environment. They can also adsorb to the exterior surface of the organism (Handy and Eddy 1991, Handy et al. 2008). Researchers compared toxicity response of Cu ions ($CuSO_4$) and nanoparticles towards zebrafish (*Danio rerio*) in dechlorinated tap water (hardness of 142 mg of $CaCO_3$ per liter and a pH of 8.2) and showed that nanoparticles were less toxic than corresponding ions (Griffitt et al. 2007). Other studies reported that ZnO NPs can also cause acute effects on fish, at concentrations higher than expected in the environment (Keerthana and Kumar 2020).

Effect of Nanoparticles on Beneficial Microbes

Soil microorganisms play a pivital role in the immobilization and cycling of carbon and other nutrients and the degradation or detoxification of pollutants that enhance soil health (Saccá et al. 2017) and regulate the efficiency and accessibility of nutrients to crop plants (Mahawar and Prasanna 2018). It needs to be highlighted that reports on the effect of nanoparticles on plant beneficial microbes are still scanty and not well understood (Mahawar and Prasanna 2018, Ameen et al. 2021b). It is illustrated in Figure 5. However, due to the increasing release of nano-based agricultural products (nanoparticle-based pesticides, fertilizers, and herbicides) that are main source of soil entry (Chhipa 2017), the interactions between nanoparticles and beneficial microorganisms are subjected to evaluation (Duhan et al. 2017). For example, nanoparticles increase the solubility of essential ingredients, when used as additives in pesticides (Ameen et al. 2021b). Many authors reported that, nanoparticles in soils have shown effects on soil fertility, microbiota in soil and agricultural crops (Fayiga 2017, Yanga et al. 2017, Pittol et al. 2017). Nanoparticles such as ZnO-NPs, CuO-NPs, AgNPs, FeO-NPs and TiO_2-NPs have shown variable chronic and acute toxic effects on soil microbes. Nanoparticle-microbe interactions depends not only on nanoparticle properties, including size, surface charges, capping agent and the presence of divalent anions/cations, but also on the composition and charge of the bacterial cell wall (Acharya et al. 2018). However, translocation of different size and also crystalline structure TiO_2 NPs may vary in soil (Klingenfuss

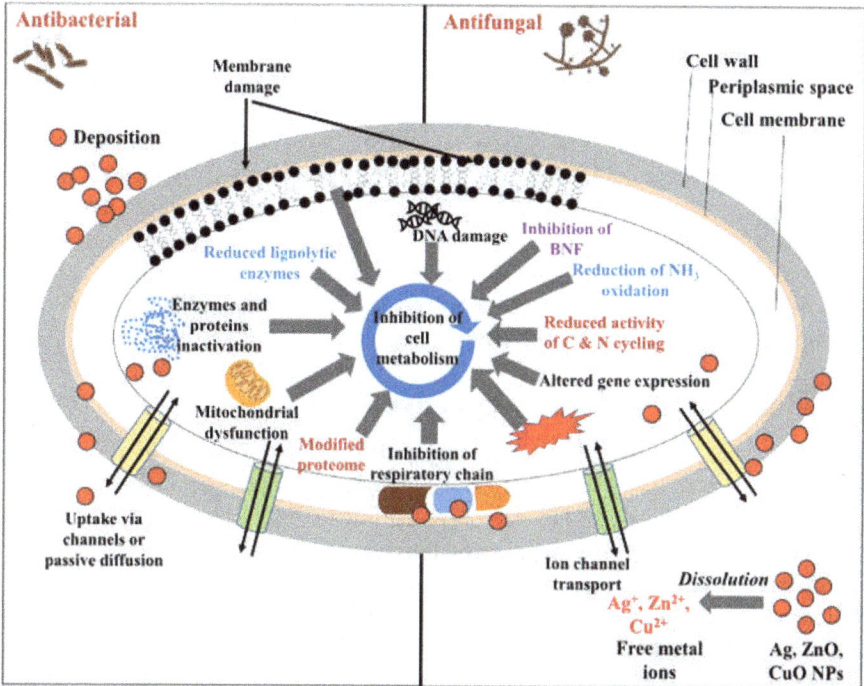

Figure 5. The toxic impact of different metal-based nanoparticles such as Ag-NPs, ZnO-NPs, CuO-NPs and their ions (Ag^+, Zn^{2+}, Cu^{2+}) on beneficial soil bacteria and fungi. The toxicity mechanism is common for both fungi and bacteria which varies with varying physicochemical features of metal nanoparticles, applied concentration, and soil environment (Ameen et al. 2021b).

2014). The soil properties like buffering capacity and natural organic matter, and also aggregation, immobilization and deposition of nanoparticles, and environmental corona formation around nanoparticles may significantly affect interactions between nanoparticles and beneficial soil microorganisms (Zhang et al. 2020).

Interestingly, it is claimed that bacterial cells could become partially or fully resistant to some nanoparticles (Ahmed et al. 2020). The four plant growth-promoting rhizobacterial species such as *Bacillus thuringiensis*, *Pseudomonas mosselii*, *Azotobacter chroococcum* and *Sinorhizobium meliloti* tolerated CuO, TiO_2, and Al_2O_3 nanoparticles in a dose up to 3 mg mL^{-1} (Ahmed et al. 2020) and were sensitive towards Ag and ZnO nanoparticles at concentration < 1500 μg mL^{-1} (Ahmed et al. 2020). Moreover, in the presence of Ag and ZnO, their surface roughness enhanced, while adherence on a solid surface, synthesis of extracellular polymeric substances and bacterial colonization were reduced as compared to untreated cells (Ahmed et al. 2020). Similarly, it was found that ZnO-NPs induced morphological changes in *Rhizobium leguminosarum* cells and damaged bacterial surface (Fan et al. 2014). Furthermore, ZnO-NPs caused early senescence of nodules, delayed the onset of nitrogen fixation, and disturbed the root nodulation process (Fan et al. 2014). It was noticed that production of some bioactive molecules (e.g., indole-3-acetic-acid) vital for plant growth and soil fertility by PGPR bacteria can be significantly reduced or completely inhibited in the presence of silver and zinc oxide NPs (Seneviratne et al. 2016). Recently, research found that ZnO-NPs suppressed activities of N-acetylglucosaminidase, glycine-aminopeptidase, aryl-sulfatase, polyphenol oxidase, and peroxidase in aquatic fungi significantly reduced the litter decomposition rate (Du et al. 2020). Moreover, studies reported that nanoparticles of CuO reduced production of siderophore in *P. chlororaphis* O6 and also increased the pyoverdine siderophore synthesis (Dimkpa et al. 2012). On the other hand, some reports showed the stimulatory impact of NPs on root-microbe symbiosis (Feng et al. 2013, Chen et al. 2017). Moreover, nanofertilizers based on ZnO nanoparticles produced by using soil fungi are known to enhance the nutrient mobilization and their uptake in plants (Mahawar and Prasanna 2018).

Conclusion and Future Perspectives

To sum up, with increasing knowledge about the synthesis of nanoparticles by fungi, myconanotechnology has been continuously progressing because the synthesis is bio-inspired, cost-effective, and has demonstrated low toxicity compared to chemically and physically synthesised nanoparticles. Furthermore, the nanoparticles thus formed have shown enhanced biocompatibility and bioactivities such as antimicrobial, anticancerous, antioxidant, applications in agriculture (for the development of nanobiosensors used in diagnosis), delivery of agrichemicals, and for therapy of the diseases caused by plant pathogens. The mycogenic nanoparticles are also useful in food packaging and enhancing the shelf-life of the fruits by forming thin films/nanoencapsulation, etc. The use of mycogenic nanoparticles as catalysts, in cosmetics and textiles, has revolutionized the world. In the future, myconanotechnology will flourish and enrich the newer technologies towards early detection of cancer, plant diseases, removal of multidrug-resistance in microbes, enhanced synergist activity of antibiotics/agrochemicals, slow and site-specific delivery of drugs. However, there is a greater need for intensive studies to understand the toxicity to humans and the environment.

Acknowledgements

Mahendra Rai is thankful to the Polish National Agency for Academic Exchange (NAWA) for financial support (Project No. PPN/ULM/2019/1/00117/A/DRAFT/00001) to visit the Department of Microbiology, Nicolaus Copernicus University, Toruń, Poland.

References

Abd-Alla, M.H., Nafady, N.A. and Khalaf, D.M. 2016. Assessment of silver nanoparticles contamination on faba bean-*Rhizobium leguminosarum* bv. viciae-*Glomus aggregatum* symbiosis: Implications for induction of autophagy process in root nodule. Agri. Ecosys. Environ. 218: 163–177.

Acharya, D., Singha, K.M., Pandey, P., Mohanta, B., Rajkumari, J. and Singha, L.P. 2018. Shape dependent physical mutilation and lethal effects of silver nanoparticles on bacteria. Sci. Rep. 8: 1–11.

Adebayo, E.A., Azeez, M.A., Alao, M.B., Oke, A.M. and Aina, D.A. 2021. Fungi as veritable tool in current advances in nanobiotechnology. Heliyon 7: e08480.

Ahmed, B., Ameen, F., Rizvi, A., Ali, K., Sonbol, H., Zaidi, A., Khan, M.S. and Musarrat, J. 2020. Destruction of cell topography, morphology, membrane, inhibition of respiration, biofilm formation, and bioactive molecule production by nanoparticles of Ag, ZnO, CuO, TiO$_2$, and Al$_2$O$_3$ toward beneficial soil bacteria. ACS Omega 5: 7861–7876.

Alzahrani, E., Sharfalddin, A. and Alamodi, M. 2015. Microwave-hydrothermal synthesis of ferric oxide doped with cobalt. Adv. Nanopart. 4: 53–60.

Ameen, F., AlHomaidan, A.A., AlSabri, A., Almansob, A. and Alnadhari, S. 2021a. Antioxidant, antifungal and cytotoxic effects of silver nanoparticles synthesized using marine fungus *Cladosporium halotolerans*. Appl. Nanoscience 148.

Ameen, F., Alsamhary, K., Alabdullatif, J.A. and ALNadhari, S. 2021b. A review on metal-based nanoparticles and their toxicity to beneficial soil bacteria and fungi. Ecotoxicol. Env. Safe. 213: 112027.

Arms, L., Smith, D.W., Flynn, J., Palmer, W., Martin, A., Woldu, A. and Hua, U. 2018. Advantages and limitations of current techniques for analyzing the biodistribution of nanoparticles. Front. Pharmacol. 9: 802.

Awad, A.M., El-Abbasi, I.H., Shoala, T., Youssef, S.A., Shaheen, D.M. and Amer, G.A. 2019. PCR and nanotechnology unraveling detection problems of the seed-borne pathogen cephalosporium maydis, the causal agent of late wilt disease in maize. Int. J. Nanotechnol. Allied Sci. 3: 30–39.

Azeez, L. 2021. Detection and evaluation of nanoparticles in soil environment. Nanomat. Soil Remed. 33–63.

Balakumaran, M.D., Ramachandran, R. and Kalaicheilvan, P.T. 2015. Exploitation of endophytic fungus, *Guignardia mangiferae* for extracellular synthesis of silver nanoparticles and their *in vitro* biological activities. Microbiol. Res. 178: 9–17.

Ballottin, D., Fulaz, S., Cabrini, F., Tsukamoto, J., Duran, N., Alves, O.L. and Tasic, L. 2017. Antimicrobial textiles: Biogenic silver nanoparticles against *Candida* and *Xanthomonas*. Mater. Sci. Eng. 75: 582–589.

Bansod, S., Bonde, S., Tiwari, V., Bawaskar, M., Deshmukh, S., Gaikwad, S. and Rai, M. 2013. Bioconjugation of gold and silver nanoparticles synthesized by *Fusarium oxysporum* and their use in rapid identification of *Candida* species by using bioconjugate-nano-polymerase chain reaction. J. Biomed. Nanotech. 9: 1962–1971.

Banu, A.N. and Balasubramanian, C. 2014. Optimization and synthesis of silver nanoparticles using Isaria fumosorosea against human vector mosquitoes. Parasitol. Res. 113: 3843–3851.

Barbosa, A.C.M.S., Silva, L.P.C., Ferraz, C.M., Tobias, F.L., de Araújo, J.V., Loureiro, B. and Braga, F.R. 2019. Nematicidal activity of silver nanoparticles from the fungus *Duddingtonia flagrans*. Int. J. Nanomedicine 14: 2341.

Bayat, M., Zargar, M., Chudinova, E., Astarkhanova, T. and Pakina, E. 2021. *In vitro* evaluation of antibacterial and antifungal activity of biogenic silver and copper nanoparticles: The first report of applying biogenic nanoparticles against *Pilidium concavum* and *Pestalotia* sp. fungi. Molecules 26: 5402.

Begum, R., Najeeb, J., Sattar, A., Naseem, K., Irfan, A., Al-Sehemi, A.G. and Farooqi, Z.H. 2020. Chemical reduction of methylene blue in the presence of nanocatalysts: A critical review. Rev. Chem. Eng. 36: 749–770.

Bijay-Singh and Craswell, E. 2021. Fertilizers and nitrate pollution of surface and ground water: An increasingly pervasive global problem. SN Appl. Sci. 3: 518.

Carrola, J., Bastos, V., Jarak, I., Oliveira-Silva, R., Malheiro, E., Daniel-da-Silva, A.L., Oliveira, H., Santos, C., Gil, A.M. and Duarte, I.F. 2016. Metabolomics of silver nanoparticles toxicity in HaCaT cells: Structure-activity relationships and role of ionic silver and oxidative stress. Nanotoxicol. 10: 1105–17.

Chauhan, A., Zubair, S., Tufail, S., Sherwani, A., Sajid, M., Raman, S.C., Azam, A. and Owais, M. 2011. Fungus-mediated biological synthesis of gold nanoparticles: Potential in detection of liver cancer. Int. J. Nanomed. 6: 2305–2319.

Chen, C., Tsyusko, O.V., McNear, D.H., Judy, J., Lewis, R.W. and Unrine, J.M. 2017. Effects of biosolids from a wastewater treatment plant receiving manufactured nanomaterials on *Medicago truncatula* and associated soil microbial communities at low nanomaterial concentrations. Sci. Total. Environ. 609: 799–806.

Chen, H. and Yada, R. 2011. Nanotechnologies in agriculture: New tools for sustainable development. Trends Food Sci. Technol. 22: 585–594.

Chhipa, H. 2017. Nanofertilizers and nanopesticides for agriculture. Environ. Chem. Lett. 15: 15–22.

Clarance, P., Luvankar, B., Sales, J., Khusro, A., Agastian, P., Tack, J.-C., Al Khulaifi, M.M., Al-Shwaiman, H.A., Elgorban, A.M. and Syed, A. 2020. Green synthesis and characterization of gold nanoparticles using endophytic fungi *Fusarium solani* and its *in-vitro* anticancer and biomedical applications. Saudi J. Biol. Sci. 27: 706–712.

Datkhile, K.D., Durgawale, P.P. and Patil, M.N. 2017. Biogenic silver nanoparticles are equally cytotoxic as chemically synthesized silver nanoparticles. Biomed. Pharmacol. J. 10: 337–344.

Deshmukh, S.P., Patil, S.M., Mullani, S.B. and Delekar, S.D. 2019. Silver nanoparticles as an effective disinfectant: A review. Mater. Sci. Eng. 97: 954–965.

Dimkpa, C.O., McLean, J.E., Britt, D.W. and Anderson, A.J. 2012. CuO and ZnO nanoparticles differently affect the secretion of fluorescent siderophores in the beneficial root colonizer *Pseudomonas chlororaphis* O6. Nanotoxicol. 6: 635–642.

Dong, L., Lai, Y., Zhou, H., Yan, B. and Liu, J. 2020. The biodistribution and transformation of nanoparticulate and ionic silver in rat organs *in vivo*. NanoImpact 20: 100265.

Du, J., Zhang, Y., Yin, Y., Zhang, J., Ma, H., Li, K. and Wan, N. 2020. Do environmental concentrations of zinc oxide nanoparticle pose ecotoxicological risk to aquatic fungi associated with leaf litter decomposition? Water Res. 178: 115840.

Du, L., Xu, Q., Huang, M., Xian, L. and Feng, J.X. 2015. Synthesis of small silver nanoparticles under light radiation by fungus *Penicillium oxalicum* and its application for the catalytic reduction of methylene blue. Mater. Chem. Phys. 160: 40–47.

Duhan, J.S., Kumar, R., Kumar, N., Kaur, P., Nehra, K. and Duhan, S. 2017. Nanotechnology: The new perspective in precision agriculture. Biotechnol. Rep. 15: 11–23.

Dutta, S., Gupta, B., Srivastava, S.K. and Gupta, A.K. 2021. Recent advances on the removal of dyes from wastewater using various adsorbents: A critical review. Mater. Adv. 2: 4497–4531.

El-Abbasi, I.H., Khalil, A.A., Awad, H.M. and Shoala, T. 2020. Nano-diagnostic technique for detection of rice pathogenic fungus *Pyricularia oryzae*. Indian Phytopathol. 73: 673–682.

Elegbede, J.A., Lateef, A., Azeez, M.A., Asafa, T.B., Yekeen, T.A., Oladipo, I.C. and Gueguim-Kana, E.B. 2018. Fungal xylanases-mediated synthesis of silver nanoparticles for catalytic and biomedical applications. IET Nanobiotechnol. 12: 857–863.

Fan, R., Huang, Y.C., Grusak, M.A., Huang, C.P. and Sherrier, D.J. 2014. Effects of nano-TiO_2 on the agronomically-relevant *Rhizobium*-legume symbiosis. Sci. Total Environ. 503–512.

Fang, Y. and Ramasamy, R.P. 2015. Current and prospective methods for plant disease detection. Biosensors 5: 537–561.

FAO. 2018. Food and Agriculture Organization of the United Nations (FAO). Statistical Databases. Accessed on 02/2022. In: https://www.fao.org/news/archive/news-by-date/2018/en/.

FAO. 2020. Food and Agricultural Organization of the United Nations (FAO). Sustainable Food and Agriculture. Accessed on 02/2022. In: https://www.fao.org/sustainability/news/detail/en/c/1274219/.

FAO. 2021. Food and Agricultural Organization of the United Nations (FAO). Pesticides use, pesticides trade and pesticides indicators. Global, regional and country trends, 1990–2019. FAOSTAT Analytical Brief Series No. 29. Roma. Accessed on 02/2022. In: https://www.fao.org/3/cb6034en/cb6034en.pdf.

Fayiga, J. 2017. Nanoparticles in biosolids: Effect on soil health and crop growth. Peertechz J. Environ. Sci. Toxicol. 2: 059–067.

Federici, G., Shaw, B.J. and Handy, R.D. 2007. Toxicity of titanium dioxide nanoparticles to rainbow trout, (*Oncorhynchus mykiss*): Gill injury, oxidative stress, and other physiological effects. Aquat. Toxicol. 84: 415–430.

Feng, Y.Z., Cui, X.C., He, S.Y., Dong, G., Chen, M. and Wang, J.H. 2013. The role of metal nanoparticles in influencing arbuscular mycorrhizal fungi effects on plant growth. Environ. Sci. Technol. 47: 9496–9504.

Ferdous, Z. and Nemmar, A. 2020. Health impact of silver nanoparticles: A review of the biodistribution and toxicity following various routes of exposure. Int. J. Mol. Sci. 21: 2375.

Feroze, N., Arshad, B., Younas, M., Afridi, M.I., Saqib, S. and Ayaz, A. 2020. Fungal mediated synthesis of silver nanoparticles and evaluation of antibacterial activity. Microsc. Res. Tech. 83: 72–80.

Ferraz, C.M., Campodonico de Oliveira, M.L., Barbosa de Assis, J.P., Silva, L.P.C., Tobias, F.L., Lima, T.F. and Braga, F.R. 2021. *In vitro* evaluation of the nematicidal effect of *Duddingtonia flagrans* silver nanoparticles against strongylid larvae (L3). Biocontrol Sci. Technol. 1–5.

Garcia-Rubio, R., de Oliviera, H.C., Rivera, J. and Trevijano-Contador, N. 2020. The fungal cell wall: *Candida, Cryptococcus* and *Aspergillus* species. Front. Microbiol. 10: 2993.

Gerloff, K., Albrecht, C. and Boots, A.W. 2009. Cytotoxicity and oxidative DNA damage by nanoparticles in human intestinal Caco-2 cells. Nanotoxicology 3: 355–364.

Gherbawy, Y.A., Shalaby, I.M., El-sadek, M.S.A., Elhariry, H.M. and Banaja, A.A. 2013. The anti-fasciolasis properties of silver nanoparticles produced by *Trichoderma harzianum* and their improvement of the anti-fasciolasis drug triclabendazole. Int. J. Mol. Sci. 14: 21887–21898.

Griffitt, J., Weil, R. and Hyndman, K.A. 2007. Exposure to copper nanoparticles causes gill injury and acute lethality in zebrafish (*Danio rerio*). Env. Sci. Technol. 41: 8178–8186.

Guilger, M., Pasquoto-Stigliani, T., Bilesky-Jose, N., Grillo, R., Abhilash, P.C. and Fraceto, L.F. 2017. Biogenic silver nanoparticles based on *Trichoderma harzianum*: Synthesis, characterization, toxicity evaluation and biological activity. Sci. Rep. 7: 44421.

Guilger-Casagrande, M. and de Lima, R. 2019. Synthesis of silver nanoparticles mediated by fungi: A review. Front. Bioeng. Biotechnol. 7: 287.

Guo, H. 2016. Intravenous administration of silver nanoparticles causes organ toxicity through intracellular ROS-related loss of interendothelial junction. Part. Fibre. Toxicol. 13: 21.

Handy, R.D. and Eddy, F.B. 1991. Effects of inorganic cations on Na+ adsorption to the gill and body surface of rainbow trout, *Oncorhynchus mykiss*, in dilute solutions. Can. J. Fish. Aquat. Sci. 48: 1829–1837.

Handy, R.D., Kammer, F., Lead, J.R., Hassello, M., Owen, R. and Crane, M. 2008. The ecotoxicology and chemistry of manufactured nanoparticles. Ecotoxicology 17: 287–314.

Hillyer, J.F. and Albrecht, R.M. 2001. Gastrointestinal persorption and tissue distribution of differently sized colloidal gold nanoparticles. J. Pharm. Sci. 90: 1927–1936.

Hossen, S., Hossain, M.K., Basher, M.K., Mia, M.N.H., Rahman, M.T. and Uddin, M.J. 2019. Smart nanocarrier-based drug delivery systems for cancer therapy and toxicity studies: A review. J. Adv. Res. 15: 1–18.

Hu, X., Saravanakumar, K., Jin, T. and Wang, M.-H. 2019. Mycosynthesis, characterization, anticancer and antibacterial activity of silver nanoparticles from endophytic fungus *Talaromyces purpureogenus*. Int. J. Nanomed. 14: 3427–3438.

Hussain, S.M., Hess, K.L. and Gearhart, J.M. 2005. *In vitro* toxicity of nanoparticles in BRL3A rat liver cells. Toxicol. *In Vitro* 19: 975–983.

Ingle, A., Gade, A., Pierrat, S., Sönnichsen, C. and Rai, M. 2008. Mycosynthesis of silver nanoparticles using the fungus *Fusarium acuminatum* and its activity against some human pathogenic bacteria. Curr. Nanosci. 4: 141–144.

Ingle, A.P., Gupta, I., Jogee, P. and Rai, M. 2020. Role of nanotechnology in the detection of mycotoxins: A smart approach. pp. 11–33. *In*: Rai, M. and Abd-Elsalam, K.A. (eds.). Nanomycotoxicology Academic Press, Elsevier.

Islam, S.N., Naqvi, S.M.A., Parveen, S. and Ahmad, A. 2021. Application of mycogenic silver/silver oxide nanoparticles in electrochemical glucose sensing; alongside their catalytic and antimicrobial activity. 3 Biotech. 11: 1–11.

Jaswal, T. and Gupta, J. 2021. A review on the toxicity of silver nanoparticles on human health. Mat. Today: Proceedings 5: 2214–7853.

Jin, C., Wang, K., Oppong-Gyebi, A. and Hu, J. 2020. Application of nanotechnology in cancer diagnosis and therapy—A mini-review. Int. J. Med. Sci. 17: 2964–2973.

Jing, X.X., Su, Z.Z., Xing, H.E., Wang, F.Y., Shi, Z.Y. and Liu, X.Q. 2016. Biological effects of ZnO nanoparticles as influenced by arbuscular mycorrhizal inoculation and phosphorus fertilization. Environ. Sci. 37: 3208–3215.

Jogee, P. and Rai, M. 2020. Application of nanoparticles in inhibition of mycotoxin-producing fungi. pp. 239–250. *In*: Rai, M. and Abd-Elsalam, K.A. (eds.). Nanomycotoxicology, Academic Press, Elsevier.

Joner, E.J., Hartnik, T. and Amundsen, C.E. 2008. Environmental fate and ecotoxicity of engineered nanoparticles. *In*: Norwegian Pollution Control Authority Report no. TA 2304/2007, Bioforsk, Norway, pp. 1–64.

Kalaiselvan, V. and Rajasekaran, A. 2009. Biosynthesis of silver nanoparticles from *Aspergillus niger* and evaluation of its wound healing activity in experimental rat model. Int. J. Pharm. Tech. Res. 4: 1523–1529.

Kalpana, V.N. and Devi Rajeswari, V. 2018. A review on green synthesis, biomedical applications, and toxicity studies of ZnO NPs. Bioinorg. Chem. Appl. 18: 1–12.

Keerthana, S. and Kumar, A. 2020. Potential risks and benefits of zinc oxide nanoparticles: A systematic review. Crit. Rev. Toxicol. 50: 47–71.

Kermanizadeh, A., Gaiser, B.K. and Hohnston, H. 2014. Toxicological effect of engineered nanomaterials on the liver. Br. J. Pharmacol. 171: 3980–3987.

Khandel, P. and Shahi, S.K. 2018. Mycogenic nanoparticles and their bio-prospective applications: Current status and future challenges. J. Nanostructure Chem. 8: 369–391.

Khlebtsov, N. and Dykman, L. 2011. Biodistribution and toxicity of engineered gold nanoparticles: A review of *in vitro* and *in vivo* studies. Chem. Soc. Rev. 40: 1647–1671.

Kim, S.W., Kim, K.S., Lamsal, K., Kim, Y.J., Kim, S.B., Jung, M.Y. and Lee, Y.S. 2009. An *in vitro* study of the antifungal effect of silver nanoparticles on oak wilt pathogen Raffaelea sp. J. Microbiol. Biotechnol. 19: 760–764.

Klingenfuss, F. 2014. Testing of TiO_2 nanoparticles on wheat and microorganisms in a soil microcosm. Master's Thesis, University of Gothenburg, Gothenburg, Sweden, 2014.

Lancet Editorial. 2018. GLOBOCAN 2018: Counting the toll of cancer. Lancet 392(10152): 985. Doi: 10.1016/S0140-6736(18)32252-9.

Lassaletta, L., Billen, G., Grizzetti, B., Anglade, J. and Garnier, J. 2014. 50 year trends in nitrogen use efficiency of world cropping systems: The relationship between yield and nitrogen input to cropland. Environ. Res. Lett. 9: 105011.

Li, G., He, D., Qian, Y., Guan, B., Gao, S., Cui, Y. and Wang, L. 2012. Fungus-mediated green synthesis of silver nanoparticles using *Aspergillus terreus*. Int. J. Mol. Sci. 13: 466–476.

Li, Q., Liu, F., Li, M., Chen, C. and Gadd, G.M. 2021. Nanoparticle and nanomineral production by fungi. Fung. Biol. Rev. Doi: https://doi.org/10.1016/j.fbr.2021.07.003.

Liao, C., Li, Y. and Tjong, S.C. 2019. Bactericidal and cytotoxic properties of silver nanoparticles. Int. J. Mol. Sci. 20: 449.

Liao, M. and Liu, H. 2012. Gene expression profiling of nephrotoxicity from copper nanoparticles in rats after repeated oral administration. Environ. Toxicol. Pharmacol. 34: 67–80.

Liu, F., Wen, L.X., Li, Z.Z., Yu, W. and Sun, H.Y. 2006. Porous hollow silica nanoparticles as controlled delivery system for water-soluble pesticide. Mater. Res. Bull. 41: 2268–2275.

Mahawar, H. and Prasanna, R. 2018. Prospecting the interactions of nanoparticles with beneficial microorganisms for developing green technologies for agriculture. Env. Nanotechnol. Monitoring Manag. 10: 477–485.

Mani, V.M., Kalaivani, S., Sabarathinam, S., Vasuki, M., Soundari, A.J.P.G., Das, M.P.A., Elfasakhany, A. and Pugazhendhi, A. 2021. Copper oxide nanoparticles synthesized from an endophytic fungus *Aspergillus terreus*: Bioactivity and anti-cancer evaluations. Environmental Research 201: 111502.

Mansoor, S., Zahoor, I., Baba, T.R., Padder, S.A., Bhat, Z.A., Koul, A.M. and Jiang, L. 2021. Fabrication of silver nanoparticles against fungal pathogens. Front. Nanotechnol. 3: 679358.

Mao, L., Chen, Z., Wang, Y. and Chen, C. 2021. Design and application of nanoparticles as vaccine adjuvants against human corona virus infection. J. Inorg. Biochem. 219: 111454.

Marambio-Jones, C.E.M. 2010. A review of the antibacterial effects of silver nano materials and potentials implications for humans health and the environment. J. Nanopart. Res. 12: 1531–151.

Marcato, P.D., Nakasato, G., Brocchi, M., Melo, P.S., Huber, S.C., Ferreira, I.R. and Durán, N. 2012. Biogenic silver nanoparticles: Antibacterial and cytotoxicity applied to textile fabrics. J. Nano. Res. 20: 69–76.

Merkoçi, A. 2021 Smart nanobiosensors in agriculture. Nat. Food 2: 920–921.

Mora-Aguilera, G., Acevedo-Sánchez, G., Guzmán-Hernández, E., Flores-Colorado, O.E., Coria-Contreras, J.J., Mendoza-Ramos, C., Martínez-Bustamante, V.I., López Buenfil, A., González-

Gómez, R. and Javier-López, M.A. 2021. Web-based epidemiological surveillance systems and applications to coffee rust disease. Mexican J. Phytopathol. 39: 452–492.

Narayanan, K.B., Park, H.H. and Sakthivel, N. 2013. Extracellular synthesis of mycogenic silver nanoparticles by *Cylindrocladium floridanum* and its homogeneous catalytic degradation of 4-nitrophenol. Spectrochim. Acta A Mol. Biomol. Spectrosc. 116: 485–490.

Noori, A., White, J.C. and Newman, L.A. 2017. Mycorrhizal fungi influence on silver uptake and membrane protein gene expression following silver nanoparticle exposure. J. Nanopart. Res. 19: 66.

Osama, E., El-Sheikh, S.M., Khairy, M.H. and Galal, A.A. 2020. Nanoparticles and their potential applications in veterinary medicine. J. Adv. Vet. Anim. Res. 10: 268–273.

Osmond, M.J. and Mccall, M.J. 2010. Zinc oxide nanoparticles in modern sunscreens: An analysis of potential exposure and hazard. Nanotoxicol. 4: 15–41.

Othman, A.M., Elsayed, M.A., Al-Balakocy, N.G., Hassan, M.M. and Elshafei, A.M. 2019. Biosynthesis and characterization of silver nanoparticles induced by fungal proteins and its application in different biological activities. J. Genet. Eng. Biotechnol. 17: 1–13.

Patel, P.D. 2002. (Bio) sensors for measurement of analytes implicated in food safety: A review. Trend. Analyt. Chem. 21: 96–115.

Petkar, K.C., Patil, S.M., Chavhan, S.S., Kaneko, K., Sawant, K.K., Kunda, N.K. and Saleem, I.Y. 2021. An overview of nanocarrier-based adjuvants for vaccine delivery. Pharmaceutics 13: 455.

Pimentel, D. and Burgess, M. 2014. Environmental and economic costs of the application of pesticides primarily in the United States. pp. 47–71. *In*: Pimentel, D. and Peshin, R. (eds.). Integrated Pest Management. Springer, Switzerland.

Pittol, M., Tomacheski, D., Simões, D.N., Ribeiro, V.F. and Santana, R.M.C. 2017. Macroscopic effects of silver nanoparticles and titanium dioxide on edible plant growth. Environ. Nanotechnol. Monit. Manag. 8: 127–133.

Pratsinis, A., Hervella, P., Leroux, J.C., Pratsinis, S.E. and Sotiriou, G.S. 2013. Toxicity of silver nanoparticles in marcrophages. Small 12: 2576–84.

Radetić, M. 2013. Functionalization of textile materials with silver nanoparticles. J. Mat. Sci. 48: 95–107.

Rahman, K., Khan, S.U., Fahad, S., Chang, M.X., Abbas, A., Khan, W.U., Rahman, L., Ul Haq, Z., Nabi, G. and Khan, D. 2019. Nano-biotechnology: A new approach to treat and prevent malaria. Int. J. Nanomed. 14: 1401–1410.

Rai, M., Bonde, S., Golinska, P., Trzcińska-Wencel, J., Gade, A., Abd-Elsalam, K., Shende, S., Gaikwad, S. and Ingle, A. 2021b. *Fusarium* as a novel fungus for the synthesis of nanoparticles: Mechanism and applications. J. Fungi 7: 139.

Rai, M., Ingle, A.P., Gaikwad, S., Padovani, F.H. and Alves, M. 2016. The role of nanotechnology in control of human diseases: perspectives in ocular surface diseases. Crit. Rev. Biotechnol. 36: 777–787.

Rai, M., Ingle, A.P., Paralikar, P., Gupta, I., Medici, S. and Santos, C.A. 2017. Recent advances in use of silver nanoparticles as antimalarial agents. Int. J. Pharm. 526: 254–270.

Rai, M., Ingle, A.P., Trzcinska-Wencel, J., Wypij, M., Bonde, S., Yadav, A., Kratosova, G. and Golinska, P. 2021a. Biogenic silver nanoparticles: What we know and what do we need to know? Nanomaterials 11: 2901.

Rai, M., Yadav, A. and Gade, A. 2009a. Silver nanoparticles as a new generation of antimicrobials. Biotechnol. Adv. 27: 76–83.

Rai, M., Yadav, A. Bridge, P. and Gade, A. 2009b. Myconanotechnology: A new and emerging science. pp. 258–267. *In*: Rai, M. and Bridge, P. (eds.). Applied Mycology, CABI, UK.

Rai, M., Yadav, A., Ingle, A.P., Reshetilov, A., Blanco-Prieto, M.J. and Feitosa, C.M. 2019. Neurodegenerative diseases: The real problem and nanobiotechnological solutions. pp. 1–17. *In*: Rai, M. and Yadav, A. (eds.). Nanobiotechnology in Neurodegenerative Diseases, Springer, Switzerland.

Raies, A.B. and Bajic, V.B. 2016. In silico toxicology: Computational methods for the prediction of chemical toxicity. Wiley Interdiscip Rev. Comput. Mol. Sci. 6: 147–172.

Rana, S. and Kalaichelvan, P.T. 2013. Ecotoxicity of nanoparticles. Int. Sch. Res. Notices 2013: 574648.

Rao, P.V., Nallappan, D., Madhavi, K., Rahman, S., Jun, L.W. and Gan, S.H. 2016. Phytochemicals and biogenic metallic nanoparticles as anticancer agents, oxidative medicine and cellular longevity. Oxidat. Med. Cell. Longevity 23: 15.

Renganathan, S., Subramaniyan, S., Karunanithi, N., Vasanthakumar, P., Kutzner, A., Kim, P.-S. and Heese, K. 2021. Antibacterial, antifungal, and antioxidant activities of silver nanoparticles biosynthesized from *Bauhinia tomentosa* Linn. Antioxidants 10: 1959. https://doi.org/10.3390/ antiox10121959.

Rezvantalab, S., Drude, N.I., Moraveji, M.K., Güvener, N., Koons, E.K., Shi, Y., Lammers, T. and Kiessling, F. 2018. PLGA-based nanoparticles in cancer treatment. Front. Pharmacol. 9: 1260.

Rojas, C.A.A., Jex, A.R., Gasser, R.B. and Scheerlinck, J.P.Y. 2014. Techniques for the diagnosis of Fasciola infections in animals: Room for improvement. Adv. Parasitol. 85: 65–107.

Sabella, S., Carney, R.P., Brunetti, V., Malvindi, M.A., Al-Juffali, N., Vecchio, G., Janes, S.M., Bakr, O.M., Cingolani, R., Stellacci, F. and Pompa, P.P. 2014. A general mechanism for intracellular toxicity of metal-containing nanoparticles. Nanoscale 6: 7052–61.

Saccá, M.L., Caracciolo, A.B., Lenola, D. and Grenni, M.P. 2017. Ecosystem services provided by soil microorganisms. pp. 9–24. *In*: Lukac, M., Grenni, P. and Gamboni, M. (eds.). Soil Biological Communities and Ecosystem Resilience, Springer, Cham.

Santos, T.S., Silva, T.M., Cardoso, J.C., de Albuquerque-Júnior, R.L.C., Zielinska, A., Souto, E.B., Severino, P. and da Costa Mendonça, M. 2021. Biosynthesis of silver nanoparticles mediated by entomopathogenic fungi: Antimicrobial resistance, nanopesticides, and toxicity. Antibiotics 10: 852.

Saravanakumar, K. 2019. Biosynthesis and characterization of copper oxide nanoparticles from indigenous fungi and its effect of photothermolysis on human lung carcinoma. J. Photochem. Photobiol. B Biol. 7.

Seneviratne, M., Gunaratne, S., Bandara, T., Weerasundara, L., Rajakaruna, N., Seneviratne, G. and Vithanage, M. 2016. Plant growth promotion by *Bradyrhizobium japonicum* under heavy metal stress. S. Afr. J. Bot. 105: 19–24.

Sharma, A., Sagar, A., Rana, J. and Rani, R. 2022. Green synthesis of silver nanoparticles and its antibacterial activity using fungus Talaromyces purpureogenus isolated from Taxus baccata Linn. Micro Nanosyst. Lett. 10: 1–12.

Sugunan, A., Warad, H.C., Boman, M. and Dutta, J. 2006. Zinc oxide nanowires in chemical bath on seeded substrates: Role of hexamine. J. Sol-Gel Sci. Technol. 39: 49–56.

Sundaravadivelan, C. and Padmanabhan, M.N. 2014. Effect of mycosynthesized silver nanoparticles from filtrate of *Trichoderma harzianum* against larvae and pupa of dengue vector *Aedes aegypti* L. Environ. Sci. Pollut. Res. Int. 21: 4624–33.

Vale, G., Mehennaoui, K., Cambier, S., Libralato, G., Jomini, S. and Domingos, R.F. 2015. Manufactured nanoparticles in the aquatic environment-biochemical responses on freshwater organisms: A critical overview. Aquat. Toxicol. 170: 162–174.

van der Zande Sahu, M., Vandebriel, R.J. and Van Doren, E. 2012. Distribution, elimination, and toxicity of silver nanoparticles and silver ions in rats after 28-day oral exposure. ACS Nano 6: 7427–7442.

Velhal, S.G., Kulkarni, S.D. and Latpate, R.V. 2016. Fungal mediated silver nanoparticle synthesis using robust experimental design and its application in cotton fabric. Int. Nano Lett. 6: 257–264.

Vigneshwaran, N., Varadarajan, P.V. and Balasubramanya, R.H. 2007. Application of metallic nanoparticles in textiles. Nanotechnologies for the Life Sciences. https://doi.org/10.1002/9783527610419. NTLS0136.

Wang, F.Y., Liu, X.Q., Shi, Z.Y., Tong, R.J., Adams, C.A. and Shi, X.J. 2016. Arbuscular mycorrhizae alleviate negative effects of zinc oxide nanoparticle and zinc accumulation in maize plants—A soil microcosm experiment. Chemosphere 147: 88–97.

Wang, Z., Wei, F., Liu, S.Y., Xu, Q., Huang, J.Y., Dong, X.Y. and Chen, H. 2010. Electrocatalytic oxidation of phytohormone salicylic acid at copper nanoparticles-modified gold electrode and its detection in oilseed rape infected with fungal pathogen Sclerotinia sclerotiorum. Talanta 80: 1277–1281.

Worrall, E.A., Hamid, A., Mody, K.T., Mitter, N. and Pappu, H.R. 2018. Nanotechnology for plant disease management. Agronomy 8: 285.

Wu, L., Wang, M. and Wei, D. 2021. Advances in gold nanoparticles for mycotoxin. Analyst 146: 1793–1806.

Xiong, L., Zhang, X., Huang, Y.X., Liu, W.J., Chen, Y.L. and Yu, S.S. 2018. Biogenic synthesis of Pd-based nanoparticles with enhanced catalytic activity. ACS Appl. Nano Mater. 1: 1467–1475.

Yadav, A., Kon, K., Kratosova, G., Duran, N., Ingle, A.P. and Rai, M. 2015. Fungi as an efficient mycosystem for the synthesis of metal nanoparticles: Progress and key aspects of research. Biotechnol. Lett. 7: 2099–2120.

Yang, D. and Watts, J. 2005. Particle surface characteristics may play an important role in phytotoxicity of alumina nanoparticles. Toxicol. Lett. 158: 122–132.

Yanga, W. and Cao, Y. 2017. Interactions between nanoparticles and plants: Phytotoxicity and defense mechanisms. J. Plant Interact. 12: 158–169.

Yilmaz, B., Terekeci, H., Sandal, S. and Kelestimur, F. 2020. Endocrine disrupting chemicals: Exposure, effects on human health, mechanism of action, models for testing and strategies for prevention. Rev. Endocr. Metab. Disord. 21: 127–147.

Zhang, P., Guo, Z., Zhang, Z., Fu, H., White, J.C. and Lynch, I. 2020. Nanomaterial transformation in the soil–plant system: Implications for food safety and application in agriculture. Small 16: 2000705.

Zhang, S., Gao, H. and Bao, G. 2015. Physical process of nanoparticle cellular endocytosis. ACS Nano 9: 8655–8671.

Section II
Biomedical Applications

2

Synthesis of Biogenic Nanoparticles by Fungi and Cancer Applications

Nelson Durán,[1,2,]* Marcelo B. de Jesus,[3] Queila C. Dias,[1] Gerson Nakazato[4] and Wagner J. Fávaro[1,]*

Introduction

Microorganisms are known to efficiently produce different types of unique nanostructures. This property is used to synthesize nanomaterials for many applications. For instance, metallic molecules can be produced by different microorganisms, such as bacteria and fungi, based on biogenic methods and induced synthesis, in order to control the biogenic synthesis of nanostructures presenting specific morphologies and assemblies. However, the biogenic synthesis of nanoparticles is manageable in terms of particle geometry and processing at an industrial scale, regardless of the nanoparticle production accuracy (Fang et al. 2019, Ghosh et al. 2021, Koul et al. 2021).

Nevertheless, biogenically induced synthesis has encouraged researchers in this field to synthesize inorganic nanoparticles by using regular metal precursors (Grasso et al. 2020, Gupta and Seema 2021), since this process also produces a wide variety of compositions. Organisms such as bacteria, fungi, and algae are cultured to generate nanoparticles. However, the fungi-based biogenic synthesis

[1] Laboratory of Urogenital Carcinogenesis and Immunotherapy, Department of Structural and Functional Biology, Universidade Estadual de Campinas (UNICAMP), Campinas, São Paulo, Brazil.
[2] Nanomedicine Research Unit (Nanomed), Center for Natural and Human Sciences (CCNH), Universidade Federal do ABC (UFABC), Santo André, São Paulo, Brazil.
[3] Department of Biochemistry and Tissue Biology, Institute of Biology, Universidade Estadual de Campinas, UNICAMP, Campinas, São Paulo, Brazil.
[4] Laboratory of Basic and Applied Bacteriology, Department of Microbiology, Biology Sciences Center, Universidade Estadual de Londrina (UEL), Londrina, PR, Brazil.
* Corresponding authors: nelsonduran1942@gmail.com; wjfavaro@gmail.com

of nanostructures presents undeniable advantages over bacteria-based biogenic synthesis. Therefore, research in this field has been encouraged over the past decade. The benefits of using fungi lie in the fact that the mycelium is easily scaled up and that the protein extraction process is economically feasible and ecologically green (Durán et al. 2005, 2016, Gaikwad et al. 2013, How et al. 2021). Fungi presenting enzymes in both the cytoplasm and cell wall that can turn metal cations into nanoparticles (Durán et al. 2005, 2010, Chatterjee et al. 2020, Bruna et al. 2021). Metal cations have a positive charge, which is captured by fungi to start biosynthesis processes. Furthermore, fungi can secrete large amounts of protein and it enhances the biogenic production of nanoparticles (Mughal et al. 2021).

Among biogenic nanoparticles, mainly the gold and silver metal nanoparticles have been widely investigated in the nanomedicine field (Durán et al. 2016, Santos et al. 2019). Some of the examples of the use of nanoparticles concern the following applications: magnetic nanoparticles, for imaging purposes (Heo et al. 2020, Billings et al. 2021); silver nanoparticles, for photodetectors (Paysen et al. 2020, Bouafia et al. 2021, Phummirat et al. 2021); barium titanate nanoparticles, for breast cancer cell treatment (Yoon et al. 2020, Ahamed et al. 2020); zinc oxide nanoparticles, for lung cancer cells (Chabattula et al. 2021); copper and copper oxide (Noor et al. 2020), cobalt oxide (Vijayanandan and Balakrishnan 2018), magnesium oxide (Saied et al. 2021), selenium nanoparticles (Wadhwani et al. 2016) in cell cultures; gold nanoparticles, for light-induced magnetism (Cheng et al. 2020, Iqbal et al. 2021); gold, for antimicrobial activity (Lee et al. 2020, Mikhailova 2021); and gold nanoparticles, for SARS-CoV-2 virus monitoring throughout the pandemic period (Moitra et al. 2020, Zhang et al. 2021). These studies have emphasized the usefulness of biogenic nanoparticles as a versatile and cost-effective tool for therapeutic and technological applications.

The aim of the current chapter is to review the biogenic synthesis of metallic nanoparticles and the current state of art of these nanoparticles' applications, both *in vitro* and *in vivo*. Finally, it focused on the fungal-mediated synthesis of different nanoparticles and on their applications in cancer cases, mainly in cancer treatment applied *in vitro* and *in vivo*.

Biogenic Nanoparticles by Fungi

Although there has not been an increase in the number of studies about the biogenic process of gold nanoparticles as compared to silver nanoparticles, a relatively larger number of studies are focusing on investigating the use of gold in nanomedicine applications. Hulikere et al. (2017) have synthesized gold nanoparticles by using fungal species *Cladosporium cladosporioides* isolated from *Sargassum wightii* seaweeds (Hulikere and Joshi 2019). They have shown that nanoparticle synthesis mechanism was based on NADPH-dependent reductase enzymes, as well as that the phenolic moieties of compounds worked as nanoparticles' synthesis inductors. This mechanism was previously suggested for *Fusarium oxysporum* in silver nanoparticles (Durán et al. 2005). Several biogenic syntheses use fungal species *F. oxysporum* to synthesize nanoparticles, as seen in different studies conducted with silver nanoparticles. Srivastava and co-researchers used fungal species *F. oxysporum* to synthesize silver nanoparticles and showed its antibacterial effects on *Escherichia coli* and *Pseudomonas aeruginosa* (Srivastava et al. 2019). The literature has already described different nanoparticles produced by *F. oxysporum*, such as copper (Pham et al. 2019, Noor et al. 2020, Mani et al. 2021), cobalt oxide (Abdel-Aziz et al. 2020), magnesium oxide (Saied et al. 2021), and gold (Naimi-Shamel et al. 2019).

Thus, previously described fungi can also synthesize metal oxide nanoparticles (Durán and Seabra 2012, Seabra et al. 2013). Magnetite is a common iron oxide with magnetic properties. These nanoparticles have been used to eliminate heavy metals and other types of materials from the environment due to their high surface area/ volume ratio (Crane et al. 2011). Magnetite nanoparticles are highly efficient in this elimination process in comparison to other metal nanoparticles. It happens because their superparamagnetic behavior enables separating them from wastewater due to surface charge and complexation processes. The use of fungus-Fe_3O_4 nanocomposites for nuclear waste management was reported in literature (Ding et al. 2015). Magnetite nanoparticles have been widely used for medical applications such as MRI (Peigneux et al. 2016), oscillation damping and position sensing (Yan et al. 2017), and recording machines. Similar to bacteria, fungi also show major disadvantages in biosafety. Fungal species such as *F. oxysporum* are unsafe due to their pathogenicity and health concerns. Fortunately, many non-pathogenic fungi available in nature are suitable to be used in nanoparticle synthesis. Different *Trichoderma* fungal species are safe examples, since they have been used in food applications (Hu et al. 2019), as well as in the medical and paper industries (Abd-Elsalam 2021, Bahrulolum et al. 2021).

The following examples can be listed as fungal-based biogenic synthesis of silver nanoparticles, with antimicrobial activity and no application in cancer treatment: *Cladosporium cladosporioides* (Hulikere and Joshi 2019), *Fusarium oxysporum* (Srivastava et al. 2019), *Fusarium oxysporum* (Marcato et al. 2012, Santos et al. 2019, Ballottin et al. 2017), *Phoma gardeniae* (Rai et al. 2015), *Aspergillus flavus* (Sulaiman et al. 2015), *Trichoderma harzianum* (Guilger-Casagrande et al. 2019), *Fusarium scirpi* (Rodriguez-Serrano et al. 2020), *Trichoderma longibrachiatum* (Elamawi et al. 2018), *Nigrospora oryzae* (Dawoud et al. 2021), *Alternaria* sp. (Singh et al. 2017), *Phomopsis helianthi* (Gond et al. 2020), *Colletotrichum* sp. (Azmath et al. 2016), and *Aspergillus tubingensis* (Rodrigues et al. 2021). Those used for the biogenic synthesis of copper nanoparticles comprise of *Fusarium oxysporum* (Majumder 2012), *Aspergillus niger* (Naqvi et al. 2017), from copper oxide: *Penicillium chrysogenum* (El-Batal et al. 2020), for selenium: *Mariannaea* sp. HJ (Zhang et al. 2019), *Fusarium* sp. and *Trichoderma reesi* (Gharieb et al. 1995); for cobalt oxide: *Aspergillus nidulans* (Vijayanandan and Balakrishnan 2018), for magnesium oxide: *Aspergillus terreus* (Saied et al. 2021), and for gold: *Cladosporium cladosporioides* (Hulikere et al. 2017), *Cladosporium oxysporum* (Bhargava et al. 2016), *Trichoderma harzianum* (Abdel-Kareem et al. 2018) and *Cantharellus* sp. (Jha et al. 2021).

Table 1 shows some of the most significant metallic nanoparticles and oxides investigated about cancer in the literature. Based on this table, it is possible to see that silver and gold nanoparticles are the metal nanoparticles mainly produced by fungi.

Table 1. Some examples of recent studies on biogenic nanoparticles synthesis using fungi for cancer treatment.

Nanoparticles	Fungus	Application	Reference
Silver	*Ganoderma neo-japonicum Imazeki*	Anticancer	Gurunathan et al. 2013
Silver	*Calocybe indica*	Anticancer	Gurunathan et al. 2015
Silver	*Fusarium oxysporum*	Antimicrobial and anticancer	Husseiny et al. 2015
Silver	*Pleurotus djamo* var. *roseus*	Anticancer	Raman et al. 2015
Silver	*Guignardia mangiferae*	Anticancer	Balakumaran et al. 2015
Silver	*Pestalotiopsis microspora*	Antioxidant and anticancer	Netala et al. 2016
Silver	*Cunninghamella echinulata*	Anticancer	Anbazhagan et al. 2017
Silver	*Trichoderma viride*	Anticancer, immunomodulatory activity	Adebayo-Tayo et al. 2019
Silver	*Fusarium oxysporum*	Antitumoral	Ferreira et al. 2019
Silver	*Fusarium oxysporum*	Anticancer	Ferreira et al. 2020
Silver	*Talaromyces purpurogenus*	Antimicrobial, anticancer	Bhatnagar et al. 2019
Silver	*Aspergillus fumigatus*	Antimicrobial, anticancer	Othman et al. 2019
Silver	*Aspergillus sydowii*	Antifungal, antiproliferative	Wang et al. 2021
Silver	*Aspergillus terreus*	Antimicrobial, anticancer	Lotfy et al. 2021
Silver	*Rhizopus stolonifera*	Anticancer	Banu et al. 2021
Copper	*Aspergillus terreus*	Antioxidant, antimicrobial, anticancer	Noor et al. 2020
CuO	*Trichoderma asperellum*	Anticancer	Saravanakumar et al. 2019
CuO	*Aspergillus terreus*	Antimicrobial, antifungal, anticancer	Mani et al. 2021
ZnO	*Aspergillus terreus*	Anticancer	Baskar et al. 2015
ZnO	*Alternaria tenuissima*	Antimicrobial, anticancer	Abdelhakim et al. 2020
ZnO	*Trichoderma harzianum*	Antimicrobial, anticancer	Saravanakumar et al. 2020
Selenium	*Fusarium semitectum*	Antioxidant, antimicrobial, anticancer	Abbas and Baker 2020
Gold	*Pleurotus ostreatus*	Antimicrobial and anticancer	El Domany et al. 2018
Gold	*Aspergillus flavus*	Anticancer	Abu-Tahon et al. 2020
Gold	*Fusarium solani*	Anticancer	Clarance et al. 2020

Biogenic Nanoparticles Synthesized by Fungi and their Effect on Cancer in Cell Culture

(i) Silver Nanoparticles

Recent reviews have described the current state of knowledge about the processes and mechanisms involved in the biogenic production of silver nanoparticles, as well as about their potential to be used in the medical field. However, the reported biogenic silver nanoparticles are mainly derived from phyto-sources and have a remarkably small number of fungal origins (Ratan et al. 2020, Rai et al. 2021). MDA-MB-231 breast cancer cells treated for 24 hours with different concentrations of bio-silver nanoparticles (1–10 µg/mL) deriving from *Ganoderma neo-japonicum* (*Imazeki*, spherical, 4.4 ± 2 nm in size) have shown decreased cell viability, dose-dependent membrane damage (Figure 1A), as well as cleaved/active caspase-3 upregulation. Furthermore, silver cations released from bio-silver nanoparticles are the leading cause of caspase-3 activation (Figure 1B) and the ultimate cause of oxidative stress (over 50% of the control) (Gurunathan et al. 2013). These findings have shown that silver nanoparticles had cytotoxic effects and apoptotic properties, and suggested that reactive oxygen species produced by silver nanoparticles play a key role in apoptosis. Thus, these nanoparticles could contribute to the development of adequate anticancer drugs capable of contributing to advancements in nanomedicine applied to cancer therapy.

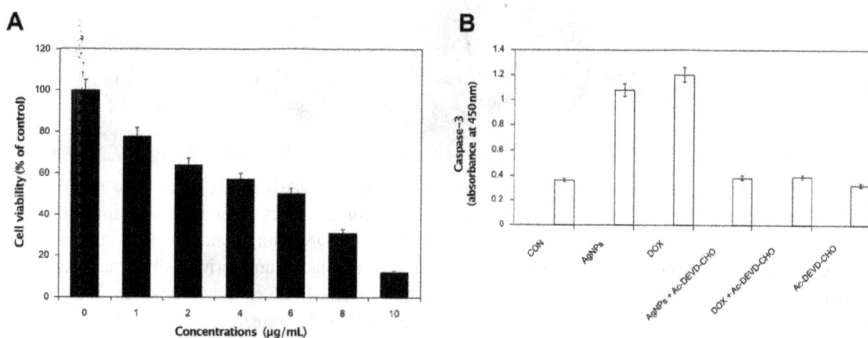

Figure 1. (A) Cytotoxic effect of silver nanoparticles (AgNPs) on MDA-MB-231 human breast cancer cells. Cells were treated with AgNPs at various concentrations for 24 hours, and cytotoxicity was determined by the MTT (3-[4,5-dimethylthiazol-2-yl]-2,5-diphenyltetrazolium bromide) method. Notes: The results represent the means of three separate experiments, and error bars represent the standard error of the mean. Treated groups showed statistically significant differences from the control group with Student's t-test (P < 0.05). (B) Silver nanoparticles (AgNPs) induce the activity of caspase 3 in MDA-MB-231 human breast cancer cells. MDA-MB-231 cells were treated with AgNPs, Doxorubicin (DOX), and/or a caspase-3 inhibitor (Ac-DEVD-CHO) for 24 hours. Notes: The concentration of the p-nitroaniline released from the substrate was calculated from the absorbance values at 405 nm. The results represent the means of three separate experiments, and error bars represent the standard error of the mean. Treated groups showed statistically significant differences from the control group with Student's t-test (P < 0.05). Abbreviation: CON, control (Gurunathan et al. 2013, by Dove Medical Press Limited, and licensed under Creative Commons Attribution – Non-Commercial (unported, v3.0) License).

The literature reported the inhibitory effect of silver nanoparticles deriving from *Calocybe indica* (20 nm size) on MDA-MB-231 breast cancer cell lines (Gurunathan et al. 2015). Silver nanoparticles produced by this fungus (F-AgNPs) were compared to silver nanoparticles produced by bacteria (*Bacillus tequilensis*) (B-AgNPs) in the aforementioned study. Cell viability experiments have indicated the dose-dependent toxic effect of silver nanoparticles, which was evidenced by Lactate dehydrogenase (LDH) leakage, reactive oxygen species (ROS) activation, and terminal deoxynucleotidyl transferase dUTP nick end labeling (TUNEL)-positive cells in MDA-MB-231 breast cancer cells (Figure 2).

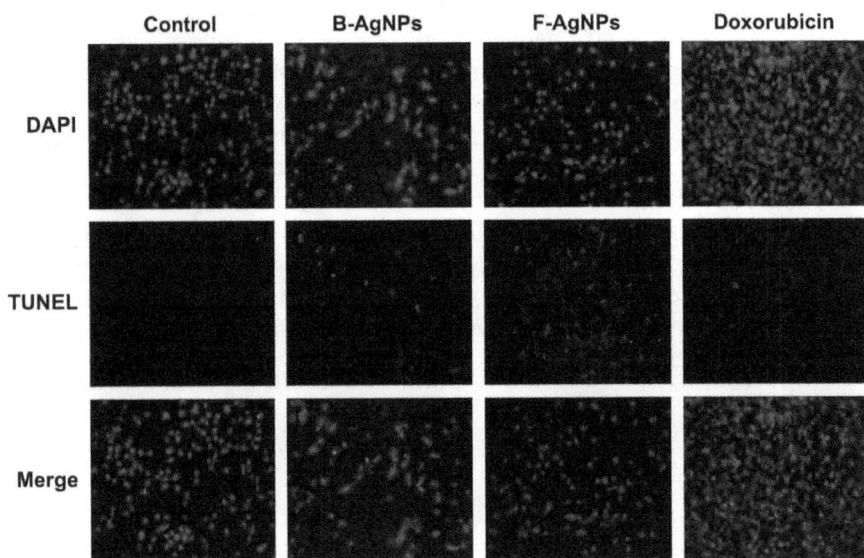

Figure 2. B-AgNPs and F-AgNPs promote apoptosis. Notes: MDA-MB-231 cells were treated with respective IC50 concentrations of B-AgNPs or F-AgNPs for 24 hours. Fluorescent staining of cells was recorded. Representative images are shown for apoptotic DNA fragmentation (red staining) and corresponding nuclei (blue staining). Abbreviations: B-AgNPs, bacterium-derived silver nanoparticles; DAPI, 4',6-diamidino-2-phenylindole; F-AgNPs, fungus-derived silver nanoparticles; IC50, half-maximal inhibitory concentration; TUNEL, terminal deoxynucleotidyl transferase dUTP nick end labeling (from Gurunathan et al. 2015, by Dove Medical Press Limited, and licensed under Creative Commons Attribution – Non-Commercial (unported v3.0) License).

Furthermore, Western blot measurements have shown that silver nanoparticles have caused cell apoptosis by inducing p53, p-Erk1/2, and caspase-3 signaling, as well as Bcl-2 down-regulation (Figure 3).

Therefore, a simple approach was adopted to synthesize silver nanoparticles based on using *C. indica*, as well as to investigate its p53-dependent cell death mechanism in MDA-MB-231 human breast cancer cells. These data were indicative of new nanotherapeutic molecules that play important role in cancer treatment.

After screening approximately 13 tested fungal species, *Guignardia mangiferae* was the one capable of producing well-dispersed and stabilized silver nanoparticles (spherical and 5–30 nm, in size). Furthermore, in addition to presenting excellent antifungal activity against pathogenic plant fungi, it also showed cytotoxic effect

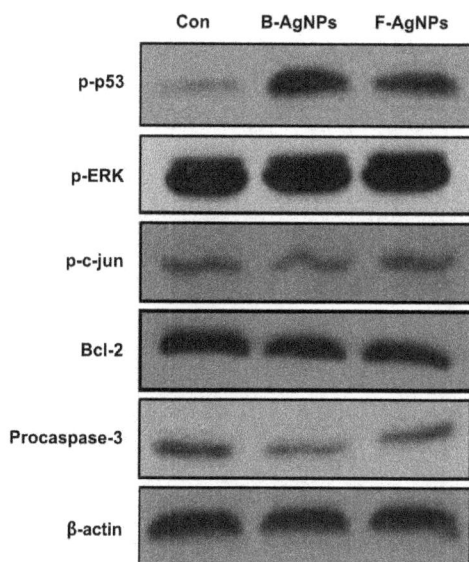

Figure 3. Western blot analysis of p-p53, p-Erk1/2, p-c-Jun, Bcl-2, procaspase-3, and actin expression in MDA-MB-231 cells exposed to B-AgNPs or F-AgNPs. Notes: MDA-MB-231 cells were treated with respective IC50 concentrations of B-AgNPs or F-AgNPs for 24 hours. Western blot analysis determined the expression of p-p53, p-Erk1/2, p-c-Jun, Bcl-2, and procaspase-3 protein levels. Both B-AgNPs and F-AgNPs led to increased levels of p-p53, p-Erk1/2, and decreased levels of procaspase-3, whereas no alteration in expression was observed for p-cJun. Bcl-2 expression was significantly reduced. Equal protein loading was confirmed by analysis of β-actin protein levels. The results are representative of three independent experiments. Abbreviations: B-AgNPs, bacterium-derived AgNPs; Con, control; F-AgNPs, fungus-derived AgNPs; IC50, half-maximal inhibitory concentration (Gurunathan et al. 2015 by Dove Medical Press Limited, and licensed under Creative Commons Attribution – Non-Commercial (unported v3.0) License).

on cell cultures, as well as recorded IC_{50} values of 63.37, 27.54, and 23.84 µg/mL against normal Vero (African monkey kidney), cervical (HeLa) and breast (MCF-7) cancer cells, respectively. Thus, these data have suggested that silver nanoparticles deriving from *G. mangiferae* were biocompatible with animals and likely had broader biomedical/pharmaceutical applicability (Balakumaran et al. 2015).

Based on Water-Soluble Tetrazolium salts (WST-1) assay results, silver nanoparticles synthesized by fungal species *F. oxysporum* have shown antitumor activity against the MCF-7 (human breast carcinoma) cell line. The viability of the assayed compounds in response to their concentration has efficiently decreased. The mean cytotoxic measurements led to IC_{50} value of 121.23 µg/mL. The IC_{50} value recorded for silver nanoparticles was indicative of strong cytotoxic and anticancer drugs (Husseiny et al. 2015).

According to Raman and the co-authors, viable PC3 cells presenting polygonal and round shapes after treatment with biogenic silver nanoparticles deriving from *Pleurotus djamor* var. *roseus* have changed their morphology, formed irregular confluent aggregates, and presented cell condensation and agglomeration (Raman et al. 2015). After these transformations, biogenic silver nanoparticles have inhibited PC3 cell proliferation due to IC_{50} value of 10 µg/mL, after 24-hour incubation.

Morphological measurements based on propidium iodide (PI) staining have evidenced shrunken nuclei, as well as DNA condensation and deterioration, a fact that indicated the cytotoxic effect of silver nanoparticles on PC3 cells. These results have shown that silver nanoparticles presented an effective antiproliferative effect on PC3 cells, since they inhibited their growth, decreased DNA synthesis, and improved apoptosis.

Silver nanoparticles produced from *Pestalotiopsis microspore* have shown efficient cytotoxic effect against mouse melanoma (B16F10) - IC_{50} value of 26.43 ± 3.41 µg/mL; human ovarian carcinoma (SKOV3) - IC_{50} value of 16.24 ± 2.48 µg/mL; human lung adenocarcinoma (A549) - IC_{50} value of 39.83 ± 3.74 µg/mL; and human prostate carcinoma (PC3) - IC_{50} value of 27.71 ± 2.89 µg/mL cells. Notably, silver nanoparticles were biocompatible to healthy cells of the Chinese hamster ovary cell line (CHO) - IC_{50} value of 438.53 ± 4.2 µg/mL. The most significant effect of these particles was observed in SKOV3 cells, which indicated concentration-dependent apoptotic changes, comprising irregular protrusions in the plasma membrane, cell shrinkage, irreversible chromatin condensation in the cell nucleus and fragmented nuclei (Netala et al. 2016).

Anbazhagan and co-researchers synthesized silver nanoparticles (spherical and 20–50 nm, in size) based on using *Cunninghamella echinulata*. They tested their cytotoxicity against Vero cell lines (African green monkey kidney cells) and these nanoparticles recorded IC_{50} value of 62.8 µg/mL (Anbazhagan et al. 2017). This effect diverged from the cytotoxic activity of $AgNO_3$ (~ 25 µg/mL), and suggested that myco-synthesized nanoparticles were lesser cytotoxic to healthy cells than bulk salts. Taken together, these results have evidenced less toxic silver nanoparticles to be explored in several biomedical applications.

The cytotoxicity of silver nanoparticles deriving from *Trichoderma viride* against hepatitis-2C (Hep-2C) and rotavirus cell lines recorded IC_{50} value of 54.27 ± 0.02 µg/mL and 38.33 ± 0.04, respectively. In addition, a dose-dependent value of cytotoxic activity against Hep-2C cell lines was observed *in vitro* (Adebayo-Tayo et al. 2019).

Extracellular pigment deriving from *Talaromyces purpurogenus* (previously known as *Penicillium purpurogenum*) was applied as a reducing agent to produce silver nanoparticles (both spherical and with other structures showing a size distribution of approximately 40 nm and zeta potential of −24.8 mV). The cytotoxicity of silver nanoparticles against HeLa (human cervical cancer), HepG2 (human liver cancer), and HEK-293 (human embryonic kidney) cell lines (5-fluorouracil was used as positive control) has shown important HepG2 cell line activity at IC_{50} value of 11.1 µg/mL (Figure 4) (Bhatnagar et al. 2019).

Figure 4A shows that cell survival has decreased in all cell lines, as nanoparticle concentration increased. This outcome suggests the cytotoxic effect of silver nanoparticles on the cell population. Results have also shown that HEK-293 was more resistant to these interactions than HeLa and HepG2, since more than 60% of HEK-293 survived the exposure to silver nanoparticle concentration of 100 µg/mL. These findings have indicated that silver nanoparticles presented selective behavior towards cancer cells. In this case, nanoparticles had the strongest effect on HepG2.

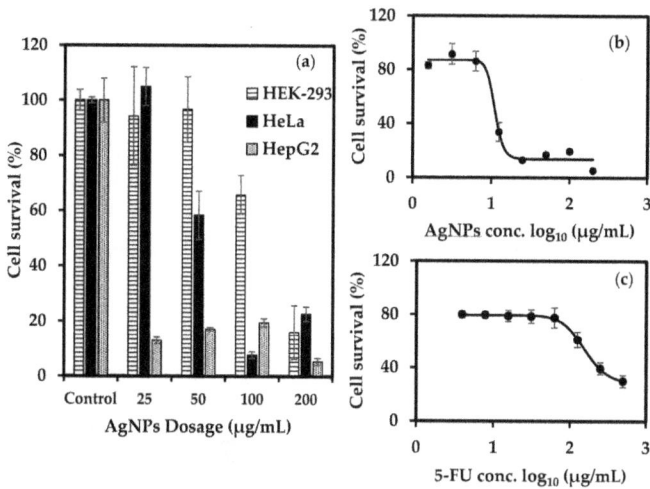

Figure 4. (A) Anticancer activity of silver nanoparticles against various cell lines. (B) Effect of silver nanoparticles on HepG2 cell line. (C) Effect of 5-FU on HepG2 cell line. Error bars represent standard deviation (n = 3) (Bhatnagar et al. 2019, open-access Creative Common CC BY license from MDPI).

Figures 4B and 4C show that silver nanoparticles were more effective anticancer agents than the standard anticancer agent 5-FU.

Silver nanoparticles' green synthesis by fungal proteins deriving from *Aspergillus fumigatus* used against cell cultures, recorded IC_{50} values against human colon cancer cell lines *in vitro* (HCT116), lung adenocarcinoma human alveolar basal epithelial cells (A549), human breast cancer cell line (MCF7) and human Caucasian prostate adenocarcinoma (PC3), as shown in Table 2 (Othman et al. 2019). In addition, biosynthesized silver nanoparticles were capable of preventing tumor cell line growth at low concentrations, although they effectively stopped HCT116 (84.6%), A549 (60.8%), MCF7 (68.9%), and PC3 (69.5%) cell line growth at a concentration of 54 µg/mL.

The antitumor ability of biogenic silver nanoparticles was recently reported in the literature, since they presented IC_{50} against the 5637 (bladder carcinoma) cell line (Ferreira et al. 2019, 2020). This assay used two approaches, namely: the

Table 2. Antitumoral activities (IC_{50}) of silver nanoparticles from *Aspergillus fumigatus*.

Cell Line	Antitumor Activity	
	IC_{50} (µg/mL)	IC_{90} (µg/mL)
BJ1 (normal skin fibroblast)	–	–
HCT116 (colon cell line)	31.1	52.3
A549 (lung carcinoma cell line)	45.4	72.8
MCF7 (human Caucasian breast adenocarcinoma cell line)	40.9	65.7
PC3 (prostate cell line)	33.5	60.7

IC_{50} lethal concentration of the sample causes the death of 50% of cells in 48 hours, IC90 lethal concentration of the sample causes the death of 90% of cells in 48 h (modified from Othman et al. 2019).

MTT [3-(4,5-Dimethylthiazol-2-yl)-2,5-Diphenyltetrazolium Bromide] and calcein/
PI assays, which were based on different chemical principles. It was done in order
to increase the validity of results and to avoid artifacts. Figure 5A shows that all
assays reported comparable dose-response associations and similar IC_{50} values of:
10.57 µM, for MTT; 9.79 µM, for calcein; and 13.72 µM, for PI. Then, the cytotoxicity
of biosynthetic silver nanoparticles was assessed on time (Figure 5B). Data have
shown that the cytotoxicity profile of these nanoparticles after 6-hour treatment was
similar to that observed after 24-hour treatment (24-hours vs. 6-hours : $p > 0.05$).

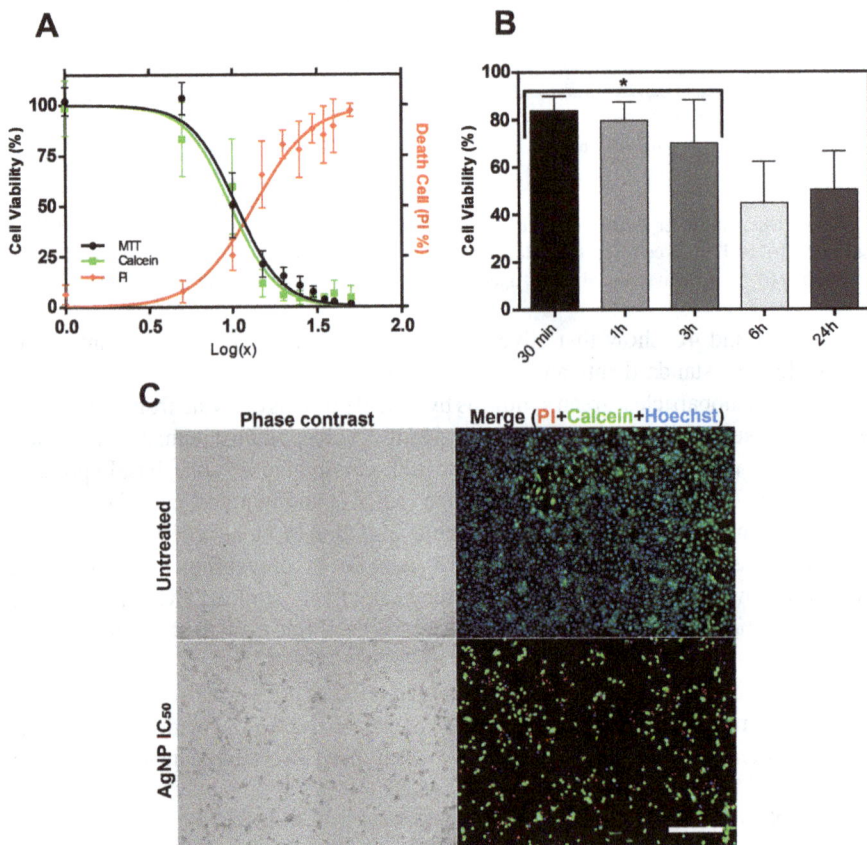

Figure 5. AgNP cytotoxicity in urinary bladder carcinoma 5636 cells. (A) Cells were treated with
increasing concentrations of AgNP (1–50 µM) for 24 h. Each value represents the mean ± S.D. of three
independent experiments (n = 3), cell viability was normalized to untreated control). Cell viability
was measured by MTT formazan absorbance (ex. 570 nm) and Calcein fluorescence (~ 492/517
nm). For cytotoxicity evaluation, counts of PI-positive nuclei were used in relation to 100% labeled
nuclei with Hoechst 33342. (B) Cytotoxicity of the AgNP was evaluated in time variation. Cells were
treated with AgNP IC50 for 1 h, 3 h, 6 h, and 24 h. The results were normalized to the untreated group
(100% viable cells). Each value represents the mean ± S.D. of three independent experiments. For statistical
tests, p < 0.05 for ANOVA was used, followed by Tukey's test (24 h vs. 1 h, 3 h, 6 h). (C) Representative
images of IC50 by Calcein/PI assay obtained by phase contrast and the images below represent the merge
of the DAPI, GFP, and PI light cubes. Column: I – Control group, II – IC50 treatment. Phase contrast,
Hoechst + Calcein + PI. Scale bar: 200 µm (Ferreira et al. 2020, by permission of Elsevier B.V.).

No significant viability decrease was observed at shorter time periods (1 hour and 3 hours) in comparison to the 24-hour treatment group ($p < 0.05$). The viability assay has shown that the cytotoxicity of silver nanoparticles was dose and time dependent under the evaluated conditions. In addition, the influence of silver nanoparticle treatment on cell morphology is depicted in Figure 5C. Representative Calcein/ PI assay images have shown that control cells were mostly attached and presented cuboidal epithelial-like morphology. Cells subjected to silver nanoparticles-based treatment were lysed and presented morphological changes (cell rounding and shrinkage). Moreover, there was an increase in the number of calcein-negative cells, which evidenced loss of cellular viability, as well as PI-positive cells, which indicated cell death. In addition, low silver nanoparticle concentration induced DNA damage and chromosomal aberrations due to low toxicity (Durán et al. 2005, Lima et al. 2013). Lima et al. (2013) did not observe genotoxicity response in several human culture cells subjected to up to 10 mg/mL of capped silver nanoparticles with the mean size of approximately 40 nm. Silver nanoparticles' interaction with cells has triggered several cellular responses, in passive and/or active manners (De Lima et al. 2012).

Subsequently, the underlying mechanisms of silver nanoparticle cell death effect on 5637 cells were investigated. Likely cytotoxicity was defined based on nanoparticles' physicochemical properties, i.e., on surface features such as corona protein, size, and shape (De Lima et al. 2012, Durán et al. 2015). Several studies have described apoptosis as the main cell death pathway enabled by silver nanoparticles (Cheng et al. 2013). However, biogenic nanoparticles are likely to show specific cytotoxicity mechanisms, since protein corona deriving from the generation process can change cell internalization and present cytotoxic effects associated with its properties (Mirshafiee et al. 2013, Durán et al. 2015). Therefore, the study assessed DNA fragmentation based on TUNEL analysis, since TUNEL is representative of apoptosis (Figure 6). Doxorubicin used as a positive control (Figure 6A) has clearly shown 5637 fluorescein-dUTP-labeled cells after 24-hours of incubation. Furthermore, cells treated with silver nanoparticles for 6 hours (Figure 6A) and 24 hours (Figure 6A) have also significantly enhanced dUTP fluorescence, a fact that indicated DNA fragmentation induction ($p < 0.05$, Figure 6B).

In addition to TUNEL assay, apoptosis was also determined based on caspase-3 activation by cleaved Caspase-3. Figure 6C shows typical images of cleaved Caspase-3 measured through confocal microscopy. Doxorubicin has significantly increased cleaved Caspase-3 levels in this experiment. Cells exposed to silver nanoparticles have shown a gradual increase in cleaved caspase-3 levels over 1 to 6 hours, as well as expression levels similar to those observed for cells treated with doxorubicin. Caspase-3 activation triggers the cleavage of caspase-activated DNAse (ICAD) inhibitor; this process releases caspase-activated DNAse (CAD), which translocates to the nucleus and fragments the DNA (Arora et al. 2008).

Data about silver nanoparticles deriving from *Rhizopus stolonifera* (spherical shape, 5–50 nm, in size) used in anticancer experiment (MTT assay) have shown efficient anti-cancer activity against both EAC (esophageal adenocarcinoma cancer) (IC_{50} of 2.15 µg/mL; at 8 µg/mL - 68.13% inhibition) and HT-29 (colon cancer) (IC_{50} of 2.0 µg/mL; at 8.0 µg/mL - 68.46% inhibition) cell lines. Moreover, biogenic

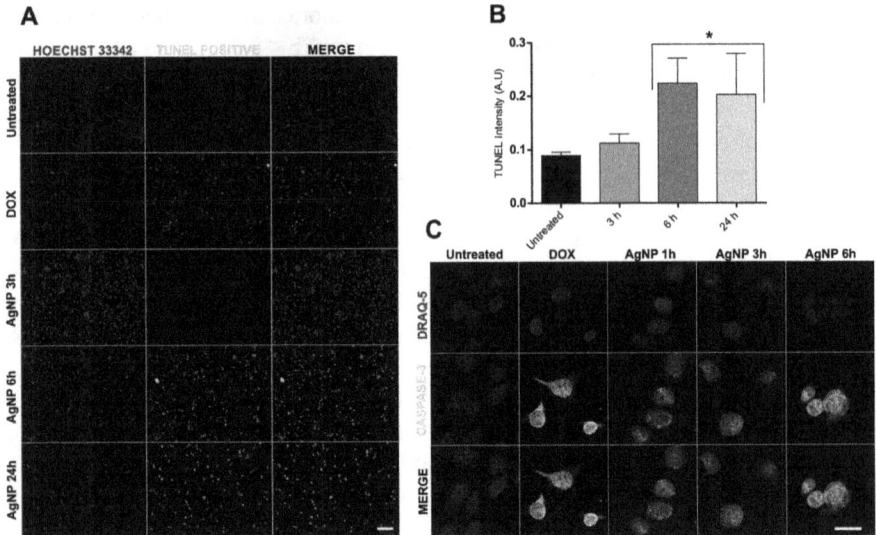

Figure 6. AgNP induces apoptosis in a time-dependent manner in 5637 cells. (A) Cells were exposed to IC50 AgNP for 3, 6, and 24 h. Cells containing the terminal deoxynucleotidyl transferase dUTP nick end-labeling were labeled with Click-iT TUNEL Alexa Fluor Imaging Assay kit (Molecular Probes, Invitrogen) following the manufacturer's instructions. (A) Hoechst 33342 (blue) and TUNEL (green) double-positive cells are shown. Groups: Untreated; Doxorubicin; AgNP 3 h; AgNP 6 h; AgNP 24 h. The images were obtained with fluorescence microscopy (Cytation 5) in the objective of ×10, scale bar of 50 μm. (B) Fluorescence intensity (A.U.) of images labeled with Click-iT TUNEL Alexa Fluor 488 was quantitated by BioTek software Gen5 (Cytation 5). For statistical tests, $p < 0.05$ for ANOVA was used, followed by Tukey's test (untreated vs. 3 h, 6 h, and 24 h). (C) Activation of Caspase-3 in 5637 cells in response to AgNP exposure. Cells were treated for 1 h, 3 h, 6 h with AgNP IC50 and 24 h with Doxorubicin (apoptosis-positive control). After the treatment, cells were fixed with PFA 4%, immunostained with primary antibody cleaved Caspase-3 (Asp175) in green, and counterstained with DRAQ5 (nuclei, blue). Confocal images were obtained in the objective of 63× with and AIRY fixed in 1. Scale bar of 25 μm. (For interpretation of the references to color in this figure legend, the reader is referred to the web version of this article) (Ferreira et al. 2020, by permission of Elsevier).

synthesis using *R. stolonifera* has shown excellent anticancer activity in cell cultures (Banu et al. 2021).

Silver nanoparticles deriving from *Aspergillus sydowii* have shown antiproliferative effects against cancer cells. Based on the MTT assay, silver nanoparticles have shown a dose-dependent antiproliferation effect on HeLa and MCF-7 cells (IC$_{50}$ ~ 25 μg/mL and ~ 7 μg/mL, respectively). Due to their size and surface charges, the antiproliferation mechanism of silver nanoparticles has mainly focused on the following aspects: (1) oxidative stress induction, (2) DNA damage, (3) mitochondrial damage, (4) immune system activation, (5) cell cycle arrest induction, (6) apoptotic process activation (Wang et al. 2021).

An assessment of IC$_{50}$ of *A. terreus* silver nanoparticles on MCF-7 and Vero cell lines showed that MCF-7 cells were susceptible to silver nanoparticles at IC$_{50}$ of 87.5 μg/mL, as well as that Vero cells presented IC$_{50}$ of 350 μg/mL, the concentration dependence in the injury observed in MCF-7 cells was suggested. Furthermore, the maximum non-toxic silver nanoparticle concentration observed for Vero and MCF-7 cell lines was 43.75 and 21.87 μg/mL, respectively (Lotfy et al. 2021).

(ii) Copper/CuO Nanoparticles

Copper nanoparticles produced from *Aspergillus niger* (500 nm, in size; Z-average of 398.2 nm, and PDI-poly dispersion index of 0.246) have shown significant cytotoxic effect on Huh-7 (human hepatocellular carcinoma) cell lines at IC_{50} of 3.09 μg/mL, based on the MTT assay (Noor et al. 2020).

Copper oxide nanoparticles deriving from *Trichoderma asperellum* have shown a cytotoxic effect on human lung carcinoma cells (A549) at IC_{50} of 40.62 μg/mL. However, results were more significant in the presence of NIR light (IC_{50} of 24.7 μg/mL), which may have happened due to the combination of copper oxide nanoparticles to the photothermal effect, which induced heat by cancer cells ablation (Saravanakumar et al. 2019). The apoptosis index recorded for untreated cells was 15.26 and the one recorded for the copper oxide nanoparticles-NIR association was 79.82. This analysis has shown evidence that the Bcl-2 (protein % of 30.12) expression was more significant in untreated cells than in the treated ones. It also showed evidence that Caspase 3 expression was higher (protein % of 48.95) in cells treated with copper oxide nanoparticles-NIR association. The identified signaling apoptotic proteins comprised Bcl-2 (observed in the outer mitochondrial membrane, induces cell survival and inhibits pro-apoptotic-related proteins), and Cas-3 (a pro-apoptotic protein capable of causing cancer cell death). Western blot data have shown up-regulation of both Bcl-2 in non-treated cells and Cas-3 in cells treated with copper oxide nanoparticles. This outcome has indicated that nanoparticles induced apoptosis, and it also corroborated previously reported data (Saravanakumar et al. 2018).

The anticancer effect of copper oxide nanoparticles produced by *Aspergillus terreus* on colon cancer cell lines (HT-29) has shown IC_{50} of 22 μg/mL in the MTT assay and 32.11% of total cells were in the S phase of the cell cycle, in fluorescence-activated cell sorting (FACS) analysis. Angiogenesis suppression in tumor cells was evaluated based on Hen's Egg Test on Chorioallantoic Membrane (HET-CAM) *in vivo*, the higher concentration of copper oxide used in this test was 60 μL (stock solution 1 mg/mL) and it recorded blood vessel suppression of 31.36% (after 2 hours) and 81.81% (after 18 hours), respectively. Copper oxide nanoparticles presented dose-dependent anticancer activity. The results of this research have evidenced the significant role played by copper oxide nanoparticles in cancer treatment (Mani et al. 2021).

(iii) Selenium Nanoparticles

Selenium nanoparticles (*Fusarium semitectum*, 92.33 ± 48.5 nm, in size) have shown selective anticancer potential against Caco-2 Human Colon cancer cells (IC_{50} of 10.24 μg/mL), SNU-16 gastric carcinoma (IC_{50} of 13.27 μg/mL) and A431 skin cancer cells (IC_{50} of 20.44 μg/mL). On the other hand, they did not show cytotoxic effects on healthy liver cells (THLE2) and presented low toxicity on healthy kidney cells (Vero) (Abbas and Baker 2020). These findings have evidenced the positive effects of selenium nanoparticles as likely candidates to be used in future therapeutic applications.

(iv) Zinc Oxide Nanoparticles

Zinc oxide nanoparticles deriving from *Aspergillus terreus* were capped by L-asparaginase (called CuO nanoparticles-metalloprotein) found in the fungal extract (Baskar et al. 2015). MCF-7 viability has decreased from 64.3% to 35.02% in cells treated with 50 μL (0.4 μg/mL) of Zn–MP solution. Decreased MCF-7 cell viability observed after treatment with metalloproteins has suggested that biosynthesized metalloproteins can be used as an efficient anticancer drug.

Zinc oxide nanoparticles synthesized by *Trichoderma harzianum* (spherical shape and 30.34 nm in size) did not show cytotoxicity in embryonic fibroblast cells (NIH3T3). This outcome has indicated zinc oxide nanoparticles' concentration-dependent suppressive effect on A549 human lung carcinoma cells at IC_{50} of 158 μg/mL, which was also confirmed by fluorescent cytochemistry. In addition, Annexin V-FITC staining results have indicated the incidence of apoptotic cells in nanoparticles-treated A549 cells. Apoptotic cells were not observed in non-treated A549 cells in the control group. These data have suggested that ZnO nanoparticles induced necrotic cell death (Saravanakumar et al. 2020).

Based on the MTT assay, ZnO nanoparticles derived from *Alternaria tenuissima* have inhibited cancer cell proliferation at IC_{50} of 55.76 μg/mL (HFB-4) (human melanocytes), 16.87 μg/mL (HepG-2) (hepatocellular carcinoma), and 18.02 μg/mL (MCF-7) (human breast carcinoma). Thus, ZnO nanoparticles were effective in the non-malignant HBF-4 cell line, as well as in malignant MCF-7 and HepG-2 cancer cell lines. Interestingly, ZnO nanoparticles have shown specificity against malignant cells rather than the healthy ones. In addition, ZnO nanoparticles presented anticancer activity similar to that of Taxol (Abdelhakim et al. 2020).

(v) Gold Nanoparticles

Gold nanoparticles generated by *Pleurotus ostreatus* (El Domany et al. 2018) were more efficient than chemically synthesized gold nanoparticles against the cytotoxic activity of human liver cancer (HepG2) and human colon carcinoma (HCT-116) cell lines *in vitro*, at a concentration of 25 μg/mL, but they recorded lower efficiency against the human Caucasian prostate adenocarcinoma cell line (PC3). One of the most significant effects of gold nanoparticles was observed in cell morphology/shape. Anticancer activity based on the investigated doses was expressed in percentage, as follows: HepG2 (33.5%), HCT-116 (22.7%), and PC3 (14.6%). Gold nanoparticles have shown strong antiproliferation activity in different cancer cell lines at the same concentration range applied in the current study (Bhat et al. 2013). This outcome can be explained by gold nanoparticles' irregular shape and by their association with other organic compounds (Lee et al. 2015). Thus, the comparison between antiproliferation activity of biogenic and commercial gold nanoparticles against several cancer cell lines has confirmed the effectiveness of using biogenic nanoparticles in cancer treatment (El Domany et al. 2018).

Gold nanoparticles (*Fusarium solani*, 40–45 nm in size) have shown cytotoxicity in cervical cancer HeLa cells (IC_{50} 1.3 ± 0.5 μg/mL), as well as against human breast cancer MCF-7 cells (IC_{50} of 0.8 ± 0.5 μg/mL). They presented dose-dependent cytotoxic effect on, and induced apoptosis in, both cancer cells. Furthermore, apoptotic cell accumulation has decreased in the sub-G0/G1 phase cell cycle of MCF-7 cancer cells. Altogether, these data have suggested a versatile biomedical therapy based on a safer chemotherapeutic drug with low systemic toxicity (Clarance et al. 2020).

Abu-Tahon et al. have confirmed that biosynthesized gold nanoparticles deriving from *Aspergillus flavus* are a strong anticancer agent against human lung carcinoma (A549) (IC_{50} of 53.5 μg/mL), human liver cancer (HepG2) (IC_{50} of 60.7 μg/mL), and human breast cancer (MCF7) cell lines (IC_{50} of 100 μg/mL) (Abu-Tahon et al. 2020). Another research reported similar results after using gold nanoparticles deriving from *Justicia adhatoda* leaf extract in A549 tumor cell lines (IC_{50} of 80 μg/mL) (Latha et al. 2018). Gold nanoparticles used at the concentration of 53.5 μg/mL were capable of reducing the percentage of cells in the G0/G1 and S phases in comparison to untreated cells. The percentage of treated cells decreased from 47.59% to 31.28% in the G0/G1 phase and from 44.23% to 32.74% in the S phase. On the other hand, the percentage of treated cells in the G2/M phase recorded a sharp increase to 35.98% in comparison to the increase to 8.18% observed in control cells. The aforementioned authors concluded that the Annexin-V FITC technique and cell cycle have confirmed cell death via apoptosis and cell cycle arrest in the G2/M phase for the A549 cell line (Abu-Tahon et al. 2020).

Biogenic Nanoparticles (Fungi) and Cancer Experiments *in vivo*

A study about immunomodulatory activity was carried out with Swiss albino female mice (6 weeks old) weighing approximately 20 g (Adebayo-Tayo et al. 2019). Groups 1 and 2 were used as control, Group 3 was treated with sheep red blood cells, Group 4 was treated with silver nanoparticles, and Group 5 was treated with *T. viride* fungal filtrate. The results are shown in Table 3. Groups treated with biosynthesized silver nanoparticles and fungal filtrate recorded immunomodulatory activity (Table 3) significantly different from that of other groups. Group 3, which comprised mice treated with sheep red blood cells, recorded high IgG (118 ± 0.23) and IgA (230 ± 0.45 mg/dL) levels. On the other hand, Group 4 (mice treated with silver

Table 3. Immunomodulatory activity of silver nanoparticles from *T. viride*.

S/N	Group	SRBC	SNPS	FF	IgG (mg/dL)	IgA (mg/dL)	IgM (mg/dL)
1	GRP1a	–	–	–	0.000	0.000	0.000
2	GRP1b	–	–	–	79 ± 0.65^b	171 ± 0.97^c	38 ± 0.12^c
3	GRP2	+	–	–	118 ± 0.23^a	230 ± 0.45^b	73 ± 0.26^b
4	GRP3	–	+	–	48 ± 0.27^d	258 ± 0.73^a	96 ± 0.22^a
5	GRP4	–	–	+	63 ± 0.39^c	75 ± 0.81^d	24 ± 0.17^d

n = 6; p < 0.05 – significant difference; GRP1a – mice not exposed to cigarette smoke and not treated, GRP1b – mice exposed to cigarette smoke and not treated, GRP2 – mice administered sheep red blood cells, GRP3 – mice administered silver nanoparticles, GRP4 – mice administered *T. viride* filtrate, SNPs – silver nanoparticles FF – fungal filtrate, SN – serial number, SRBC – sheep red blood cell (modified from Adebayo-Tayo et al. 2019 and by Wroclaw Medical University).

nanoparticles) recorded the highest IgG (258 ± 0.73 mg/dL), whereas Group 5 (mice treated with fungal filtrate) recorded the lowest IgG (63 ± 0.39 mg/dL). In addition, silver nanoparticle-treated mice (Group 4) recorded the highest IgM (96 ± 0.22 mg/dL), whereas Group 5 (treated with fungal filtrate) recorded the lowest IgM (24 ± 0.17 mg/dL). Further, immunological activity *in vivo* with silver nanoparticles has shown significant IgA and IgM immunostimulation.

It was suggested that it may have happened because nanoparticles stimulated macrophage activity in the immune system to protect hosts from cancer, in order to eliminate neoplastic cells and reject non-self-components (Adebayo-Tayo et al. 2019). According to research, chemically synthesized nanoparticles had immunomodulatory effect on these cell lines, either alone or associated with other immunomodulatory agents (Swarnakar et al. 2013). They also acted as targeted drug/vaccine delivery vehicles for macrophages (Adebayo-Tayo et al. 2019).

Although there are several anticancer treatments *in vivo* based on silver nanoparticles available in the literature (Miranda et al. 2022), so far, no biogenic synthesis by fungi was reported. According to one of the first studies conducted in this field, the antitumor effectiveness of silver nanoparticles deriving from *Fusarium oxysporum*—spherical and 20–50 nm, in size (Durán et al. 2005), was evaluated in female C57BL/6Junib mice (N-methyl - N-nitrosourea - MNU), who were chemically induced to non-muscle invasive bladder cancer (NMIBC) and subsequently, treated with these nanoparticles (Ferreira et al. 2020). These mice were treated through intravesical route with biogenic silver nanoparticles at concentrations of 500, 200, and 50 µg/mL (0.5, 0.2, and 0.05 mg/mL). Histopathological analyses (Figure 7) (Table 4) have shown that treatment with silver nanoparticles at the highest concentration (0.5 mg/mL), was not effective in treating NMIBC, since 100% of animals belonging to this group still presented malignant neoplastic lesions after

Table 4. Histopathological changes (%) of mice's urinary bladder from different experimental groups.

Groups					
Histopathology	Control (n = 7)	NMIBC (Cancer) (n = 7)	NMIBC/AgNPs (0.05 mg/mL) (n = 7)	NMIBC/AgNPs (0.2 mg/mL) (n = 7)	NMIBC/AgNPs (0.5 mg/mL) (n = 7)
Normal	7(100%)	–	1(14.28%)	–	–
Flat hyperplasia	–	–	3(42.85%)	2(28.57%)	–
Flat carcinoma *in situ* (pTis)	–	–	2(28.57%)	4(57.15%)	2(28.57%)
Papillary urothelial carcinoma (pTa)	–	3(42.85%)	1(14.28%)	1(14.28%)	3(42.85%)
High-grade urothelian cancer invading *lamina propia*	–	4(57.15%)	–	–	2(28.57%)

Benign Lesions: Flat hyperplasia; Malignant lesions: pTis; pTa; pT1. The histopathological alterations were normalized by the number of mice (n) examined in each group. *$P < 0.0001$ (proportions test). (from Ferreira et al. 2020, by permission of Elsevier B.V.).

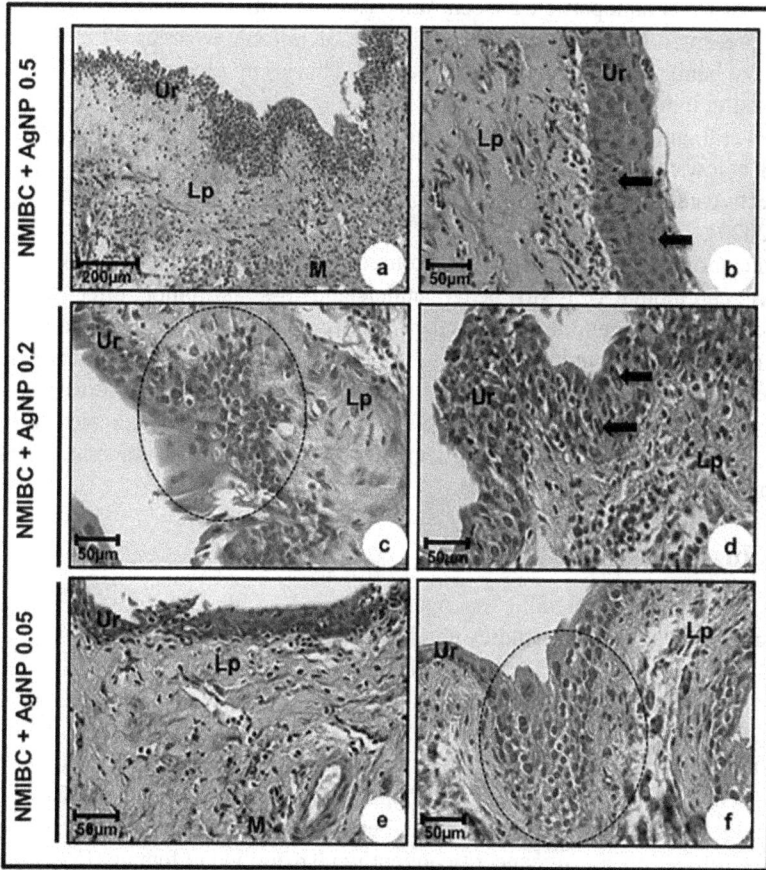

Figure 7. Photomicrographs of urinary bladder from NMIBC + AgNP 0.5 (a, b), NMIBC + AgNP 0.2 (c, d) and NMIBC + AgNP 0.05 (e, f) groups. (a) pTa tumor characterized by extensive papillary lesions, urothelial cells with a disordered arrangement and loss of polarity, intense cellular pleomorphism, and numerous mitosis figures. (b), (d) pTis tumor characterized by cellular atypia: bulky nuclei with reduced cytoplasm and prominent nucleoli (arrows). (c), (f) Flat hyperplasia (circle) characterized by thickening of the urothelium and absence of cytological atypia. (e) Normal urothelium comprises 2–3 layers: a layer of basal cells, an intermediate layer of cells, and a superficial or apical layer composed of cells in an umbrella. a - f: Lp - lamina propria, M - muscular layer, Ur –urothelium (Ferreira et al. 2020, by permission of Elsevier B.V.).

treatment. If one takes into consideration that chemical NMIBC (N-methyl-N-nitrosourea - MNU) induction led to 100% of malignant lesions, it can be observed for that experimental group that the tumor progression inhibition promoted by the adopted treatment was evidenced by the percentage of non-malignant lesions (benign and/or lack of lesion).

Thus, treatment with 0.2 mg/mL of silver nanoparticles recorded tumor progression inhibition of 28.57%, which was represented by the presence of flat hyperplasia (benign lesion) in 28.57% of animals in this group (Figure 7). However, the highest tumor progression inhibition rate was observed in the group treated with the lowest silver nanoparticle concentration (0.05 mg/mL). 14.28% of animals in this group presented healthy urothelium (absence of lesion), whereas 42.85% of them presented benign urothelial lesions (Figure 7). This group recorded 57.13% of tumor progression inhibition.

Several authors have confirmed that silver nanoparticles can inhibit cancer cell growth at low concentrations (mg/ml). These nanoparticles can also be functionalized as effective carrier systems for the administration of cancer cell targeted drugs (Huy et al. 2020).

The molecular mechanism of cytotoxicity was investigated in 5637 human bladder carcinoma cells to better understand the antitumor effect of silver nanoparticles. Time and dose dependent cytotoxicity and exhaustive analysis have evidenced cell death induction due to apoptosis. In addition, silver nanoparticles induced a decrease in cell migration and proliferation. These observations have corroborated the antitumor quality of silver nanoparticles and indicated that they could be a cost-effective and promising candidate to be used for bladder cancer therapy (Ferreira et al. 2019, 2020).

Conclusions

Fungal syntheses of metallic nanoparticles, and of their oxides, are excellent anticancer agents. So far, studies conducted in this field have mainly focused on using cell cultures *in vitro* (cancer cells: HeLa, MCF-7, MDA-MB-231, HepG2, HCT116, A549, PC3, UBC-5636. HT-20, EAC, A431, Huh-7 and healthy cells: HEK-293, BJ1, THLE2, NIH3T3, Vero).

The most relevant nanoparticles with anticancer activity generated by fungi comprised of, silver, gold, selenium, copper, and oxides, such as CuO, ZnO, CoO, and MgO. To the best of our knowledge, Ferreira et al. is the only study available in the literature reporting anticancer treatment *in vivo* applied to female C57BL/6Junib mice, based on biogenetic silver nanoparticles produced by fungus species *Fusarium oxysporum* (Ferreira et al. 2020). From the therapeutic perspective, there was very little knowledge about the use of biogenic or biosynthetic silver nanoparticles to treat bladder cancer. However, relevant results observed in studies about the antitumor activity of these nanoparticles on 5637 bladder cancer cell cultures, led to migration inhibition and antiproliferation activity due to decreased colony formation. This encouraged the researchers to investigate the antitumor activity of biogenic/biosynthesized silver nanoparticles deriving from fungi *in vivo* (Ferreira et al. 2020). The main results have shown that low nanoparticle concentrations, inhibited tumor progression in approximately 58% of treated animals, 14% of animals presented healthy urothelium, whereas 43% of them only presented benign lesions in the urinary bladder. Thus, biogenic silver nanoparticles should be investigated as a new therapeutic alternative for NMIBC treatment.

Acknowledgements

Support from FAPESP, CNPq, INOMAT (MCTI/CNPq) are acknowledged.

References

Abbas, H.S. and Baker, D.H.A. 2020. Biological evaluation of selenium nanoparticles biosynthesized by *Fusarium semitectum* as and anticancer agents. Egypt. J. Chem. 63(4): 1119–1133.

Abd-Elsalam, K.A. 2021. Special issue: Fungal nanotechnology. J. Fungi. 7(8): 583.

Abdel-Aziz, M.M., Emam, T.M. and Elsherbiny, E.A. 2020. Bioactivity of magnesium oxide nanoparticles synthesized from cell filtrate of endobacterium *Burkholderia rinojensis* against *Fusarium oxysporum*. Mater. Sci. Eng. C 109(0): 110617.

Abdelhakim, H.K., El-Sayed, E.R. and Rashidi, F.B. 2020. Biosynthesis of zinc oxide nanoparticles with antimicrobial, anticancer, antioxidant and photocatalytic activities by the endophytic *Alternaria tenuissima*. J. Appl. Microbiol. 128(6): 1634–1646.

Abdel-Kareem, M.M. and Zohri, A.A. 2018. Extracellular mycosynthesis of gold nanoparticles using *Trichoderma hamatum*: Optimization, characterization andantimicrobial activity. Lett. Appl. Microbiol. 67(5): 465–475.

Abu-Tahon, M.A., Ghareib, M. and Abdallah, W.E. 2020. Environmentally benign rapid biosynthesis of extracellular gold nanoparticles using *Aspergillus flavus* and their cytotoxic and catalytic activities. Process Biochem. 95(0): 1–11.

Adebayo-Tayo, B.C., Ogunleye, G.E. and Ogbole, O. 2019. Biomedical application of greenly synthesized silver nanoparticles using the filtrate of *Trichoderma viride*: Anticancer and immunomodulatory potentials. Polim. Med. 49(2): 57–63.

Ahamed, M., Akhtar, M.J., Khan, M.A.M., Alhadlaq, H.A. and Alshamsan, A. 2020. Barium titanate (BaTiO3) nanoparticles exert cytotoxicity through oxidative stress in human lung carcinoma (A549) cells. Nanomaterials 10(11): 2309.

Anbazhagan, S., Azeez, S., Morukattu, G., Rajan, R., Venkatesan, K. and Thangavelu, K.P. 2017. Synthesis, characterization and biological applications of mycosynthesized silver nanoparticles. 3 Biotech. 7(0): 333.

Arora, S., Jain, J., Rajwade, J.M. and Paknikar, K.M. 2008. Cellular responses induced by silver nanoparticles: *In vitro* studies. Toxicol. Lett. 179(2): 93–100.

Azmath, P., Baker, S., Rakshith, D. and Satish, S. 2016. Mycosynthesis of silver nanoparticles bearing antibacterial activity. Saudi Pharm. J. 24(2): 140–146.

Balakumaran, M.D., Ramachandran, R. and Kalaichelvan, P.T. 2015. Exploitation of endophytic fungus, Guignardia mangiferae for extracellular synthesis of silver nanoparticles and their *in vitro* biological activities. Microbiol. Res. 178(0): 9–17.

Ballottin, D., Fulaz, S., Cabrini, F., Tsukamoto, J., Durán, N., Alves, O.L. and Tasic, L. 2017. Antimicrobial textiles: Biogenic silver nanoparticles against *Candida* and *Xanthomonas*. Mat. Sci. Eng. C 75(0): 582–589.

Bahrulolum, H., Javanshir, S.N.N., Tarrahimofrad, H., Mirbagheri, V.S., Easton, A.J. and Ahmadian, G. 2021. Green synthesis of metal nanoparticles using microorganisms and their application in the agrifood sector. J. Nanobiotechnol. 19(0): 86.

Bhargava, A., Jain, N., Khan, M.A., Pareek, V., Dilip, R.V. and Panwar, J. 2016. Utilizing metal tolerance potential of soil fungus for efficient synthesis of gold nanoparticles with superior catalytic activity for degradation of rhodamine B. J. Environ. Manag. 183(0): 22–32.

Bhat, R., Sharanabasava, V.G., Deshpande, R., Shetti, U., Sanjeev, G. and Venkataraman, A. 2013. Photo-biosynthesis of irregular shaped functionalized gold nanoparticles using edible mushroom *Pleurotus Florida* and its anticancer evaluation. J. Photochem. Photobiol. B: Biol. 125(0): 63–69.

Banu, A., Gousuddin, M. and Yahya, E.B. 2021. Green synthesized monodispersed silver nanoparticles' Characterization and their efficacy against cancer cells. Biomed. Res. Ther. 8(8): 4476–4482.

Baskar, G., Chandhuru, J., Fahad, K.S., Praveen, A., Chamundeeswari, M. and Muthukumar, T. 2015. Anticancer activity of fungal Lasparaginase conjugated with zinc oxide nanoparticles. J. Mat. Sci.: Mater. Med. 26(1): 43.

Billings, C., Langley, M., Warrington, G., Mashali, F. and Johnson, J.A. 2021. Magnetic particle imaging: Current and future applications, magnetic nanoparticle synthesis methods and safety measures. Int. J. Mol. Sci. 22(14): 7651.

Bhatnagar, S., Kobori, T., Ganesh, D., Ogawa, K. and Aoyagi, H. 2019. Biosynthesis of silver nanoparticles mediated by extracellular pigment from *Talaromyces purpurogenus* and their biomedical applications. Nanomaterials 9(7): 1042.

Bouafia, A., Laouini, S.E., Ahmed, A.S.A., Soldatov, A.V., Algarni, H., Feng Chong, K. and Ali, G.A.M. 2021. The recent progress on silver nanoparticles: Synthesis and electronic applications. Nanomaterials 11(9): 2318.

Bruna, T., MaldonadoBravo, F., Jara, P. and Caro, N. 2021. Silver nanoparticles and their antibacterial applications. Int. J. Mol. Sci. 22(13): 7202.

Chabattula, S.C., Gupta, P.K., Tripathi, S.K., Gahtori, R., Padhi, P., Mahapatra, S., Biswal, B.K., Singh, S.K., Dua, K., Ruokolainen, J., Mishra, Y.K., Jha N.K., Bishi, D.K. and Kesari, K.K. 2021. Anticancer therapeutic efficacy of biogenic Am-ZnO nanoparticles on 2D and 3D tumor models. Mat. Today Chem. 22(12): 100618.

Chatterjee, S., Mahanty, S., Das, P., Chaudhuri, P. and Das, S. 2020. Biofabrication of iron oxide nanoparticles using manglicolous fungus *Aspergillus niger* BSC-1 and removal of Cr(VI) from aqueous solution. Chem. Eng. J. 385(0): 123790.

Cheng, O.H.C., Son, D.H. and Sheldon, M. 2020. Light-induced magnetism in plasmonic gold nanoparticles. Nat. Photonics 14(0): 365–368.

Cheng, X., Zhang, W., Ji, Y., Meng, J., Guo, H., Liu, J., Wu, X. and Xu, H. 2013. Revealing silver cytotoxicity using Au nanorods/Ag shell nanostructures: Disrupting cell membrane and causing apoptosis through oxidative damage. RSC Adv. 3(7): 2296–2305.

Clarance, P., Luvankar, B., Sales, J., Khusro, A., Agastian, P., Tack, J.-C., Al Khulaifi, M.M., AL-Shwaiman, H.A., Elgorban, A.M., Syed, A. and Kim, H.-J. 2020. Green synthesis and characterization of gold nanoparticles using endophytic fungi Fusarium solani and its *in-vitro* anticancer and biomedical applications. Saudi J. Biol. Sci. 27(0): 706–712.

Crane, R.A., Dickinson, M., Popescu, I.C. and Scott, T.B. 2011. Magnetite and zero-valent iron nanoparticles for the remediation of uranium contaminated environmental water. Water Res. 45(0): 2931–2942.

Dawoud, T.M., Yassin, M.A., El-Samawaty, A.R.M. and Elgorban, A.M. 2021. Silver nanoparticles synthesized by *Nigrospora oryzae* showed antifungal activity. Saudi J. Biol. Sci. 28(3): 1847–1852.

De Lima, R., Seabra, A.B. and Durán, N. 2012. Silver nanoparticles: A brief review of cytotoxicity and genotoxicity of chemically and biogenically synthesized nanoparticles. J. Appl. Toxicol. 32(11): 867–879.

Ding, C., Cheng, W., Sun, Y. and Wang, X. 2015. Novel fungus-Fe_3O_4 bio-nanocomposites as high performance adsorbents for the removal of radionuclides. J. Hazard. Mater. 295(0): 127–137.

Durán, N. and Seabra, A.B. 2012. Metallic oxide nanoparticles: State of the art in biogenic syntheses and their mechanisms. Appl. Microbiol. Biotechnol. 95(0): 275–288.

Durán, N., Durán, M., de Jesus, M.B., Seabra, A.B., Fávaro, W.J. and Nakazato, G. 2016. Silver nanoparticles: A new view on mechanistic aspects on antimicrobial activity. Nanomedicne: NBM 12(0): 789–799.

Durán, N., Marcato, P.D., Alves, O.L., De Souza, G.I.H. and Esposito, E. 2005. Mechanistic aspects of biosynthesis of silver nanoparticles by several *Fusarium oxysporum* strains. J. Nanobiotechnol. 3(8): 1–7.

Durán, N., Marcato, P.D., De Conti, R., Alves, O.L., Costa, F.T.M. and Brocchi, M. 2010. Potential use of silver nanoparticles on pathogenic bacteria, their toxicity and possible mechanisms of action. J. Braz. Chem. Soc. 21(6): 949–959.

Durán, N., Silveira, C.P., Durán, M. and Martinez, D.S.T. 2015. Silver nanoparticle protein corona and toxicity: A mini-review. J. Nanobiotechnol. 13(0): 55.

Elamawi, R.M., Al-Harbi, R.E. and Hendi, A.A. 2018. Biosynthesis and characterization of silver nanoparticles using *Trichoderma longibrachiatum* and their effect on phytopathogenic fungi. Egypt. J. Biol. Pest. Control. 28(0): 28.

El-Batal, A.I., El-Sayyad, G.S., Mosallam, F.M. and Fathy, R.M. 2020. *Penicillium chrysogenum*-mediated mycogenic synthesis of copper oxide nanoparticles using gamma rays for *in vitro* antimicrobial activity against some plant pathogens. J. Clust. Sci. 31(0): 79–90.

El Domany, E.B., Essam, T.M., Ahmed, A.E. and Farghali, A.A. 2018. Biosynthesis physico-chemical optimization of gold nanoparticles as anti-cancer and synergetic antimicrobial activity using *Pleurotus ostreatus* fungus. J. Appl. Pharm. Sci. 8(5): 119–128.

Fang, X., Wang, Y., Wang, Z., Jiang, Z. and Dong, M. 2019. Microorganism assisted synthesized nanoparticles for catalytic applications. Energies 12(1): 190.

Ferreira, L.A.B., Fóssa, F.G., Durán, N., de Jesus, M.B. and Fávaro, W.J. 2019. Cytotoxicity and antitumor activity of biogenic silver nanoparticles against non-muscle invasive bladder. J. Phys. Conf. Ser. 1323(0): 012020.

Ferreira, L.A.B., Garcia-Fossa, F., Radaic, A., Durán, N., Fávaro, W.J. and de Jesus, M.B. 2020. Biogenic silver nanoparticles: *In vitro* and *in vivo* antitumor activity in bladder cancer. Eur. J. Pharm. Biopharm. 151(0): 162–170.

Gaikwad, S.C., Birla, S.S., Ingle, A.P., Gade, A.K., Marcato, P.D., Rai, M. and Durán, N. 2013. Screening of different *Fusarium* species to select potential species for the synthesis of silver nanoparticles. J. Braz. Chem. Soc. 24(0): 1974–1982.

Gharieb, M., Wilkinson, S. and Gadd, G. 1995. Reduction of selenium oxyanions by unicellular, polymorphic and filamentous fungi: Cellular location of reduced selenium and implications for tolerance. J. Ind. Microbiol. 14(0): 300–311.

Ghosh, S., Ahmad, R., Zeyaullah, M. and Khare, S.K. 2021. Microbial nano-factories: Synthesis and biomedical applications. Front. Chem. 9(0): 626834.

Gond, S.K., Mishra, A., Verma, S.K., Sharma, V.K. and Kharwar, R.N. 2020. Synthesis and characterization of antimicrobial silver nanoparticles by an endophytic fungus isolated from *Nyctanthes arbor-tristis*. Proc. Natl. Acad. Sci. India Sect. B Biol. Sci. 90(0): 641–645.

Grasso, G., Zane, D. and Dragone, R. 2020. Microbial nanotechnology: Challenges and prospects for green biocatalytic synthesis of nanoscale materials for sensoristic and biomedical applications. Nanomaterials 10(1): 11.

Guilger-Casagrande, M., Germano-Costa, T., Pasquoto-Stigliani, T., Fraceto, L.F. and De Lima, R. 2019. Biosynthesis of silver nanoparticles employing Trichoderma harzianum with enzymatic stimulation for the control of *Scleotinia sclerotiorum*. Sci. Rep. 9(0): 14351.

Gupta, M. and Seema, K. 2021. Living nano-factories: An eco-friendly approach towards medicine and environment. *In*: Pal, K. (ed.). Bio-manufactured Nanomaterials. Springer, Cham. https://doi.org/10.1007/978-3-030-67223-2_6.

Gurunathan, S., Raman, J., Malek, S.N.A., John, P.A. and Vikineswary, S. 2013. Green synthesis of silver nanoparticles using *Ganoderma neo-japonicum* Imazeki: A potential cytotoxic agent against breast cancer cells. Int. J. Nanomed. 8(0): 4399–4413.

Gurunathan, S., Park, J.H., Han, J.W. and Kim, J.H. 2015. Comparative assessment of the apoptotic potential of silver nanoparticles synthesized by Bacillus tequilensis and Calocybe indica in MDA-MB-231 human breast cancer cells: Targeting p53 for anticancer therapy. Int. J. Nanomed. 10(0): 4203.

Heo, S., Lee, J., Lee, G.H., Heo, C.J., Kim, S.H., Yun, D.J., Park, J.B., Kim, K., Kim, Y., Lee, D., Park, G.S., Cho, H.Y., Shin, T., Yun, S.Y., Kim, S., Jin, Y.W. and Park, K.-B. 2020. Surface plasmon enhanced organic color image sensor with Ag nanoparticles coated with silicon oxynitride. Sci. Rep. 10(0): 1–8.

How, X.W., Ong, Y.S., Low, S.S., Pandey, A., Show, P.L. and Foo, J.B. 2021. How far have we explored fungi to fight cancer? Seminars in Cancer Biology 2021. https://doi.org/10.1016/j.semcancer.2021.03.009.

Hu, D., Yu, S., Yu, D., Liu, N., Tang, Y., Fan, Y., Wang, C. and Wu, A. 2019. Biogenic Trichoderma harzianum-derived selenium nanoparticles with control functionalities originating from diverse recognition metabolites against phytopathogens and mycotoxins. Food Control. 106(0): 106748.

Hulikere, M.M. and Joshi, C.G. 2019. Characterization, antioxidant and antimicrobial activity of silver nanoparticles synthesized using marine endophytic fungus—Cladosporium cladosporioides. Process Biochem. 82(0): 199–204.

Hulikere, M.M., Joshi, C.G., Danagoudar, A., Poyya, J., Kudva, A.K. and Dhananjaya, B.L. 2017. Biogenic synthesis of gold nanoparticles by marine endophytic fungus—Cladosporium cladosporioides isolated from seaweed and evaluation of their antioxidant and antimicrobial properties. Process Biochem. 63(0): 137–144.

Husseiny, S.M., Salah, T.A. and Anter, H.A. 2015. Biosynthesis of size controlled silver nanoparticles by Fusarium oxysporum, their antibacterial and antitumor activities. Beni-Suef Univ. J. Basic Appl. Sci. 4(3): 225–231.

Huy, T.Q., Huyen, P.T.M., Le, A.T. and Tonezzer, M. 2020. Recent advances of silver nanoparticles in cancer diagnosis and treatment. Anticancer Agents Med. Chem. 20(11): 1276–1287. Doi: 10.2174/1 871520619666190710121727. PMID: 31291881.

Iqbal, M.Z., Ali, I., Khan, W.S., Kong, X. and Dempsey, E. 2021. Reversible self-assembly of gold nanoparticles in response to external stimuli. Mat. Des. 205(0): 109694.

Jha, P., Saraf, A. and Soh, J.K. 2021. Antimicrobial activity of biologically synthesized gold nanoparticles from wild mushroom *Cantharellus* species. J. Scientific Res. 65(3): 78–83.

Koul, B., Poonia, A.K., Yadav, D. and Jin, J.-O. 2021. Microbe-mediated biosynthesis of nanoparticles: Applications and future prospects. Biomolecules 11(6): 886.

Latha, D., Prabu, P., Arulvasu, C., Manikandan, R., Sampurnam, S. and Narayanan, V. 2018. Enhanced cytotoxic effect on human lung carcinoma cell line (A549) by gold nanoparticles synthesized from Justicia adhatoda leaf extract. Asian Pac. J. Trop. Biomed. 8(11): 540.

Lee, K.X., Shameli, K., Yew, Y.P., Teow, S.-Y., Jahangirian, H., Moghaddam, R.R. and Webster, T.J. 2020. Recent developments in the facile bio-synthesis of gold nanoparticles (AuNPs) and their biomedical applications. Inter. J. Nanomed. 2020(15): 275–300.

Lee, K.D., Nagajyothi, P.C., Sreekanth, T.V.M. and Park, S. 2015. Eco-friendly synthesis of gold nanoparticles (AuNPs) using *Inonotus obliquus* and their antibacterial, antioxidant and cytotoxic activities. J. Ind. Eng. Chem. 26(0): 67–72.

Lim, H.K., Gurung, R.L. and Hande, M.P. 2017. DNA-dependent protein kinase modulates the anticancer properties of silver nanoparticles in human cancer cells. Mutat. Res. Toxicol. Environ. Mutagen 824(0): 32–41.

Lima, R., Feitosa, L.O., Ballottin, D., Marcato, P.D., Tasic, L. and Durán, N. 2013. Cytotoxicity and genotoxicity of biogenic silver nanoparticles. J. Phys. Conf. Ser. 429(0): 012020.

Lotfy, W.A., Alkersh, B.M., Sabry, S.A. and Ghozlan, H.A. 2021. Biosynthesis of silver nanoparticles by *Aspergillus terreus*: Characterization, optimization, and biological activities. Front. Bioeng. Biotechnol. 9: 633468.

Majumder, D.R. 2012. Bioremediation: Copper nanoparticles from electronic-waste. Inter. J. Eng. Sci. Technol. 4(10): 4380–4389.

Mani, V.M., Kalaivani, S., Sabarathinam, S., Vasuki, M., Soundari, A.J.P.G., Das, M.P.A., Elfasakhany, A. and Pugazhendhi, A. 2021.Copper oxide nanoparticles synthesized from an endophytic fungus Aspergillus terreus: Bioactivity and anticancer evaluations. Environ. Res. 201(0): 111502.

Marcato, P.D., Durán, M., Huber, S., Rai, M., Melo, P.S., Alves, O.L. and Durán, N. 2012. Biogenic silver nanoparticles and its antifungal activity as a new topical transungual drug delivery. J. Nano Res. 20(0): 99–107.

Mikhailova, E.O. 2021. Gold nanoparticles: Biosynthesis and potential of biomedical. J. Funct. Biomater. 12(4): 70.

Miranda, R.R., Sampaio, I. and Zucolotto, V. 2022. Exploring silver nanoparticles for cancer therapy and diagnosis. Colloids Surf. B: Biointerf. 210(0): 112254.

Mirshafiee, V., Mahmoudi, M., Lou, K., Cheng, J. and Kraft, M.L. 2013. Protein corona significantly reduces active targeting yield. Chem. Commun. 49(25): 2557.

Moitra, P., Alafeef, M., Dighe, K., Frieman, M. and Pan, D. 2020. Selective naked-eye detection of SARS-CoV-2 mediated by n gene targeted antisense oligonucleotide capped plasmonic nanoparticles. ACS Nano. 14(6): 7617–7627.

Mughal, B., Zaidi, S.Z.J., Zhang, X. and Hassan, S.U. 2021. Biogenic nanoparticles: Synthesis, characterisation and applications. Appl. Sci. 11(6): 2598.

Naimi-Shamel, N., Pourali, P. and Dolatabadi, S. 2019. Green synthesis of gold nanoparticles using *Fusarium oxysporum* and antibacterial activity of its tetracycline conjugant. J. Mycol. Med. 29(1): 7–13.

Naqvi, S.T.Q., Shah, Z., Fatima, N., Qadir, M.I., Ali, A. and Muhammad, S.A. 2017. Characterization and biological studies of copper nanoparticles synthesized by *Aspergillus niger*. J. Bionanosci. 11(2): 136–140.

Netala, V.R., Bethu, M.S., Pushpalatah, B., Baki, V.B., Aishwarya, S., Rao, J.V. and Tartte, V. 2016. Biogenesis of silver nanoparticles using endophytic fungus Pestalotiopsis microspora and evaluation of their antioxidant and anticancer activities. Int. J. Nanomed. 11(0): 5683–5696.

Noor, S., Shah, Z., Javed, A., Ali, A., Hussain, S.B., Zafar, DS., Ali, H. and Muhammad, S.A. 2020. A fungal based synthesis method for copper nanoparticles with the determination of anticancer, antidiabetic and antibacterial activities. J. Microbiol. Meth. 174(0): 105966.

Othman, A.M., Elsayed, M.A., Al-Balakocy, N.G., Hassan, M.M. and Elshafei, A.M. 2019. Biosynthesis and characterization of silver nanoparticles induced by fungal proteins and its application in different biological activities. J. Genet. Eng. Biotechnol. 17(0): 8.

Paysen, H., Loewa, N., Stach, A., Wells, J., Kosch, O., Twamley, S., Makowski, M.R., Schaeffter, T., Ludwig, A. and Wiekhorst, F. 2020. Cellular uptake of magnetic nanoparticles imaged and quantified by magnetic particle imaging. Sci. Rep. 10(0): 1922.

Peigneux, A., Valverde-Tercedor, C., López-Moreno, R., Pérez-González, T., Fernández-Vivas, M.A. and Jiménez-López, C. 2016. Learning from magnetotactic bacteria: A review on the synthesis of biomimetic nanoparticles mediated by magnetosome-associated proteins. J. Struct. Biol. 196(2): 75–84.

Pham, N.D., Duong, M.M., Le, M.V., Hoang, H.A. and Pham, L.K.O. 2019. Preparation and characterization of antifungal colloidal copper nanoparticles and their antifungal activity against Fusarium oxysporum and *Phytophthora capsici*. C. R. Chim. 22(11-12): 786–793.

Phummirat, P., Mann, N. and Preece, D. 2021. Applications of optically controlled gold nanostructures in biomedical engineering. Front. Bioeng. Biotechnol. 8(0): 602021.

Rai, M., Ingle, A.P., Gade, A. and Durán, N. 2015. Synthesis of silver nanoparticles by *Phoma gardenia* and *in vitro* evaluation of their efficacy against human disease-causing bacteria and fungi. IET Nanobiotechnol. 9(2): 71–75.

Rai, M., Ingle, A.P., Trzcinska-Wencel, J., Wypij, M., Bonde, S., Yadav, A., Kratošová, G. and Golinska, P. 2021. Biogenic silver nanoparticles: What we know and what do we need to know? Nanomaterials 11(0): 2901.

Raman, J., Reddy, G.R., Lakshmanan, H., Selvaraj, V., Gajendran, B., Nanjian, R., Chinnasamy, A. and Sabaratnam, V. 2015. Mycosynthesis and characterization of silver nanoparticles from *Pleurotus djamor* var. *roseus* and their *in vitro* cytotoxicity effect on PC3 cells. Process Biochem. 50(1): 140–147.

Ratan, Z.A., Haidere, M.F., Nurunnabi, M., Shahriar, S.M., Ahammad, A.J.S., Shim, Y.Y., Reaney, M.J.T. and Cho, J.Y. 2020. Green chemistry synthesis of silver nanoparticles and their potential anticancer effects. Cancers 12 (0): 855.

Rodrigues, A.G., RdC, R., Selari, P.J.R.G., de Araujo, W.L. and de Souza, A.O. 2021. Anti-biofilm action of biological silver nanoparticles produced by *Aspergillus tubingensis* and antimicrobial activity of fabrics carrying it. Biointerface Res. Appl. Chem. 11(6): 14764–14774.

Rodriguez-Serrano, C., Guzman-Moreno, J., Angelez-Chavez, C., Rodriguez-Gonzalez, V., Ortega-Sigala, J.J., Ramírez-Santoyo, R.M. and Vidales-Rodríguez, L.E. 2020. Biosynthesis of silver nanoparticles by Fusarium scirpi and its potential as an antimicrobial agent against uropathogenic *Escherchia coli* biofilms. PloS One 15(13): e0230275.

Saied, E., Eid, A.M., Hassan, S.E.-D., Salem, S.S., Radwan, A.A., Halawa, M., Saleh, F.M., Saad, H.A., Saied, E.M. and Fouda, A. 2021. The catalytic activity of biosynthesized magnesium oxide nanoparticles (MgO-NPs) for inhibiting the growth of pathogenic microbes, tanning effluent treatment, and chromium ion removal. Catalysts 11(0): 821.

Santos, L.M., Stanisic, D., Menezes, U.J., Mendonça, M.A., Barral, T.D., Seyffert, N., Azevedo, V., Durán, N., Meyer, R., Tasic, L. and Portela, R.W. 2019. Biogenic Silver nanoparticles as a post-surgical treatment for *Corynebacterium pseudotuberculosis* infection in small ruminants. Front. Microbiol. 10(0): 824.

Saravanakumar, K., Shanmugam, S. and Babu, N. 2019. Biosynthesis and characterization of copper oxide nanoparticles from indigenous fungi and its effect of photothermolysis on human lung carcinoma. J. Photochem. Photobiol B: Biol. 190(0): 103–109.

Saravanakumar, K., Chelliah, R., Shanmugam, S., Varukattu, N.B., Oh, D.-H., Kathiresan, K. and Wang, M.H. 2018. Green synthesis and characterization of biologically active nanosilver from seed extract of *Gardenia jasminoides* Ellis. J. Photochem. Photobiol. B: Biol. 185(0): 126–135.

Saravanakumar, K., Jeevithan, E., Hu, X., Chelliah, R., Ho, D-H. and Wang, M-H. 2020. Enhanced anti-lung carcinoma and anti-biofilm activity of fungal molecules mediated biogenic zinc oxide nanoparticles conjugated with β-D-glucan from barley. J. Photochem. Photobiol. B: Biol. 203(0): 111728.

Seabra, A.B., Haddad, P. and Durán, N. 2013. Biogenic synthesis of nanostructurated iron compounds: Applications and perspectives. IET-Nanobiotechnol. 7(3): 90–99.

Singh, T., Jyoti, K., Patnaik, A., Singh, A., Chauhan, R. and Chandel, S.S. 2017. Biosynthesis, characterization and antibacterial activity of silver nanoparticles using an endophytic fungal supernatant of *Raphanus sativus*. J. Genet. Eng. Biotechnol. 15(1): 31–39.

Srivastava, S., Bhargava, A., Pathak, N. and Srivastava, P. 2019. Production, characterization and antibacterial activity of silver nanoparticles produced by fusarium oxysporum and monitoring of protein-ligand interaction through *in-silico* approaches. Microb. Pathog. 129(0): 136–145.

Sulaiman, G.M., Hussein, T.H. and Saleem, M.M. 2015. Biosynthesis of silver nanoparticles synthesized by Aspergillus flavus and their antioxidant, antimicrobial and cytotoxicity properties. Bull. Mater. Sci. 38(0): 639–644.

Swarnakar, N., Thanki, K. and Jain, S. 2013. Effect of co-administration of CoQ10-loaded nanoparticles on the efficacy and cardio-toxicity of doxorubicin-loaded nanoparticles. RSC Adv. 3(34): 146–171.

Vijayanandan, A.S. and Balakrishnan, R.M. 2018. Biosynthesis of cobalt oxide nanoparticles using endophytic fungus *Aspergillus nidulans*. J. Environ. Manag. 218(0): 442–450.

Wadhwani, S.A., Shedbalkar, U.U., Singh, R. and Chopade, B.A. 2016. Biogenic selenium nanoparticles: Current status and future prospects. Appl. Microbiol. Biotechnol. 100(6): 2555–2566.

Wang, D., Xue, B., Wang, L., Zhang, Y., Liu, L. and Zhou, Y. 2021. Fungus-mediated green synthesis of nano-silver using Aspergillus sydowii and its antifungal/antiproliferative activities. Sci. Rep. 11(0): 10356.

Yan, L., Da, H., Zhang, S., López, V.M. and Wang, W. 2017. Bacterial magnetosome and its potential application. Microbiol. Res. 203(0): 19–28.

Yoon, Y.N., Lee, D.S., Park, H.J. and Kim, J.S. 2020. Barium titanate nanoparticles sensitise treatment-resistant breast cancer cells to the antitumor action of tumour-treating fields. Sci. Rep. 10(0): 2560.

Zhang, Y., Chen, M., Liu, C., Chen, J., Luo, X., Xue, Y., Liang, Q., Zhou, L., Tao, Y., Li, M., Wang, D., Zhou, J. and Wang, J. 2021. Sensitive and rapid on-site detection of SARS-CoV-2 using a gold nanoparticle-based high-throughput platform coupled with CRISPR/Cas12-assisted RT-LAMP. Sens Actuators: B Chem. 345(0): 130411.

Zhang, H., Zhou, H., Bai, J., Li, Y., Yang, J., Ma, Q. and Qu, Y. 2019. Biosynthesis of selenium nanoparticles mediated by fungus *Mariannaea* sp. HJ and their characterization. Colloids Surf. A Physicochem. Eng. Asp. 571(0): 9–16.

Zhao, X., Zhou, L., Rajoka, M.S.R., Yan, L., Jiang, C., Shao, D., Zhu, J., Shi, J., Huang, Q., Yang, H. and Jin, M. 2018. Fungal silver nanoparticles: Synthesis, application and challenges. Crit. Rev. Biotechnol. 38(6): 817–835.

3

Myconanotechnology for Cancer
Current Status and Future Prospects

Marcia Regina Salvadori

Introduction

Nanomedicine is a constantly evolving science that seeks to overcome and improve the limitations of traditional anti-cancer treatments. Due to its nanometric size, the NPs allow the retention of medicines in the tumor tissue by passive targeting, whereas the functionalized NPs perform active targeting. Their targets are molecules expressed in cancer cells, enabling the limitation of side effects in increased concentrations of drugs in the site of acting. The possibility of modification of the surface of the NPs extends the circulation time of medicines, provides a controlled release of the same and prevents its elimination by the immune system (van der Meel et al. 2019, Paus et al. 2021). These characteristics enable unique mechanisms that can greatly improve the pharmacokinetics of therapeutic agents aggregated to NPs (Salvioni et al. 2019).

In the 1960s, a new cycle in materials science was initiated with the appearance of nanotechnology. The synthesis of biogenic nanomaterials emerged as green nanotechnology, opening production routes for environmentally safe NPs, in relation to physical and chemical methods, thus avoiding the high energy consumption and use of toxic products, employing reducing agents and natural stabilizers. The production of biogenic NPs occurs through a bottom-up approach, which covers organisms such as fungi, yeasts, plants, algae, bacteria, among others (Mughal et al. 2021, Patil and Chandrasekaran 2020).

Biogenic metallic NPs act as cytotoxic agents in combating various cancers, causing autophagy and cell death. Due to their large surface area, as well as their nano size, biogenic NPs have shown to be very promising in anti-cancer therapy. In

Department of Microbiology, Biomedical Institute – II, University of São Paulo, São Paulo, 05508000, Brazil.
Emails: mrsal@usp.br, mrsalvadori@yahoo.com.br

addition to promoting the effective delivery of drugs, they present great specificity to tumor cells (Jabeen et al. 2021).

In 2009, the term myconanotechnology was introduced by Dr. Mahendra Rai (Rai et al. 2009). It is a branch of nanotechnology that defines the study of the synthesis and use of nanomaterials via fungi. Myconanotechnology is mainly applied in the areas of biomedical science (Salvadori et al. 2019), environmental science (Salvadori et al. 2018), agricultural commodities and food industry (Jagtap et al. 2021) among many others.

The studies of mycosynthesized NPs have become a promising field in nanomedicine for anti-cancer treatment. The mycosynthesized silver and gold NPs have notably revealed great anti-cancer potential. Research with silver NPs has established that they can lead to apoptosis or cytotoxicity of cancer cells, due to their interaction with several biomolecules, especially carbohydrates, proteins and DNA, possibly due to reactive oxygen species (ROS) (Barabadi et al. 2020).

In this chapter, we have conceptualized myconanotechnology for cancer in a versatile approach to the synthesis of metallic NPs via fungi, used as a potential anti-cancer agent, as well as to understand the possible anti-cancer mechanisms of action of these NPs.

Anti-Cancer Nanomedicine

The conventional anti-cancer treatments encompass several therapies that have serious limitations, such as chemotherapy, which is the most traditional treatment, and that causes side effects by harming normal cell function (Bracci et al. 2014). Hormonal treatment, although effective, is restricted to a small variety of cancers (Yang et al. 2013), and the therapy with monoclonal antibodies and tyrosine kinase inhibitors have their effectiveness limited only to cancers that have certain hyperactive receptors (Li et al. 2013).

Nanomedicine has played a significant role in oncology over the past several decades (Figure 1). A growing number of clinical studies have described several

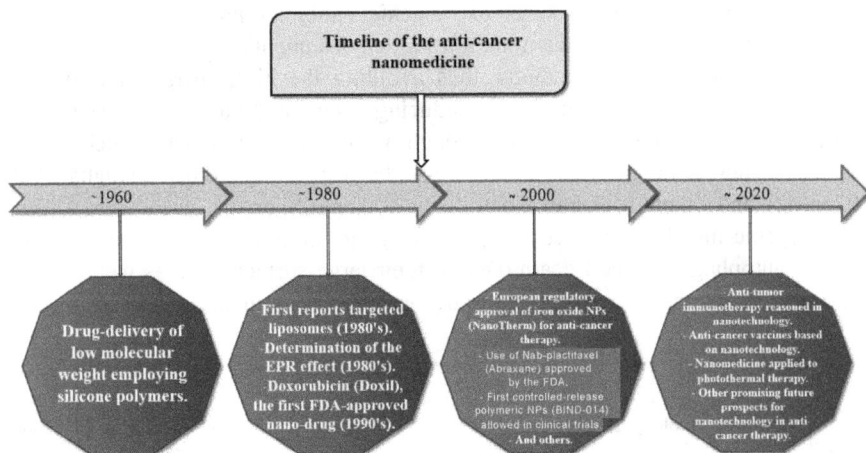

Figure 1. Brief timeline of the advancement of nanomedicine in anti-cancer therapy.

nanoformulations that were developed on the basis of unique physical and chemical properties in order to provide specific and efficient chemotherapeutic agents for tumor cells (Murciano-Goroff et al. 2020).

Nanodrug delivery systems are widely used in cancer treatments and diagnostics due to their high tumor-targeting capability, low toxicity, and controlled release characteristics (Liu et al. 2020).

The accurate development of nanomedicines in relation to their physicochemical properties allows increasing their anti-tumor activity. In addition to expanding the ability to evade the cancer cell efflux pumps and improve the intracellular endosomal delivery of drugs, enabling the selective destruction of tumor cells (Kim and Khang 2020).

The nanocarriers are NPs used as transport modules for other substances and constitute the fundamental pillar of the smart drug delivery system, using the physicochemical differences between the tumor and healthy cells to locate cancer (Hossen et al. 2019). The mechanism of anti-cancer therapy of nanocarriers comprises two types of drug delivery targeting: active and passive targeting. Active targeting, where the nanocarriers equipped with drugs, that are associated with the targeting ligands, detect specific targets overexpressed on the surface of tumor cells. As cancer cells show the overexpression of many cell surface receptors such as folate, aptamers, transferrin, peptides, antibodies, and others, cancer cells are specifically distinguished from normal cells during therapy (Attia et al. 2019). According to Nakamura et al. 2016, in passive targeting, the phenomenon of enhanced permeability effect occurs, which consists of the increased concentration of drug-loaded nanocarriers in the tumor tissue, due to the endothelial permeability of the tumor vasculature. The enhanced retention effect is a deficiency in the lymphatic system that leads to the retention of NPs in cancer. These two phenomena are called the Enhanced Permeability and Retention (EPR) effect. The accumulation of anti-cancer medicines in tumor cells increases several times compared to healthy cells in the body when the EPR effect is applied. Some of the main anti-cancer nanocarriers are mentioned below.

The dendrimers are nanocarriers composed of polymers. They are of great interest in oncology due to their monodisperse nature, branched structure and water solubility. The conventional dendrimers are susceptible to rapid expulsion by the immune system and consequently lead to lower retention by tumor cells. To solve this problem and actively target the tumor site, its chemical modification is used through copolymerization with a linear polymer and hybridization with other carriers. The dendritic surface can be modified by aptamers, proteins and antibodies, among others. The modifications can also be related to systems responsive to stimuli, such as changes in pH, heat, light, proteins, etc. (Dubey et al. 2020).

The liposomes are nanocarriers that become intelligent materials when functionalized. The structures are grafted onto liposomes so that they actively act on tumor sites, such as peptides, monoclonal antibodies, vitamins, proteins, and others. These intelligent nanocarriers respond to external and internal stimuli, such as redox reaction changes, light, pH, ultrasound, enzymatic transformations, etc. Radiolabeled liposomes can be employed to establish their distribution in the body, as well as to diagnose cancer in conjunction with treatment (Beltrán-Gracia et al. 2019).

In anti-cancer therapy, metallic nanocarriers like gold NPs can be employed in the delivery of medicines or molecules such as DNA, RNA and proteins. The gold NPs make it possible to convert light into heat and scatter the heat produced to destroy tumor cells through the surface plasmonic resonance phenomenon (Hossen et al. 2019).

To act as intelligent nanocarriers, micelles have their surfaces functionalized with aptamers, peptides, folic acid, carbohydrates, antibodies, and others. This helps to destroy tumor cells. For a controlled release of the anti-cancer drug, the micelle nucleus or crown must be functionalized. Employing these multifunctional micelles, the co-delivery scheme is very interesting to the synergistic effects in anti-tumor therapy (Majumder et al. 2020).

The nanocarrier quantum dots are employed to detect cancer cells during drug release at the target site (Liang et al. 2021). These nanocarriers reach tumor cells when functionalized actively. For example, a complex of folic acid and quantum dots is used to diagnose ovarian cancer (Zhao and Zhu 2016).

The mesoporous materials are classified as smart nanocarriers, due to their adjustable particle and pore size. They allow different drugs to be carried easily and, in addition to compatibility, the binding of these carriers to tumor cells makes them an ideal option (Gao et al. 2020).

The carbon nanotubes are another type of nanocarrier made of fullerene, which is a graphene sheet wrapped in a cylindrical tube, acquiring the shape of a nanotube. To make carbon nanotubes smart nanocarriers, it is necessary to functionalize them chemically or physically. Recent research has shown that functionalized carbon nanotubes have been able to overcome the blood-brain barrier (You et al. 2019). These nanocarriers have proven to be favorable in the delivery of plasmid DNA, aptamers, small-interfering ribonucleic acid and antisense oligonucleotides (Son et al. 2016). In addition to gene transport, it can be employed for thermal ablation of a tumor site and also for early cancer diagnosis (Chen et al. 2017).

The superparamagnetic iron oxide NPs (SPIONs) are widely studied magnetic materials. Examples of SPIONs are maghemite and magnetite NPs, mixed iron oxides with transition metals such as nickel, copper, and cobalt. The functionalization of SPIONs leads to the formation of intelligent nanocarriers, reducing their aggregation, preserving their surfaces from oxidation, increasing blood circulation avoiding the reticuloendothelial system, providing a surface for grouping drugs and targeting ligands and reducing non-specific targets. The SPIONs are nanocarriers that also have theranostic properties. Since they are magnetic, SPIONs can be located by an external magnetic field, a technique used in anti-tumor therapies (Hossen et al. 2019).

Biogenic NPs as Anti-Cancer Agents

The synthesis of biologically mediated NPs, through green nanotechnology, results in the formation of biogenic NPs. The biological techniques used for the production of NPs allow the use of reducing agents and natural stabilizers, constituting an ecologically and economically sustainable option, in relation to physical and chemical methods.

The production process of biogenic NPs involves a bottoms-up approach, which includes uni and multicellular organisms, such as bacteria, yeasts, fungi, cyanobacteria, algae, plant extracts and viruses (Hernández-Díaz et al. 2021, Kumari et al. 2021, Salvadori et al. 2017, Salvadori et al. 2015, Mukherjee et al. 2021, Tsekhmistrenko et al. 2020). The synthesis of biogenic NPs based on microorganisms constitutes a methodology different from the conventional physical and chemical methods. This is because the synthesis is performed by microbial culture filtrates (extracellular and intracellular), or dead cells from them as reducing agents (Salvadori et al. 2016). The use of microorganisms in the production of biogenic NPs is of great interest due to their immense potential for tolerance to metals and versatility to develop in different environmental conditions (Figure 2).

General Scheme of Synthesis of Biogenic Metallic Nanoparticles

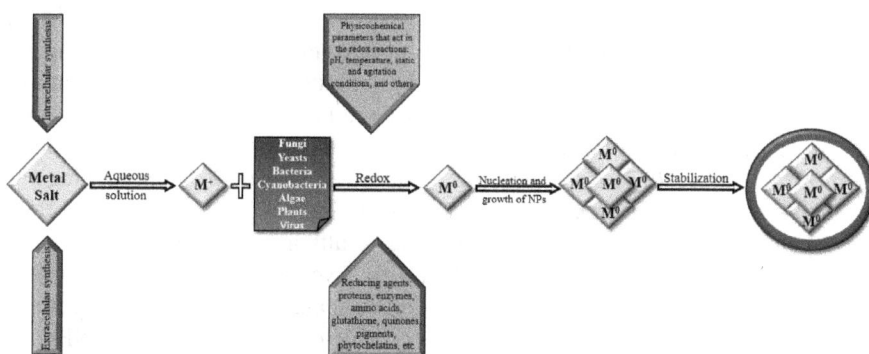

Figure 2. General scheme of synthesis of biogenic NPs.

Due to their nano size and large surface area that enables effective drug delivery and specificity to cancer cells, biogenic NPs have become a promising field in cancer treatment. Several studies of biogenic metallic NPs have been successfully conducted to determine their autophagy and cell death induction activities. These NPs have shown to be promising cytotoxic agents in combating several types of cancer. Three mechanisms were presented for the anti-tumor activity of biogenic NPs. First through apoptosis, which is directly related to the high level of ROS, which leads to oxidative stress and DNA fragmentation in the tumor cell. The second mechanism proposes the influence of proteins/DNA, resulting in alterations of the cellular chemical functions. And finally, the action of biogenic NPs in the cell membranes modifies cell permeability and mitochondrial dysfunction (Ratan et al. 2020).

The mechanisms used in the synthesis of NPs by microorganisms requires further studies for its full elucidation. However, it is known that the production of NPs by microbes covers bioadsorption, bioaccumulation, solubility, metal precipitation, toxicity and efflux systems (Liu and Catchmark 2019).

The bacteria are good candidates for the production of NPs. Due to their rapid cultivation, they are easily accessible in the environment and adapt easily to different culture conditions (temperature and oxygenation). The bacteria are able to synthesize NPs from magnetic metals, called magnetotactic bacteria, which produce magnetosomes, that are suitable for the synthesis of magnetic radicals. Magnetosomes are iron oxide nanocrystals covered by iron sulfide, and can be applied in anti-cancer treatment (Taher et al. 2021). In 2016, Rajeshkumar reported the extracellular synthesis of gold NPs employing culture supernatant of *Enterococcus* sp., obtaining high anti-cancer activity in liver cancer (HepG2) and lung cancer (A549) cell lines (Rajeshkumar 2016). Almalki and Khalifa described the synthesis of silver NPs with spherical shape with an approximate size between 5 and 15 nm by the bacterium *Bacillus* sp. (Almalki and Khalifa 2020). KFU36, a marine strain, showed anti-cancer activity in breast cancer cell line (MCF-7), inducing apoptosis in 61% of these cells.

According to Lachance, around 1500 species of yeasts, which are eukaryotic microorganisms belonging to the fungal kingdom, have already been described (Lachance 2016). Currently, yeasts have become promising in the field of nanotechnology, specifically in the synthesis of biogenic NPs, due to the advantages offered in their synthesis process, such as: easy process and handling of biomass, growth in extreme temperature conditions, pH and nutrients, production of various enzymes, easy control of laboratory conditions and cost-effectiveness (Salvadori et al. 2014a, Salem 2022). Kaler and collaborators described the synthesis of biogenic silver NPs by *Saccharomyces boulardii* cell-free extract with size ranging from 3–10 nm approximately. It showed promising results in anti-cancer activity when tested in MCF-7 cell line (Kaler et al. 2013). Two new yeasts samples named HX-YS and LPP-12Y, related to species of the *Metschnikowia* genus were described in 2021 (Liu et al. 2021). These strains are capable of synthesizing silver NPs and have shown anti-cancer activity against human lung cancer cell lines A549 and H1975.

The mycogenic pathway employed by filamentous fungi in the synthesis of NPs has been promising due to this eukaryotic, multicellular and spore-producing totipotent microorganism. It has an extremely adequate profile as it presents molecules in the cell wall that play a pivotal role in the sorption of various metals, easy cultivation, great potential to secrete extracellular enzymes, growth control, easy handling and synthesis of extracellular NPs (Rai et al. 2009, Salvadori et al. 2014b). The endophytic strain of *Fusarium solani* ATLOY-8 isolated from the plant *Chonemorpha fragrans*, has been reported to produce gold NPs with an approximate diameter of 40–45 nm, exhibiting an anti-cancer potential against cervical cancer cell lines (HeLa) and MCF-7 (Clarance et al. 2020). In 2019, Panchangam and Upputuri reported the synthesis of silver NPs that were ellipsoidal to spherical in shape with average diameter of 79–107 nm, using the filamentous fungus *Scedosporium* sp. with anti-cancer capacity against MCF7 and human prostate cancer (PC-3) cell lines (Panchangam and Upputuri 2019).

The cyanobacterial colonies (blue-green algae) can form leaves, filaments or hollow balls, fixing atmospheric nitrogen and carbon dioxide in the period of photosynthesis. Certain strains develop in the dark in organotrophic/lithotrophic/chemotrophic environments, providing a wide range of nutrition. As they are photoautotrophic organisms, they can develop under chemoautotrophic conditions in the dark and light. The cyanobacteria are ecologically viable and low-cost instruments for NPs synthesis. The cyanobacterial strains commonly used in the production of NPs range from unicellular to colonial species (Zhang et al. 2020). Hamouda et al. reported the synthesis of silver NPs using an aqueous extract of *Oscillatoria limnetica* fresh biomass with sizes ranging from 3.30–17.97 nm (Hamouda et al. 2019). The silver NPs showed cytotoxic effects against both MCF-7 and HCT-116 cell lines.

Belonging to the Protista kingdom, algae comprise autotrophic and aquatic photosynthetic eukaryotic organisms, which may be unicellular or multicellular. The microalgae are capable of synthesizing NPs through green chemistry, converting metal ions into NPs (Mukherjee et al. 2021). The ability of algae to produce NPs is due to the presence of active compounds in its cell walls such as alginate and laminarin (Sharma et al. 2016), while stabilization and capping of NPs are done by their amino acids or cysteine amides. In general terms, it can be said that the synthesis of metallic NPs by algae comprises the following steps: (a) elaboration of algae extracts in solution at high temperatures, and (b) mixing of the extract with solutions of the precursor metal, followed by stirring by a specified time. The color change at the beginning of the reaction evidences the nucleation and subsequent growth of the NPs where the nearby nucleonic particles come together and form the thermodynamically stable NPs under different geometries (Kumaresan et al. 2018). Biogenic ruthenium NPs were synthesized from marine algae *Dictyota dichotoma* with an average size of 30 nm, and have been described for presenting the anti-cancer activity to HeLa and MCF-7N cell lines (Ali et al. 2017). In 2021, Acharya and collaborators obtained silver NPs using marine algae *Chaetomorpha linum* (Acharya et al. 2021). These NPs were shown to be an efficient candidate in anti-cancer activity, which was demonstrated using human colorectal carcinoma cells (HCT-116).

There are several plants species for the production of NPs via green chemistry. They have advantages in the synthesis of biogenic NPs as they do not require cell culture maintenance, support the production of NPs on a large scale, are low cost, easy and safe to handle, easily found in nature and provide natural stabilizing agents (Zhang et al. 2020). The plants have phytoconstituents that include primary and secondary metabolites that act as reducing agents used in the synthesis of NPs, such as, vitamins, nucleic acids and alkaloids, proteins, terpenoids, amino acids, saponies, phenols and flavonoids (Jain et al. 2021). In 2016, researchers described the employment of various parts of plants to produce NPs, such as stems, latex, seeds, callus, fruits and bark (Anjum et al. 2016). The plants synthesize metallic NPs intracellularly or extracellularly. Intracellular synthesis includes the development of plants in a hydroponic solution rich in metals, soils abundant in metals, among others (Harris and Bali 2007), while the extracellular synthesis of NPs includes the use of the extract of leaves or leaves in water or ethanol (Parashar et al. 2009). Generally, the plant extract reduction technique includes mixing the aqueous solution of the metallic salt with the extract at room temperature, and the production of NPs is affected in a few minutes (Mittal et al. 2013). The *Diospyros malabarica* tree is widely cultivated in India, whose fruit was used as a synthesizer of silver NPs with anti-cancer potential against the human primary glioblastoma cell line (U87-MG) (Bharadwaj et al. 2021). Recently, it was reported that the fresh peel extracts of *Benincasa hispida* acted as reducing and stabilizing agents in the synthesis of gold NPs with spherical shape and the average size of 22.18 ± 2 nm, showing *in vitro* cytotoxic effect against the HeLa cell line (Al Saqr et al. 2021).

Viruses have been used in the synthesis of quantum dots for the production of nanomaterials in the last decade. The presence of the external viral capsid protein has important applicability in the production of NPs, offering a highly reactive surface with metallic ions (Makarov et al. 2014). The viral NPs have the ability to deliver active drugs to cancer cells, employing the unique pathophysiology of tumors such as retention effect, carcinogenic microenvironment, and enhanced permeability (Wu et al. 2021).

Myconanotechnology for Cancer: Current Status

Dr. Mahendra Rai in 2009, established the term myconanotechnology, as an area of nanotechnology that defines the study of the synthesis and use of nanomaterials via fungi. Recently, the mycogenesis of NPs is considered a relevant way of participation of fungi in nanotechnology. The synthesis of NPs by fungi can be carried out with the desired shape and size, intracellularly and/or extracellularly. The mycogenic synthesis of NPs constitutes the fundamental pillar of myconanotechnology, directing towards an original and feasible multidisciplinary science with promising results (Hamida et al. 2021).

The fungi produce a wide range of enzymes and proteins that act as reducing agents. This characteristic makes them important in the synthesis of metallic NPs. Other advantages of NPs' mycosynthesis are faster growth of the fungi compared to bacteria, under the same conditions. The fungi have mycelia with a large surface area, providing a greater interaction with metals in the synthesis of metallic NPs, and being more advantageous than bacteria (Khandel and Shahi 2018). Some examples of studies reporting mycogenic metallic NPs are listed in Table 1.

The general mechanism of mycosynthesis of metallic NPs has not been fully elucidated. However, in general terms, the intra- and extracellular mechanisms are described as follows. A possible mechanism of intracellular mycosynthesis of NPs would involve the electrostatic interaction between metal cations and amide groups present in fungal cell wall enzymes. Subsequent bioreduction of ions by enzymes present within the cell wall, would lead to aggregation of metal ions and the formation of NPs. The probable extracellular mechanism of mycogenic NPs involves the interaction between metallic cations and amide groups located in the fungal cell wall, followed by bioreduction, probably due to the presence of extracellular enzymes in the fungal cell wall. These same proteins would act as capping and stabilizing agents of the NPs (Salvadori et al. 2022). The extracellular synthesis of NPs offers the advantages of not requiring downstream processing for NPs recovery and fungal cell wall lysis (Saxena et al. 2014) (Figure 3).

The mycogenic NPs have been used extensively in the pharmaceutical industry, nanomedicine, environmental cleaning, agriculture, textile industries and many other applications (Table 2).

Table 1. Mycosynthesized metallic nanoparticles.

Fungi	Metal NPs	Size (nm)	Shape	References
Alternaria alternata	Fe	9 ± 3	Cubic	Mohamed et al. 2015
Aspergillus aculeatus	NiO	5.89	Spherical	Salvadori et al. 2014
Aspergillus flavus	TiO_2	62–74	Spherical, Oval	Rajakumar et al. 2012
Aspergillus niger	CeO_2	5–20	Spherical	Gopinath et al. 2015
Aspergillus terreus	Ag	7–23	Spherical	Lofty et al. 2021
Aspergillus tubingensis	$Ca_3P_2O_8$	28.2	Spherical	Tarafdar et al. 2012
Candida albicans	CdS	50–60	Spherical	Venkat et al. 2019
Fusarium acuminatum	Au	17	Spherical	Tidke et al. 2014
Fusarium oxysporum	Pt	10–100	Hexagons, Pentagons, Circles, Squares, Rectangles	Riddin et al. 2006
Geotrichum sp	U(VI)	50–100	Needles	Zhao et al. 2016
Hypocrea lixii	Cu	24.5	Spherical	Salvadori et al. 2013
Penicillium chrysogenum	Te	50.16	Spherical	Barabadi et al. 2018
Penicillium sp.	Zr	100	Spherical	Golnaraghi Ghomi et al. 2019
Phaffia rhodozyma	Ag, Au	4.1 ± 1.44, 2.22 ± 0.7	Quasi-Spherical, Spherical	Rónavári et al. 2018
Rhizopus oryaze	MgO	20.38 ± 9.9	Spherical	Hassan et al. 2021
Saccharomyces cerevisiae	$BaTiO_3$, Pd	8–21, 10–20	Both Spherical	Jha and Prasad 2010, Saitoh et al. (2020)
Trichoderma asperellum	CuO	110	Spherical	Saravanakumar et al. 2019
Xylaria acuta	ZnO	34–55	Hexagonal	Sumanth et al. 2020

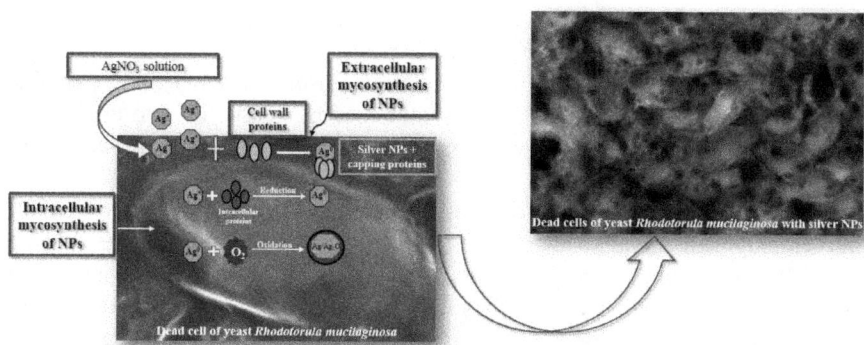

Figure 3. Schematic illustration of the mechanism of synthesis of extra and intracellular silver NPs.

One of the most promising applications of NPs synthesized by fungi is environmental cleaning, i.e., myconanobioremediation and water treatment. The main techniques used in the processes mentioned above are adsorption of toxic

Table 2. Applications of mycosynthesized metal nanoparticles.

Fungi	Mycosynthesized NPs	Potential Application	References
Aspergillus japonicus AJP01	Au	Catalytic activity	Bhargava et al. 2015
Aspergillus niger	Ag	Wound healing activity	Sundaramoorthi et al. 2009
Aspergillus niger	Cu	Anti-cancer, antidiabetic and antibacterial activities	Noor et al. 2020
Aspergillus sydowii	Ag	Antifungal/antiproliferative activities	Wang et al. 2021
Beauveria bassiana	Ag	Antibacterial activity	Tyagi et al. 2019
Lecanicillium lecanii	Ag	Textile fabrics	Namasivayam et al. 2011
Metarhizium anisopliae	Ag	Mosquitocidal activity	Amerasan et al. 2016
Penicillium oxalicum	Ag	Larvicidal activity against the larvae of *Culex quinquefasciatus*	Seetharaman et al. 2021
Pycnoporus sanguineus	Au	Biodegradation	Shi et al. 2015
Rhizopus oryzae	Au	Water hygiene management	Das et al. 2009
Saccharomyces pombe	CdS	Electric diode	Kowshik et al. 2002
Tricholoma crassum	Au	Anti-cancer activity	Basu et al. 2018
Trichoderma viride	Ag	Vegetable and fruit preservation	Fayaz et al. 2009

metals and other pollutants, and the alteration of toxic materials to non-toxic or less toxic forms (Salvadori et al. 2022).

The mycosynthesized NPs have high potential mosquitocidal properties (Hassan et al. 2021). In 2014, it was found that silver NPs synthesized by the entomopathogenic fungus, *Beauveria bassiana*, showed an effective action against the dengue vector *Aedes aegypti* (Banu and Balasubramanian 2014).

The mycogenic NPs can also be applied as biosensors. Zheng et al. reported that the alloy of Au-Ag NPs extracellularly synthesized by yeasts, used to produce an electrochemical vanillin sensor, was able to increase the response electrochemistry of vanillin five times (Zheng et al. 2010).

The myconanofunctional agricultural products constitute a new thriving route for direct employment in fields and farms, for sensing systems, pre-and post-harvest crop preservation (insecticides and herbicides), precision agriculture, targeted intelligent distribution systems, advanced diagnostics and management of mycotoxins. Thus, NPs synthesized by a wide range of fungal species can be used to improve agricultural production, optimizing growth and preservation against infections (Alghuthaymi et al. 2021, Kalia et al. 2020).

Pharmaceutical and biomedical sciences are the areas where mycogenic NPs is applied the most. The NPs synthesized by macrofungi are widely recognized for their excellent anti-cancer, nutritional, antimicrobial, immunomodulatory, antioxidant activities (Bharadwaj et al. 2020). Recently, some literature suggests that NPs synthesized by fungi have promising antibacterial activities, such as against pathogenic bacteria, *Staphylococcus aureus, Escherichia coli, Pseudomonas aeruginosa, Salmonella typhae, Bacillus cereus*, etc. (Khandel and Shahi 2018, Salvadori et al. 2019). The mycogenic NPs also have excellent antifungal activity against pathogenic fungi such as *Cryptococcus* sp., *Candida* sp., *Aspergillus* sp., *Mucor* sp., and others (Salvadori et al. 2019, Wang et al. 2021).

Every year, millions of individuals are diagnosed with cancer with a high mortality rate. It is usually triggered as a result of disturbances in various physiological processes, such as cell signaling and apoptosis. Due to the resistance of cancer cells to various anti-cancer drugs, the therapy of this disease has become a great stimulus in the search for new drugs. Some reasons such as the unspecific spread of anti-cancer drugs leading to insufficient dosages to tumor cells, severe adverse reactions, ineffective control of drug responses, especially in developing countries, are turning cancer into an incurable disease (Waseem and Nisar 2016).

The mycosynthesized metallic NPs play an expressive role in anti-cancer myco-nanotherapy, leading to a new tactic in medicine. The pristine mycofabricated NPs, without carrying drugs, were and still are used in the development of medicines with anti-cancer potential (Table 3).

The copper oxide NPs produced by *Aspergillus terreus* were effective against colon cancer cell line (HT-29) (Mani et al. 2021). Borse and co-authors observed that platinum NPs synthesized by *Saccharomyces boulardii* had good anti-cancer activity against the epidermoid carcinoma (A431) and MCF-7 cell lines (Borse et al. 2015). The gold NPs mycosynthesized by *Pleurotus ostreatus*, showed anti-cancer activity against the HepG2 and HCT-116 cell lines (El Domany et al. 2018). It was observed that the endophytic fungus *Botryosphaeria rhodina* isolated from a medicinal plant *Catharanthus roseus* (Linn.), was able to synthesize silver NPs and showed cytotoxic activity against the A549 cell line (Akther et al. 2019). Iram et al. reported the use of *Fusarium oxysporum* for the synthesis of terbium oxide NPs, which were cytotoxic

Table 3. Mycosynthesized NPs with potential anti-cancer activity.

Fungi	Mycosynthesized NPs with Anticancer Activity	References
Agaricus bisporus	Pd	Mohana and Sumathi 2020
Aspergillus niger	CdS	Alsaggaf et al. 2020
Aspergillus niger	Cu	Noor et al. 2020
Cladosporium perangustum	Ag	Govindappa et al. 2020
Fusarium solani	Au	Clarance et al. 2020
Ganoderma enigmaticum and *Trametes ljubarskyi*	Ag	Krishna et al. 2021
Penicillium sp.	MgO	Majeed et al. 2018
Polyporus umbellatus	Se	Gao et al. 2020
Trichoderma crassum	Au	Basu et al. 2018
Xylaria acuta	ZnO	Sumanth et al. 2020

to osteosarcoma cell lines (MG-63 and Saos-2) (Iram et al. 2016). In 2014, it was reported that potential anti-cancer activity using taxol-conjugated gadolinium oxide NPs was mycosynthesized by *Humicola* sp. (Khan et al. 2014).

The mycogenic NPs, especially silver NPs, have shown great potential for anti-cancer activity with mechanisms of action not yet fully elucidated (Figure 4).

The silver NPs can induce apoptosis and necrosis, damaging the ultrastructure of tumor cells, leading to ROS production and DNA damage. Through transmission electron microscopy, it was found that tumor cells in the presence of silver NPs, depending on the time and concentration of NPs, showed damage to the cell ultrastructure, such as morphological modifications, injuries of cytoplasmic organelles and different death pathways: apoptosis, necrosis and autophagy (Sooklert et al. 2019). In the cytoplasm of cancer cells exposed to silver NPs, autophagosomes that are inherent to the processes of apoptosis and necrosis are produced. The free Ag^+ cations from silver NPs are linked to the destruction of cell membranes, inducing glutathione oxidation and consequently an increase in lipid peroxidation in cell membranes, leading to the leakage of cytoplasmic constituents of injured cells (George et al. 2018). The increase of ROS leads to oxidative stress. This phenomenon causes lipid peroxidation, oxidation of amino acids into proteins, inactivation of enzymes, damages mitochondrial functions and damage to the DNA/RNA, which provides autophagy, apoptosis and necrosis of tumor cells. Cancer cells exposed to silver NPs through endocytosis can cause autophagy and apoptosis due to ROS-brokered stress responses. The production of ROS due to the presence of silver NPs can interfere with cellular signal transduction pathways that can act in apoptosis, where the mitochondrial function can be impeded by silver NPs through suspension of the mitochondrial respiratory chain (Prasad et al. 2017). The tumor cells in contact with silver NPs may have DNA repair defects, DNA methylation,

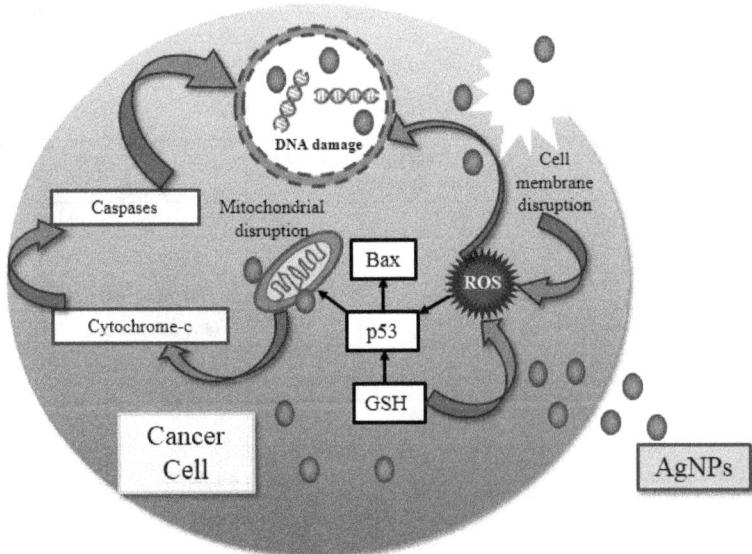

Figure 4. Schematic illustration of a potential anti-cancer mechanism of mycogenic AgNPs.

increased chromosomal aberrations and DNA base pairing errors. The silver NPs may exercise an important function in the regulation of cellular gene expression, through the up- or down-regulation of the expression of key genes, such as p53. We can also cite the up-regulation of the apoptotic precursor protein Bax, suggesting that silver NPs begin apoptosis via mitochondria. In general terms, silver NPs can present anti-cancer activity through several routes. They cause damage to the DNA via ROS, and also through the activation of cellular morphological alterations and disruption of the DNA structure directly via Ag^0 and Ag^+ provided by silver NPs (Ishida 2017). It is noteworthy that silver NPs can also impede the migration of cancer cells, reduce distant metastasis and angiogenesis (Ratan et al. 2020).

Myconanotechnology for Cancer: Future Prospects

Cancer is apparently a pivotal agent of global morbidity and mortality. Approximately 14 million new cases of cancer are expected to occur by 2035, which will generate a considerable impact on the economy and society worldwide (McDaniel et al. 2019). In view of this, there is great urgency to devise efficient and progressive therapy techniques to decrease the adverse effects of the occurrence of cancer. Conventional anti-cancer therapies include surgery, chemotherapy and radiotherapy, but the side effects interfere with the results. Conventional chemotherapy can lead to thrombophlebitis, tissue necrosis, liver and kidney dysfunction, myelosuppression, immunosuppression and occasionally multiple drug resistance, which results in chemotherapy failure. To overcome these challenging obstacles, a new branch of oncology called cancer nanomedicine has emerged (Schirrmacher 2019).

The metallic NPs can be used as new therapeutics and functionalized drug carriers with possible or existing anti-cancer drug candidates, whose side effects can be prevented with the use of this targeted conduct (Machado et al. 2019). Among the NPs synthesis methods, such as physical, chemical and biological, the biological process, more precisely the mycosynthesis of NPs, has shown to be promising. This is primarily because it is benign in nature, economically viable, easy to handle, requires simple nutrients, possesses a high capacity for binding metals to the cell wall, and intracellular metal uptake. This is besides their enormous secretory components of enzymes and proteins, which are involved in the ion reduction and capping of NPs. Promising studies are being carried out with mycogenic NPs, which exhibit efficient anti-cancer activity, against kidney, breast, prostate, lung, hepatocellular cancer, etc., in preliminary studies (Table 3).

However, some challenges in the NPs' mycosynthesis strategy require resolutions. These involve the stability, morphology and size of the crystal, monodispersity of the NPs, as well as the management of its growth. Other crucial parameters that must be studied would be the separation and purification of NPs. The use of genetically modified organisms makes it possible to enhance the production of proteins and enzymes vital for the synthesis and stabilization of NPs, increasing the capacity for metal accumulation and thus the biosynthesis of NPs (Zhang et al. 2020).

The mycosynthesized NPs have broad perspectives for application in nanomedicine, especially the anti-tumor activity that has received great attention. The mechanisms by which mycosynthesized NPs act as anti-cancer are not fully elucidated. For a more successful future application of these NPs, studies are being carried out in order to improve their biocompatibility and surface functionalization. The surface of the NPs allow the insertion of several ligands, making them a powerful nanocarrier, that would allow the optimization of its biological safety and simultaneous delivery of several anti-cancer drugs (Xu et al. 2020). Research is being carried out in order to address challenges in clinical trials regarding the dosage, route of administration and biodegradability of NPs produced by green synthesis (Ratan et al. 2020).

Conclusions

The employment of mycosynthesized metallic NPs in the field of oncological myco-nanotherapy gives a glimpse in to near future showing the union between technology and nature. During the development of this chapter, it was sought to describe the current use trends of several fungi in the synthesis of NPs with potential application in anti-cancer therapies, as well as the possible mechanisms of action of mycogenic NPs in cancer cells. It was delineated how the mycogenic NPs efficiently expressed their anti-tumor activity against several cancer cell lines, via impediment of cell proliferation, ROS production, inhibition of DNA synthesis, and apoptosis. The future perspectives of biogenic metallic NPs, with strong recommendations for their use in oncological myco-nanotherapy has been highlighted. Although more studies are needed in terms of bioavailability, cell toxicity and biocompatibility, to adopt these NPs as a safe biocompatible agent. The use of biogenic metallic NPs also requires more research to evaluate their effect singly or with other bioactive agents to treat and control the various forms of tumors, promoting a new era of cancer treatment.

Acknowledgement

The author thanks Lée Hyppolito Salvadori for her collaboration and inestimable support during the writing of this chapter.

References

Acharya, D., Satapathy, S., Somu, P., Parida, U.K. and Mishra, G. 2021. Apoptotic effect and anticancer activity of biosynthesized silver nanoparticles from marine algae *Chaetomorpha linum* extract against human colon cancer cell HCT-116. Biol. Trace. Elem. Res. 199: 1812–1822.

Akther, T., Mathipi, V., Kumar, N.S., Davoodbasha, M.A. and Srinivasan, H. 2019. Fungal-mediated synthesis of pharmaceutically active silver nanoparticles and anticancer property against A549 cells through apoptosis. Environ. Sci. Pollut. Res. Int. 26: 13649–13657.

Al Saqr, A., Khafagy, El-S., Alalaiwe, A., Aldawsari, M.F., Alshahrani, S.M., Anwer, Md. K., Khan, S., Lila, A.S.A., Hany, H.A. and Hegazy, W.A.H. 2021. Synthesis of gold nanoparticles by using green machinery: Characterization and *in vitro* toxicity. Nanomaterials 11: 808. https://doi.org/10.3390/nano11030808.

Alghuthaymi, M.A., Kalia, A., Bhardwaj, K., Bhardwaj, P., Abd-Elsalam, K.A., Valis, M. and Kuca, K. 2021. Nanohybrid antifungals for control of plant diseases: Current status and future perspectives. J. Fungi. 7: 48. https://doi.org/10.3390/jof7010048.

Ali, M.Y.S., Anuradha, V., Abishek, R., Yogananth, N. and Sheeba, H. 2017. *In vitro* anticancer activity of green synthesis ruthenium nanoparticle from *Dictyota dichotoma* marine algae. NanoWorld J. 3: 66–71.

Almalki, M.A. and Khalifa, A.Y.Z. 2020. Silver nanoparticles synthesis from *Bacillus* sp. KFU 36 and its anticancer effect in breast cancer MCF-7 cells via induction of apoptotic mechanism. J. Photochem. Photobiol. B 204: 111786. Doi: 10.1016/j.jphotobiol.2020.111786.

Alsaggaf, M.S., Elbaz, A.F., El Badawy, S. and Moussa, S.H. 2020. Anticancer and antibacterial activity of cadmium sulfide nanoparticles by *Aspergillus niger.* Adv. Polym. Technol. 2020: 4909054.

Amerasan, D., Nataraj, T., Murugan, K., Panneerselvam, C., Madhiyazhagan, P., Nicoletti, M. and Benelli, G. 2016. Myco-synthesis of silver nanoparticles using *Metarhizium anisopliae* against the rural malaria vector *Anopheles culicifacies* Giles (Diptera: Culicidae). J. Pest. Sci. 89: 249–256.

Anjum, S., Abbasi, B.H. and Shinwari, Z.K. 2016. Plant-mediated green synthesis of silver nanoparticles for biomedical applications: Challenges and opportunities. Pak. J. Bot. 48: 1731–1760.

Attia, M.F., Anton, N., Wallyn, J., Omran, Z. and Vandamme, T.F. 2019. An overview of active and passive targeting strategies to improve the nanocarriers efficiency to tumour sites. J. Pharm. Pharmacol. 71: 1185–1198.

Banu, A.N. and Balasubramanian, C. 2014. Myco-synthesis of silver nanoparticles using *Beauveria bassiana* against dengue vector, *Aedes aegypti* (Diptera: Culicidae). J. Parasitol. Res. 113: 2869–2877.

Barabadi, H., Kobarfard, F. and Vahidi, H. 2018. Biosynthesis and characterization of biogenic tellurium nanoparticles by using *Penicillium chrysogenum* PTCC 5031: A novel approach in gold biotechnology. Iran. J. Pharm. Sci. 17: 87–97.

Barabadi, H., Vahidi, H., Rashedi, M., Mahjoud, M.A., Nanda, A. and Saravanan, M. 2020. Recent advances in biological mediated cancer research using silver nanoparticles as a promising strategy for hepatic cancer therapeutics: A systematic review. Nanomed. J. 7: 251–262.

Basu, A., Ray, S., Chowdhury, S., Sarkar, A., Mandal, D.P., Bhattacharjee, S. and Kundu, S. 2018. Evaluating the antimicrobial, apoptotic, and cancer cell gene delivery properties of protein-capped gold nanoparticles synthesized from the edible mycorrhizal fungus *Tricholoma crassum.* Nanoscale Res. Lett. 13: 154.

Beltrán-Gracia, E., López-Camacho, A., Higuera-Ciapara, I., Velázquez-Fernández, J.B. and Vallejo-Cardona, A.A. 2019. Nanomedicine review: Clinical developments in liposomal applications. Cancer Nanotechnol. 10: 11. https://doi.org/10.1186/s12645-019-0055-y.

Bharadwaj, K.K., Rabha, B., Pati, S., Choudhury, B.K., Sarkar, T., Gogoi, S.K., Kakati, N., Baishya, D., Kari, Z.A. and Edinur, H.A. 2021. Green synthesis of silver nanoparticles using *Diospyros malabarica* fruit extract and assessments of their antimicrobial, anticancer and catalytic reduction of 4-nitrophenol (4-NP). Nanomaterials 11: 1999. https://doi.org/10.3390/nano11081999.

Bhardwaj, K., Sharma, A., Tejwan, N., Bhardwaj, S., Bhardwaj, P., Nepovimova, E., Shami, A., Kalia, A., Kumar, A. and Abd-Elsalam, K.A. 2020. *Pleurotus* macrofungi-assisted nanoparticles synthesis and its potential applications: A review. J. Fungi. 6: 351. Doi: 10.3390/jof6040351.

Bhargava, A., Jain, N., Gangopadhyay, S. and Panwar, J. 2015. Development of gold nanoparticle-fungal hybrid based heterogeneous interface for catalytic applications. Pro. Biochem. 50: 1293–1300.

Borse, V., Kaler, A. and Banerjee, U.C. 2015. Microbial synthesis of platinum nanoparticles and evaluation of their anticancer activity. Int. J. Emerg. Trends Electr. Electron. 11: 26–31.

Bracci, L., Schiavoni, G., Sistigu, A. and Belardelli, F. 2014. Immune-based mechanisms of cytotoxic chemotherapy: Implications for the design of novel and rationale-based combined treatments against cancer. Cell Death Differ. 21: 15–25.

Chen, Z., Zhang, A., Wang, X., Zhu, J., Fan, Y., Yu, H. and Yang, Z. 2017. The advances of carbon nanotubes in cancer diagnostics and therapeutics. J. Nanomater. 2017: 1–13.

Clarance, P., Luvankar, B., Sales, J., Khusro, A., Agastian, P., Tack, J.C., Al Khulaifi, M.M., Al-Shwaiman, H.A., Elgorban, A.M., Syed, A. and Kim, H.J. 2020. Green synthesis and characterization of gold nanoparticles using endophytic fungi *Fusarium solani* and its *in vitro* anticancer and biomedical applications. Saudi J. Biol. Sci. 27: 706–712.

Das, S.K., Das, A.R. and Guha, A.K. 2009. Gold nanoparticles: Microbial synthesis and application in water hygiene management. Langmuir 25: 8192–8199.

Dubey, S.K., Salunkhe, S., Agrawal, M., Kali, M., Singhvi, G., Tiwari, S., Saraf, S., Saraf, S. and Alexander, A. 2020. Understanding the pharmaceutical aspects of dendrimers for the delivery of anticancer drugs. Curr. Drug Targets 21: 528–540.

El Domany, E.B., Essam, T.M., Ahmed, A.E. and Farghali, A.A. 2018. Biosynthesis physico-chemical optimization of gold nanoparticles as anti-cancer and synergetic antimicrobial activity using *Pleurotus ostreatus* fungus. J. Appl. Pharm. Sci. 8: 119–128.

Fayaz, A.M., Balaji, K., Girilal, M., Kalaichelvan, P.T. and Venkatesan, R. 2009. Mycobased synthesis of silver nanoparticles and their incorporation into sodium alginate films for vegetable and fruit preservation. J. Agric. Food. Chem. 57: 6246–6252.

Gao, X., Li, X., Um, J., Ho, C.T., Su, J., Zhang, Y., Lin, X., Chen, Z., Li, B. and Xie, Y. 2020. Preparation, physicochemical characterization, and anti-proliferation of selenium nanoparticles stabilized by *Polyporus umbellatus* polysaccharide. Int. J. Biol. Macromol. 152: 605–615.

Gao, Y., Gao, D., Shen, J. and Wang, Q. 2020. A review of mesoporous sílica nanoparticles delivery systems in chemo-based combination cancer therapies. Front. Chem. 8: 598722. Doi: 10.3389/fchem.2020.598722.

George, B.P.A., Kumar, N., Abrahamse, H. and Ray, S.S. 2018 Apoptotic efficacy of multifaceted biosynthesized silver nanoparticles on human adenocarcinoma cells. Sci. Rep. 8: 14368. https://doi.org/10.1038/s41598-018-32480-5.

Golnaraghi Ghomi, A.R., Mohammadi-Khanaposhti, M., Vahidi, H., Kobarfard, F., Ameri Shah Reza, M. and Barabadi, H. 2019. Fungus-mediated extracellular biosynthesis and characterization of zirconium nanoparticles using standard *Penicillium* species and their preliminary bactericidal potential: A novel biological approach to nanoparticle synthesis. Iran J. Pharm. Res. 18: 2101–2110.

Gopinath, K., Karthika, V., Sundaravadivelan, C., Gowri, S. and Arumugam, A. 2015. Mycogenesis of cerium oxide nanoparticles using *Aspergillus niger* culture filtrate and their applications for antibacterial and larvicidal activities. J. Nanostruct. Chem. 5: 295–303.

Govindappa, M., Lavanya, M., Aishwarya, P., Pai, K., Lunked, P., Hemashekhar, B., Arpitha, B.M., Ramachandra, Y.L. and Raghavendra, V.B. 2020. Synthesis and characterization of endophytic fungi, *Cladosporium perangustum* mediated silver nanoparticles and their antioxidant, anticancer and nano-toxicological study. BioNano. Sci. 10: 928–941.

Hamida, R.S., Ali, M.A., Abdelmeguid, N.E., Al-Zaban, M.I., Baz, L. and Bin-Meferij, M.M. 2021. Lichens—A potential source for nanoparticles fabrication: A Review on nanoparticles biosynthesis and their prospective applications. J. Fungi. 7: 291. https://doi.org/10.3390/jof7040291.

Hamouda, R.A., Hussein, M.H., Abo-elmagd, R.A. and Bawazir, S.S. 2019. Synthesis and biological characterization of silver nanoparticles derived from the cyanobacterium *Oscillatoria limnetica*. Sci. Rep. 9: 13071.

Harris, A.T. and Bali, R. 2007. On the formation and extent of uptake of silver nanoparticles by live plants. J. Nanopart. Res. 10: 691–695.

Hernández-Díaz, J.A., Garza-García, J.J.O., Zamudio-Ojeda, A., León-Morales, J.M., López-Velázquez, J.C. and García-Morales, S. 2021. Plant-mediated synthesis of nanoparticles and their antimicrobial activity against phytopathogens. J. Sci. Food Agric. 101: 1270–1287.

Hassan, S.E.-D., Fouda, A., Saied, E., Farag, M.M.S., Eid, A.M., Barghoth, M.G., Awad, M.A., Hamza, M.F. and Awad, M.F. 2021. *Rhizopus oryzae*-mediated green synthesis of magnesium oxide nanoparticles (MgO-NPs): A promising tool for antimicrobial, mosquitocidal action, and tanning effluent treatment. J. Fungi. 7: 372.

Hossen, S., Hossain, M.K., Basher, M.K., Mia, M.N.H., Rahman, M.T. and Uddin, M.J. 2019. Smart nanocarrier-based drug delivery systems for cancer therapy and toxicity studies: A review. J. Adv. Res. 15: 1–18.

Iram, S., Khan, S., Ansary, A.A., Arshad, M., Siddiqui, S., Ahmad, E., Khan, R.H. and Khan, M.S. 2016. Biogenic terbium oxide nanoparticles as the vanguard against osteosarcoma. Spectrochim. Acta Part A Mol. Biomol. Spectrosc. 168: 123–131.

Ishida, T. 2017. Anticancer activities of silver ions in cancer and tumor cells and DNA damages by Ag^+-DNA base-pairs reactions. MOJ Tumor Res. 1: 8–16.

Jabeen, S., Qureshi, R., Munazir, M., Maqsood, M., Munir, M., Shah, S.S.H. and Rahim B.Z. 2021. Application of green synthesized silver nanoparticles in cancer treatment—A critical review. Mater. Res. Express 8: 092001.

Jagtap, P., Nath, H., Kumari, P.B., Dave, S., Mohanty, P., Das, J. and Dave, S. 2021 Mycogenic fabrication of nanoparticles and their applications in modern agricultural practices & food industries. pp. 475–488. *In*: Sharma, V.K., Shah, M.P., Parmar, S. and Kumar, A. (eds.). Fungi Bio-prospects in Sustainable Agriculture, Environment and Nano-technology. Elsevier: Amsterdam, The Netherlands.

Jain, N., Jain, P., Rajput, D. and Patil, U.K. 2021. Green synthesized plant-based silver nanoparticles: Therapeutic prospective for anticancer and antiviral activity. Micro and Nano Syst. Lett. 9. https://doi.org/10.1186/s40486-021-00131-6.

Jha, A.K. and Prasad, K. 2010. Synthesis of BaTiO$_3$ nanoparticles: A new sustainable green approach. Integr. Ferroelectr. 117: 49–54.

Kaler, A., Jain, S. and Banerjee, U.C. 2013. Green and rapid synthesis of anticancerous silver nanoparticles by *Saccharomyces boulardii* and insight into mechanism of nanoparticle synthesis. Biomed. Res. Int. 2013: 872940. Doi: 10.1155/2013/872940.

Kalia, A., Abd-Elsalam, K.A. and Kuca, K. 2020. Zinc-based nanomaterials for diagnosis and management of plant diseases: Ecological safety and future prospects. J. Fungi. 6: 222. https://doi.org/10.3390/jof6040222.

Khan, S.A., Gambhir, S. and Ahmad, A. 2014. Extracellular biosynthesis of gadolinium oxide (Gd$_2$O$_3$) nanoparticles, their biodistribution and bioconjugation with the chemically modified anticancer drug taxol. Beilstein J. Nanotechnol. 5: 249–257.

Khandel, P. and Shahi, S.K. 2018. Mycogenic nanoparticles and their bio-prospective applications current status and future challenges. J. Nanostructure Chem. 8: 369–391.

Kim, K. and Khang, D. 2020. Past, present, and future of anticancer nanomedicine. Int. J. Nanomedicine 15: 5719–5743.

Kowshik, M., Deshmukh, N., Vogel, W., Urban, J., Kulkarni, S.K. and Paknikar, K.M. 2002 Microbial synthesis of semiconductor CdS nanoparticles, their characterization, and their use in the fabrication of an ideal diode. Biotechnol. Bioeng. 78: 583–588.

Krishna, G., Srileka, V., Charya, M.A.S., Serea, E.S.A. and Shalan, A.E. 2021. Biogenic synthesis and cytotoxic effects of silver nanoparticles mediated by white rot fungi. Heliyon 7: e06470. DOI: https://doi.org/10.1016/j.heliyon.2021.e06470.

Kumaresan, M., Vijai, A.K., Govindaraju, K., Tamilselvan, S. and Ganesh, K.V. 2018. Seaweed *Sargassum wightii* mediated preparation of zirconia (ZrO$_2$) nanoparticles and their antibacterial activity against Gram positive and Gram negative bacteria. Microb. Pathog. 124: 311–315.

Kumari, S., Tehri, N., Gahlaut, A. and Hooda, V. 2021. Actinomycetes mediated synthesis, characterization, and applications of metallic nanoparticles. Inorg. Nano-Met. Chem. 51: 1386–1395.

Lachance, M.A. 2016. Paraphyly and (yeast) classification. Int. J. Syst. Evol. Microbiol. 66: 4924–4929.

Li, G.N., Wang, S.P., Xue, X., Qu, X.J. and Liu, H.P. 2013. Monoclonal antibody-related drugs for cancer therapy. Drug Discov. Ther. 7: 178–184.

Liang, Z., Khawar, M.B., Liang, J. and Sun, H. 2021. Bio-conjugated quantum dots for cancer research: Detection and imaging. Front. Oncol. 11: 749970. Doi: 10.3389/fonc.2021.749970.

Liu, C.G., Han, Y.H., Kankala, R.K., Wang, S.B. and Chen, A.Z. 2020. Subcellular performance of nanoparticles in cancer therapy. Int. J. Nanomedicine 15: 675–704.

Liu, K. and Catchmark, J.M. 2019. Enhanced mechanical properties of bacterial cellulose nanocomposites produced by co-culturing *Gluconacetobacter hansenii* and *Escherichia coli* under static conditions. Carbohydr. Polym. 219: 12–20.

Liu, X., Chen, J.L., Yang, W.Y., Qian, Y.C., Pan, J.Y., Zhu, C.N., Liu, L., Ou, W.B., Zhao, H.X. and Zhang, D.P. 2021. Biosynthesis of silver nanoparticles with antimicrobial and anticancer properties using two novel yeasts. Sci. Rep. 11: 15795. https://doi.org/10.1038/s41598-021-95262-6.

Lofty, W.A., Alkersh, B.M., Sabry, S.A. and Ghozlan, H.A. 2021. Biosynthesis of silver nanoparticles by *Aspergillus terreus*: Characterization, optimization, and biological activities. Front. Bioeng. Biotechnol. 9: 633468. Doi: 10.3389/fbioe.2021.633468.

Machado, R., Pironi, A., Alves, R., Dragalzew, A., Dalberto, I. and Chorilli, M. 2019. Recent advances in the use of metallic nanoparticles with antitumoral action-review. Curr. Med. Chem. 26: 2108–2146.

Majeed, S., Danish, M. and Muhadi, N.F.B.B. 2018. Genotoxicity and apoptotic activity of biologically synthesized magnesium oxide nanoparticles against human lung A-549 cell line. Adv. Nat. Sci. Nanosci. Nanotechnol. 9: 025011. DOI: 10.1088/2043-6254/AAC42C.

Majumder, N., Das, N.G. and Das, S.K. 2020. Polymeric micelles for anticancer drug delivery. Ther. Deliv. 11: 613–635.

Makarov, V., Love, A.J., Sinitsyna, O., Makarova, S., Yaminsky, I., Taliansky, M. and Kalinina, N.O. 2014. "Green" nanotechnologies: Synthesis of metal nanoparticles using plants. Acta Nat. 6: 35–44.

Mani, V.M., Kalaivani, S., Sabarathinam, S., Vasuki, M., Soundari, A.J.P.G., Das, M.P.A., Elfasakhany, A. and Pugazhendhi, A. 2021. Copper oxide nanoparticles synthesized from an endophytic fungus *Aspergillus terreus*: Bioactivity and anti-cancer evaluations. Environ. Res. 201: 111502. https://doi.org/10.1016/j.envres.2021.111502.

McDaniel, J.T., Nuhu, K., Ruiz, J. and Alorbi, G. 2019. Social determinants of cancer incidence and mortality around the world: An ecological study. Glob. Health Promot. 26: 41–49.

Mittal, A.K., Chisti, Y. and Banerjee, U.C. 2013. Synthesis of metallic nanoparticles using plant extracts. Biotechnol. Adv. 31: 346–356.

Mohamed, Y.M., Azzam, A.M., Amin, B.H. and Safwat, N.A. 2015. Mycosynthesis of iron nanoparticles by *Alternaria alternata* and its antibacterial activity. Afr. J. Biotechnol. 14: 1234–1241.

Mohana, S. and Sumathi, S. 2020. Multi-functional biological effects of palladium nanoparticles synthesized using *Agaricus bisporus*. J. Clust. Sci. 31: 391–400.

Mughal, B., Zaidi, S.Z.J., Zhang, X. and Ul Hassan, S. 2021. Biogenic nanoparticles: Synthesis, characterisation and applications. Appl. Sci. 11: 2598. https://doi.org/10.3390/app11062598.

Mukherjee, A., Sarkar, D. and Sasmal, S. 2021. A review of green synthesis of metal nanoparticles using algae. 2021. Front. Microbiol. 12: 693899. Doi: 10.3389/fmicb.2021.693899.

Murciano-Goroff, Y.R., Taylor, B.S., Hyman, D.M. and Schram, A.M. 2020. Toward a more precise future for oncology. Cancer Cell. 37: 431–442.

Nakamura, Y., Mochida, A., Choyke, P.L. and Kobayashi, H. 2016. Nanodrug delivery: Is the enhanced permeability and retention effect sufficient for curing cancer? Bioconjug. Chem. 27: 2225–2238.

Namasivayam, S.K.R. and Avimanyu. 2011. Silver nanoparticle synthesis from *Lecanicillium lecanii* and evalutionary treatment on cotton fabrics by measuring their improved antibacterial activity with antibiotics against *Staphylococcus aureus* (ATCC 29213) and *E. coli* (ATCC 25922) strains. Int. J. Pharm. Pharm. Sci. 3: 190–195.

Noor, S., Shah, Z., Javed, A., Ali, A., Hussain, S.B., Zafar, S., Ali, H. and Muhammad, S.A. 2020. A fungal based synthesis method for copper nanoparticles with the determination of anticancer, antidiabetic and antibacterial activities. J. Microbiol. Methods. 174: 105966. Doi: 10.1016/j.mimet.2020.105966.

Panchangam, R.L. and Upputuri, R.T.P. 2019. *In vitro* biological activities of silver nanoparticles synthesized from *Scedosporium* sp. isolated from soil. Braz. J. Pharm. Sci. 55: e00254. http://dx.doi.org/10.1590/s2175-97902019000200254.

Parashar, V., Parashar, R., Sharma, B. and Pandey, A.C. 2009. *Parthenium* leaf extract mediated synthesis of silver nanoparticles: A novel approach towards weed utilization. Dig. J. Nanomater. Biostruct. 4: 45–50.

Patil, S. and Chandrasekaran, R. 2020. Biogenic nanoparticles: A comprehensive perspective in synthesis, characterization, application and its challenges. J. Genet. Eng. Biotechnol. 18: 67. https://doi.org/10.1186/s43141-020-00081-3.

Paus, C., van der Voort, R. and Cambi, A. 2021. Nanomedicine in cancer therapy: Promises and hurdles of polymeric nanoparticles. Explor. Med. 2: 167–185.

Prasad, S., Gupta, S.C. and Tyagi, A.K. 2017. Reactive oxygen species (ROS) and cancer: Role of antioxidative nutraceuticals. Cancer Lett. 387: 95–105.

Rai, M., Yadav, A., Bridge, P. and Gade, A. 2009. Myconanotechnology: A new and emerging science. pp. 258–267. *In*: Rai, M. and Bridge, P.D. (eds.). Applied Mycology. CAB International, Oxfordshire UK.

Rajakumar, G., AbdulRahuman, A., Roopan, S.M., Khanna, V.G., Elango, G., Kamaraj, C., Zahir, A.A. and Velayutham, K. 2012. Fungus-mediated biosynthesis and characterization of TiO_2 nanoparticles and their activity against pathogenic bacteria. Spectrochim. Acta A 91: 23–29.

Rajeshkumar, S. 2016. Anticancer activity of eco-friendly gold nanoparticles against lung and liver cancer cells. J. Genet. Eng. Biotechnol. 14: 195–202.

Ratan, Z.A., Haidere, M.F., Nurunnabi, Md., Shahriar, S.Md., Ahammad, A.J.S., Shim, Y.Y., Reaney, M.J.T. and Cho, J.Y. 2020. Green chemistry synthesis of silver nanoparticles and their potential anticancer effects. Cancers 12: 855. Doi: 10.3390/cancers12040855.

Riddin, T.L., Gericke, M. and Whiteley, C.G. 2006. Analysis of the inter- and extracellular formation of platinum nanoparticles by *Fusarium oxysporum* f. sp. *lycopersici* using response surface methodology. Nanotechnology 17: 3482–3489.

Rónavári, A., Igaz, N., Gopisettty, M.K., Szerencsés, B., Kovács, D., Papp, C., Vágvölgyi, C., Boros, I.M., Kónya, Z., Kiricsi, M. and Pfeiffer, I. 2018. Biosynthesized silver and gold nanoparticles are potent antimycotics against opportunistic pathogenic yeasts and dermatophytes. Int. J. Nanomedicine 13: 695–703.

Saitoh, N., Fujimori, R., Yoshimura, T., Tanaka, H., Kondoh, A., Nomura, T. and Konishi, Y. 2020. Microbial recovery of palladium by baker's yeast through bioreductive deposition and biosorption. Hydrometallurgy 196: 105413. Doi: 10.1016/j.hydromet.2020.105413.

Salem, S.S. 2022. Bio-fabrication of selenium nanoparticles using Baker's yeast extract and its antimicrobial efficacy on food borne pathogens. Appl. Biochem. Biotechnol. Doi: 10.1007/s12010-022-03809-8. PMID: 34994951.

Salvadori, M.R., Ando, R.A. and Corrêa, B. 2022. Bio-separator and bio-synthesizer of metallic nanoparticles—A new vision in bioremediation. Mater. Lett. 306: 130878. https://doi.org/10.1016/j.matlet.2021.130878.

Salvadori, M.R., Ando, R.A., Nascimento, C.A.O. and Corrêa, B. 2014a. Intracellular biosynthesis and removal of copper nanoparticles by dead biomass of yeast isolated from the wastewater of a mine in the brazilian Amazonia. PLoS One 9(1): e87968. https://doi.org/10.1371/journal.pone.0087968.

Salvadori, M.R., Ando, R.A., do Nascimento C.A.O. and Corrêa, B. 2014b. Bioremediation from wastewater and extracellular synthesis of copper nanoparticles by the fungus *Trichoderma koningiopsis*. J. Environ. Sci. Health A Tox. Hazard. Subst. Environ. Eng. 49: 1286–1295.

Salvadori, M.R., Ando, R.A., Nascimento, C.A.O. and Corrêa, B. 2015. Extra and intracellular synthesis of nickel oxide nanoparticles mediated by dead fungal biomass. PLoS One 10(6): e0129799. https://doi.org/10.1371/journal.pone.0129799.

Salvadori, M.R., Ando, R.A., Muraca, D., Knobel, M., Nascimento, C.A.O. and Corrêa, B. 2016. Magnetic nanoparticles of Ni/NiO nanostructured in film form synthesized by dead organic matrix of yeast. RSC Adv. 6: 60683–60691.

Salvadori, M.R., Ando, R.A., Nascimento, C.A.O. and Corrêa, B. 2017. Dead biomass of Amazon yeast: A new insight into bioremediation and recovery of silver by intracellular synthesis of nanoparticles. J. Environ. Sci. Health A Tox. Hazard. Subst. Environ. Eng. 52: 1112–1120.

Salvadori, M.R., Ando, R.A., Nascimento, C.A.O. and Corrêa, B. 2018. Biosynthesis of metal nanoparticles via fungal dead biomass in industrial bioremediation process. pp. 165–199. *In*: Prasad, R., Kumar, V., Kumar, M. and Wang, S. (eds.). Fungal Nanobionics: Principles and Applications. Springer Nature, Singapore.

Salvadori, M.R., Lepre, L.F., Ando, R.A., do Nascimento, C.A.O. and Corrêa, B. 2013. Biosynthesis and uptake of copper nanoparticles by dead biomass of *Hypocrea lixii* isolated from the metal mine in the Brazilian Amazon region. PLoS One 8(11): e80519. Doi: 10.1371/journal.pone.0080519.

Salvadori, M.R., Monezi, T.A., Mehnert, D.U. and Corrêa, B. 2019. Antimicrobial activity of Ag/Ag$_2$O nanoparticles synthesized by dead biomass of yeast and their biocompatibility with mammalian cell lines. IJRSMB 5: 7–15.

Salvadori, M.R., Nascimento, C.A.O. and Corrêa, B. 2014. Nickel oxide nanoparticles film produced by dead biomass of filamentous fungus. Sci. Rep. 4: 6404. https://doi.org/10.1038/srep06404.

Salvioni, L., Rizzuto, M.A., Bertolini, J.A., Pandolfi, L., Colombo, M. and Prosperi, D. 2019. Thirty years of cancer nanomedicine: Success, frustration, and hope. Cancer 11: 1855. Doi: 10.3390/cancers11121855.

Saravanakumar, K., Shanmugam, S., Varukattu, N.B., Ali, D.M., Kathiresan, K. and Wang, M.H. 2019. Biosynthesis and characterization of copper oxide nanoparticles from indigenous fungi and its effect of photothermolysis on human lung carcinoma. J. Photochem. Photobiol. B 190: 103–109.

Saxena, J., Sharma, M.M., Gupta, S. and Singh, A. 2014. Emerging role of fungi in nanoparticle synthesis and their applications. J. Pharm. Pharm. Sci. 3: 1586–1613.

Schirrmacher, V. 2019. From chemotherapy to biological therapy: A review of novel concepts to reduce the side effects of systemic cancer treatment (Review). Int. J. Oncol. 54: 407–419.

Seetharaman, P.K., Chandrasekaran, R., Periakaruppan, R., Gnanasekar, S., Sivaperumal, S., Abd-Elsalam, K.A., Valis, M. and Kuca, K. 2021. Functional attributes of myco-synthesized silver nanoparticles from endophytic fungi: A new implication in biomedical applications. Biology 10: 473. https://doi. org/ 10.3390/biology10060473.

Sharma, A., Sharma, S., Sharma, K., Chetri, S.P.K., Vashishtha, A., Singh, P., Kumar, R., Rathi, B. and Agrawal, V. 2016. Algae as crucial organisms in advancing nanotechnology: A systematic review. J. Appl. Phycol. 28: 1759–1774.

Shi, C., Zhu, N., Cao, Y. and Wu, P. 2015. Biosynthesis of gold nanoparticles assisted by the intracellular protein extract of *Pycnoporus sanguineus* and its catalysis in degradation of 4-nitroaniline. Nanoscale Res. Lett. 10: 147. Doi: 10.1186/s11671-015-0856-9.

Son, K.H., Hong, J.H. and Lee, J.W. 2016. Carbon nanotubes as cancer therapeutic carriers and mediators. Int. J. Nanomed. 11: 5163–5185.

Sooklert, K., Wongjarupong, A., Cherdchom, S., Wongjarupong, N., Jindatip, D., Phungnoi, Y., Rojanathanes, R. and Sereemaspun, A. 2019. Molecular and morphological evidence of hepatotoxicity after silver nanoparticle exposure: A systematic review, *in silico*, and ultrastructure investigation. Toxicol. Res. 35: 257–270.

Sumanth, B., Lakshmeesha, T.R., Ansari, M.A., Alzohairy, M.A., Udayashankar, A.C., Shobha, B., Niranjana, S.R., Srinivas, C. and Almatroudi, A. 2020. Mycogenic synthesis of extracellular zinc oxide nanoparticles from *Xylaria acuta* and its nanoantibiotic potential. Int. J. Nanomedicine 15: 8519–8536.

Sundaramoorthi, C., Kalaivani, M., Mathews, D.M., Palanisamy, S., Kalaiselvan, V. and Rajasekaran, A. 2009. Biosynthesis of silver nanoparticles from *Aspergillus niger* and evaluation of its wound healing activity in experimental rat model. Int. J. Pharmtech. Res. 4: 1523–1529.

Taher, Z., Legge, C., Winder, N., Lysyganicz, P., Rawlings, A., Bryant, H., Muthana, M. and Staniland, S. 2021. Magnetosomes and magnetosome mimics: Preparation, cancer cell uptake and functionalization for future cancer therapies. Pharmaceutics 13: 367. https://doi.org/10.3390/pharmaceutics13030367.

Tarafdar, J.C., Raliya, R. and Rathore, I. 2012. Microbial synthesis of phosphorous nanoparticle from tri-calcium phosphate using *Aspergillus tubingensis* TFR-5. J. Bionanosci. 6: 84–89.

Tidke, P.R., Gupta, I., Gade, A.K. and Rai, M. 2014. Fungus-mediated synthesis of gold nanoparticles and standardization of parameters for its biosynthesis. IEEE. Trans. Nanobioscience 13: 397–402.

Tsekhmistrenko, S.I., Bityutskyy, V.S., Tsekhmistrenko, O.S., Horalskyi, L.P., Tymoshok, N.O. and Spivak, M.Y. 2020. Bacterial synthesis of nanoparticles: A green approach. Biosyst. Divers. 28: 9–17.

Tyagi, S., Tyagi, P.K., Gola, D., Chauhan, N. and Bharti, R.K. 2019. Extracellular synthesis of silver nanoparticles using entomopathogenic fungus: Characterization and antibacterial potential. SN Appl. Sci. 1: 1545. https://doi.org/10.1007/s42452-019-1593-y.

van der Meel, R., Sulheim, E., Shi, Y., Kiessling, F., Mulder, W.J.M. and Lammers, T. 2019. Smart cancer nanomedicine. Nat. Nanotechnol. 14: 1007–1017.

Venkat, K.S., Sowmya, B., Geetha, R., Karpagambigai, S., Jacquline, R.P., Rajesh, K.S. and Lakshmi, T. 2019. Preparation of yeast mediate semiconductor nanoparticles by *Candida albicans* and its bactericidal potential against *Salmonella typhi* and *Staphylococcus aureus*. Int. J. Res. Pharm. Sci. 10: 861–864.

Wang, D., Xue, B., Wang, L., Zhang, Y., Liu, L. and Zhou, Y. 2021. Fungus-mediated green synthesis of nano-silver using *Aspergillus sydowii* and its antifungal/antiproliferative activities. Sci. Rep. 11: 10356. https://doi.org/10.1038/s41598-021-89854-5.

Waseem, M. and Nisar, M.A. 2016. Fungal-derived nanoparticles as novel antimicrobial and anticancer agents. *In*: Farrukh, M.A. (ed.). Functionalized Nanomaterials. IntechOpen. Doi: 10.5772/66922. Available from: https://www.intechopen.com/chapters/53484.

Wu, Y., Li, J. and Shin, H.J. 2021. Self-assembled viral nanoparticles as targeted anticancer vehicles. Biotechnol. Bioprocess. Eng. 26: 25–38.

Xu, L., Wang, Y.Y., Huang, J., Chen, C.Y., Wang, Z.X. and Sie, H. 2020. Silver nanoparticles: Synthesis, medical applications and biosafety. Theranostics 10: 8996–9031.

Yang, G., Nowsheen, S., Aziz, K. and Georgakilas, A.G. 2013. Toxicity and adverse effects of Tamoxifen and other anti-strogen drugs. Pharmacol. Ther. 139: 392–404.

You, Y., Wang, N., He, L., Shi, C., Zhang, D., Liu, Y., Luo, L. and Chen, T. 2019. Designing dual-functionalized carbon nanotubes with high blood-brain-barrier permeability for precise orthotopic glioma therapy. Dalton Trans. 48: 1569–1573.

Zhang, D., Ma, X.L., Gu, Y., Huang, H. and Zhang, G.W. 2020. Green synthesis of metallic nanoparticles and their potential applications to treat cancer. Front. Chem. 8: 799. Doi: 10.3389/fchem.2020.00799.

Zhao, C., Li, X., Ding, C., Liao, J., Du, L., Yang, J., Yang, Y., Zhang, D., Tang, J., Liu, N. and Sun, Q. 2016. Characterization of uranium bioaccumulation on a fungal isolate *Geotrichum* sp. dwc-1 as investigated by FTIR, TEM and XPS. J. Radioanal. Nucl. Chem. 310: 165–175.

Zhao, M.X. and Zhu, B.J. 2016. The research and applications of quantum dots as nanocarriers for targeted drug delivery and cancer therapy. Nanoscale Res. Lett. 11: 207. Doi: 10.1186/s11671-016-1394-9.

Zheng, D., Hu, C., Gan, T., Dang, X. and Hu, S. 2010. Preparation and application of a novel vanillin sensor based on biosynthesis of Au-Ag alloy nanoparticles. Sens. Actuators B Chem. 148: 247–252.

4

Mycogenic Nanoparticles as Novel Antimicrobial Agents

Sougata Ghosh,[1,2,]* Khalida Bloch[1] and Sirikanjana Thongmee[2]

Introduction

Advances in the field of nanotechnology have helped in extending the horizon of their application from catalysis to sensors, food, agriculture, environment, and pharmaceutics. A plethora of bioactive nanoparticles can be used for biomedical applications that include diagnosis, sensing, bioimaging, targeted drug delivery and triggered release (Rokade et al. 2018, Saravanan et al. 2021). The activity of nanoparticles is largely dependent on the size and shape which is also a key player in determining their physico-chemical and optoelectronic properties (Jamdade et al. 2019). However, the hazardous and toxic chemicals used in the physical and chemical synthesis of nanoparticles often pose a challenge by rendering the resulting nanoparticles toxic and biologically incompatible (Rokade et al. 2017, Sharma et al. 2019). Hence, there is a continuously growing need to develop environmentally benign methods for the synthesis of biocompatible nanoparticles for biomedical applications.

Although several bacteria, algae, medicinal plants and their metabolites are used for the biogenic synthesis of nanoparticles, mycogenic synthesis using either live or dead fungus seems to be very attractive (Shende et al. 2018, Ghosh 2018, Bhagwat et al. 2018). Several fungal metabolites serve as reducing and stabilizing agents attributing to the stability of the biogenic nanoparticles (Ghosh and Webster 2021). Hence, myco-assisted nanoparticles synthesis is considered a rapid, efficient, green, economical, and sustainable process compared to others (Moghaddam et al. 2015). Fungal cultures are easy to handle and maintain. A substantial quantity of fungal biomass can be generated in a short time even using a culture medium with cheaper raw materials as a nutrient source. The enzymes present in fungi have

---

[1] Department of Microbiology, School of Science, RK University, Rajkot, Gujarat, India.
[2] Department of Physics, Faculty of Science, Kasetsart University, Bangkok 10900, Thailand.
* Corresponding author: ghoshsibb@gmail.com

high redox potential which aids in the reduction of metal ions into nanoparticles (Guilger-Casagrande and Lima 2019). Fungi are preferred because they have the ability to produce a high amount of secretory proteins which helps in the formation of nanoparticles. This chapter gives a comprehensive detail of the use of fungi for the synthesis of several nanoparticles such as silver (AgNPs), gold (AuNPs), copper (CuNPs), selenium (SeNPs) for antimicrobial activity.

Mycogenic Nanoparticles

Rai et al. (2009a) coined the term 'Myconanotechnology (myco = fungi, nanotechnology = the creation and exploitation of materials in the size range of 1–100 nm)' for the first time. This branch of science deals with the fungi assisted synthesis of metal nanoparticles that covers both fungal diversity and the new applied interdisciplinary science. The fungal cells or myconanofactories are exploited for the synthesis of metal nanoparticles of silver, gold, copper, platinum, palladium, zirconium, silica, titanium, iron (magnetite) and even zinc oxide (Gade et al. 2010). There are many advantages of using fungi in fabrication of nanoparticles. This process doesn't require toxic chemicals for reduction of metal ions to respective nanoparticles. Instead, fungus derived NADH transfer and NADH-dependent enzymes are involved in the reduction of metal ions (Rai et al. 2009b). Also no hazardous stabilizing agents are needed in this process. Fungus secreted proteins are responsible for capping or stabilizing the mycogenic nanoparticles. Moreover, the nanoparticles synthesized using fungi present good polydispersity, dimensions and stability as compared to other available methods. Being nontoxic, the mycosynthesized nanoparticles have promising applications in the health care, textile and agricultural industries. Several fungi are reported to synthesize bioactive nanoparticles with promising antimicrobial activity that is summarized in Table 1. The following section gives an elaborate account of the myco-assisted synthesis of antimicrobial nanoparticles.

Silver Nanoparticles

Biosynthesis of silver nanoparticles (AgNPs) was carried out using endophytic fungi isolated from ethnomedicinal plant *Gloriosa superba* L. The fungus was isolated from the plant material collected from the Department of Forests, Government of Meghalaya. The plant material was surface sterilized and was placed on potato dextrose agar (PDA) plates supplemented with streptomycin. Endophytic fungal isolates GS1 and GS2 were identified as *Alternaria* sp. and *Penicillium funiculosum*, respectively. The isolates were inoculated in 100 mL potato dextrose broth and were agitated at 120 rpm for 96 hours at 25°C. Biomass was harvested by filtration and was rinsed with sterile distilled water. Wet biomass (10 g) was transferred to 100 mL sterile double distilled water and was agitated at 120 rpm for 48 hours at 25°C. The cell filtrate was then collected by filtering biomass with Whatman filter paper No. 1. An aqueous solution of 1 mM silver nitrate ($AgNO_3$) was reacted with the filtrate. The reaction mixture was incubated in darkness at room temperature. Synthesized nanoparticles were characterized using UV-Vis spectroscopy and the surface plasmon absorption band for GS1 and GS2 was found to be 415 and 403 nm respectively. The morphology of the nanoparticles was determined using transmission electron microscopy (TEM). The AgNPs synthesized from GS1 were 5–20 nm in size while GS2 was 5–10 nm in size with a spherical shape. Antimicrobial activity of AgNPs was checked against pathogenic microorganisms *Streptococcus pyogenes* MTCC1925, *Escherichia coli* MTCC730, *Enterococcus faecalis* MTCC2729 and *Candida albicans* MTCC183. A zone of inhibition of 13 mm was observed in AgNPs synthesized from GS2 while it was 10 mm in the case of AgNPs synthesized from GS1. The nanoparticles synthesized from GS2 showed good inhibitory effect against *E. coli*, *E. faecalis* and *C. albicans* with a zone of inhibition of 12 mm whereas a moderate zone of 10 mm was observed against *S. pyogenes*. The AgNPs synthesized from GS1 showed a maximum zone of 10 mm against *C. albicans*, *E. faecalis* and *E. coli*. The synergistic effect of AgNPs synthesized from *P. funiculosum* GS2 in combination with chloramphenicol and gentamycin was studied against *S. pyogenes* MTCC1925, *E. coli* MTCC730 and *E. faecalis* MTCC2729. AgNPs along with gentamycin showed a zone of inhibition of 23 mm which was 0.19 fold increase against *E. coli*. AgNPs along with chloramphenicol showed a zone of inhibition of 37 mm while for *E. faecalis* it was 27 mm when combined with gentamycin (Devi et al. 2014).

AgNPs were synthesized using a *Amylomyces rouxii* strain KSU-09. The fungus was isolated from the roots of the *Phoenix dactylifera* (date palm) and grown on PDA followed by which the mycelia were harvested by filtration. The extracellular synthesis of AgNPs was achieved using 10 g of wet biomass. The biomass was added to Milli-Q water and was placed for 72 hours. After centrifugation at 3000 rpm for 10 min, the mycelia-free extract was recovered that was mixed with 1 mM of AgNO$_3$ and was allowed to incubate at $28 \pm 0.5°C$, 200 rpm for 72 hours. The development of yellowish-brown color indicated the formation of AgNPs. The strong surface plasmon resonance (SPR) band was obtained at 420 nm. The X-ray diffraction pattern showed four intense peaks at 38°, 44.18°, 64.29° and 77.19° indexed to (1 1 1), (2 0 0), (2 2 0) and (3 1 1) planes which suggested that the AgNPs were crystalline in nature. The shapes of AgNPs were spherical. Atomic force microscopy exhibited that the average size of nanoparticles was 14–20 nm. The antimicrobial activity of synthesized nanoparticles was tested against *Shigella dysenteriae* type I, *Staphylococcus aureus*, *Citrobacter* sp., *Escherichia coli*, *Pseudomonas aeruginosa*, *Bacillus subtilis*, *Candida albicans* and *Fusarium oxysporum*. AgNPs inhibited the test pathogens whereas no zone of inhibition was observed for the mycelia-free extract alone (Musarrat et al. 2010).

Likewise, mycogenic synthesis of AgNPs was carried out using *Aspergillus clavatus* that was isolated from the stem of *Azadirachta indica* A. Juss. The endophyte was grown aerobically in liquid medium and was incubated for 72 hours at 150 rpm, 25°C. The biomass was filtered and mixed with Milli-Q deionized water. The fungal biomass was agitated and was again filtered with Whatman filter paper No. 1. Wet fungal biomass (20 g) was added into 1 mM of AgNO$_3$ and reacted under shaking condition (200 rpm). The cell-free extract (10 mL) was also used for the synthesis of AgNO$_3$. The rapid color change to brown indicated the formation of AgNPs. The excitation of SPR was obtained at 415 nm which was determined using UV-Vis spectroscopy. The size of the nanoparticles was in the range of 10–25 nm with some hexagonal and non spherical polyhedral shape. AFM revealed that the NPs were spherical in shape with a height of 2–6 nm and width 10–60 nm. Antimicrobial activity of AgNPs was tested against *C. albicans*, *P. fluorescens* and *E. coli*. The maximum zone of inhibition (16 mm) was observed against the *C. albicans*. The minimum inhibitory concentration (MIC) was 5.83 µg/mL and the minimum fungicidal concentration (MFC) was 9.7 µg/mL against *C. albicans* (Verma et al. 2010).

In another study, the synthesis of AgNPs was carried out from marine *Aspergillus niger*. The fungal biomass (5 g) was mixed with 1 mM $AgNO_3$ and the pH of the reaction was optimized between 3 to 10. The reaction mixture was incubated for 72 hours at 27°C. The gradual color change from colorless to brown after 24 hours indicated the formation of AgNPs. At pH 10, the color change was obtained within 1 min, while at pH 8 and 9 it took 2 min. TEM analysis revealed that the size of spherical AgNPs was in range of 25–26 nm. The antimicrobial activity was tested at pH 5, 8, 9 and 10 against *Bacillus megaterium*, *S. aureus*, *Proteus vulgaris* and *Shigella sonnei*. A zone of inhibition was observed against all test organisms. Synergistic antibacterial activity was also evaluated in combination with gentamycin (10 µg/disc). When nanoparticles were tested along with antibiotics, an increase in the area against all test organisms was noted. Bacterial growth kinetics was studied. The maximum sensitivity of AgNPs was observed at pH 8.5, 8, 7.4 and 6 against *S. aureus*, *B. megaterium*, *P. vulgaris* and *S. sonnei*, respectively (Vala and Shah 2012).

Extracellular synthesis of AgNPs was achieved using *Aspergillus sydowii* where the fungus was initially isolated from soil and cultured on SDA medium. The cell filtrate was mixed with $AgNO_3$ and a brown color was developed that indicated the formation of AgNPs. Synthesis was carried out at room temperature in darkness. After 1 hour, UV-Vis absorption peak at 420 nm was obtained. The size of the AgNPs was in the range of 1–24 nm with polydispersed spherical shape. Several reaction parameters were optimized that include temperature (20–60°C), pH (5.0–8.0) and substrate concentration (0.5–2.5 mM). The best results were obtained at 50°C, pH 8.0 and 1.5 mM substrate concentration. Antifungal activity of synthesized AgNPs was checked against *Aspergillus* spp. (*A. fumigatus*, *A. flavus*, and *A. terreus*), *Fusarium* spp. (*F. solani*, *F. moniliforme*, *F. oxysporum*), *Candida* spp. (*C. albicans*, *C. glabrata*, *C. parapsilosis* and *C. tropicalis*), *Cryptococcus neoformans* and *Sporothrix schenckii*. In comparison to itraconazole and fluconazole, AgNPs exhibited a broad-spectrum antifungal activity at very low concentrations (0.25 µg/mL). It was also observed that in the presence of AgNPs, the growth of *Candida* spp. and *Aspergillus* spp. was significantly reduced (Wang et al. 2021).

Extracellular synthesis of AgNPs was also carried out using entomopathogenic fungus *Beauveria bassiana*. The fungus was cultivated on PDA medium for 72 hours at 25°C. The fungi were grown on a liquid composite medium for 72 hours followed by which the wet biomass was harvested and further incubated for 48 hours in a flask containing sterile distilled water. It was then filtered and the filtrate was allowed to mix with 1 mM of $AgNO_3$. The reaction mixture was incubated at 25°C for 30 min at

100 rpm in darkness after which a brown color was developed marking the synthesis of AgNPs. The absorption peak in UV-Vis spectroscopy was obtained at 450 nm. TEM micrograph revealed that the synthesized AgNPs were triangular, circular and hexagonal in shape. The size of AgNPs was in the range of 10–50 nm. Optimum pH and temperature for synthesis were 6.0 and 25°C, respectively when the filtrate of *B. bassiana* was used. The zeta potential of the AgNPs was –22 mV with a sharp and single peak. Antimicrobial activity and growth kinetics of the AgNPs against different microorganisms were studied. Maximum reduction in growth was observed in *E. coli* (67.2%), while it was 63.3% in the case of *P. aeruginosa*. This indicated that AgNPs have a deleterious effect on the reproductive capacity of microbes. MIC value of AgNPs synthesized from fungus was 2.5, 3 and 4.5 ppm against *E. coli*, *P. aeruginosa* and *S. aureus*, respectively. The synergistic effect of AgNPs along with ciprofloxacin was also tested. The MIC of AgNPs along with antibiotics was 0.4, 0.4 and 0.5 ppm against *E. coli*, *P. aeruginosa*, and *S. aureus* respectively (Tyagi et al. 2019).

In another study, *Bipolaris tetramera* KF934408 was isolated from rhizospheric soil on Pikovskaya's medium and the colonies were further subcultured for 72 hours at $28 \pm 2°C$. The isolated fungus was then grown on maltose glucose yeast peptone broth and incubated at 27°C. The fungal biomass was harvested after 120 hours. Wet biomass (15 g) was added into deionized water and was kept for 48 hours at 27°C, 150 rpm followed by reaction with $AgNO_3$(1 mM). The reaction mixture was kept in dark conditions. The synthesized nanoparticles showed an absorption peak at 350 nm indicating the formation of AgNPs. The dynamic light scattering (DLS) spectra showed an intense peak at 109.4 nm while the TEM images revealed that the AgNPs were spherical, triangular and hexagonal in shape. The size of NPs was in the range of 54.78–73.49 nm. Antibacterial activity against *B. subtilis*, *Bacillus cereus*, *S. aureus*, *E. coli* and *E. aerogenes* using agar well diffusion method showed clear zones of inhibition. The AgNPs also showed an inhibitory effect against *A. niger* and *Trichoderma*. A zone of inhibition of 1 cm diameter was recorded against *A. niger* and 1.2 cm against *Trichoderma* (Fatima et al. 2015).

Extracellular synthesis of AgNPs was carried out using the supernatant of *Candida glabrata*. The fungus was isolated from the oropharyngeal mucosa of human immunodeficiency virus (HIV) patients with oral candidiasis using a sterile cotton swab. The swab was then dipped into 5 mL of Sabouraud dextrose broth (SDB), 1 mL from which was inoculated on HiCHROM medium plates and it was allowed to incubate at 37°C for 48 hours. The pure colonies were suspended into SDB and incubated at 28°C for 48–72 hours. The culture was then centrifuged at 12,000 rpm for 10 min and the supernatant was collected that was added into 1 mM $AgNO_3$ and was kept at room temperature. The change in color of the reaction mixture to brown from colorless indicated the formation of AgNPs. The synthesized AgNPs showed an absorption peak at 460.64 nm. FTIR analysis suggested that the proteins and enzymes attributed to the extracellular synthesis of the AgNPs. TEM micrographs showed that the NPs were spherical, oval and well dispersed which were mostly uniform in shape with 2–15 nm in size. Antimicrobial activity of AgNPs was tested against various bacterial spp., e.g., *S. aureus*, *E. coli*, *S. typhimurium*, *S. flexneri*,

K. pneumoniae, P. aeruginosa, Candida spp., e.g., *C. albicans, C. dubliniensis, C. parapsilosis, C. tropicalis, C. krusei,* and *C. glabrata* using well diffusion method. AgNPs had an excellent inhibitory activity against all test organisms. The MIC and MFC values were 31.25–125 µg/mL and 62.5–250 µg/mL respectively for bacterial strains, while it was 62.5–250 µg/mL and 125–500 µg/mL respectively for fungal strains (Jalal et al. 2018).

The AgNPs was also synthesized using endophytic fungi *Cladosporium cladosporioides.* The fungus was isolated from brown algae *Sargassum wightii* and cultured on PDA medium at 28°C for 30 days. The fungal biomass was separated by filtration and the filtrate (10 mL) was reacted with 1 mM of $AgNO_3$ (90 mL). The synthesis of NPs was confirmed using UV-Vis spectroscopy that showed a strong absorption peak at 440 nm. The shape and size of the nanoparticles were determined using FEGSEM that revealed that the particles were spherical with size ranging from 30–60 nm. FTIR measurements showed that the proteins, enzymes and polyphenols present in the fungal filtrate played an active role in reduction of metal ions to AgNPs and their capping. The antimicrobial activity of synthesized NPs was tested against pathogenic bacteria and a fungus. AgNPs exhibited zone of inhibition against *S. aureus, S. epidermis, B. subtilis, E. coli* and *C. albicans.* The inhibitory effect of AgNPs against Gram-positive bacteria was comparatively less than Gram-negative. However, the mycogenic AgNPs inhibited the bacteria as well as the fungus significantly, indicating its broad spectrum antimicrobial activity (Hulikere and Joshi 2019).

Synthesis of AgNPs was also carried out using endophytic fungi *Cryptosporiopsis ericae* isolated from ethnomedicinal plant *Potentilla fulgens* L. The fungal biomass was developed by culturing the fungi on the PDA medium. The culture was allowed to agitate at 120 rpm for 96 hours at 25 ± 2°C followed by which the biomass was harvested by filtration. The wet biomass was added into the sterile double distilled water and kept for incubation followed by filtration. The recovered filtrate was then reacted with 1 mM $AgNO_3$ under room temperature in dark conditions. The appearance of the brown color indicated the formation of AgNPs that was further confirmed by UV-Vis spectroscopy. After 72 hours the absorption was maximum at 440 nm indicating the formation of small and spherical AgNPs. The SEM and TEM analysis was carried out to determine the morphology and size of NPs. The average size obtained was 2–15 nm. The antimicrobial synergy of AgNPs was tested in combination with antibacterial and antifungal agents such as chloramphenicol and fluconazole against pathogenic microorganisms, e.g., *S. aureus* MTCC96, *S. enterica* MTCC735, *E. coli* MTCC730, *E. faecalis* MTCC2729 and *C. albicans* MTCC183. AgNPs along with antibacterial and antifungal agents showed zone of inhibition with a diameter of 37, 35, 36, 38, and 35 mm against *S. enterica* MTCC735, *E. coli* MTCC730, *S. aureus* MTCC96, *E. faecalis* MTCC2729 and *C. albicans* respectively. MIC and MBC of AgNPs were 15 µM and 20 µM against *E. coli* MTCC730 and *S. enterica* MTCC735, respectively while it was 20 µM and 25 µM against *S. aureus* MTCC96 and *E. faecalis* MTCC2729, respectively (Devi and Joshi, 2014).

Curvularia lunata, an endophytic fungus isolated from the leaves of *Catharanthus roseus* was used for the synthesis of AgNPs. The wet fungal biomass was suspended in sterile double distilled water and was filtered. The cell filtrate was reacted with 1 mM of AgNO$_3$ and was incubated at room temperature under dark conditions for 24 hours. After 2 hours, development of the dark brown color indicated the formation of AgNPs. The surface plasmon resonance was obtained at 422 nm in UV-Vis spectroscopy. FTIR analysis showed that the presence of protein played an important role in the synthesis and stabilization of the AgNPs. The mycogenic AgNPs were spherical with size ranging from 10–50 nm in diameter. Dynamic light scattering (DLS) analysis showed that the average size of AgNPs was 64.3 nm. Zeta potential exhibited single peak at –26.6 mV. The antimicrobial effect of AgNPs was evaluated against human pathogens such as *E. coli*, *P. aeruginosa*, *S. paratyphi*, *B. subtilis*, *S. aureus* and *B. cereus*. Antimicrobial activity could be significantly enhanced on combining the AgNPs with different antibiotics (Ramalingmam et al. 2015).

In another study, AgNPs were synthesized using *Fusarium oxysporum*. The fungus was initially cultivated on PDA medium followed by MYPG medium. The mycelia were harvested and mixed with distilled water. Wet biomass (10 g) was reacted with 100 mL of 1 mM of AgNO$_3$. The reaction mixture was incubated at 28°C in darkness for 190 hours at 150 rpm. The visual color change to the brown color indicated the formation of AgNPs. The absorption peak was obtained between 408 and 411 nm indicating the SPR of AgNPs. SEM analysis showed that AgNPs were of spherical shape while TEM analysis revealed the size of AgNPs ranged from 21.3–37.3 nm with uniform distribution. The antimicrobial property of AgNPs was tested against *E. coli*, *Enterobacter cloacae*, *Klebsiella pneumoniae*, *Proteus mirabilis*, *P. aeruginosa*, *S. aureus*, and *S. epidermidis*. AgNPs (25% v/v) showed significant inhibition of *E. coli*, *P. aeruginosa* and *K. pneumoniae* whereas there was no inhibition in the case of *P. mirabilis*. AgNPs at concentration of 50% exhibited the highest antimicrobial effect against *E. coli*, *P. mirabilis*, *C. albicans* and *Candida krusei* (Ahmed et al. 2018).

Biomass filtrate of *Fusarium solani* was used for the synthesis of AgNPs. Cultivation of *F. solani* was carried out in a fermentation medium followed by harvesting of the biomass after 72 hours by filtration. The biomass was then mixed with distilled water and kept for 72 hours at 30–32°C and the aqueous phase was separated by filtration. The biomass filtrate was mixed with AgNO$_3$ (0.085 g/100 mL) and incubated for 48 hours at 25°C. The SPR was obtained at 420 nm. TEM micrograph showed that the size of NPs was in the range of 3–8 nm. The AgNPs were loaded on cotton fabrics and its durability was determined. The efficiencies of the antibacterial finish on the cotton fabric (bacterial reduction) were 97% and 91% against *S. aureus* and *E. coli*, respectively (El-Rafie et al. 2010).

Endophytic fungus *Lasiodiplodia theobromae* was isolated from *Cinnamomum zeylanicum* after surface sterilization of the petioles and placing them on the PDA medium. Further, the isolated fungus was inoculated on PDB and incubated for 12 days at 28°C. The fungal mycelium was further treated incubated in distilled water and the biomass was filtered followed by collection of the cell filtrate. The cell filtrate was further reacted with AgNO$_3$ (1 mM) in the proportion of 1:1 in darkness for 24 hours for the synthesis of AgNPs. UV-Vis spectroscopy showed the absorption peak at 420 nm due to SPR of the AgNPs. DLS analysis revealed that the NPs were

136.3 nm in diameter with an intercept of 0.98 and a low polydispersity index of 0.433. The zeta potential of AgNPs was −17.7. FEGSEM confirmed that the particles were spherical and oval in shape with an average particle size of 76 nm. The antibacterial effect was tested against *P. aeruginosa* ATCC27853 and clinical sample-driven *P. aeruginosa*. AgNPs showed antibacterial effect against both the organisms. A zone of inhibition was obtained at concentrations of 25 μg and 50 μg of AgNPs. The MIC value was 15 μg/mL in case of *P. aeruginosa* ATCC27853 while the growth rate of the clinical isolate was reduced to 96% (Ranjani et al. 2020).

AgNPs were synthesized using *Macrophomina phaseolina*. The fungus was grown on PDA medium at 28°C followed by which the mycelium was transferred into PDB and incubated for another 5 days. Further, the mycelia were harvested and washed with sterile deionized water and agitated for 72 hours at 120 rpm. The extract was then collected and filtered. The cell-free filtrate was mixed with $AgNO_3$ (1 mM) and was kept on a shaker (120 rpm) at 28°C in darkness for 24, 48 and 72 hours. The UV-Vis spectroscopy showed single absorption with a maximum peak at 450 nm. TEM micrograph showed that the AgNPs were mostly spherical and polydispersed in nature as evident from Figure 1. The particles were in a range from 3.33–40.15 nm with an average of 17.26 ± 1.87 nm. The morphology was confirmed by atomic force microscopy (AFM) which revealed that the particles were symmetrical and spherical. SDS-PAGE analysis showed an 85 kDa protein band which was speculated to be responsible for the capping and stabilization of AgNPs. The antimicrobial activity of AgNPs was tested against human and plant pathogenic bacteria and multidrug-resistant bacteria (MDR). AgNPs exhibited a zone of inhibition against *E. coli* and MDR *E. coli* with an increase in concentration. Wild type and MDR *A. tumefaciens* were also inhibited with the increase in the concentration of AgNPs. AgNPs also showed an inhibitory effect on the growth of human and plant bacteria which was evaluated using growth kinetics (Chowdhury et al. 2014).

In another study, the endophytic fungus was isolated from the leaves of *Nyctanthes arbour-tristis*. *Phomopsis helianthi* was grown on PDB and the fungal biomass was filtered followed by which it was suspended in the deionized distilled water and agitated at 200 rpm for 24 hours. Further, it was filtered, and centrifuged to get cell-free extract, 50 mL of which was added to 1 mM of $AgNO_3$. The reaction mixture was further incubated for 24 hours at 200 rpm. The synthesis of AgNPs by the fungal extract was confirmed by UV-Vis spectroscopy that showed a SPR excitation peak at 422 nm. The synthesized AgNPs had the size of 40 nm with spherical, pentagonal and hexagonal shape. The mycogenic AgNPs showed a maximum zone of inhibition (14 mm) against *E. coli* and *P. aeruginosa*. Hence, the AgNPs acted as a potent antimicrobial agent (Gond et al. 2020).

Fungal endophyte was also isolated from *Phlogacanthus thyrisflorus* that were inoculated in the malt extract broth and incubated for 7 days at 25°C. The mycelium free extract was filtered and used for the synthesis of AgNPs. Aqueous solution of $AgNO_3$ (1 mM) was mixed with 5 mL of fungal extract and incubated on a shaker for 3 days at 150 rpm. The formation of brown color indicated the formation of AgNPs. The UV-Vis spectroscopy showed an absorption peak at 456 nm. The particles were spherical in shape which was determined using TEM and AFM. FTIR analysis

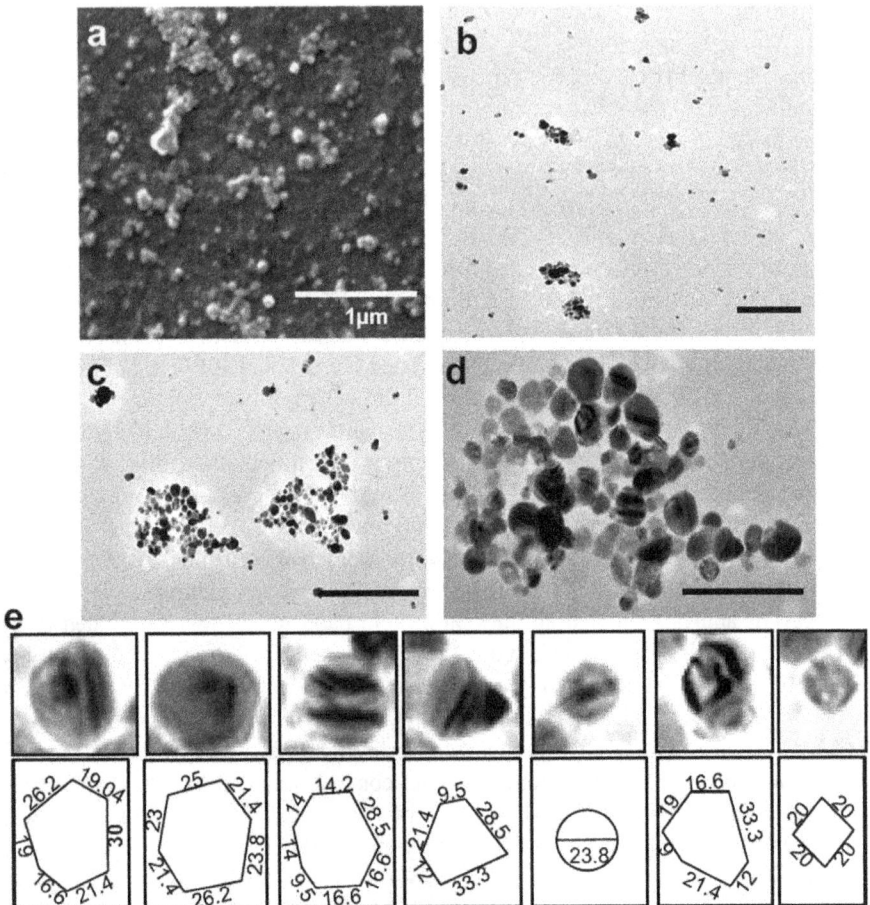

Figure 1. Electron micrographs of AgNPs (a) Scanning electron microscopy micrograph of AgNPs produced with *M. phaseolina* at 50,000 magnification (bar = 1 μm). (b, c, d) Transmission electron micrograph of AgNPs at different magnifications (bar = 100 nm). (e) Measurement of nanoparticles of different shapes. Reprinted from Chowdhury, S., Basu, A. and Kundu, S. 2014. Green synthesis of protein capped silver nanoparticles from phytopathogenic fungus *Macrophomina phaseolina* (Tassi) Goid with antimicrobial properties against multidrug-resistant bacteria. Nanoscale Res. Lett. 9: 365.

suggested that several reducing compounds aided the biosynthesis of AgNPs. Antibacterial activity was tested against *S. aureus*, *B. subtilis*, *P. aeruginosa* and *E. coli*. AgNPs along with streptomycin showed zone of inhibition with a diameter of 18.25 ± 0.75, 18.50 ± 1.00, 17.25 ± 2.25 and 17.75 ± 2.75 against *P. aeruginosa*, *E. coli*, *S. aureus* and *B. subtilis* (Bhattacharjee et al. 2017).

AgNPs were synthesized from the fungal extract that was prepared by inoculating 72 hour old culture *Penicillium radiatolobatum* on PDB and incubation for 10 days at $26 \pm 2°C$. The mycelia were filtered and washed with distilled water. Further 20 g of fungal biomass was incubated for 72 hours at 180 rpm in distilled water. The cell filtrate (40 mL) was mixed with 60 mL of 3 mM $AgNO_3$ and was kept under darkness for 24 hours at 26 ± 2°C. The visual observation indicated the formation of the brown color from the light yellow color indicating the formation of AgNPs. This was confirmed by UV-Vis spectroscopy. The absorption peak was obtained in the range of 410–420 nm. The zeta potential of AgNPs was –22.2 ± 0.87 mV. XRD analysis confirmed the crystalline nature of AgNPs. Several proteins and alkynes were involved in the synthesis of AgNPs which was revealed by FTIR analysis. The size of synthesized NPs was in the range of 5.09 to 24.52 nm. The particles were spherical, triangular and hexagonal in shape. The antimicrobial assay was carried out against various drug-resistant bacteria and fungus with various concentrations of the AgNPs varying between 0.1 to 1 mg/mL. Zones of inhibition were in a range from 8.2 ± 0.25 to 1.47 ± 0.36 mm, 18.9 ± 0.25 to 25.1 ± 0.32 mm, 12 ± 0.25 to 19 ± 0.30 mm, 20 ± 0.25 to 25 ± 0.15 mm and 10.3 ± 0.81 to 17.8 ± 0.41 mm against *B. cereus, S. auerus, E. coli, S. enterica* and *Listeria monocytoegenes*, respectively (Naveen et al. 2021).

In yet another study, mycogenic synthesis of AgNPs was carried out using extracellular filtrate of *Phomopsis liquidambaris* strain SA1 that was isolated from the fresh and healthy leaves of *Salacia chinensis*. The leaves were surface sterilized and inoculated on PDA medium supplemented with antibiotics. After two weeks, the fungus was collected, transferred onto PDA medium and was incubated for 10 days. The mycelia were aseptically ground to prepare the fungal extract in distilled water that was further filtered. The filtrate was centrifuged at 5000 rpm for 15 min and the supernatant was used for the synthesis of AgNPs. SA1 (10 mL) was added into 90 mL of 1 mM of $AgNO_3$ and incubated at 37°C at 200 rpm for 24 hours in darkness. After incubation, brown color formation was observed with a maximum absorption due to SPR at 430 nm. A high yield of NPs was obtained at alkaline pH. Polydispersed AgNPs were synthesized at 37°C with pH 8. FTIR revealed that several proteins, flavonoids and triterpenoids acted as reducing and capping agents during synthesis of AgNPs. The zeta potential was –25.7 mV. TEM micrographs showed that the AgNPs were spherical in shape with a size of 18.7 nm. The antibacterial activity of AgNPs was tested at different concentrations of 10 μg/mL, 20 μg/mL and 40 μg/mL. Maximum zone of inhibition with a diameter of 17 mm, 18 mm, 17 mm, 16 mm, 15 mm, 15 mm, 15 mm and 10 mm was obtained against *E. coli* MTCC1687, *P. mirabilis* MTCC425, *Micrococcus luteus* MTCC1809, *Shigella flexneri* MTCC9543, *K. pneumoniae* MTCC4031, *Vibrio cholerae* MTCC3906, *Pseudomonas putida* MTCC1194, *S. typhi* MTCC531, respectively (Seetharaman et al. 2018).

Synthesis of AgNPs was carried out from white-rot fungi *Schizophyllum commune* and *Pycnoporus sanguineus*. Both Malaysian white-rot fungi were cultivated for 4 days after which the mycelia were harvested, centrifuged and pellets were collected. Further, the pellets were washed with deionized water. Thereafter the pellets (1% w/v) and supernatant (1% v/v) were mixed with 0.001 M $AgNO_3$ and incubated for 5 days, at 200 rpm and 30°C. The color change from pale yellow to brown indicated the formation of AgNPs. The SPR band obtained for the AgNPs synthesized from both the fungus was at 420 nm. The TEM analysis revealed that the average particles size of AgNPs synthesized from *P. sanguineus* and *S. commune* were 15.8 nm and 20.9 nm, respectively. Intracellular synthesis of AgNPs was observed.

The formation of AgNPs was at the circumference of the fungi with the average size of 61.0 nm and 30.1 nm for *P. sanguineus* and *S. commune*, respectively. AgNPs were further tested for antimicrobial and antifungal activity. AgNPs synthesized from *S. commune* showed a maximum zone of inhibition of 2.0 cm against *S. aureus*. The MIC value was in between 0.03 to 47 µg/mL while MBC/MFC values were in a range from 0.09 to 47 µg/mL in all cases. Hence AgNPs synthesized from white rot fungi have excellent bactericidal and fungicidal activities (Chan and Don 2012).

Gold Nanoparticles

Mycogenic gold nanoparticles (AuNPs) were derived using *Bipolaris tetramera*. The rhizospheric soil was used for the isolation of fungi. The fungi were grown on MGYP medium. Wet biomass (15 g) of fungus was mixed with 1 mM HAuCl$_4$ and was kept under dark conditions. Average particles size of AuNPs was 73.82 nm which was determined using DLS analysis. TEM micrograph in Figure 2b revealed that the size of AuNPs was nearly 58.4 nm for the spherical shape, 110.13 nm for the triangular ones and 261.73 nm when the shape was hexagonal. AuNPs showed moderate antibacterial activity against *B. cereus*, *S. aureus* and *E. aeruginosa* while it didn't show any antifungal activity (Fatima et al. 2015).

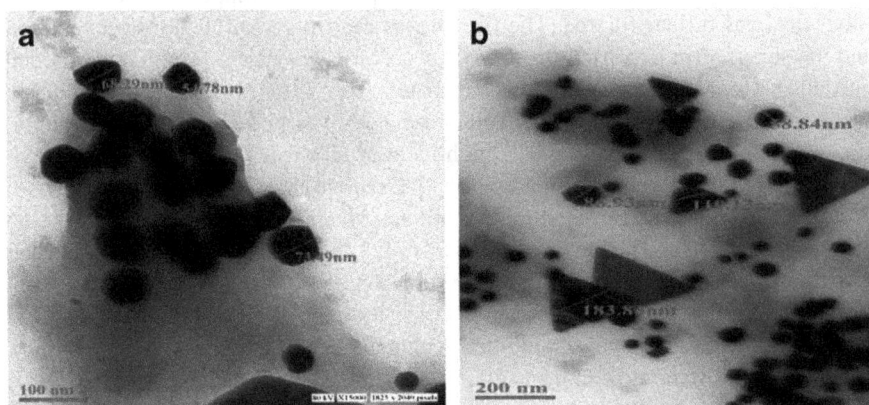

Figure 2. Transmission electron microscopic analysis of nanoparticles. (a) AgNPs; (b) AuNPs. Reprinted from Fatima, F., Bajpai, P., Pathak, N., Singh, S., Priya, S. and Verma S.R. 2015. Antimicrobial and immunomodulatory efficacy of extracellularly synthesized silver and gold nanoparticles by a novel phosphate solubilizing fungus *Bipolaris tetramera*. BMC Microbiol. 15: 52.

Synthesis of AuNPs was carried out using *Pleurotus osteratus*. The fungus was initially isolated on PDA medium and incubated for 7 days at 21°C. The biomass was allowed to grow and then it was filtered. The cell filtrate was then mixed with HAuCL$_4$ in a ratio of 10:1 for 24 hours at 37°C under shaking condition (120 rpm). The change in color from yellow to purple-pink indicated the formation of AuNPs. UV-Vis spectroscopy showed the absorption peak around 550 nm. The size and shape of AuNPs were checked using TEM analysis. The synthesized AuNPs were spherical in shape with size ranging from 10 to 30 nm. The zeta potential of AuNPs was -24 mV. Various parameters were optimized for the highest production of AuNPs. The maximum production of AuNPs was obtained at 5 mM concentration of the gold salt, at 200 rpm with a ratio of 5:1 under incubation for 48 hours. Antimicrobial synergy was also checked. AuNPs enhanced the antimicrobial activity of all antibiotics by 1–2 mm in the zone of inhibition against all test organisms (El Domany et al. 2018).

Selenium Nanoparticles

Penicillium expansum ATCC36200 was used for the synthesis of selenium nanoparticles (SeNPs). The fungus was cultivated on PDA medium and incubated at 30°C for 12 days at pH 6.0 under shaking condition (150 rpm). The biomass was filtered using Whatman filter paper No. 1. Selenium dioxide (2 mM) was mixed with the fungal biomass filtrate, followed by incubation at 30°C for 24 hours at 150 rpm. Development of red color indicated the synthesis of SeNPs that were further dried at 80°C for 48 hours. Fungal proteins were involved in the reduction of selenium ions into SeNPs. UV-Vis spectroscopy showed a sharp peak at 295 nm. TEM analysis revealed that the SeNPs were spherical in shape with the size ranging from 4 to 12.7 nm. SEM images showed that the particles were randomly distributed with agglomeration. EDX indicated the signature absorption peaks of elemental selenium at 1.379 keV, 1.419 keV, 11.222 keV and 12.496 keV. FTIR spectroscopy confirmed that the fungal metabolites played the main role in the formation, size reduction and stabilization of SeNPs. The average size of SeNPs was in a range from 3 to 82 nm as determined using X-ray diffraction (XRD) analysis. *In-vitro* antimicrobial activity of SeNPs was tested against various Gram-positive, Gram-negative bacteria, unicellular and multicellular fungi. SeNPs (2000 μg/mL) showed zone of inhibition with diameter of $36.3 \pm 0.882, 30.3 \pm 1.093, 28.3 \pm 0.333, 26 \pm 0.557$, $25.6 \pm 0.667, 23.7 \pm 0.145$ and 22.9 ± 0.493 mm against *S. aureus* ATCC23235, *B. subtilis* ATCC6051, *E. coli* ATCC8739, *P. aeruginosa* ATCC9027, *C. albicans* ATCC90028, *A. fumigatus* RCMB02568, and *A. niger* RCMB02724, respectively. MIC of SeNPs was 250 μg/mL against *A. fumigatus* RCMB02568 and *A. niger* RCMB02724 with zone of inhibition with a diameter of 10.4 ± 0.348 mm and 11 ± 0.577 nm, respectively (Hashem et al. 2021).

Metal Oxide Nanoparticles

Mycogenic synthesis of zinc oxide (ZnO) and copper oxide (CuO) nanoparticles was carried out using *Penicillium chrysogenum* MF318506 strain. The fungus was cultivated in CzapekDoX (CD) fermentative broth medium containing $NaNO_3$ (2 g/L), $MgSO_4.7H_2O$ (0.5 g/L), KCl (0.5 g/L), KH_2PO_4 (0.5 g/L), FeSO4 (0.001 g/L) and sucrose (20 g/L). The inoculated broth was incubated at 30°C for 7 days at pH 6.0. The biomass was separated by filtration. The filtrate was used for the synthesis of ZnONPs and CuONPs from zinc acetate (2 mM) and copper acetate (2 mM), respectively after mixing with the fungal filtrate followed by incubation at 30°C for 48 hours at 150 rpm. The synthesized NPs were dried at 80°C for 48 hours that resulted in formation of white and green crystalline powder of ZnONPs and CuONPs, respectively. The absorption maxima were recorded at 380 nm and 337 nm for ZnONPs and CuONPs, respectively using UV-Vis spectroscopy. The size and shape of the nanoparticles were determined using TEM. The morphology of the synthesized ZnONPs and CuONPs were hexagonal and spherical, respectively which was capped by active metabolites present in the fungal filtrate. The average size of the nanoparticles was in the range of 9.0–35.0 nm for ZnONPs and 10.5 to 59.7 nm for CuONPs. The SAED pattern indicated polycrystalline nature of both the nanoparticles. SEM analysis showed that the nanoparticles were randomly distributed. EDX analysis showed the elemental percent ratio of 58.3% Zn and 20.0% O in ZnONPs, while it was 46.4% Cu and 18.3% O in case of CuONPs. Several metabolites secreted in the filtrate of fungal biomass aided in the formation, size reduction and stabilization of ZnONPs and CuONPs.

The antimicrobial activity of the synthesized nanoparticles was tested against *S. aureus*, *B. subtilis*, *P. aeruginosa*, *E. coli* and *S. typhimurium*. ZnONPs (5 mg/mL) showed zone of inhibition with diameter of 16.33 ± 0.88, 13.5 ± 0.28, 12.43 ± 0.23, 11.06 ± 0.34, and 11.13 ± 0.41 mm against *S. aureus*, *B. subtilis*, *P. aeruginosa*, *E. coli* and *S. typhimurium*, respectively. The mycogenic CuONPs (5 mg/mL) showed zone of inhibition of 22 ± 0.57, 16.26 ± 0.63, 13.6 ± 0.4, 11.93 ± 0.52, and 11.66 ± 0.33 mm against *S. aureus*, *B. subtilis*, *P. aeruginosa*, *E. coli* and *S. typhimurium*, respectively. MIC for ZnONPs was 2 mg/mL against *S. aureus*, *E. coli* and *P. aeruginosa*, while it was 3 mg/mL against *B. subtilis* and *S. typhimurium*. MIC for CuONPs was 1.0, 1.5, 2 and 3 mg/mL against *S. aureus*, *B. subtilis*, *P. aeruginosa*, *E. coli* and *S. typhimurium*, respectively (Mohamed et al. 2021).

Mechanism

Many studies have highlighted the nanobiotechnological potential of the fungi although the exact mechanism involved in the synthesis of metal and metal oxide nanoparticles is yet to be elucidated. The extracellular synthesis might occur due to the presence of enzymes and proteins in the fungal extract which can reduce the metal ions to corresponding nanoparticles. The parameters such as pH, temperature, fungal species, metal salt concentration, as well as the presence of capping agents directly affect the size of the nanoparticles. Many biomolecules react with ions but mainly nicotinamide adenine dinucleotide (NADH) and NADH dependent nitrate reductase enzymes are considered as important factors in the biosynthesis of nanoparticles from fungi. Extracellular NADPH-dependent nitrate reductase enzymes and quinones also participate in the formation of nanoparticles (Ghosh et al. 2021). Similarly various mechanisms exist by which the biogenic nanoparticles exhibit the antimicrobial activity as depicted in Figure 3. The nanoparticles can initially adsorb on the surface of the microbial cells eventually entering by pore formation and thereby resulting in the oxidative stress due to formation of reactive oxygen species (ROS). This leads to damage of the DNA and the proteins. Alteration of the membrane permeability results in leakage of cellular metabolites that disrupts the metabolic processes. Similarly, the nanoparticles can also damage the vital enzymes leading to impairment of the pathways related to survival, cell division and gene expression (Singh et al. 2018).

Table 1. Fungi mediated synthesis of different nanoparticles with antimicrobial activity.

Nanoparticles	Fungi	Size	Shape	Activity	Target Organism	References
Ag	*Alternaria solani* GS1	5–20	Spherical	Antibacterial and antifungal	*S. pyogenes* MTCC1925, *E. coli* MTCC730, *E. faecalis* MTCC2729 and *C. albicans* MTCC183	Devi et al. 2014
	Amylomyces rouxii	14–20	Spherical	Antibacterial and antifungal	*S. dysenteriae* type I, *S. aureus*, *Citrobacter* sp., *E. coli*, *P. aeruginosa*, *B. subtilis*, *C. albicans* and *F. oxysporum*	Musarrat et al. 2010
	Aspergillus clavatus	10–25	Hexagonal and non spherical polyhedral	Antibacterial and antifungal	*C. albicans*, *P. fluorescens* and *E. coli*	Verma et al. 2010
	Aspergillus niger	25–26	Spherical	Antibacterial	*B. megaterium*, *S. aureus*, *P. vulgaris* and *S. sonnei*	Vala and Shah 2012
	Aspergillus sydowii	1–24	Spherical	Antifungal	*A. fumigatus*, *A. flavus* and *A. terreus*, *F. solani*, *F. moniliforme*, *F. oxysporum*, *C. albican*, *C. glabrata*, *C. parapsilosis* and *C. tropicalis*, *C. neoformans* and *S. schenckii*	Wang et al. 2021
	Beauveria bassiana	10–50	Traingular, circular and hexagonal	Antibacterial	*E. coli*, *P. aeruginosa* and *S. aureus*	Tyagi et al. 2019
	Bipolaris tetramera	54.78–73.49	Spherical, triangular and hexagonal	Antibacterial	*B. subtilis*, *Bacillus cereus*, *S. aureus*, *E. coli* and *E. aerogenes*	Fatima et al. 2015
	Candida glabrata	2–15	Spherical, oval	Antibacterial and antifungal	*S. aureus*, *E. coli*, *S. typhimurium*, *S. flexneri*, *K. pneumoniae*, *P. aeruginosa*, *C. albicans*, *C. dubliniensis*, *C. parapsilosis*, *C. tropicalis*, *C. krusei* and *C. glabrata*	Jalal et al. 2018
	Cladosporium cladosporioides	30–60	Spherical	Antibacterial and antifungal	*S. aureus*, *S. epidermis*, *B. subtilis*, *E. coli* and *C. albicans*	Hulikere and Joshi 2019

Table 1 contd. ...

...Table 1 contd.

Nanoparticles	Fungi	Size	Shape	Activity	Target Organism	References
	Cryptosporiopsis ericae	2–15	Spherical	Antibacterial and antifungal	*S. aureus* MTCC96, *S. enterica* MTCC735, *E. coli* MTCC730, *E. faecalis* MTCC2729, and *C. albicans* MTCC183	Devi and Joshi 2014
	Curvularia lunata	10–50	Spherical	Antibacterial	*E. coli, P. aeruginosa, S. paratyphi, B. subtilis, S. aureus* and *B. cereus*	Ramalingmam et al. 2015
	Fusarium oxysporum	21.3–37.3	Spherical	Antibacterial	*E. coli, E. cloacae, K. pneumoniae, Proteus mirabilis, P. aeruginosa, S. aureus* and *S. epidermidis*	Ahmed et al. 2018
	Fusarium solani	3–8	–	Antibacterial	*S. aureus* and *E. coli*	El-Rafie et al. 2010
	Lasiodiplodia theobromae	76	Spherical and oval	Antibacterial	*P. aeruginosa* ATCC27853	Ranjani et al. 2020
	Macrophomina phaseolina	3.33–40.15	Spherical	Antibacterial	*E. coli* and *A. tumefaciens*	Chowdhury et al. 2014
	Phomopsis helianthi	40	Spherical, pentagonal and hexagonal	Antibacterial	*E. coli* and *P. aeruginosa*	Gond et al. 2020
	Penicillium oxalicum		Spherical	Antibacterial	*S. aureus, B. subtilis, P. aeruginosa* and *E. coli*	Bhattacharjee et al. 2017
	Penicillium radiatolobatum	5.09–24.52	Spherical, triangle and hexagonal	Antibacterial	*B. cereus, S. auerus, E. coli, S. enterica* and *L. monocytoegenes*	Naveen et al. 2021
	Phomopsis liquidambaris SA1	18.7	Spherical	Antibacterial	*E. coli* MTCC1687, *P. mirabilis* MTCC425, *M. luteus* MTCC1809, *S. flexneri* MTCC9543, *K. pneumoniae* MTCC4031, *V. cholerae* MTCC3906, *P. putida* MTCC1194, *S. typhi* MTCC531	Seetharaman et al. 2018

	Fungus	Size	Shape	Activity	Microbes	Reference
Au	*Pycnoporus sanguineus* and *Schizophyllum commune*	20.9 and 15.8	Spherical	Antibacterial and antifungal	*S. aureus* and white rot fungi	Chan and Don 2012
	Bipolaris tetramera	58.4, 110.13, and 261.73	Spherical, triangular and hexagonal	Antibacterial	*B. cereus, S. aureus* and *E. aeruginosa*	Fatima et al. 2015
	Pleurotus osteratus	10–30	Spherical	Antimicrobial	–	El Domany et al. 2018
CuO	*Penicillium chrysogenum* MF318506	10.5–59.7	Hexagonal and spherical	Antibacterial	*S. aureus, B. subtilis, P. aeruginosa, E. coli* and *S. typhimurium*	Mohamed et al. 2021
Se	*Penicillium expansum*	4–12.7	Spherical	Antibacterial, antifungal	*S. aureus* ATCC23235, *B. subtilis* ATCC6051, *E. coli* ATCC8739, *P. aeruginosa* ATCC9027, *C. albicans* ATCC90028, *A. fumigatus* RCMB02568, and *A. niger* RCMB02724	Hashem et al. 2021
ZnO	*Penicillium chrysogenum* MF318506	9.0–35.0	Hexagonal and spherical	Antibacterial	*S. aureus, B. subtilis, P. aeruginosa, E. coli* and *S. typhimurium*	Mohamed et al. 2021

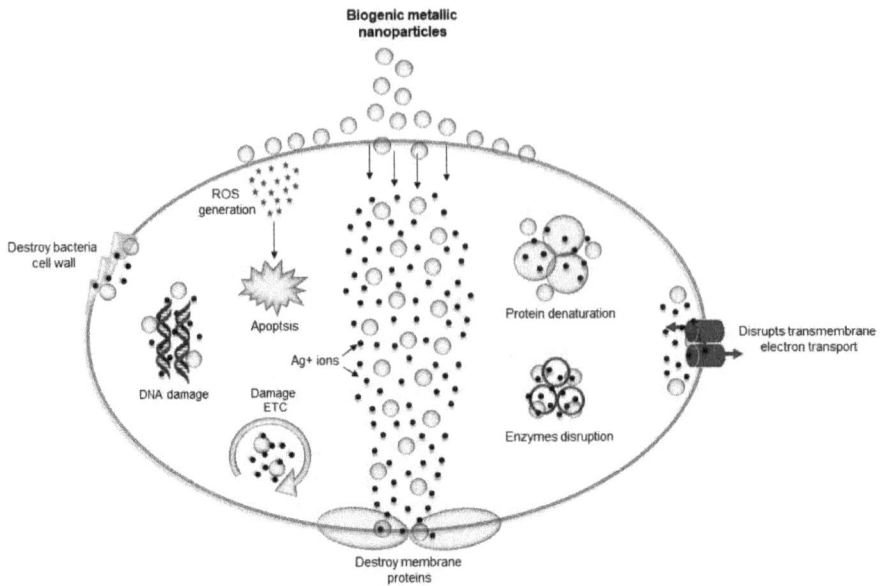

Figure 3. Various mechanisms of antimicrobial activity of biogenic metallic nanoparticles. ROS: reactive oxygen species. Reprinted from Singh, P., Garg A., Pandit S., Mokkapati, V.R.S.S. and Mijakovic, I. 2018. Antimicrobial effects of biogenic nanoparticles. Nanomaterials 8: 1009.

Conclusion

Mycogenic synthesis of nanoparticles offers various advantages as it is an environmentally benign, rapid and effective route for the synthesis of biocompatible nanoparticles with diverse biomedical applications. Fungal extracts are comprised of several metabolites that are responsible for both reduction and capping during nanoparticle synthesis. Different species of fungi, as well as endophytic fungi, are used for the synthesis of nanoparticles. However, there are many disadvantages of using fungi for the synthesis of nanoparticles which are needed to be addressed. Requirement of sterile conditions, the time required by the fungus to grow and produce sufficient biomass, and the low yield of nanoparticles are some of the limitations in mycogenic synthesis which should be addressed. Mycogenic nanoparticles are potent antimicrobial agents that can be effectively used for developing nanomedicine for controlling multidrug-resistant microorganisms.

Acknowledgement

Dr. Sougata Ghosh acknowledges Kasetsart University, Bangkok, Thailand for Post Doctoral Fellowship and funding under Reinventing University Program (Ref. No. 6501.0207/10870 dated 9th November, 2021).

References

Ahmed, A., Hamzah, H. and Maaroof, M. 2018. Analyzing formation of silver nanoparticles from the filamentous fungus *Fusarium oxysporum* and their antimicrobial activity. Turk. J. Biol. 42(1): 54–62.

Bhagwat, T.R., Joshi, K.A., Parihar, V.S., Asok, A., Bellare, J. and Ghosh, S. 2018. Biogenic copper nanoparticles from medicinal plants as novel antidiabetic nanomedicine. World J. Pharm. Res. 7(4): 183–196.

Bhattacharjee, S., Debnath, G., Das, A.R., Saha, A.K. and Das, A.R. 2017. Characterization of silver nanoparticles synthesized using an endophytic fungus, *Penicillium oxalicum* having potential antimicrobial activity. Adv. Nat. Sci.: Nanosci. Nanotechnol. 8(4): 1–6.

Chan, Y.C. and Don, M.M. 2012. Characterization of Ag nanoparticles produced by white-rot fungi and its *in vitro* antimicrobial activities. Int. Arab. J. Antimicrob. Agents 2(3): 1–8.

Chowdhury, S., Basu, A. and Kundu, S. 2014. Green synthesis of protein capped silver nanoparticles from phytopathogenic fungus *Macrophomina phaseolina* (Tassi) Goid with antimicrobial properties against multidrug-resistant bacteria. Nanoscale Res. Lett. 9: 365.

Devi, L.S. and Joshi, S.R. 2014. Evaluation of the antimicrobial potency of silver nanoparticles biosynthesized by using an endophytic fungus, *Cryptosporiopsis ericae* PS4. J. Microbiol. 52(8): 667–674.

Devi, L.S., Bareh, D.A. and Joshi, S.R. 2014. Studies on biosynthesis of antimicrobial silver nanoparticles using endophytic fungi isolated from the ethno-medicinal plant *Gloriosa superba* L. Proc. Natl. Acad. Sci., India, Sect. B Biol. Sci. 84(4): 1091–1099.

El Doman, Y.E.B., Essam, T.M., Ahmed, A.E. and Farghali, A.A. 2018. Biosynthesis physico-chemical optimization of gold nanoparticles as anti-cancer and synergetic antimicrobial activity using *Pleurotus ostreatus* Fungus. J. App. Pharm. Sci. 8(5): 119–128.

El-Rafie, M.H., Mohamed, A.A., Shaheen, Th.I. and Hebeish, A. 2010. Antimicrobial effect of silver nanoparticles produced by fungal process on cotton fabrics. Carbohydr. Polym. 80(3): 779–782.

Fatima, F., Bajpai, P., Pathak, N., Singh, S., Priya, S. and Verma, S.R. 2015. Antimicrobial and immunomodulatory efficacy of extracellularly synthesized silver and gold nanoparticles by a novel phosphate solubilizing fungus *Bipolaris tetramera*. BMC Microbiol. 15: 52.

Gade, A., Ingle, A., Whiteley, C. and Rai, M. 2010. Mycogenic metal nanoparticles: Progress and applications. Biotechnol. Lett. 32: 593–600.

Ghosh, S. 2018. Copper and palladium nanostructures: A bacteriogenic approach. Appl. Microbiol. Biotechnol. 101(18): 7693–7701.

Ghosh, S. and Webster, T.J. 2021. Nanobiotechnology: Microbes and Plant Assisted Synthesis of Nanoparticles, Mechanisms and Applications. Elsevier Inc. USA.

Ghosh, S., Shah, S. and Webster, T.J. 2021. Recent trends in fungal biosynthesis of nanoparticles. pp. 403–452. *In*: Shah, M.P., Sharma, V.K., Parmar, S. and Kumar, A. (eds.). Fungi Bio-Prospects in Sustainable Agriculture, Environment and Nano-Technology. Volume 3: Fungal Metabolites and Nano-technology. Elsevier USA.

Gond, S.K., Mishra, A., Verma, S.K., Sharma, V.K. and Kharwar, R.N. 2020. Synthesis and characterization of antimicrobial silver nanoparticles by an endophytic fungus isolated from *Nyctanthes arbor-tristis*. Proc. Natl. Acad. Sci., India, Sect. B Biol. Sci. 90(3): 641–645.

Guilger-Casagrande, M. and Lima, R. 2019. Synthesis of silver nanoparticles mediated by fungi: A review. Front. Bioeng. Biotechnol. 7: 287.

Hashem, H.H., Khalil, A.M.A., Reyad, A.M. and Salem, S.S. 2021. Biomedical applications of mycosynthesized selenium nanoparticles using *Penicillium expansum* ATTC 36200. Biol. Trace Elem. Res. 199(10): 3998–4008.

Hulikere, M.M. and Joshi, C.G. 2019. Characterization, antioxidant and antimicrobial activity of silver nanoparticles synthesized using marine endophytic fungus—*Cladosporium cladosporioides*. Process Biochem. 82.

Jalal, M., Ansari, M.A., Alzohairy, M.A., Ali, S.G., Khan, H.M., Almatroudi, A. and Raees, K. 2018. Biosynthesis of silver nanoparticles from oropharyngeal *Candida glabrata* isolates and their antimicrobial activity against clinical strains of bacteria and fungi. Nanomaterials 8(8): 586.

Jamdade, D.A., Rajpali, D., Joshi, K.A., Kitture, R., Kulkarni, A.S., Shinde, V.S., Bellare, J., Babiya, K.R. and Ghosh, S. 2019. *Gnidia glauca* and *Plumbago zeylanica* mediated synthesis of novel copper nanoparticles as promising antidiabetic agents. Adv. Pharmacol. Sci. 2019: 9080279.

Moghaddam, A.B., Namvar, F., Moniri, M., Tahir, P.M., Azizi, S. and Mohamad, R. 2015. Nanoparticles biosynthesized by fungi and yeast: A review of their preparation, properties, and medical applications. Molecules 20: 16540–16565.

Mohamed, A.A., Abu-Elghait, M., Ahmed, N.E. and Salem, S.S. 2021. Eco-friendly mycogenic synthesis of ZnO and CuO nanoparticles for *in vitro* antibacterial, antibiofilm, and antifungal applications. Biol. Trace Elem. Res. 199(7): 2788–2799.

Musarrat, J., Dwivedi, S., Singh, B.R., Al-Khedhairy, A.A., Azam, A. and Naqvi, A. 2010. Production of antimicrobial silver nanoparticles in water extracts of the fungus Amylomycesrouxii strain KSU-09. Bioresour. Technol. 101(22): 8772–8776.

Naveen, K.V., Sathiyaseelan, A., Mariadoss, A.V.A., Xiaowen, H., Saravanakumar, K. and Wang, M. 2021. Fabrication of mycogenic silver nanoparticles using endophytic fungal extract and their characterization, antibacterial and cytotoxic activities. Inorg. Chem. Commun. 128: 108575.

Rai, M., Yadav, P., Bridge, P. and Gade, A. 2009a. Myconanotechnology: A new and emerging science. pp. 258–267. *In*: Rai Bridge (ed.). Applied Mycology. CABI Publication, UK.

Rai, M., Yadav, A. and Gade, A. 2009b. Silver nanoparticles as a new generation of antimicrobials. Biotech. Adv. 27: 76–83.

Ramalingmam, P., Muthukrishnan, S. and Thangaraj, P. 2015. Biosynthesis of silver nanoparticles using an endophytic fungus, *Curvularia lunata* and its antimicrobial potential. J. Nanosci. Nanoeng. 1(4): 241–247.

Ranjani, S., Shariq, A.M., Adnan, M., Senthil, K.N., Ruckmani, K. and Hemalatha, S. 2020. Synthesis, characterization and applications of endophytic fungal nanoparticles. Inorg. Nano-Met. Chem. 51(2): 1–8.

Rokade, S., Joshi, K., Mahajan, K., Patil, S., Tomar, G., Dubal, D., Parihar, V.S., Kitture, R., Bellare, J.R. and Ghosh, S. 2018. *Gloriosa superba* mediated synthesis of platinum and palladium nanoparticles for induction of apoptosis in breast cancer. Bioinorg. Chem. Appl. 2018: 4924186.

Rokade, S.S., Joshi, K.A., Mahajan, K., Tomar, G., Dubal, D.S., Parihar, V.S., Kitture, R., Bellare, J., Ghosh, S., 2017. Novel anticancer platinum and palladium nanoparticles from *Barleria prionitis*. Glob. J. Nanomedicine 2(5): 555600.

Saravanan, A., Senthil Kumar, P., Karishma, S., Vo, D-V.N., Jeevanantham, S., Yaashikaa, P.R. and George, C.S. 2021. A review on biosynthesis of metal nanoparticles and its environmental applications. Chemosphere 264: 128580.

Seetharaman, P.K., Chandrasekaran, R., Gnanasekar, S., Chandrakasan, G., Gupta, M., Babu, D. and Sivaperumal, S. 2018. Antimicrobial and larvicidal activity of eco-friendly silver nanoparticles synthesized from endophytic fungi *Phomopsis liquidambaris*. Biocatal. Agric. Biotechnol. 16: 22–30.

Sharma, D., Kanchi, S. and Bisetty, K. 2019. Biogenic synthesis of nanoparticles: A review. Arab. J. Chem. 12: 3576–3600.

Shende, S., Joshi, K.A., Kulkarni, A.S., Charolkar, C., Shinde, V.S., Parihar, V.S., Kitture, R., Banerjee, K., Kamble, N., Bellare, J. and Ghosh, S. 2018. *Platanus orientalis* leaf mediated rapid synthesis of catalytic gold and silver nanoparticles. J. Nanomed. Nanotechnol. 9: 2.

Singh, P., Garg, A., Pandit, S., Mokkapati, V.R.S.S. and Mijakovic, I. 2018. Antimicrobial effects of biogenic nanoparticles. Nanomaterials 8: 1009.

Tyagi, S., Tyagi, P.K., Gola, D., Chauhan, N. and Bharti, R.K. 2019. Extracellular synthesis of silver nanoparticles using entomopathogenic fungus: Characterization and antibacterial potential. SN Appl. Sci. 1: 1545.

Vala, A.K. and Shah, S. 2012. Rapid synthesis of silver nanoparticles by a marine-derived fungus *Aspergillus niger* and their antimicrobial potentials. Int. J. Nanosci. Nanotechnol. 8: 197–206.

Verma, V.C., Kharwar, R.N. and Gange, A.C. 2010. Biosynthesis of antimicrobial silver nanoparticles by the endophytic fungus *Aspergillus clavatus*. Nanomedicine 5(1): 33–40.

Wang, D., Xue, B., Wang, L., Zhang, Y., Liu, L. and Zhou, Y. 2021. Fungus-mediated green synthesis of nano-silver using *Aspergillus sydowii* and its antifungal/antiproliferative activities. Sci. Rep. 11: 10356.

5

Myconanotechnology in Vaccine Adjuvants, Challenges for Crises to Come

Juan Bueno

"The deviation of man from the state in which he was originally placed by nature seems to have proved to him a prolific source of diseases"

"While the vaccine discovery was progressive, the joy I felt at the prospect before me of being the instrument destined to take away from the world one of its greatest calamities [smallpox], blended with the fond hope of enjoying independence and domestic peace and happiness, was often so excessive that, in pursuing my favorite subject among the meadows, I have sometimes found myself in a kind of reverie"

—Edward Jenner (1749–1823)

Introduction

Vaccines are the greatest public health measure through which humanity counts for the prevention of infectious diseases at a global level (Andrianou et al. 2021). Vaccines work on a mechanism in which a microbial antigen is administered, which then stimulates a specific and protective immune response by the host. This response produces protective and neutralizing antibodies which increase their survival (Slifka and Amanna 2019, Pang et al. 2021). Thus, every vaccine has two important components, the antigen (part of the microorganism that will be recognized by

Research Center of Bioprospecting and Biotechnology for Biodiversity Foundation (BIOLABB) Colombia.
Email: juangbueno@gmail.com

the immune system of the host) and the adjuvant (Tan et al. 2018, Schijns et al. 2020). Vaccine adjuvants can be defined as any molecule or complex molecular or supramolecular assembly that is significantly able to augment an immune response when administered simultaneously with a vaccine (Figure 1) (Roth et al. 2021, Chowdhury et al. 2022). It is thought that many adjuvants mediate their effects by acting as immune potentiators, eliciting early innate immune responses which enhance vaccine effectiveness by greatly exacerbating the overall vaccine responses (Pulendran et al. 2021, Lodaya et al. 2022). This topic is an important research field since immunization with purified protein antigens typically results in the induction of modest antibody response with little or no T cell response. Additionally, multiple immunizations may be required to elicit sufficient antibody responses (Galen et al. 2021, McMillan et al. 2021). Developers may seek to include adjuvants in vaccine candidates to enhance the efficacy of weak antigens, to induce appropriate immune responses not sufficiently induced in the absence of adjuvant or both (Pan et al. 2021, Pogostin and McHugh 2021). There are several properties that a good adjuvant is required to have. These include, not being toxic in its effective dose to fulfill its adjuvant effect, strongly induce the cellular and humoral immune response, produce a good immunological memory, to not cause allergic or autoimmune responses, not be mutagenic, carcinogenic or teratogenic, to not be pyrogen and finally, to be pharmacologically stable in a wide range of temperatures and pH (Cao et al. 2020, Pollard and Bijker 2021, Tsakiri et al. 2021). In SARS-CoV-2 (severe acute respiratory syndrome coronavirus 2), public health crisis MF59 (an emulsion containing naturally occurring squalene oil and water) and AS03 as emulsion adjuvants, were used in vaccines (O'Hagan et al. 2020, Lodaya et al. 2022). For that reason, it is necessary to undertake extensive research and development of new vaccine adjuvants to improve the protection against infectious diseases in humans and animals. This strategy should have the ability to prevent the spread of infectious disease and zoonoses, improving the quality of life of the population (Excler et al. 2021, Krishnan et al. 2021). Equally, identification of such immunomodulatory properties will be crucial in the discovery of novel, clinically-relevant compounds for augmenting existing immunotherapy or vaccination practices (Hu et al. 2018, Li and Li 2020). In this order of ideas, the application of nanomaterials as vaccine adjuvants is a very promising strategy, because antigens delivered by nanomaterials can be protected from degradation and released in a sustainable manner, and their uptake by antigen-presenting cells (APCs) is more efficient (Chatzikleanthous et al. 2021, Mao et al. 2021). Thus, among the licensed adjuvants, MF59 (Novartis) is a nano-adjuvant that is 165 nm in diameter with the ability to recruit neutrophils, monocytes, and dendritic cells (DCs) and enhance antigen uptake (Irvine et al. 2020, Abbasi and Uchida 2021). Likewise, fungi have been employed in the production of nanomaterials to be applied in medicine, agriculture, and industry. So, it can be a source of nanocomposites with immunomodulatory activity that allows developing novel adjuvants that achieve antigenic processing of current vaccines (Sharma et al. 2019, Wang et al. 2020). In view of this, the objective of this chapter will be to determine the possibilities of nanomaterials obtained from fungi for use as nano-adjuvant with which to produce future vaccines. This will help us fight against the great challenges of public health that infectious diseases and cancer impose on us.

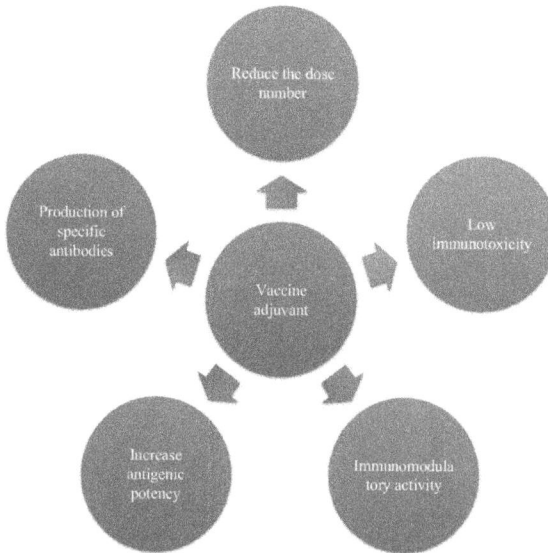

Figure 1. Characteristic properties of vaccine adjuvants.

Vaccine Adjuvants in Antigen Uptake

The processing of the antigen is a natural phenomenon of primordial importance. It allows the presentation of a microorganism to the cells of the immune system with which specific antibodies can be developed to counteract the microbial invasion (Lu et al. 2018, Schechter et al. 2019). Likewise, the uptake and presentation of the microbial antigen through the APCs such as macrophages, B lymphocytes and dendritic cells is a process that induces an adaptive immune response efficiently, except in cases in where microorganisms have evasion capacity of immunity (de Oliveira et al. 2021, Iyer et al. 2021). In this way, a vaccine adjuvant has the ability to stimulate the immune system so that it recognizes the microbial antigen. It also induces the production of both neutralizing and protective antibodies by the host, i.e., antibodies capable of protecting against infectious disease so as to neutralize the causal agent when it comes into contact with predisposed patients (Figure 2) (Haun et al. 2020, Wagner and Weinberger 2020). Thus, a good adjuvant of vaccines has the ability to stimulate the antigen presenting cells of the vaccine receptor in order to present the microbial antigen that is in the formulation more efficiently (Tsai et al. 2020, dong Zhang et al. 2021). For this reason, an adjuvant works by stimulating innate immunity to enhance processing through the activation of Toll-like receptors (TLR) (Ong et al. 2021, Roßmann et al. 2021). TLRs are transmembrane receptors that recognize infectious agents and make molecular signaling that induces activity in APCs (Mokhtari et al. 2021, Sameer and Nissar 2021). Thus, nanomaterials have submitted the ability to join the TLRs by stimulating innate immunity and antigenic recognition (Dowling and Mansell 2016, Liu et al. 2017). This makes the nanomaterials an important source of future adjuvants for the development of vaccines with high immunogenicity (Chung et al. 2020, Khalaj-Hedayati et al. 2020).

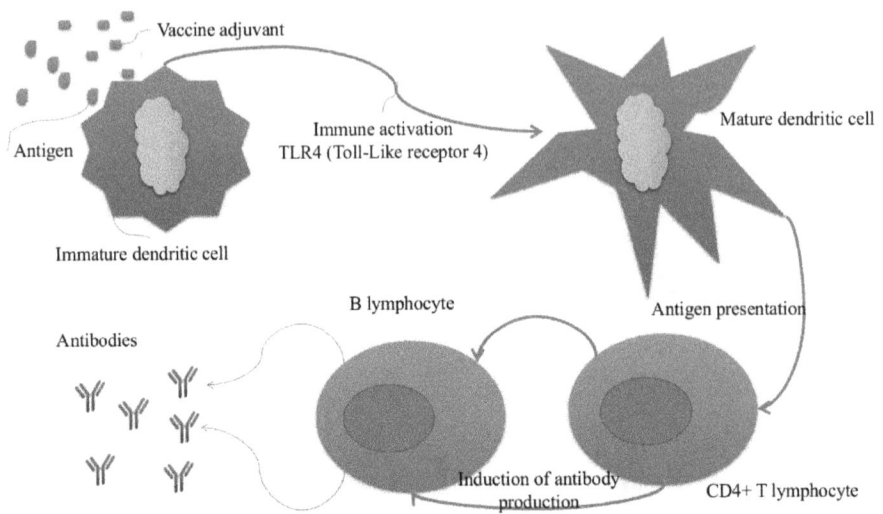

Figure 2. Mechanism of action of vaccine adjuvants.

Nanomaterials as Vaccine Adjuvants

In this way, nanomaterials interact with the receptors of innate immunity. They can also produce an inflammatory response for being agonists of toll-like receptors (Javaid et al. 2019, Rossella et al. 2021). Equally, nanomaterials can deliver the antigen to be presented properly, as well as increase the penetration of the vaccine to tissue layers containing the APCs (Tiwari et al. 2021, Trincado et al. 2021). Likewise, nanomaterials can stimulate the activity of dendritic cells to recognize the antigens present in vaccines (Khan and Khan 2021, Xu et al. 2021). Thus, nanoparticles can stimulate the clonal expansion of CD4+ lymphocytes in response to antigenic recognition to start a protective cellular and humoral immune response (Pati et al. 2018, Malachowski and Hassel 2020). In this order of ideas, the combination of nanocomposites with both microbial and tumoral antigens is a valid approximation for the development of new vaccines with high immunogenicity (Figure 3) (Baig et al. 2018, Trabbic et al. 2021). When evaluating nanotechnology for the development of vaccine adjuvants, to prevent and address the upcoming public health crisis in the world, it is necessary to develop green synthesis systems of nanoparticles that are biocompatible with the host, as well as a sustainable biotechnology system that allows producing these compounds industrially (Wen et al. 2019, Seyfoori et al. 2021, Yayehrad et al. 2021). In this way, microbial nanotechnology is a great field of research for synthesis and obtaining biotechnological derivatives for the formulation of useful vaccines for the prevention of infectious diseases that affect human beings (Zhu et al. 2014, Han et al. 2018). Finally, these bionanomaterial-based vaccine adjuvants have the ability to overcome the challenges and limitations of current vaccines, increasing antigenic presentation, as well as efficient and protective antibody production (Chauhan et al. 2020, Vu et al. 2021).

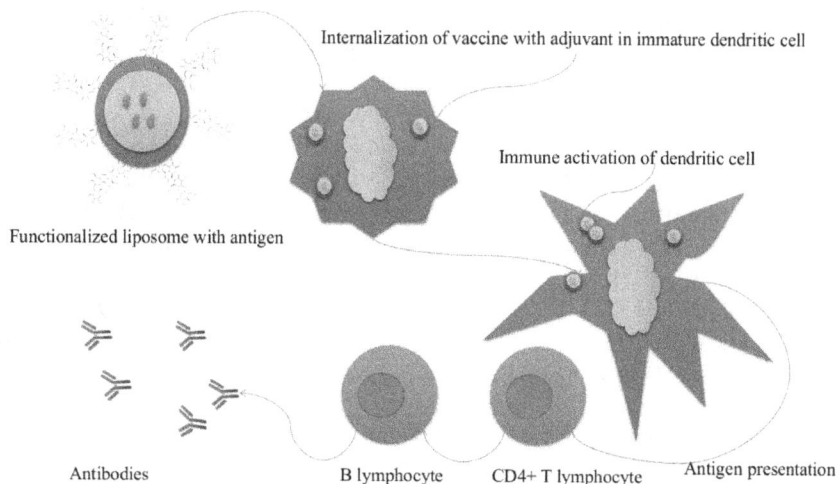

Figure 3. Nanomaterials and vaccine adjuvants.

Bioinspired and Biomimetic Nanoparticles Activating the Immune System

Bioinspired and biomimetic nanomedicines use biocompatible biological models for the development of nanoproducts with pharmacological activity (Chen et al. 2019, Liu et al. 2021). In this way, this approach seeks to transport molecules to specific sites as well as amplify antigens and achieve immunoactivation (Vijayan et al. 2019, Thorp et al. 2020). Among them are nanoparticles based on viruses and nanoparticles associated with biomolecules of plasma membranes of bacteria and/or fungi (Sundar and Kumar Prajapati 2012, Weiss et al. 2020). Thus, these nanoconjugates contain fungal biomolecules that functionalize the nanoparticles such as the chitosan, the mannan and the chitin (Nag et al. 2021, Niculescu and Grumezescu 2021). Likewise, in this order of ideas, bioinspired nanoparticles functionalized with molecules from fungi cell walls and membranes have shown to activate lectin receptors and mature dendritic cells by TLR (Chakravarty and Vora 2021, Chugh et al. 2021). Equally, the combination of antigens with mycotic components such as the natural carbohydrate polysaccharides, induces a strong cellular and humoral immune response. These compounds also have good bioavailability, biodegradability and stability, so they are considered good candidates to be vaccine adjuvants (Bashiri et al. 2020, Lang and Huang 2020). In this way, the routes of the fungal synthesis can be used for obtaining both functionalized nanoparticles and organic nanoparticles that can be employed as immune system enhancers that induce a response to different antigens in a bioinspired nanotechnology model (Sabu et al. 2019, Sushnitha et al. 2020).

Microbial Nanotechnology, Green Synthesis in Vaccine Adjuvants Development

In this order of ideas, lipid nanoparticles, protein nanoparticles, micelles, liposomes and polymer nanoparticles are nanoparticle vaccine platforms considered for new developments in nanovaccines (Figure 4) (Das and Ali 2021, Taneja et al. 2021). In this way, these organic nanoparticles can be produced by fungi on an industrial scale, in a model of microbial nanotechnology to be more biodegradable and compatible with the biological systems of the host (Dorcheh and Vahabi 2016, Meena et al. 2021). Thus, nanoparticles based on lipids have proven efficiency in the encapsulation of mRNA (messenger RNA), protecting the degradation of the antigen and allowing its presentation (Chatzikleanthous et al. 2021). Likewise, organic nanoparticles containing chitosan, β-glucan and mannan have demonstrated the ability to increase the immunogenicity of antigens in evaluated formulations (Zhang et al. 2021). Likewise, organic nanoparticles can promote the maturation of dendritic cells through TLR4 (Toll-Like receptor 4), as well as stimulation of C-type lectin receptors that are expressed in immune system cells (Vang et al. 2017, Lenders et al. 2020). Organic nanoparticles also decrease the unwanted autoimmune responses by increasing the specificity of the humoral immune response during the process of transporting the antigen (Lu et al. 2021, Mitchell et al. 2021). In this way, the green synthesis of nanomaterials involves microbial models that allow the obtaining of specific biomaterials which are used to develop more biocompatible products, such as vaccines. Among these models, the fungi are widely employed (Ghosh et al. 2021, Shreyash et al. 2021). Thus, multicellular filamentous fungi and yeast have been employed for the synthesis of lipid nanoparticles, protein nanoparticles, micelles, liposomes, and polymer nanoparticles; which have a low cost of production, easy synthesis, and low toxicity (Raveendran et al. 2017, Madamsetty et al. 2019).

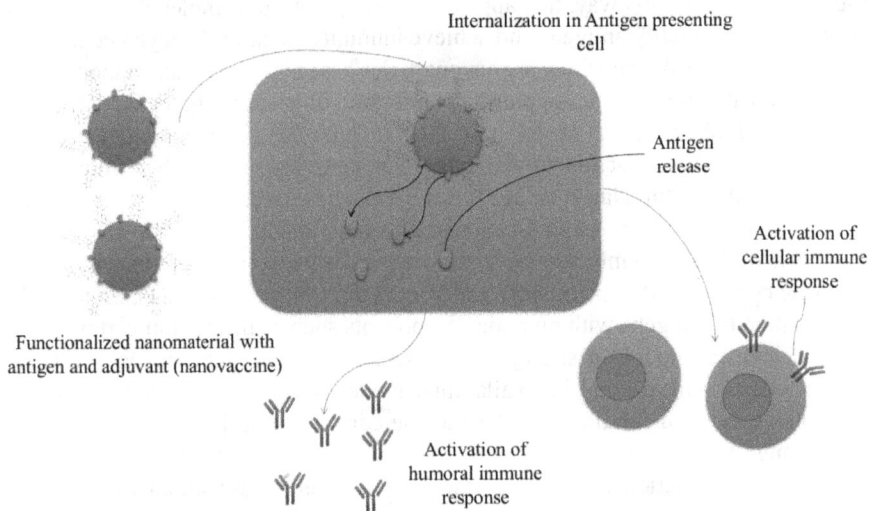

Figure 4. Nanovaccine mechanism of action.

Fungal Nanotechnology, Biosynthetic Routes in Adjuvants

Fungal nanotechnology seeks to employ biosynthetic routes of mycotic cells to obtain nanomaterials that can be employed in medicine, agriculture and industry. In this way, the use of biosynthetic organisms such as fungi, allows to develop industrial processes with low pollution rate due to the biotechnological use of biological production mechanisms. Hence, this approach is being applied in green chemical synthesis to avoid the toxicity of waste with other industrial approaches (Khan et al. 2018, Yadav et al. 2019). In this case, the biosynthetic routes of the fungal cells can be coupled with industrial processes for obtaining bionanomaterials with which to develop new adjuvants to enhance immunogenicity (Elieh-Ali-Komi and Hamblin 2016, Pan and Cui 2020). Especially in its processes of oxidoreduction, the fungi can synthesize organic nanoparticles for subsequent functionalization that can induce an inflammatory response by stimulation of innate immunity (Zolnik et al. 2010, Cronin et al. 2020). Fungi have great production capacity of functionalized organic nanoparticles. The fungal capsular polysaccharides are of special interest, among which there are immunoactive compounds such as α- and β-glucans, glycoproteins, chitin, as well as complex mannans (Złotko et al. 2019, Garcia-Rubio et al. 2020). Likewise, in the fungal biofilms, the composition of the wall acquires other characteristics, which make it change its immunogenicity and make it more attractive to the development of adjuvants (García-Carnero et al. 2020, Hameed et al. 2021). In the same way, fungal lipids (fatty acids, oxylipins, sphingolipids, phospholipids, glycolipids and sterols) can be used for the synthesis of micelles, lipid nanoparticles and liposomes, highly functional and with the ability to stimulate innate immunity to enhance antigenic recognition (Drescher and van Hoogevest 2020, Tenchov et al. 2021). Equally, stimulation of TLR4 has also been reported by fungal lipids that are recognized by the receptor and induce activation of the immune response with inflammation and stimulation of dendritic cells (Pereira-Dutra et al. 2019, Patin et al. 2019). In this way, the conjugation of organic nanoparticles functionalized with cell wall components and fungal lipids is an important approach in the search for new biomaterials for the development of nanoadjuvants that allow immunopotentiation to acquire immunity against infectious diseases and cancer (Kelly et al. 2017, Yenkoidiok-Douti and Jewell 2020).

Functionalizing Fungal Organic Nanoparticles

Functionalization is a process by which the properties and characteristics of the nanoparticles are increased by its conjugation with bioactive compounds on their surface that allow to give it greater pharmacological activity (Matur et al. 2020, Sanità et al. 2020). In this way, the functionalization allows cellular internalization of the nanomaterials as well as an increase in its biocompatibility with which a cellular and humoral immune response can be induced safely at the host (de Carvalho Lima et al. 2021, Hou et al. 2021). In this order of ideas, functionalization of nanocomposites can be achieved with conjugation of fungal biomolecules that stimulate innate immunity by the ability to be recognized by TLR (Figure 4) (Tabasum et al. 2017, Auría-Soro et al. 2019). Thus, among the most studied fungal cells, wall components for functionalizing organic nanoparticles is chitosan which is a mucopolysaccharide with promising use in immunotherapy for its immunostimulant activity (Table 1) (Jhaveri et al. 2021, Namasivayam et al. 2021). Also, the beta-glucan (polysaccharide biopolymer) nanoparticles have reported to be important for the design of microparticles for the transport of medications, as well as immune response induction (Lee et al. 2019, Li and Cheung 2019). In this order of ideas, the best nanoparticle functionalization strategy with biomolecules includes covalent methods for presenting greater stability (Bahrami et al. 2020, Marques et al. 2020). Likewise, the use of covalent methods for the functionalization of nanoparticles by conjugation with biomolecules has increased efficiency in the delivery of the antigen, as well as in the induction of a protective cellular and humoral immune response (Figure 5) (Tornesello et al. 2020, Xu et al. 2022). But it is important to note that each developed nanomaterial with covalent methods has coupling condition that guarantees its stability, penetration capacity and activating potency over innate immunity, which makes it a correct assessment of biological activity to determine if it has the characteristics of a good adjuvant (Li et al. 2021, Tang et al. 2021).

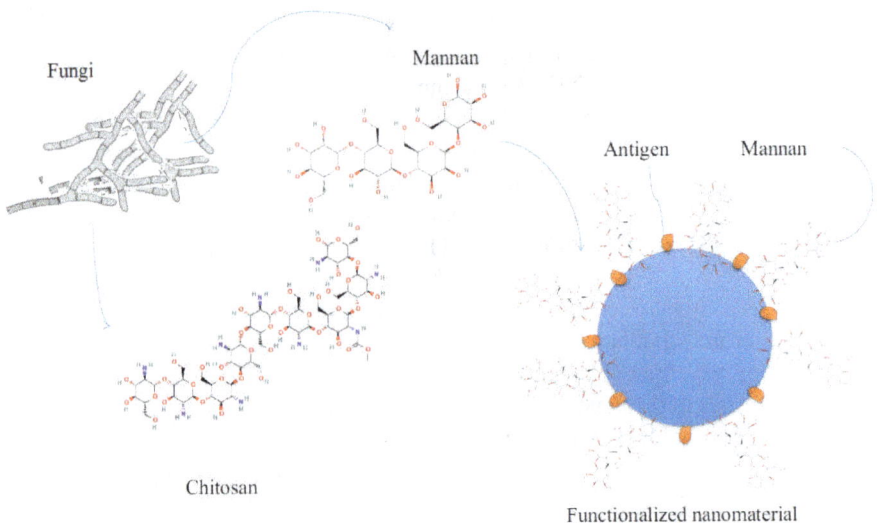

Figure 5. Functionalized nanomaterial with fungal cell wall components.

Table 1. Fungal cell wall components with biological activity on immune system.

Fungal Cell Wall Components	Biological Activity Immune System
α- and β-glucans	Activation of macrophages, neutrophils and natural killer cells
Glycoproteins	Activation of Toll-like receptors
Chitin	Induction of cytokine production, leukocyte recruitment, and macrophage activation
Complex mannans	Induce macrophage activation

Evaluation of Vaccine Adjuvants

In order to evaluate a vaccine adjuvant, firstly it is necessary to differentiate between immunostimulation and immunotoxicity, and an immune response will lead to the production of antibodies and immunological memory, an autoimmune response can be produced in the second (Bou Zerdan et al. 2021, Magadán et al. 2021). In this way, the best strategy to reduce immunotoxicity is to ensure the biocompatibility of the formulations used for vaccination, as well as nanoparticles, especially inorganic nanomaterials that have presented greater toxicity over the immune system (Kus-Liśkiewicz et al. 2021, Shams et al. 2021). It is also necessary to determine that the use of nanoadjuvants should produce a localized inflammation for the activation of the APCs and the antigenic presentation safely for the vaccinated patient (O'Neill et al. 2021, Tsakiri et al. 2021). Likewise, the evaluation of the toxicity on the dendritic cells *in vitro* should be determined to select the adjuvants to be taken into account in subsequent phases of formulation and development of vaccines (HogenEsch et al. 2018, D'Amico et al. 2021). On the other hand, it is prioritized to evaluate the capacity of the vaccine adjuvants to produce antigen-antibody reactions with immunocomplex formation, that enhances or initiates autoimmune diseases in the population (Rauch et al. 2018, Roe 2021). It is also necessary to report that organic and biocompatible nanoparticles reduce the phenomenon known as protein corona which is acquired by nanoparticles as a protein envelope during the administration *in vivo* and that can induce unwanted immune responses (Guerrini et al. 2018, Francia et al. 2019). Thus, research in new adjuvants should seek immunogenicity without inducing widespread inflammation, known as autoimmune/inflammatory syndrome induced by adjuvants (ASIA), as well as allergic or autoimmune reactions that affect the quality of life of vaccinated people (Jara et al. 2017, Pereira et al. 2020).

Conclusions

Vaccine adjuvants are an important approximation in the production of new formulations that increase the immunogenicity of weak antigens and allow development of protective antibodies against infectious diseases and cancer (Li et al. 2021). In this order of ideas, using nanotechnological tools to increase the penetration and recognition of antigens by APCs is a valid approach to obtain vaccines more rapidly and effectively (Durán-Lobato et al. 2021). In the same way the nanoparticles have presented immunotoxicity problems by the conformation of the protein corona, as well as toxicity against immune system cells. Thus, the employment of more biocompatible bionanomaterials obtained through microbial nanotechnology processes is a solution plausible to the safe introduction of nanoparticles in the design and development of vaccines (Florindo et al. 2020). For this reason, the use of fungi as a source of bionanomaterials that induce an adequate immune response without producing autoimmunity has been evaluated for the production and functionalization of new models of organic nanoparticle synthesis that can be useful for medicine in the future (Guo et al. 2020).

Acknowledgment

The author thanks Research Center of Bioprospecting and Biotechnology for Biodiversity Foundation (BIOLABB) for their collaboration and invaluable support during the writing of this chapter.

References

Abbasi, S. and Uchida, S. 2021. Multifunctional immunoadjuvants for use in minimalist nucleic acid vaccines. Pharmaceutics 13(5): 644.

Andrianou, X.D., Pronk, A., Galea, K.S., Stierum, R., Loh, M., Riccardo, F., Pezzotti, P. and Makris, K.C. 2021. Exposome-based public health interventions for infectious diseases in urban settings. Environ. Int. 146: 106246.

Auría-Soro, C., Nesma, T., Juanes-Velasco, P., Landeira-Viñuela, A., Fidalgo-Gomez, H., Acebes-Fernandez, V., Gongora, F., Almendral-Parra, M., Manzano-Roman, R. and Fuentes, M. 2019. Interactions of nanoparticles and biosystems: Microenvironment of nanoparticles and biomolecules in nanomedicine. Nanomaterials 9(10): 1365.

Bahrami, A., Delshadi, R. and Jafari, S.M. 2020. Active delivery of antimicrobial nanoparticles into microbial cells through surface functionalization strategies. Trends Food Sci. Technol. 99: 217–228.

Baig, M.S., Keservani, R.K., Ahmad, M.F. and Baig, M.E. 2018. Smart delivery of nanobiomaterials in drug delivery. In Nanobiomaterials. Apple Academic Press, pp. 241–287.

Bashiri, S., Koirala, P., Toth, I. and Skwarczynski, M. 2020. Carbohydrate immune adjuvants in subunit vaccines. Pharmaceutics 12(10): 965.

Bou Zerdan, M., Moussa, S., Atoui, A. and Assi, H.I. 2021. Mechanisms of immunotoxicity: Stressors and evaluators. Int. J. Mol. Sci. 22(15): 8242.

Cao, W., He, L., Cao, W., Huang, X., Jia, K. and Dai, J. 2020. Recent progress of graphene oxide as a potential vaccine carrier and adjuvant. Acta Biomater. 112: 14–28.

Chakravarty, M. and Vora, A. 2021. Nanotechnology-based antiviral therapeutics. Drug Deliv. Transl. Res. 11: 748–787.

Chatzikleanthous, D., O'Hagan, D.T. and Adamo, R. 2021. Lipid-based nanoparticles for delivery of vaccine adjuvants and antigens: Toward multicomponent vaccines. Mol. Pharm. 18(8): 2867–2888.

Chauhan, G., Madou, M.J., Kalra, S., Chopra, V., Ghosh, D. and Martinez-Chapa, S.O. 2020. Nanotechnology for COVID-19: Therapeutics and vaccine research. ACS Nano 14(7): 7760–7782.

Chen, Z., Wang, Z. and Gu, Z. 2019. Bioinspired and biomimetic nanomedicines. Acc Chem. Res. 52(5): 1255–1264.

Chowdhury, S., Toth, I. and Stephenson, R.J. 2022. Dendrimers in vaccine delivery: Recent progress and advances. Biomaterials 280: 121303.

Chugh, V., Vijaya Krishna, K. and Pandit, A. 2021. Cell membrane-coated mimics: A methodological approach for fabrication, characterization for therapeutic applications, and challenges for clinical translation. ACS Nano 15(11): 17080–17123.

Chung, Y.H., Beiss, V., Fiering, S.N. and Steinmetz, N.F. 2020. COVID-19 vaccine front runners and their nanotechnology design. ACS Nano 14(10): 12522–12537.

Cronin, J.G., Jones, N., Thornton, C.A., Jenkins, G.J., Doak, S.H. and Clift, M.J. 2020. Nanomaterials and innate immunity: A perspective of the current status in nanosafety. Chem. Res. Toxicol. 33(5): 1061–1073.

D'Amico, C., Fontana, F., Cheng, R. and Santos, H.A. 2021. Development of vaccine formulations: Past, present, and future. Drug Deliv. Transl. Res. 11(2): 353–372.

Das, A. and Ali, N. 2021. Nanovaccine: An emerging strategy. Expert Rev. Vaccines 20(10): 1273–1290.

de Carvalho Lima, E.N., Diaz, R.S., Justo, J.F. and Piqueira, J.R.C. 2021. Advances and perspectives in the use of carbon nanotubes in vaccine development. Int. J. Nanomedicine 16: 5411.

de Oliveira Viana, I.M., Roussel, S., Defrêne, J., Lima, E.M., Barabé, F. and Bertrand, N. 2021. Innate and adaptive immune responses toward nanomedicines. Acta Pharm. Sin. B 11(4): 852–870.

dong Zhang, Y., Li, M., Du, G.S., Chen, X.Y. and Sun, X. 2021. Advanced oral vaccine delivery strategies for improving the immunity. Adv. Drug Deliv. Rev. 113928.

Dorcheh, S.K. and Vahabi, K. 2016. Biosynthesis of nanoparticles by fungi: Large-scale production. Fungal Metabolites, 395–414.

Dowling, J.K. and Mansell, A. 2016. Toll-like receptors: The swiss army knife of immunity and vaccine development. Clin. Transl. Immunology 5(5): e85.

Drescher, S. and van Hoogevest, P. 2020. The phospholipid research center: Current research in phospholipids and their use in drug delivery. Pharmaceutics 12(12): 1235.

Durán-Lobato, M., López-Estévez, A.M., Cordeiro, A.S., Dacoba, T.G., Crecente-Campo, J., Torres, D. and Alonso, M.J. 2021. Nanotechnologies for the delivery of biologicals: Historical perspective and current landscape. Adv. Drug Deliv. Rev. 176: 113899.

Elieh-Ali-Komi, D. and Hamblin, M.R. 2016. Chitin and chitosan: Production and application of versatile biomedical nanomaterials. Int. J. Adv. Res. (Indore) 4(3): 411.

Excler, J.L., Saville, M., Berkley, S. and Kim, J.H. 2021. Vaccine development for emerging infectious diseases. Nat. Med. 27(4): 591–600.

Florindo, H.F., Kleiner, R., Vaskovich-Koubi, D., Acúrcio, R.C., Carreira, B., Yeini, E., Tiram, G., Liubomirski, Y. and Satchi-Fainaro, R. 2020. Immune-mediated approaches against COVID-19. Nat. Nanotechnol. 15(8): 630–645.

Francia, V., Yang, K., Deville, S., Reker-Smit, C., Nelissen, I. and Salvati, A. 2019. Corona composition can affect the mechanisms cells use to internalize nanoparticles. ACS Nano 13(10): 11107–11121.

Galen, J.E., Wahid, R. and Buskirk, A.D. 2021. Strategies for enhancement of live-attenuated *Salmonella*-based carrier vaccine immunogenicity. Vaccines 9(2): 162.

Garcia-Rubio, R., de Oliveira, H.C., Rivera, J. and Trevijano-Contador, N. 2020. The fungal cell wall: *Candida, Cryptococcus*, and *Aspergillus* species. Front. Microbiol. 10: 2993.

García-Carnero, L.C., Martínez-Álvarez, J.A., Salazar-García, L.M., Lozoya-Pérez, N.E., González-Hernández, S.E. and Tamez-Castrellón, A.K. 2020. Recognition of fungal components by the host immune system. Curr. Protein Pept. Sci. 21(3): 245–264.

Ghosh, S., Ahmad, R., Zeyaullah, M. and Khare, S.K. 2021. Microbial nano-factories: Synthesis and biomedical applications. Front. Chem. 9: 194.

Guerrini, L., Alvarez-Puebla, R.A. and Pazos-Perez, N. 2018. Surface modifications of nanoparticles for stability in biological fluids. Materials 11(7): 1154.

Guo, Z., Richardson, J.J., Kong, B. and Liang, K. 2020. Nanobiohybrids: Materials approaches for bioaugmentation. Sci. Adv. 6(12): eaaz0330.

Hameed, S., Hans, S., Singh, S., Dhiman, R., Monasky, R., Pandey, R.P., Thangamani, S. and Fatima, Z. 2021. Revisiting the vital drivers and mechanisms of β-glucan masking in human fungal pathogen, *Candida albicans*. Pathogens 10(8): 942.

Han, J., Zhao, D., Li, D., Wang, X., Jin, Z. and Zhao, K. 2018. Polymer-based nanomaterials and applications for vaccines and drugs. Polymers 10(1): 31.

Haun, B.K., Lai, C.Y., Williams, C.A., Wong, T.A.S., Lieberman, M.M., Pessaint, L., Andersen, H. and Lehrer, A.T. 2020. CoVaccine HT™ adjuvant potentiates robust immune responses to recombinant SARS-CoV-2 spike S1 immunization. Front. Immunol. 11: 2780.

HogenEsch, H., O'Hagan, D.T. and Fox, C.B. 2018. Optimizing the utilization of aluminum adjuvants in vaccines: You might just get what you want. NPJ Vaccines 3(1): 1–11.

Hou, X., Zaks, T., Langer, R. and Dong, Y. 2021. Lipid nanoparticles for mRNA delivery. Nat. Rev. Mater., 1–17.

Hu, Z., Ott, P.A. and Wu, C.J. 2018. Towards personalized, tumour-specific, therapeutic vaccines for cancer. Nat. Rev. Immunol. 18(3): 168–182.

Irvine, D.J., Aung, A. and Silva, M. 2020. Controlling timing and location in vaccines. Adv. Drug Deliv. Rev. 158: 91–115.

Iyer, S., Yadav, R., Agarwal, S., Tripathi, S. and Agarwal, R. 2021. Bioengineering strategies for developing vaccines against respiratory viral diseases. Clin. Microbiol. Rev. 35(1): e00123–21.

Jara, L.J., García-Collinot, G., Medina, G., del Pilar Cruz-Dominguez, M., Vera-Lastra, O., Carranza-Muleiro, R.A. and Saavedra, M.A. 2017. Severe manifestations of autoimmune syndrome induced by adjuvants (Shoenfeld's syndrome). Immunol. Res. 65(1): 8–16.

Javaid, N., Yasmeen, F. and Choi, S. 2019. Toll-like receptors and relevant emerging therapeutics with reference to delivery methods. Pharmaceutics 11(9): 441.

Jhaveri, J., Raichura, Z., Khan, T., Momin, M. and Omri, A. 2021. Chitosan nanoparticles-insight into properties, functionalization and applications in drug delivery and theranostics. Molecules 26(2): 272.

Kelly, S.H., Shores, L.S., Votaw, N.L. and Collier, J.H. 2017. Biomaterial strategies for generating therapeutic immune responses. Adv. Drug Deliv. Rev. 114: 3–18.

Khalaj-Hedayati, A., Chua, C.L.L., Smooker, P. and Lee, K.W. 2020. Nanoparticles in influenza subunit vaccine development: Immunogenicity enhancement. Influenza Other Respir. Viruses 14(1): 92–101.

Khan, M.A. and Khan, A. 2021. Role of NKT cells during viral infection and the development of NKT cell-based nanovaccines. Vaccines 9(9): 949.

Khan, S.U., Saleh, T.A., Wahab, A., Khan, M.H.U., Khan, D., Khan, W.U., Rahim, A., Kamal, S., Khan, F. and Fahad, S. 2018. Nanosilver: New ageless and versatile biomedical therapeutic scaffold. Int. J. Nanomedicine 13: 733.

Krishnan, S., Thirunavukarasu, A., Jha, N.K., Gahtori, R., Roy, A.S., Dholpuria, S., Kesari, K., Singh, S., Dua, K. and Gupta, P.K. 2021. Nanotechnology-based therapeutic formulations in the battle against animal coronaviruses: An update. J. Nanopart. Res. 23(10): 1–16.

Kus-Liśkiewicz, M., Fickers, P. and Ben Tahar, I. 2021. Biocompatibility and cytotoxicity of gold nanoparticles: Recent advances in methodologies and regulations. Int. J. Mol. Sci. 22(20): 10952.

Lang, S. and Huang, X. 2020. Carbohydrate conjugates in vaccine developments. Front. Chem. 8: 284.

Lee, K., Kwon, Y., Hwang, J., Choi, Y., Kim, K., Koo, H.J., Seo, Y., Jeon, H. and Choi, J. 2019. Synthesis and functionalization of β-glucan particles for the effective delivery of doxorubicin molecules. ACS Omega 4(1): 668–674.

Lenders, V., Koutsoumpou, X., Sargsian, A. and Manshian, B.B. 2020. Biomedical nanomaterials for immunological applications: Ongoing research and clinical trials. Nanoscale Adv., 2(11): 5046–5089.

Li, M., Kaminskas, L.M. and Marasini, N. 2021. Recent advances in nano/microparticle-based oral vaccines. J. Pharm. Investig. 1–14.

Li, Q., Shi, Z., Zhang, F., Zeng, W., Zhu, D. and Mei, L. 2022. Symphony of nanomaterials and immunotherapy based on the cancer–immunity cycle. Acta Pharm. Sin. B 12(1): 107–134.

Li, W.H. and Li, Y.M. 2020. Chemical strategies to boost Cancer vaccines. Chem. Rev. 120(20): 11420–11478.

Li, X. and Cheung, P.C.K. 2019. Application of natural β-glucans as biocompatible functional nanomaterials. Food Science and Human Wellness 8(4): 315–319.

Liu, J., Liew, S.S., Wang, J. and Pu, K. 2021. Bioinspired and biomimetic delivery platforms for cancer vaccines. Adv. Mater. 2103790.

Liu, Y., Hardie, J., Zhang, X. and Rotello, V.M. 2017, December. Effects of engineered nanoparticles on the innate immune system. In Seminars in Immunology (Vol. 34). Academic Press, pp. 25–32.

Lodaya, R.N., Gregory, S., Amiji, M.M. and O'Hagan, D.T. 2022. Overview of vaccine adjuvants. In Practical Aspects of Vaccine Development. Academic Press, pp. 9–25.

Lu, L., Duong, V.T., Shalash, A.O., Skwarczynski, M. and Toth, I. 2021. Chemical conjugation strategies for the development of protein-based subunit nanovaccines. Vaccines 9(6): 563.

Lu, L.L., Suscovich, T.J., Fortune, S.M. and Alter, G. 2018. Beyond binding: Antibody effector functions in infectious diseases. Nat. Rev. Immunol. 18(1): 46–61.

Madamsetty, V.S., Mukherjee, A. and Mukherjee, S. 2019. Recent trends of the bio-inspired nanoparticles in cancer theranostics. Front. Pharmacol. 10: 1264.

Magadán, S., Mikelez-Alonso, I., Borrego, F. and González-Fernández, Á. 2021. Nanoparticles and trained immunity: Glimpse into the future. Adv. Drug Deliv. Rev. 175: 113821.

Malachowski, T. and Hassel, A. 2020. Engineering nanoparticles to overcome immunological barriers for enhanced drug delivery. Bioeng. Transl. Med. 1: 35–50.

Mao, L., Chen, Z., Wang, Y. and Chen, C. 2021. Design and application of nanoparticles as vaccine adjuvants against human corona virus infection. J. Inorg. Biochem. 219: 111454.

Marques, A.C., Costa, P.J., Velho, S. and Amaral, M.H. 2020. Functionalizing nanoparticles with cancer-targeting antibodies: A comparison of strategies. J. Control Release 320: 180–200.

Matur, M., Madhyastha, H., Shruthi, T.S., Madhyastha, R., Srinivas, S.P., Navya, P.N. and Daima, H.K. 2020. Engineering bioactive surfaces on nanoparticles and their biological interactions. Scientific Reports 10(1): 1–14.

McMillan, C.L., Young, P.R., Watterson, D. and Chappell, K.J. 2021. The next generation of influenza vaccines: Towards a universal solution. Vaccines 9(1): 26.

Meena, M., Swapnil, P., Yadav, G. and Sonigra, P. 2021. Role of fungi in bio-production of nanomaterials at megascale. In *Fungi Bio-Prospects in Sustainable Agriculture, Environment and Nano-technology*. Academic Press, pp. 453–474.

Mitchell, M.J., Billingsley, M.M., Haley, R.M., Wechsler, M.E., Peppas, N.A. and Langer, R. 2021. Engineering precision nanoparticles for drug delivery. Nat. Rev. Drug Discov. 20(2): 101–124.

Mokhtari, Y., Pourbagheri-Sigaroodi, A., Zafari, P., Bagheri, N., Ghaffari, S.H. and Bashash, D. 2021. Toll-like receptors (TLRs): An old family of immune receptors with a new face in cancer pathogenesis. J. Cell Mol. Med. 25(2): 639–651.

Nag, M., Lahiri, D., Mukherjee, D., Banerjee, R., Garai, S., Sarkar, T., Ghosh, S., Dey, A., Ghosh, S., Pattnaik, S., Edinur, H., Kari, Z., Pati, S. and Ray, R.R. 2021. Functionalized chitosan nanomaterials: A jammer for quorum sensing. Polymers 13(15): 2533.

Namasivayam, S.K.R., Nivedh, S.K., Bharani, R.A., Rebecca, J. and Nishanth, A.N. 2021. Formulation optimization of chitosan nanoparticles incorporated rabies viral antigen and its influence on the release kinetics, immune potency and biosafety potential. Carbohydrate Polymer Technologies and Applications 100096.

Narasimhan, B., Goodman, J.T. and Vela Ramirez, J.E. 2016. Rational design of targeted next-generation carriers for drug and vaccine delivery. Annu. Rev. Biomed. Eng. 18: 25–49.

Niculescu, A.G. and Grumezescu, A.M. 2021. Polymer-based nanosystems—A versatile delivery approach. Materials 14(22): 6812.

O'Neill, C.L., Shrimali, P.C., Clapacs, Z.P., Files, M.A. and Rudra, J.S. 2021. Peptide-based supramolecular vaccine systems. Acta Biomater. S1742–7061.

Ong, G.H., Lian, B.S.X., Kawasaki, T. and Kawai, T. 2021. Exploration of pattern recognition receptor agonists as candidate adjuvants. Front. Cell Infect. Microbiol. 968.

O'Hagan, D.T., Lodaya, R.N. and Lofano, G. 2020, November. The continued advance of vaccine adjuvants–'we can work it out'. In Seminars in Immunology. Academic Press, p. 101426.

Pan, C., Yue, H., Zhu, L., Ma, G.H. and Wang, H.L. 2021. Prophylactic vaccine delivery systems against epidemic infectious diseases. Adv. Drug Deliv. Rev. 113867.

Pan, J. and Cui, Z. 2020. Self-assembled nanoparticles: Exciting platforms for vaccination. Biotechnol. J. 15(12): 2000087.

Pang, N.Y.L., Pang, A.S.R., Chow, V.T. and Wang, D.Y. 2021. Understanding neutralising antibodies against SARS-CoV-2 and their implications in clinical practice. Mil. Med. Res. 8(1): 1–17.

Pati, R., Shevtsov, M. and Sonawane, A. 2018. Nanoparticle vaccines against infectious diseases. Front. Immunol. 9: 2224.

Patin, E.C., Thompson, A. and Orr, S.J. 2019, May. Pattern recognition receptors in fungal immunity. In Seminars in Cell & Developmental Biology (Vol. 89). Academic Press, pp. 24–33.

Pereira, B., Xu, X.N. and Akbar, A.N. 2020. Targeting inflammation and immunosenescence to improve vaccine responses in the elderly. Front. Immunol. 11: 2670.

Pereira-Dutra, F.S., Teixeira, L., de Souza Costa, M.F. and Bozza, P.T. 2019. Fat, fight, and beyond: The multiple roles of lipid droplets in infections and inflammation. J. Leukoc. Biol. 106(3): 563–580.

Pogostin, B.H. and McHugh, K.J. 2021. Novel vaccine adjuvants as key tools for improving pandemic preparedness. Bioengineering 8(11): 155.

Pollard, A.J. and Bijker, E.M. 2021. A guide to vaccinology: From basic principles to new developments. Nat. Rev. Immunol. 21(2): 83–100.

Pulendran, B., Arunachalam, P.S. and O'Hagan, D.T. 2021. Emerging concepts in the science of vaccine adjuvants. Nat. Rev. Drug Discov. 20(6): 454–475.

Rauch, S., Jasny, E., Schmidt, K.E. and Petsch, B. 2018. New vaccine technologies to combat outbreak situations. Front. Immunol. 9: 1963.

Raveendran, S., Rochani, A.K., Maekawa, T. and Kumar, D.S. 2017. Smart carriers and nanohealers: A nanomedical insight on natural polymers. Materials 10(8): 929.

Roe, K. 2021. An explanation of the pathogenesis of several autoimmune diseases in immuno-compromised individuals. Scand. J. Immunol. 93(3): e12994.

Rossella, S., Trovato, M. and Manco, R. 2021. Exploiting viral sensing mediated by Toll-like receptors to design innovative vaccines. NPJ Vaccines 6(1).

Roth, G.A., Picece, V.C., Ou, B.S., Luo, W., Pulendran, B. and Appel, E.A. 2021. Designing spatial and temporal control of vaccine responses. Nat. Rev. Mater. 1–22.

Roßmann, L., Bagola, K., Stephen, T., Gerards, A.L., Walber, B., Ullrich, A., Schülke, S., Kamp, C., Spreitzer, I., Hasan, M., David-Watine, B., Shorte, S., Bastian, M. and van Zandbergen, G. 2021. Distinct single-component adjuvants steer human DC-mediated T-cell polarization via Toll-like receptor signaling toward a potent antiviral immune response. PNAS 118(39).

Sabu, C., Mufeedha, P. and Pramod, K. 2019. Yeast-inspired drug delivery: Biotechnology meets bioengineering and synthetic biology. Expert. Opin. Drug Deliv. 16(1): 27–41.

Sameer, A.S. and Nissar, S. 2021. Toll-Like Receptors (TLRs): Structure, functions, signaling, and role of their polymorphisms in colorectal cancer susceptibility. Biomed. Res. Int. 2021.

Sanità, G., Carrese, B. and Lamberti, A. 2020. Nanoparticle surface functionalization: How to improve biocompatibility and cellular internalization. Front. Mol. Biosci. 7: 381.

Schechter, M.C., Satola, S.W. and Stephens, D.S. 2019. Host defenses to extracellular bacteria. In Clinical Immunology. Elsevier, pp. 391–402.

Schijns, V., Fernández-Tejada, A., Barjaktarović, Ž., Bouzalas, I., Brimnes, J., Chernysh, S., Gizurarson, S., Gursel, I., Jakopin, Ž., Lawrenz, M., Nativi, C., Paul, S., Pedersen, G., Rosano, C., Ruiz-de-Angulo, A., Slütter, B., Thakur, A., Christensen, D. and Lavelle, E.C. 2020. Modulation of immune responses using adjuvants to facilitate therapeutic vaccination. Immunol. Rev. 296(1): 169–190.

Seyfoori, A., Shokrollahi Barough, M., Mokarram, P., Ahmadi, M., Mehrbod, P., Sheidary, A., Madrakian, T., Kiumarsi, M., Walsh, T., McAlinden, K., Ghosh, C., Sharma, P., Zeki, A., Ghawami, S. and Akbari, M. 2021. Emerging advances of nanotechnology in drug and vaccine delivery against viral associated respiratory infectious diseases (VARID). Int. J. Mol. Sci. 22(13): 6937.

Shams, F., Golchin, A., Azari, A., Mohammadi Amirabad, L., Zarein, F., Khosravi, A. and Ardeshirylajimi, A. 2021. Nanotechnology-based products for cancer immunotherapy. Mol. Biol. Rep. 1–24.

Sharma, P., Jang, N.Y., Lee, J.W., Park, B.C., Kim, Y.K. and Cho, N.H. 2019. Application of ZnO-based nanocomposites for vaccines and cancer immunotherapy. Pharmaceutics 11(10): 493.

Shreyash, N., Bajpai, S., Khan, M.A., Vijay, Y., Tiwary, S.K. and Sonker, M. 2021. Green synthesis of nanoparticles and their biomedical applications: a review. ACS Applied Nano Materials 4(11): 11428–11457.

Slifka, M.K. and Amanna, I.J. 2019. Role of multivalency and antigenic threshold in generating protective antibody responses. Front. Immunol. 10: 956.

Sundar, S. and Kumar Prajapati, V. 2012. Drug targeting to infectious diseases by nanoparticles surface functionalized with special biomolecules. Curr. Med. Chem. 19(19): 3196–3202.

Sushnitha, M., Evangelopoulos, M., Tasciotti, E. and Taraballi, F. 2020. Cell membrane-based biomimetic nanoparticles and the immune system: Immunomodulatory interactions to therapeutic applications. Front. Bioeng. Biotechnol. 8: 627.

Tabasum, S., Noreen, A., Kanwal, A., Zuber, M., Anjum, M.N. and Zia, K.M. 2017. Glycoproteins functionalized natural and synthetic polymers for prospective biomedical applications: A review. Int. J. Biol. Macromol. 98: 748–776.

Tan, K., Li, R., Huang, X. and Liu, Q. 2018. Outer membrane vesicles: Current status and future direction of these novel vaccine adjuvants. Front. Microbiol. 9: 783.

Taneja, P., Sharma, S., Sinha, V.B. and Yadav, A.K. 2021. Advancement of nanoscience in development of conjugated drugs for enhanced disease prevention. Life Sci. 268: 118859.

Tang, Z., Xiao, Y., Kong, N., Liu, C., Chen, W., Huang, X., Xu, D., Ouyang, J., Feng, C., Wang, C., Wang, J., Zhang, H. and Tao, W. 2021. Nano-bio interfaces effect of two-dimensional nanomaterials and their applications in cancer immunotherapy. Acta Pharm. Sin. B 11(11): 3447.

Tenchov, R., Bird, R., Curtze, A.E. and Zhou, Q. 2021. Lipid nanoparticles—from liposomes to mRNA vaccine delivery, a landscape of research diversity and advancement. ACS Nano. 15(11): 16982–17015.

Thorp, E.B., Boada, C., Jarbath, C. and Luo, X. 2020. Nanoparticle platforms for antigen-specific immune tolerance. Front. Immunol. 11: 945.

Tiwari, N., Osorio-Blanco, E.R., Sonzogni, A., Esporrín-Ubieto, D., Wang, H. and Calderon, M. 2021. Nanocarriers for skin applications: Where do we stand? Angew. Chem. Int. Ed. Engl. 61(3): e202107960.

Tornesello, A.L., Tagliamonte, M., Tornesello, M.L., Buonaguro, F.M. and Buonaguro, L. 2020. Nanoparticles to improve the efficacy of peptide-based cancer vaccines. Cancers 12(4): 1049.

Trabbic, K.R., Kleski, K.A. and Barchi, J.J. 2021. A stable nano-vaccine for the targeted delivery of tumor-associated glycopeptide antigens. bioRxiv.

Trincado, V., Gala, R.P. and Morales, J.O. 2021. Buccal and sublingual vaccines: A review on oral mucosal immunization and delivery systems. Vaccines 9(10): 1177.

Tsai, S.J., Black, S.K. and Jewell, C.M. 2020. Leveraging the modularity of biomaterial carriers to tune immune responses. Adv. Funct. Mater. 30(48): 2004119.

Tsakiri, M., Naziris, N. and Demetzos, C. 2021. Innovative vaccine platforms against infectious diseases: Under the scope of the COVID-19 pandemic. Int. J. Pharm. 610: 121212.

Vang, K.B., Safina, I., Darrigues, E., Nedosekin, D., Nima, Z.A., Majeed, W., Watanabe, F., Kannapardy, G., Kore, R., Casciano, D., Zharov, V., Griffin, R., Dings, R. and Biris, A.S. 2017. Modifying dendritic cell activation with plasmonic nano vectors. Scientific Reports 7(1): 1–11.

Vijayan, V., Mohapatra, A., Uthaman, S. and Park, I.K. 2019. Recent advances in nanovaccines using biomimetic immunomodulatory materials. Pharmaceutics 11(10): 534.

Vu, M.N., Kelly, H.G., Kent, S.J. and Wheatley, A.K. 2021. Current and future nanoparticle vaccines for COVID-19. EBioMedicine 74: 103699.

Wagner, A. and Weinberger, B. 2020. Vaccines to prevent infectious diseases in the older population: Immunological challenges and future perspectives. Front. Immunol. 11: 717.

Wang, W., Xue, C. and Mao, X. 2020. Chitosan: Structural modification, biological activity and application. Int. J. Biol. Macromol. 164: 4532–4546.

Weiss, C., Carriere, M., Fusco, L., Capua, I., Regla-Nava, J.A., Pasquali, M., Scott, J., Vitale, F., Unal, M., Mattevi, C., Bedognetti, D., Merkoçi, A., Tasciotti, E., Yilmazer, A., Gogotsi, Y., Stellacci, F. and Delogu, L.G. 2020. Toward nanotechnology-enabled approaches against the COVID-19 pandemic. ACS Nano. 14(6): 6383–6406.

Wen, R., Umeano, A.C., Kou, Y., Xu, J. and Farooqi, A.A. 2019. Nanoparticle systems for cancer vaccine. Nanomedicine 14(5): 627–648.

Xu, C., Xing, R., Liu, S., Qin, Y., Li, K., Yu, H. and Li, P. 2021. Loading effect of chitosan derivative nanoparticles on different antigens and their immunomodulatory activity on dendritic cells. Marine Drugs 19(10): 536.

Xu, R., Dong, Y., Zhang, Y., Wang, X., Zhang, C. and Jiang, Y. 2022. Programmed nanoparticle-loaded microparticles for effective antigen/adjuvant delivery. Particuology 60: 77–89.

Yadav, A.N., Singh, S., Mishra, S. and Gupta, A. 2019. Recent Advancement in White Biotechnology through Fungi. Springer International Publishing.

Yayehrad, A.T., Siraj, E.A., Wondie, G.B., Alemie, A.A., Derseh, M.T. and Ambaye, A.S. 2021. Could nanotechnology help to end the fight against COVID-19? Review of current findings, challenges and future perspectives. Int. J. Nanomedicine 16: 5713.

Yenkoidiok-Douti, L. and Jewell, C.M. 2020. Integrating biomaterials and immunology to improve vaccines against infectious diseases. ACS Biomater. Sci. Eng. 6(2): 759–778.

Zhang, X., Zhang, Z., Xia, N. and Zhao, Q. 2021. Carbohydrate-containing nanoparticles as vaccine adjuvants. Expert. Rev. Vaccines (just-accepted).

Zhu, X., Radovic-Moreno, A.F., Wu, J., Langer, R. and Shi, J. 2014. Nanomedicine in the management of microbial infection–overview and perspectives. Nano Today 9(4): 478–498.

Zolnik, B.S., González-Fernández, Á., Sadrieh, N. and Dobrovolskaia, M.A. 2010. Minireview: Nanoparticles and the immune system. Endocrinology 151(2): 458–465.

Złotko, K., Wiater, A., Waśko, A., Pleszczyńska, M., Paduch, R., Jaroszuk-Ściseł, J. and Bieganowski, A. 2019. A report on fungal $(1{\rightarrow}3)$-α-d-glucans: Properties, functions and application. Molecules 24(21): 3972.

6

Antioxidant Activity of Mycosynthesized Nanoparticles

Jashanpreet Kaur,[1] *Diksha Dhiman*[1] *and Anu Kalia*[2,*]

Introduction

Among the many facets of modern science, nanotechnology offers novel applications in chemical and material industries, medical and engineering disciplines (Albrecht et al. 2006, Galdiero et al. 2011). It involves control over atoms and molecules to produce novel materials called nano-materials possessing special and useful functions. Nowadays, metal nanoparticles exhibiting extraordinary catalytic and biological properties are the most promising scientific tools and have proved to be excellent heterogeneous catalysts. The diverse applications of these nanomaterials developed through nanotechnological interventions have profound effects on society, economy, and the common good (Hashem et al. 2021). Significant progress has been documented for the development of processes and protocols for nanomaterial synthesis, characterization and functionalization (Khan et al. 2015, Ali et al. 2016, Ali et al. 2018). Though the physical and chemical synthesis protocols are useful, the use of hazardous and environment-corrosive chemicals during the synthesis process limit the application of nanoparticles in most fields (Gupta et al. 2021). Therefore, a major impetus has been observed for the biological synthesis of nanomaterials.

The green synthesis protocol generally involves a reduction process in which the microbes such as bacteria and fungi emerge as electrochemically active entities and prove to be effective in the metal reduction process under wide extensions of natural conditions (Kitching et al. 2015). Moreover, these microorganisms and plants offer a

[1] Department of Microbiology, College of Basic Sciences and Humanities, Punjab Agricultural University, Punjab, 141004, India.
[2] Electron Microscopy and Nanoscience Laboratory, Department of Soil Science, College of Agriculture, Punjab Agricultural University, Punjab, 141004, India.
* Corresponding author: kaliaanu@pau.edu

faster rate of reduction and significantly higher biocompatibility for the synthesis of various nanoparticles. Thus, biological approaches of nanoparticles synthesis entice higher interest in the scientific network (BalaKumaran et al. 2020).

The use of nanomaterials is expanding over domains such as agriculture, food and medicine. It has been realized that biocompatible nanoparticles must be generated which will bear low to negligible ecological footprints and associated health hazards (Reddy et al. 2016, Choudhary et al. 2018). However, there are several fundamental challenges that have to be overcome to generate complex nanostructures through green synthesis techniques such as the need for an even distribution of size, shape and composition, reproducibility, scalability, and amenability. A variety of facile methods have been developed for rapid synthesis, allowing for control over particle size, shape, composition, and other specific properties of nanoparticles. The cost of the nanomaterials synthesized through sophisticated techniques are fairly cost-intensive (Pandey and Jain 2020, Baig et al. 2021). Biological nanoparticle synthesis has emerged as a viable alternative to physical or chemical methods, over the past two decades (Attarad et al. 2016).

Green Synthesis of Nanomaterials

Biosynthesis of nanoparticles has been successfully performed by a wide variety of plants and uni- to multi-cellular organisms such as bacteria (Tiquia-Arashiro and Rodrigues 2016), algae (Shah et al. 2015a), endophytic fungi (Govindappa et al. 2016a, Janakiraman et al. 2019, Mousa et al. 2021) and plants (Bhardwaj et al. 2020a, Kumar et al. 2020, Kalia et al. 2020). Microorganisms and plants are biological entities which exhibit synthesis of a variety of biocompatible metallic nanoparticles such as silver, gold, copper, platinum, zinc, iron and others that can lead to synthesis in an environment-friendly manner with minimal production costs (Husen and Siddiqi 2014, Singh et al. 2016, Das et al. 2018, Jaffri and Ahmad 2019, Kaur et al. 2020, Kumar et al. 2020, Kalia et al. 2020). Apart from these benefits, involvement of biocatalytic molecules or enzymes, other proteins, and secondary metabolites involved in reduction, stabilisation, and capping of biogenic nanoparticles provides a greater resistance to aggregation compared to the chemically synthesized nanoparticles (Kalia and Kaur 2018). Due to their increased production capacity, the unit cost of production can be reduced, consequently resulting in a lower price of production of the nanomaterials (Pandey and Jain 2020). However, to ascertain the eco-safety aspects, it is also necessary that toxicological research and criteria for assessment of the toxicity of nanoparticles must be delineated for greater clarity (Attarad et al. 2016).

Nanoparticle Synthesis through Microbial Cells

The two most fundamental approaches for synthesizing nanomaterials are top-down and bottom-up methods. The top-down approach for the most part works with the material in its bulk form, and the size reduces to the nanoscale and is finished by particular ablations like lithography, thermal deterioration, laser evacuation, mechanical preparing, and sputtering, etc. (Abou El-Nour et al. 2010). As an alternative, the bottom-up approach may be the most effective way of synthesizing the material from atomic or nuclear species through chemical responses, allowing for antecedent particles to grow in size. In homogeneous systems, it is more appropriate for arranging nanoparticles in which catalysts (e.g., diminishing operators, and chemicals) combine nanostructures that are compelled by catalyst properties, reaction media, and conditions (e.g., solvents, stabilizers, and temperature). For instance, the most common synthetic pathway for metal nanoparticle synthesis is chemical reduction method (Lal et al. 2011). Microorganisms are now emerging as eco-friendly nanofactories due to their cost-effectiveness and the lack of harmful chemicals and high energy demands for physiochemical union (Kashyap et al. 2013, Shah et al. 2015). Further, these microbes have the ability to reduce metal salts to metal nanoparticles with a limited size dispersion and, accordingly, less polydispersity (Kashyap et al. 2013, Singh et al. 2016).

Mycosynthesis of Nanoparticles: The Fungal Nano-Factories

Myconanosynthesis is one of the approaches for achieving rapid, one-pot and cost-effective synthesis of nanoparticles (Adebayo et al. 2021). The generated nanoparticles or bionanoparticles thus obtained are generally considered less toxic than those made of pristine metallic nanoparticles with or without chemical functionalization (Singh et al. 2019). As a result, researchers are keenly interested in mycosynthesis techniques for production of nanoparticles (Jutz and Böker 2011). Several fungal species have been shown to be capable of synthesizing a variety of nanoparticles, including Ag, Au, TiO_2, and Pt. There are more than 30 different fungi that have been showcased to be capable of acting as myco-nanofactories for efficient and easy synthesis of nanoparticles (Castro-Longoria et al. 2011, Syed and Ahmad 2013). Since fungi are easy to handle and have a high tolerance to metals, these are often considered attractive agents in biogenic synthesis of metallic nanoparticles (Siddiqi and Husen 2016). Also, on cultivation, fungus leads to production of appreciable amounts of biomass which also possess higher resistance to pressure and agitation (Adebayo et al. 2021). No additional steps are required to extract the culture filtrate as opposed to bacteria (Siddiqi and Husen 2016).

Also, fungi secrete high amounts of enzymes and other extracellular proteins that offer more stability to the 'bare' nanoparticles (Guilger-Casagrande and Lima 2019). Several fungal proteins, such as nitrate reductases, and NADPH-dependent reductases for example in *Penicillium* species and *Fusarium oxysporum* have a crucial role for nanoparticle synthesis (Kumar et al. 2007, Kitching et al. 2015). The choice of the fungal species for the mycosynthesis of nanoparticles may just require fungi possessing a high bioaccumulation capacity, must contain certain secondary and other metabolites and should be amenable to basic downstream preparation (Kashyap et al. 2013, Alghuthaymi et al. 2015, Kitching et al. 2015). Additionally, fungi are more likely than bacteria to acquire metals, especially since metal salts exhibit a higher ability to bind to cell walls of the fungal hyphae resulting in high-yield of nanoparticles (Singh et al. 2016). Also, there is a requirement for optimization of only few critical factors such as pH and temperature of the reaction, amount of fungal biomass, metal ion concentration, and culture medium that can produce nanoparticles with desired characteristics (Khandel and Shahi 2018a). It is thus highly feasible to produce inorganic nanoparticles at large scales through mycogenic synthesis technique. Further, the biosynthetic pathway for the generation of mycogenic nanoparticles with better-defined nano-sale size and shape characters is gaining popularity due to the single-step, green, cost-effective, safe, and clean nature of the synthesis protocol (Abd-Elsalam 2021, Bahrulolum et al. 2021). This protocol is thus, highly feasible for the production of inorganic nanoparticles on a large scale (Saxena et al. 2016, BalaKumaran et al. 2020).

Among the various types of fungal genera evaluated for the synthesis of nanoparticles, *Fusarium* species has emerged as a novel microorganism for the mycosynthesis of various types of nanoparticles with better size and monodispersity. The genus *Fusarium* has been the prime choice of many researchers because of its various expedient properties (Rai et al. 2021). *Fusarium* can be grown, identified, and maintained easily and exhibits extracellular production of enzymes which leads to feasible downstream processing and easy collection of synthestized nanoparticles. Moreover, it can be used for the rapid and sustainable production of various nanoparticles. *Fusarium* species have the ability to produce nanoparticles both extracellularly and intracellularly. Nanoparticles synthesized using *Fusarium* strains act as robust antimicrobial agents. A high-temperature tolerant *F. oxysporum* exhibited production of a higher amount of protein at a temperature of 60–80°C. On utilization of this fungal culture for NP biosynthesis, it was observed that such high-temperature induced enhanced protein production led to progressively increased the rate of nanoparticle synthesis. Zirconia, titanium and cadmium sulfide nanoparticles have also been synthesized by *Fusarium oxysporum*. These metal sulfide nanoparticles such as CdS, ZnS, PbS, and MoS_2 can be formed extracellularly when *Fusarium oxysporum* was exposed to respective sulfates in aqueous solutions (Rai et al. 2021).

Types of Mycosynthesized Nanoparticles

Various types of nanoparticles can be synthesized using biological approaches following the categories of metal, metal oxide, and other miscellaneous nanoparticles.

Metal Nanoparticles: Metallic nanoparticles include gold, silver, and other alloys. Several fungi such as *Verticillium*, *Fusarium oxysporum* and *Aspergillus flavus* are capable of producing silver nanoparticles (AgNPs) in solution, as a thin film, or as a bulk material that accumulates on the surface of cells. Many studies have been published on biogenic synthesis of AgNPs (Guilger et al. 2017, Al Abboud 2018, Danagoudar et al. 2020), but the specific mechanisms involved are still to be identified. Certain research publications have reported the role of biomolecules of fungal origin. These biomolecules are involved in complex electron transfer pathways among others, such as those responsible for the conversion of NADPH/NADH into $NADP^+/NAD^+$, can react with metal ions and play a significant role in synthesis of nanoparticles (Thakkar et al. 2010, Gudikandula et al. 2017). The biogenic production of metallic nanoparticles is anticipated to have significant effects on NADH-dependent nicotinamide adenine dinucleotide (NADH) and nitrate reductase enzymes (Zomorodian et al. 2016, Baymiller et al. 2017).

The mycogenic synthesis of AgNPs get initiated when the silver ions are reduced by certain extracellularly secreted enzymes in fungal filtrate, resulting in the production of nanoparticles of elemental silver (Ag^0) as part of the extracellular synthesis of nanoparticles. The color of the filtrate changes as the reaction completes, and UV-visible spectroscopy can be used to investigate the alteration of optical properties of the material using surface plasmon resonance bands (Ahmad et al. 2003, Kumar et al. 2007). These bands vary in wavelength between 400 to 450 nm. The occurrence of peaks at longer wavelengths indicates the formation of larger nanoparticles (Elamawi et al. 2018). Apart from the fungus species, temperature, pH, and the dispersion medium, as well as the structure of the biomolecules capping the nanoparticles can affect the morphology of the generated nanoparticles. Therefore, the synthesis conditions influence the size of the nanoparticles (Khandel and Shahi 2018b, Lee and Jun 2019). Furthermore, the optical colour displayed by the nano-dispersion can be linked to the surface plasmon resonance, which is affected by the nanoparticle size (Adeeyo and Odiyo 2018, Lee and Jun 2019, Bhangale et al. 2019).

Nanoparticles of gold (AuNPs) are a promising material because of their low toxicity, compatibility with the body, high surface area to volume ratios, and the ability to apply ligands easily. The AuNPs can be produced by use of fungal extracts as a single-step, one pot easy synthesis, low cost, and in an environment-friendly manner. Gold nanoparticles (AuNPs) have been synthesized extracellularly by use of *Fusarium oxysporum* culture filtrates and intracellularly in case of fungus *Verticillium* sp. Similarly, *Aspergillus terreus* hyphal extracts on incubation in the dark at room temperature for 48 hours with 1 mM gold chloride led to synthesis of AuNPs. Bimetallic Au-Ag alloys can also be synthesized from *Fusarium oxysporum*. *Fusarium semitectum* has been used to produce core-shell Au-Ag alloy nanoparticles, which proved to be relatively stable for a number of weeks (Li et al. 2011).

Metal Oxide Nanoparticles: A number of oxide nanoparticles can be synthesized through mycosynthesis, such as TiO_2, Sb_2O_3, SiO_2, $BaTiO_3$, and ZrO_2. *Fusarium oxysporum* has been used to synthesize nanoparticles of SiO_2 and TiO_2, respectively, from aqueous anionic complexes of SiF_6^{2+} and TiF_6^{2+}. Furthermore, tetragonal $BaTiO_3$ and quasi-spherical ZrO_2 nanoparticles were obtained from *F. oxysporum* hyphal extracts with sizes ranging from 4–5 nm to 3–11 nm (Li et al. 2011).

Other Nanoparticles: Biological systems utilize biopolymers such as protein and microbial cells as well as organic/inorganic compounds to form composites of ordered structures. Microbes have been reported to synthesize nanoparticles such as $PbCO_3$, $CdCO_3$, $SrCO_3$, PHB, $Zn_3(PO_4)_2$, and CdSe, in addition to those listed above. The aqueous Sr^{2+} ions were used to grow $SrCO_3$ crystals on challenging fungi (Rautaray et al. 2004). As *Fusarium oxysporum* grows, proteins modulate strontianite crystal morphology and direct its hierarchical assembly into higher-order structures. Biotemplates derived from fungal cells were used to synthesize zinc phosphate nanopowders (Ram Kumar Pandian et al. 2009). Complex structured nanoparticles such as $Zn_3(PO_4)_2$ powders exhibiting butterfly-like microstructures have also been reported to be synthesized (Yan et al. 2009). Their dimensions ranged from 10 to 80 nm in width and from 80 to 200 nm in length. Researchers demonstrated that a highly luminescent CdSe quantum dot can be synthesized by *Fusarium oxysporum* at room temperature.

Mechanism of Mycosynthesized Nanoparticles

A fungus may synthesize nanoparticles either intracellularly or extracellularly. The intracellular NP synthesis involves administering a metal precursor to the mycelial culture, which eventually gets internalized in the generated fungal biomass. In order to extract the nanoparticles from the biomass, chemical treatment, centrifugation, and filtration methods must be performed (Castro-Longoria et al. 2011, Rajput et al. 2016, Molnár et al. 2018). Extracellular synthesis occurs by addition of a metal precursor to an aqueous filtrate containing only fungal biomolecules, leading to the formation of free nanoparticles in the dispersion (Figure 1). This method is advantageous as it is not necessary to perform any procedures to release the nanoparticles from the cells (Sabri et al. 2016, Azmath et al. 2016, Gudikandula et al. 2017, Costa Silva et al. 2017). Nanoparticle dispersions must be purified to remove fungal residues and impurities, as well as to make the nanoparticles virtually nontoxic. Some methods to accomplish this aspect include simple filtration, membrane filtration, gel filtration, dialysis, and ultracentrifugation (Ashrafi et al. 2013, Qidwai et al. 2018, Yahyaei and Pourali 2019).

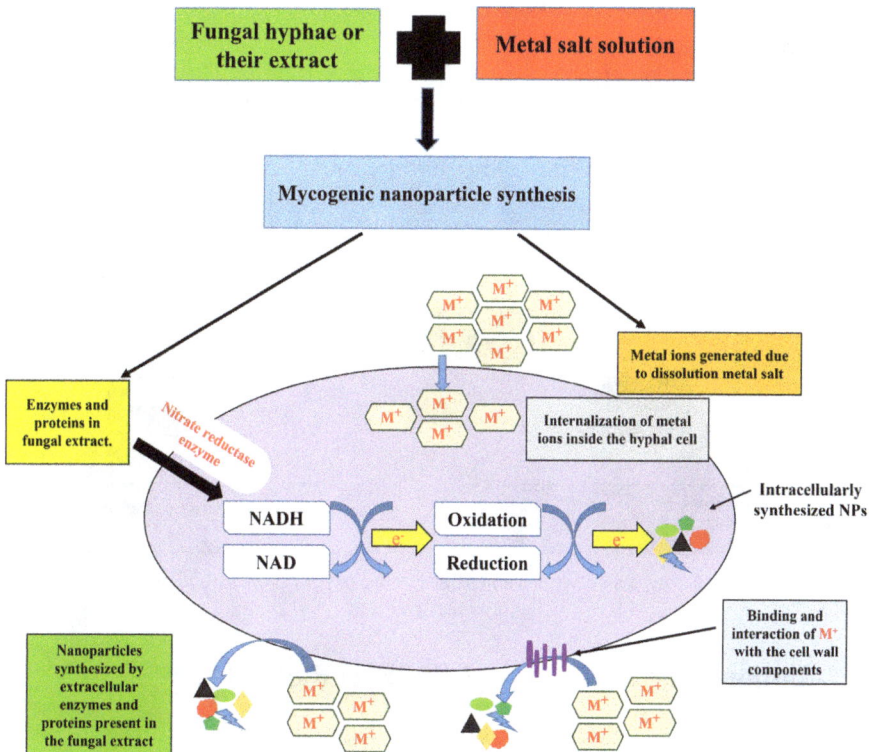

Figure 1. Mechanism of synthesis of metal/metal oxide nanoparticles due to interaction of metal ions with a variety of fungal cell macro and biomolecules.

Myconanoparticles and their Antioxidant Properties

Free radicals cause damage to cells by stealing electrons from damaged cells. Antioxidants prevent this damage and stabilize it as well. In addition to converting free radicals into waste byproducts, antioxidants also help to remove waste from the body. Natural antioxidant research has recently grown more active in a variety of sectors. Plants and food ingredients have been tested for antioxidant activity through several methods. Selecting an appropriate assay based on the chemical(s) of interest is critically important to investigate the antioxidant activity of a chemical(s). The antioxidant assays can be performed as direct and indirect assays. There are several methods which are used for assessing one of most common aspect of oxidative stress, i.e., lipid peroxidation. It can be estimated through thiobarbituric acid assay (TBA), malonaldehyde/high-performance liquid chromatography (MA/HPLC) assay, malonaldehyde/gas chromatography (MA/GC) assay, and carotene bleaching assay. Other assays consist of 2,2-diphenyl-1-picrylhydrazyl (DPPH) assay, 2,2-bimethylbenzothiazoline-6-sulfonic acid (2,2,2'-ABTS) assay, ferric reducing antioxidant power (FRAP) assay, ferrous oxidation-xylenol orange (FOX) assay, ferric thiocyanate (FTC) assay, and aldehyde/carboxylic acid (ACA) assay (Moon and Shibamoto 2009). The general methods that can be utilized to evaluate the antioxidant potential of the mycogenic nanoparticles have been presented in Figure 2.

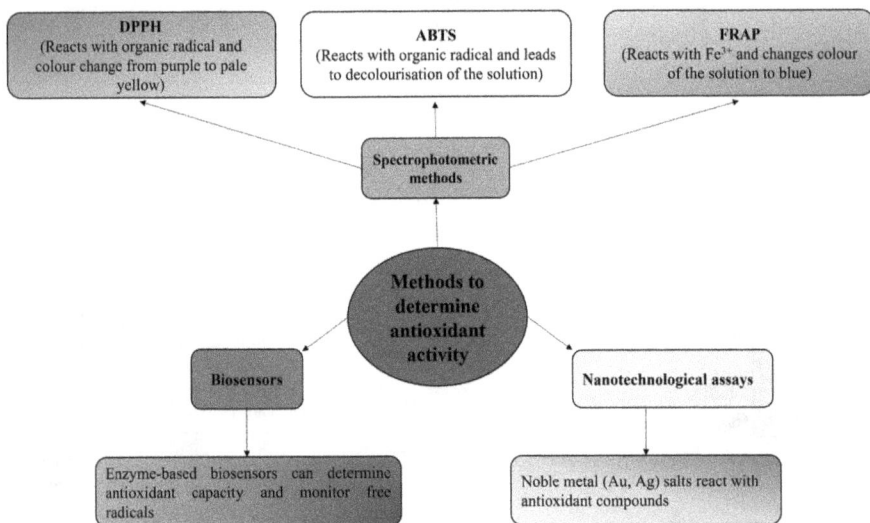

Figure 2. Antioxidant assays for determination of the reactive oxygen species (ROS) mitigating potential of myconanoparticles.

2, 2-diphenyl-1-picrylhydrazyl (DPPH) Radical Scavenging Assay

DPPH assay measures antioxidants' capacity to scavenge free radicals. A hydrogen atom from antioxidants reduces an electron of nitrogen in DPPH to form its corresponding hydrazine. To evaluate antioxidant activity, and to assess its ability to scavenge free radicals or donate hydrogen, this method is fast, simple, inexpensive, and widely used. Furthermore, solid or liquid samples of complex biological systems can be quantified with this technique. As a result, the total antioxidant effectiveness of a system can be determined. Since the radical compound is stable and need not be generated, the DPPH assay is considered a valid, accurate, easy, and economical method to test antioxidants' capacity to scavenge radicals (Kedare and Singh 2011). In addition to its antioxidative properties, certain biomolecules such as phenolic compounds can act as reducing agents and hydrogen donors, quenching singlet and triplet oxygen, or decomposing peroxides to neutralize free radicals (Huang et al. 2007).

A hydrogen atom donor and electron transfer are responsible for the reaction of DPPH with antioxidants. Thus, the substitution of 2,2-diphenyl-1-picrylhydrazyl radical (2,2-diphenyl-1-picrylhydrazines) for non-free radical DPPHHS (2,2-diphenyl-1-picrylhydrazine) results in a change from purple to yellow in color. When nanoparticles are evaluated for their antioxidant properties, for example, AgNPs exhibit a hydrogen transfer to free radicals, stable products are formed. The DPPH radicals are scavenged by conjugating bio compounds to silver ions. Thus, synthesized AgNPs could serve as antioxidative agents (Patra and Baek 2016).

The DPPH radical scavenging assay has been used to evaluate the antioxidative properties of a variety of mycosynthesized nanoparticles. The protocol involves preparation of methanolic solution of 0.1 mM DPPH which is added to sample and incubated for 30 minutes at 37°C in the dark. To determine the reduction of the DPPH radical, the material containing the same concentration of DPPH is compared with a blank containing material without DPPH and the change in absorbance at 517 nm is continuously observed. The standards used for comparing the DPPH radical scavenging activity include L-ascorbic acid, gallic acid (Drăgan et al. 2016).

The percentage of antioxidant activity can be evaluated by using the following standard formula:

$$\% \ inhibition = \frac{control \ abs. - sample \ abs.}{control \ abs.} \times 100$$

The most unique characteristic of the fungal biomolecules is that these not only function as reducing agents but can also work as biotemplating agents particularly the catalytic and structural proteins. *Aspergillus fumigatus* extracts were found to act as effective DPPH radical neutralizers, either by transferring electrons or hydrogen atoms to yellow-colored diamagnetic molecules, which is consistent with phenolic compounds in the extract exhibiting hydrogen donating properties. The positive correlation between TPCs and DPPHs ($r = 0.817$) also support this hypothesis (Arora and Chandra 2011).

ABTS Radical Cation Decolorization Activity

Antioxidants can be evaluated on their ability to scavenge ABTS and create the ABTS radical (ABTS$^{\cdot+}$) in the ABTS assay. These compounds donate hydrogen as a result of their antioxidant properties. As a result of the reaction between ABTS and a strong oxidizing agent (potassium persulphate), ABTS$^{\cdot+}$ exhibits a blue or green color (Prior et al. 2005).

General procedure for ABTS includes dissolving ABTS (7 mM) in distill water. The solution can then be mixed with potassium persulphate (140 mM) to produce the ABTS cation radical solution. This solution can be further stored at room temperature in the dark for 16 hours before use. The working concentration of the ABTS solution can be prepared by mixing it with methanol before experimentation followed by incubation with the test sample at room temperature for 10 minutes (dark condition). After incubation, measurement of the absorbance at 734 nm against a blank can be utilized to determine the percentage of ABTS scavenging with Trolox serving as a standard (Yang et al. 2011).

$$\text{ABTS} + \text{scavenging effect (\%)} = \frac{AB - AA}{AB} \times 100$$

where, AB = absorbance of ABTS radical + methanol
AA = absorbance of ABTS radical + sample extract/standard

Ferric Reducing Antioxidant Power or FRAP Assay

According to FRAP assay, ferric 2,4,6-tripyridyltriazine (Fe(III)-TPTZ) is reduced to ferrous (Fe(II)-TPTZ) complexes by electrons donated by the antioxidant compounds. In the end product (Fe(II)-TPTZ), the colour intensity represents the antioxidant capacity (Iloki-Assanga et al. 2015). This method involves reducing the Fe^{3+} TPTZ complex (colorless complex) by electron-donating antioxidants under ambient conditions to form the Fe^{2+}-tripyridyltriazine complex (blue complex). The ferric reducing antioxidant power assay (FRAP) can be utilised to analyse nanoparticles generated in the fungal culture filtrate. The procedure involves mixing TPTZ (10 mM TPTZ in 40 mM HCl) and FeCl$_3$ (20 mM) in acetate buffer (0.3 M, pH 3.6). This is then mixed with the test NPs specimens (0.5 mL) at various concentrations (0.25–1.0 mg/mL) to produce FRAP solution (4.5 mL). Deionized water and ethanol can be used as controls. After 30 minutes of incubation at 37°C, the absorbance at 593 nm is measured. As Fe^{3+}–TPTZ get reduced to Fe^{2+}–TPTZ, a dark blue colour develops. A freshly produced aqueous ascorbic acid solution (0.2–1.0 mg/mL) can be used as the standard in the tests (Pulido et al. 2000).

Hydrogen Peroxide Scavenging Assay

The ability of nanoparticles to scavenge hydrogen peroxide (H_2O_2) moieties can also be a useful method to assess the antioxidant activity. This method involves preparation of a solution (pH 7.4) of hydrogen peroxide (2.0 mM) in phosphate buffer (50 mM). Hydrogen peroxide concentration can be determined spectrophotometrically by taking absorbance at 230 nm wavelength. The standard curve can be obtained by use of ascorbic acid solution (ranging from 25 to 250 µg/ml in the respective solvent). The final volume of the ascorbic acid aliquots can be prepared with 50 mM phosphate buffer (pH 7.4) or solvent (methanol). Tubes can then be vortexed after addition of hydrogen peroxide solution (0.6 mL), and the absorbance of the hydrogen peroxide at 230 nm can be measured after 10 minutes against a blank. As a control, 50 mM phosphate buffer without hydrogen peroxide can be utilized (Khan et al 2012).

Mycosynthesized NPs and their Antioxidant Activity

Metal and Metal Oxide Nanoparticles

Nanoparticles of gold and silver are the most popular metal nanoparticles. Silver nanoparticles exhibit a variety of activities such as antibacterial and antifungal properties, biocatalytic and photocatalytic properties, chemical stability, non-toxicity, ability to absorb visible light and far-IR light besides being low cost as compared to noble metals such as gold and platinum (Danagoudar et al. 2020). Mycological synthesis is the easiest and most utilized method to synthesize the AgNPs because fungi are easier to culture in the lab and growth occurs all year round. Endophytic fungi can also produce AgNPs with the possibility of easy growth and preparation, which have become the latest trend in nanotechnology (Danagoudar et al. 2020).

The antioxidant potential of myco-nanoantioxidants are anticipated to be affected by diverse factors including the morphological features such as size dimensions, particle shape (isometric/anisometric), besides the surface charge or zeta potential that governs the agglomeration of the nucleating or nucleated nanoparticles, occurrence of surface ligands or biomolecular functionalization or coating, and finally the dissolution of the metal ions from the surface. Several studies have been published claiming silver nanoparticles to be the most effective antioxidants and antimicrobials (Govindappa et al. 2016a). In a study conducted by BalaKumaran et al. (2020) the antioxidant activity of *A. terreus* myco-derived Ag and Au nanoparticles was compared. The AgNPs showcased higher antifungal, antioxidant and cytotoxic properties as compared to AuNPs. The AgNPs and AuNPs were screened individually for free radical scavenging activity by using DPPH with minor modifications. As per this study with increasing NP concentrations, AgNPs and AuNPs displayed a greater scavenging activity. A maximum of 92.31% scavenging activity was observed for AgNPs and AuNPs at a concentration of 20 g/mL. While the standard phenolic, gallic acid, on the other hand, had a 97.58% scavenging activity. Interestingly, the present study observed a twofold increase in antioxidant activity compared with the previous study in which IC50 values for mycosynthesized AgNPs were found to be 76.95 g/mL (BalaKumaran et al. 2020). The researchers have also evaluated the modified ABTS cation radical decolourisation assay to determine the ABTS radical scavenging activity of AgNPs and AuNPs. They observed that at 100 g/mL concentration of AgNPs and AuNPs 91.48% and 85.39% of DPPH scavenging activity was recorded respectively. Similarly, Bagur et al. (2020) also demonstrated that endophytic fungus mediated AgNPs possess dose-dependent free radical scavenging activity. Moreover, Nagajyothi et al. (2014) found that AgNPs became more effective in scavenging free radicals as their concentrations increased. The maximum inhibitory activity was observed at 1 mM AgNPs; and the minimum inhibition was observed at 0.125 mM. Taha et al. (2019) have used three different experiments to demonstrate that mycosynthesized AgNPs' antioxidant activity varied with the amount of NPs. Their study demonstrated that 30 µg/mL of AgNPs exhibit significantly more antioxidant activity than the other two concentrations (10 and 20 µg/mL). A comparative study was planned for identification of the FRAP activity of the biosynthesized AgNPs with respect to the culture filtrate. The results showed that biosynthesized AgNPs exhibited a stronger FRAP activity (Pulido et al. 2000). The antioxidant properties of various mycogenic nanoparticles have been demonstrated in Table 1.

In a study conducted by Govindappa et al. (2016a) fungi as a model system were used and the antioxidant, antibacterial, anti-inflammatory and tyrokinase effects associated with silver nanoparticles synthesized from fungal endophytes were evaluated. *Penicillium* species, an endophytic fungus from *Glycosis mauritiana*, was used to make the silver nanoparticles. The method produced silver nanoparticles within 10 minutes as indicated by change in colour. The produced AgNPs exhibited antioxidant activity appreciably comparable to ascorbic acid on DPPH radical scavenging assay. Also, they reported a lower FRAP reduction activity of ascorbic acid compared to the mycosynthesized AgNPs (Govindappa et al. 2016a).

Danagoudar et al. (2020) used *P. citrinum* CGJ-C1 isolated from *Tragia involucrata*, an ethnomedicinal plant, to generate AgNPs in their research. They have evaluated the antioxidant activity of the generated AgNPs in MCF-7 breast carcinoma cells. Myco-nanoparticles were reported to have excellent cytotoxic effects, primarily against malignant cells, as well as demonstrated DNA fragmentation and larvicidal properties. Thus, the researchers proposed that these AgNPs can be used to develop novel drugs to treat various ailments caused by a variety of causative agents (BalaKumaran et al. 2020).

Another study compared the antioxidant potential of AgNPs synthesized using plant and fungal extracts as biomediators, i.e., two different green synthesis methods were compared. The authors claimed synthesis of AgNPs of variable size and morphology from the two sources, i.e., the plant and fungal extracts used in this study (Mohammed et al. 2018). They further used DPPH scavenging method to examine the free radical scavenging activity of AgNPs synthesized using two different bio-mediators. A concentration-dependent effect on DPPH scavenging activity was identified for the synthesized AgNPs. Among the two biomediators, the plant-derived AgNPs possessed higher DPPH radical scavenging activity compared to the fungus derived AgNPs. It could be attributed to the more negative charge on plant-synthesised nanoparticles compared to mycosynthesized ones (Mohammed et al. 2018). This trend was also observed by Azizi et al. (2017).

Using a *P. italicum* strain isolated from Iraqi citrus lemon fruits, Taha and co-workers (2019) developed an eco-friendly and convenient method of synthesizing AgNPs. The AgNPs showed significant anti-oxidant and anti-microbial properties. AgNPs induced cell death via apoptosis rather than necrosis, showed

Table 1. Antioxidant properties of various myco-nanoparticles.

Type of Nanoparticle Synthesized	Fungal Strain Used	Morphology/ Size (nm)	Antioxidant Activity	References
Chitosan NPs	*Trichoderma harzianum* SKCGW008	10 to 314 nm, mean diameter of 46 nm	• DPPH radical scavenging activity of CSNPs equivalent to ascorbic acid • Dose-dependent antioxidant activity	(Saravanakumar et al. 2018)
Gold NPs	*C. cladosporioides*	60 nm	• Dose-dependent radical scavenging capacity against DPPH radicals	(Joshi et al. 2017)
	Rhizopus stolonifer	1 to 5 nm	• DPPH radicals scavenged in dose-dependent manner % DPPH radical inhibition linearly correlated	(Binupriya et al. 2010)
Silver NPs	*Aspergillus brunneoviolaceus*	0.72 to 15.21 nm, spherical shaped	DPPH radical scavenging activity: 63.97%	(Mistry et al. 2021)
	C. cladosporoides	100 nm	Radical scavenging ability increased with increasing concentration	(Manjunath Hulikere and Joshi 2019)
	Penicillium species	18 nm	• AgNPs at 0.1 µg/mL equivalent to ascorbic acid Higher FRAP reducing ability of AgNPs over standard ascorbic acid	(Govindappa et al. 2016b)
	Pestalotiopsis microspora	2 to 10 nm, spherical shaped	• DPPH IC50: 76.95 ± 2.96 µg/mL • H_2O_2 radical IC50: 94.95 ± 2.18 µg/mL	(Netala et al. 2016)
	Pleurotus florida	20 to 50 nm	DPPH radical inhibition (IC50): 85 µg/mL	(Ragunath et al. 2017)
	Trichoderma harzianum	72 nm	DPPH scavenging activity IC_{50}: 0.79 µg/mL	(Konappa et al. 2021)
Zinc Oxide NPs	*Aspergillus niger*	80 to 130 nm	At 100 g/mL ZnONPs • Maximum DPPH scavenging efficiency (57.74%), • ABTS scavenging efficiency (73.58%)	(Gao et al. 2019)

superior inhibitory activity against MCF-7 cells (Taha et al. 2019). They have assessed the antioxidant activity of synthesized AgNPs by the bleaching of three different radicals, i.e., DPPH, hydroxyl and resazurin radicals. AgNPs exhibited concentration-dependent antioxidant activities at three different concentrations of 10, 20 and 30 mg mL^{-1}. In contrast to ascorbic acid, the 30 g mL^{-1} concentration showed a maximum inhibition of 60%. The results of the research demonstrated the high antioxidant activity potential of the synthesised AgNPs and also that these can be employed as antioxidants or constituents in antioxidant formulations for possible biomedical and pharmaceutical applications.

The physiological versatility of the fungal cells can help in the synthesis of multiple types of NPs from a single fungal species. Endophytic fungus isolates were obtained from several Egyptian plants and cell-free filtrates of these isolate were tested for the production of five different NPs, namely Co_3O_4NPs, CuONPs, Fe_3O_4NPs, NiONPs, and ZnONPs. By screening the endophytes, it was found that *A. terreus* strain ORG-1 isolated from *O. majorana* leaves was capable of reducing all five salt types (Mousa et al. 2021). A comparison of the antioxidant activities of synthesized nanoparticles with ascorbic acid revealed their promising antioxidant potential. Compared to ascorbic acid, the NPs synthesized were found to exhibit significant antioxidant activity at different concentrations. Ascorbic acid and formed nanoparticles were also least inhibitory at 25 µg mL^{-1}. Furthermore, all five types of synthesised NPs decreased DPPH free radicals in a dose-dependent manner, with an increase in NPs concentration leading to a significant increase in the scavenging activity. The order of the IC50 value was ZnONPs > Fe_3O_4NPs > CuONPs > Co_3O_4NPs > NiONPs, whereas standard ascorbic acid recorded IC50 value (71.64 µg mL^{-1}) lower than all the tested NPs (Mousa et al. 2021). The cerium oxide nanoparticles have been considered to be most effective metal oxide nano-antioxidant that have a great promise for treatment of the ROS related and age-specific chronic neurodegenerative disorders such as Parkinson's and Alzheimer's disease (Weaver and Stabler 2015, Rajeshkumar and Naik 2018). Mycosynthesis of cerium oxide NPs has been performed through incubation of the mycelial growth or culture filtrate with cerium salt. Khan and Ahmad (2013) have reported the synthesis of the cerium oxide NPs by incubation of mycelia of thermophilic fungus *Humicola* sp. with cerium nitrate hexahydrate solution (10^{-3} M) at 50°C under shaking conditions. Other reports involved CeO NPs synthesis from *Aspergillus niger* (Gopinath et al. 2015), *Fusarium solani* (Venkatesh et al. 2016) and *Curvularia lunata* (Munusamy et al. 2014). However, none of these studies reported the antioxidant properties of these NPs.

Other than the asco- and zygomycetes, basidiomycetes have also been evaluated for their ability to synthesize nanoparticles (Kalia and Kaur 2018, Kaur et al. 2020). Due to great diversity of bioactive metabolites with a variety of biological functions, several research studies have been performed for examining the feasibility of synthesis of nanoparticles (NP) from edible and medicinal mushrooms (Bhardwaj et al. 2020b). Published reports document synthesis of both intracellular and extracellular gold (Au) and silver (Ag) nanoparticles using various proteins and polysaccharides found in mushrooms (Owaid and Ibraheem 2017). As a result of the compounds secreted by medicinal mushrooms, the NPs are stable, long lasting, water-soluble, and possess excellent dispersion properties (Bhardwaj et al. 2020b). The antioxidant activity of AgNPs derived from the hyphal or fruit body extracts of the fungal cultures, viz., *Inonotus obliquus, Ganoderma lucidum*, and *Phellinus igniarius* was evaluated in a study (Owaid and Ibraheem 2017). *Agaricus bisporus*-derived AgNPs exhibited positive antioxidant activity, besides the metal NPs obtained from the *G. lucidum, P. igniarius*, and *I. obliquus*. These NPs have also been found to exhibit potent anticancer activities (Owaid and Ibraheem 2017).

Metalloid Nanoparticles

Hashem and co-workers (2021) conducted a study for green and eco-friendly biogenic synthesis of selenium nanoparticles (Se-NPs) with *P. expansum* and also examined the antimicrobial, antioxidant, anticancer, and hemocompatibility properties of mycosynthesized Se-NPs. The Se-NPs exhibited antioxidant properties which were determined for concentrations ranging from 4 to 12.5 µg/mL. Biosynthesised Se-NPs from *P. expansum* showed potent antioxidant and anticancer activities in prostate cancer (PC3) cells while causing no harm to normal cells. At concentrations of up to 250 µg/mL, these Se-NPs exhibited nonhemolytic activity on human RBCs (Hashem et al. 2021).

The antioxidant activity of mycosynthesized Se-NPs has been determined by using a DPPH free radical analysis. These nanoparticles have been shown to possess anti-atherosclerotic, anti-inflammatory, antitumor, anticancer, anti-mutagenic, and anti-microbial properties. As compared to ascorbic acid, the results revealed that Se-NPs have high antioxidant properties, with antioxidant activity above 50% at concentrations around 30 ng/L. There is significant evidence that Se NPs have good antioxidant activity. Other studies have examined Se-NPs that have been synthesized by different techniques (Zhang et al. 2018). These results support the use of mycosynthesized Se-NPs as natural antioxidants to protect humans from oxidative stress-related chronic illnesses (Hashem et al. 2021).

Polymeric Nanoparticles

Chitosan is a natural linear polymer which is widely utilized in the medicine industry because of its unique features including the biocompatibility, degradability, nontoxicity, and permeability. Antimicrobial, wound healing, drug carrier, biosensor agents, and water purification are just a few of the biomedical applications of this biopolymer (Saravanakumar et al. 2018). Manufacturing polymeric nanoparticles with increased bioactivity, such as antioxidant and antibacterial characteristics, is required for the fabrication of polymer-based biomaterials. Fungal cultures have been used for the synthesis of the chitosan nanoparticles. The crude enzymes of *Trichoderma harzianum* SKCGW008 were used to prepare chitosan nanoparticles (CSNPs) and spherical shaped nano-scale chitosan particles of uniform size were obtained (Saravanakumar et al. 2018). On evaluation of the antioxidant potential of chitosan nanoparticles through four different antioxidant assays the antioxidant potency of the CSNPs was recorded to be significantly higher when compared to the positive standard ascorbic acid. The CSNPs were found to possess an antioxidant activity of 40 µg.mL^{-1}, which was equivalent to standard ascorbic acid (60 µg.mL^{-1}). The presence of amino groups and amines in mycogenic CSNPs adorn higher DPPH radical scavenging activity than the CSNPs developed through ionotropic gelation method (Saravanakumar et al. 2018).

Conclusions and Future Prospects

The control over the interactions among the atoms and molecules in order to produce nano-materials with various useful functions has been prerequisite for different nanotechnological interventions (Hashem et al. 2011). Myconanotechnology strategies and their commercial applications in agriculture and allied sectors are likely to occur in the coming years, based on recent technological advancements and ongoing research aimed at improving nanomaterial synthesis efficiency and exploring their applications in agriculture (Banerjee and Ravishankar Rai 2018, Hu et al. 2019, Mousa et al. 2021). The fungal mycelia secerete copious amounts of extracellular proteins and enzymes which help in quick synthesis of nanoparticles besides confering stability to the developed nanoparticles. Further, the high biomass production capability of fungal cultures does not require any additional steps to extract the culture filtrate as opposed to the bacteria cultures. It is thus highly feasible to produce inorganic nanoparticles at large scales with this method (Attarad et al. 2016) and to conduct the noble synthesis of nanostructures of other metal oxides, nitrides, carbides, etc. Furthermore, research interventions critical to optimize the various reaction conditions to achieve better control over the nanoparticles' size, shape, and monodispersity are required. The unique optical, catalytic and biological properties that these myco-nanoparticles exhibit for instance the antioxidant properties are gaining recent interest. Antioxidant potentials of the myconanoparticles have been utilized for biomedical applications including the development of new drug delivery systems to combat different types of cancer disorders. Further, these nano-anti-oxidant formulations will find economic significance in cosmetics and nutraceutical industry. More focused studies and experimental trials are required to identify the relative potential and mechanism of antioxidant action of the nano-antioxidants. Besides, the biomolecules responsible for reducing and stabilizing nanoparticles (enzymes or proteins) with greater antioxidant potential also must be deciphered. Further, to make the production process of these myco-nano-antioxidants commercially viable, a new formulation of low-cost recovery techniques is also required to be identified.

References

Abd-Elsalam, K.A. 2021. Special issue: Fungal nanotechnology. J. Fungi. 7: 4–6 . Doi: 10.3390/jof7080583.

Abou El-Nour, K.M.M., Eftaiha, A., Al-Warthan, A. and Ammar, R.A.A. 2010. Synthesis and applications of silver nanoparticles. Arab. J. Chem. 3: 135–140.

Adebayo, E.A., Azeez, M.A., Alao, M.B., Oke, A.M. and Aina, D.A. 2021. Fungi as veritable tool in current advances in nanobiotechnology. Heliyon 7: e08480. Doi: 10.1016/j.heliyon.2021.e08480.

Adeeyo, A.O. and Odiyo, J.O. 2018. Biogenic synthesis of silver nanoparticle from mushroom exopolysaccharides and its potentials in water purification. Open Chem. J. 5: 64–75. Doi: 10.2174/1874842201805010064.

Ahmad, A., Mukherjee, P., Senapati, S., Mandal, D., Islam Khan, M., Kumar, R. and Sastry, M. 2003. Extracellular biosynthesis of silver nanoparticles using the fungus *Fusarium oxysporum*. Colloids Surfaces B: Biointerfaces 28(4): 313–318.

Al Abboud, M.A. 2018. Fungal biosynthesis of silver nanoparticles and their role in control of Fusarium wilt of sweet pepper and soil-borne fungi *in vitro*. Int. J. Pharmacol. 14: 773–780. Doi: 10.3923/ijp.2018.773.780.

Albrecht, M.A., Evans, C.W. and Raston, C.L. 2006. Green chemistry and the health implications of nanoparticles. Green Chem. 8: 417–432. Doi: 10.1039/b517131h.

Alghuthaymi, M.A., Almoammar, H., Rai, M., Said-Galiev, E. and Abd-Elsalam, K.A. 2015. Myconanoparticles: Synthesis and their role in phytopathogens management. Biotechnol. Biotechnol. Equip. 29: 221–236. Doi: 10.1080/13102818.2015.1008194.

Ali, A., Ambreen, S., Maqbool, Q., Naz, S., Shams, M.F., Ahmad, M., Phull, A.R. and Zia, M. 2016. Zinc impregnated cellulose nanocomposites: Synthesis, characterization and applications. J. Phys. Chem. Solids 98: 174–182. Doi: 10.1016/j.jpcs.2016.07.007.

Ali, A., Phull, A.R. and Zia, M. 2018. Elemental zinc to zinc nanoparticles: Is ZnO NPs crucial for life? Synthesis, toxicological, and environmental concerns. Nanotechnol. Rev. 7: 413–441. Doi: 10.1515/ntrev-2018-0067.

Arora, D.S. and Chandra, P. 2011. Antioxidant activity of Aspergillus fumigatus. ISRN Pharmacol. 2011: 1–11. Doi: 10.5402/2011/619395.

Ashrafi, S.J., Rastegar, M.F., Ashrafi, M., Yazdian, F., Pourrahim, R. and Suresh, A.K. 2013. Influence of external factors on the production and morphology of biogenic silver nanocrystallites. J. Nanosci. Nanotechnol. 13: 2295–2301. Doi: 10.1166/jnn.2013.6791.

Attarad, A., Zafar, H., Zia, M., Ihsan ul Haq, Abdul Rehman Phull, Joham Sarfraz Ali and Altaf Hussain. 2016. Synthesis, characterization, applications, and challenges of iron oxide nanoparticles. 49–67.

Azmath, P., Baker, S., Rakshith, D. and Satish, S. 2016. Mycosynthesis of silver nanoparticles bearing antibacterial activity. Saudi Pharm. J. 24: 140–146. Doi: 10.1016/j.jsps.2015.01.008.

Bagur, H., Medidi, R.S., Somu, P., Choudhury, P.W.J., karua, C.S., Guttula, P.K., Melappa, G. and Poojari, C.C. 2020. Endophyte fungal isolate mediated biogenic synthesis and evaluation of biomedical applications of silver nanoparticles. Mater. Technol. 00: 1–12. Doi: 10.1080/10667857.2020.1819089.

Bahrulolum, H., Nooraei, S., Javanshir, N., Tarrahimofrad, H., Mirbagheri, V.S., Easton, A.J. and Ahmadian, G. 2021. Green synthesis of metal nanoparticles using microorganisms and their application in the agrifood sector. J. Nanobiotechnology 19: 1–26. Doi: 10.1186/s12951-021-00834-3.

Baig, N., Kammakakam, I., Falath, W. and Kammakakam, I. 2021. Nanomaterials: A review of synthesis methods, properties, recent progress, and challenges. Mater. Adv. 2: 1821–1871. Doi: 10.1039/d0ma00807a.

BalaKumaran, M.D., Ramachandran, R., Balashanmugam, P., Jagadeeswari, S. and Kalaichelvan, P.T. 2020. Comparative analysis of antifungal, antioxidant and cytotoxic activities of mycosynthesized silver nanoparticles and gold nanoparticles. Mater. Technol. Doi: 10.1080/10667857.2020.1854518.

Banerjee, K. and Ravishankar Rai, V. 2018. A review on mycosynthesis, mechanism, and characterization of silver and gold nanoparticles. Bionanoscience 8: 17–31.

Baymiller, M., Huang, F. and Rogelj, S. 2017. Rapid one-step synthesis of gold nanoparticles using the ubiquitous coenzyme NADH. Matters. Doi: 10.19185/matters.201705000007.

Bhangale, H.G., Bachhav, S.G., Nerkar, D.M., Sarode, K.M. and Patil, D.R. 2019. Study on optical properties of green synthesized silver nanoparticles for surface plasmon resonance. J. Nanosci. Technol. 5: 658–661. Doi: 10.30799/jnst.230.19050203.

Bhardwaj, K., Dhanjal, D.S., Sharma, A., Nepovimova, E., Kalia, A., Thakur, S., Bhardwaj, S., Chopra, C., Singh, R., Verma, R., Kumar, D., Bhardwaj, P. and Kuča, K. 2020a. Conifer-derived metallic nanoparticles: Green synthesis and biological applications. Int. J. Mol. Sci. 21: 1–22. Doi: 10.3390/ijms21239028.

Bhardwaj, K., Sharma, A., Tejwan, N., Bhardwaj, S., Bhardwaj, P., Nepovimova, E., Shami, A., Kalia, A., Kumar, A., Abd-Elsalam, K.A. and Kuča, K. 2020b. Pleurotus macrofungi-assisted nanoparticle synthesis and its potential applications: A review. J. Fungi. 6: 1–21. Doi: 10.3390/jof6040351.

Binupriya, A.R., Sathishkumar, M. and Yun, S.I. 2010. Biocrystallization of silver and gold ions by inactive cell filtrate of Rhizopus stolonifer. Colloids Surfaces B Biointerfaces 79: 531–534. Doi: 10.1016/j.colsurfb.2010.05.021.

Castro-Longoria, E., Vilchis-Nestor, A.R. and Avalos-Borja, M. 2011. Biosynthesis of silver, gold and bimetallic nanoparticles using the filamentous fungus Neurospora crassa. Colloids Surfaces B Biointerfaces 83: 42–48. Doi: 10.1016/j.colsurfb.2010.10.035.

Choudhary, M.A., Manan, R., Aslam Mirza, M., Rashid Khan, H., Qayyum, S. and Ahmed, Z. 2018. Biogenic synthesis of copper oxide and zinc oxide nanoparticles and their application as antifungal agents. Int. J. Mater. Sci. Eng. 4: 1–6. Doi: 10.14445/23948884/ijmse-v4i1p101,

Costa Silva, L.P., Pinto Oliveira, J., Keijok, W.J., da Silva, A.R., Aguiar, A.R., Guimarães, M.C.C., Ferraz, C.M., Araújo, J.V., Tobias, F.L. and Braga, F.R. 2017. Extracellular biosynthesis of silver nanoparticles using the cell-free filtrate of nematophagous fungus Duddingtonia flagrans. Int. J. Nanomedicine 12: 6373–6381. Doi: 10.2147/IJN.S137703.

Danagoudar, A., Pratap, G.K., Shantaram, M., Ghosh, K., Kanade, S.R. and Joshi, C.G. 2020. Characterization, cytotoxic and antioxidant potential of silver nanoparticles biosynthesised using endophytic fungus (Penicillium citrinum CGJ-C1). Mater. Today Commun. 25: 101385. Doi: 10.1016/j.mtcomm.2020.101385.

Das, S., Chakraborty, J., Chatterjee, S. and Kumar, H. 2018. Prospects of biosynthesized nanomaterials for the remediation of organic and inorganic environmental contaminants. Environ. Sci. Nano. 5: 2784–2808. Doi: 10.1039/C8EN00799C.

Drăgan, M., Daniela Stan, C., Iacob, A. and Profire, L. 2016. Assessment if *in vitro* antioxidant and anti-inflammatory activities of new azetidin-2-one derivatives of ferulic acid. Farmacia 64(5): 717–721.

Elamawi, R.M., Al-Harbi, R.E. and Hendi, A.A. 2018. Biosynthesis and characterization of silver nanoparticles using Trichoderma longibrachiatum and their effect on phytopathogenic fungi. Egypt. J. Biol. Pest Control 28: 1–11. Doi: 10.1186/s41938-018-0028-1.

Galdiero, S., Falanga, A., Vitiello, M., Cantisani, M., Marra, V. and Galdiero, M. 2011. Silver nanoparticles as potential antiviral agents. Molecules 16: 8894–8918. Doi: 10.3390/molecules16108894.

Gao, Y., Arokia Vijaya Anand, M., Ramachandran, V., Karthikkumar, V., Shalini, V., Vijayalakshmi, S. and Ernest, D. 2019. Biofabrication of zinc oxide nanoparticles from aspergillus niger, their antioxidant, antimicrobial and anticancer activity. J. Clust. Sci. 30: 937–946. Doi: 10.1007/s10876-019-01551-6.

Gopinath, K., Karthika, V., Sundaravadivelan, C., Gowri, S. and Arumugam, A. 2015. Mycogenesis of cerium oxide nanoparticles using Aspergillus niger culture filtrate and their applications for antibacterial and larvicidal activities. J. Nanostructure Chem. 5: 295–303. Doi: 10.1007/s40097-015-0161-2.

Govindappa, M., Farheen, H., Chandrappa, C.P., Channabasava Rai, R.V. and Raghavendra, V.B. 2016a. Mycosynthesis of silver nanoparticles using extract of endophytic fungi, Penicillium species of Glycosmis mauritiana, and its antioxidant, antimicrobial, anti-inflammatory and tyrokinase inhibitory activity. Adv. Nat. Sci. Nanosci. Nanotechnol. 7. Doi: 10.1088/2043-6262/7/3/035014.

Govindappa, M., Farheen, H., Chandrappa, C.P., Channabasava Rai, R.V. and Raghavendra, V.B. 2016b. Mycosynthesis of silver nanoparticles using extract of endophytic fungi, Penicillium species of Glycosmis mauritiana, and its antioxidant, antimicrobial, anti-inflammatory and tyrokinase inhibitory activity. Adv. Nat. Sci. Nanosci. Nanotechnol. 7. Doi: 10.1088/2043-6262/7/3/035014.

Gudikandula, K., Vadapally, P. and Singara Charya, M.A. 2017. Biogenic synthesis of silver nanoparticles from white rot fungi: Their characterization and antibacterial studies. OpenNano 2: 64–78. Doi: 10.1016/j.onano.2017.07.002.

Guilger-Casagrande, M. and Lima, R. de. 2019. Synthesis of silver nanoparticles mediated by fungi: A review. Front. Bioeng. Biotechnol. 7: 1–16. Doi: 10.3389/fbioe.2019.00287.

Guilger, M., Pasquoto-Stigliani, T., Bilesky-Jose, N., Grillo, R., Abhilash, P.C., Fraceto, L.F. and De Lima, R. 2017. Biogenic silver nanoparticles based on trichoderma harzianum: Synthesis, characterization, toxicity evaluation and biological activity. Sci. Rep. 7: 1–13. Doi: 10.1038/srep44421.

Gupta, K., Chundawat, T.S. and Malek, N.A.N.N. 2021. Antibacterial, antifungal, photocatalytic activities and seed germination effect of mycosynthesized silver nanoparticles using fusarium oxysporum. Biointerface Res. Appl. Chem. 11: 12082–12091. Doi: 10.33263/BRIAC114.1208212091.

Hashem, A.H., Khalil, A.M.A., Reyad, A.M. and Salem, S.S. 2021. Biomedical applications of mycosynthesized selenium nanoparticles using Penicillium expansum ATTC 36200. Biol. Trace. Elem. Res. 199: 3998–4008. Doi: 10.1007/s12011-020-02506-z.

Hashem, A.H., Mohamed, A., Khalil, A., Reyad, A.M. and Salem, S.S. 2011. Biomedical applications of mycosynthesized selenium nanoparticles using Penicillium expansum ATTC 36200. Doi: 10.1007/s12011-020-02506-z/Published.

Hu, X., Saravanakumar, K., Jin, T. and Wang, M.H. 2019. Mycosynthesis, characterization, anticancer and antibacterial activity of silver nanoparticles from endophytic fungus Talaromyces purpureogenus. Int. J. Nanomedicine 14: 3427–3438. Doi: 10.2147/IJN.S200817.

Huang, W.Y., Cai, Y.Z., Hyde, K.D., Corke, H. and Sun, M. 2007. Endophytic fungi from Nerium oleander L (Apocynaceae): Main constituents and antioxidant activity. World J. Microbiol. Biotechnol. 23: 1253–1263. doi: 10.1007/s11274-007-9357-z.

Husen, A. and Siddiqi, K.S. 2014. Phytosynthesis of nanoparticles: Concept, controversy and application. Nanoscale Res. Lett. 9: 1–24. Doi: 10.1186/1556-276X-9-229.

Iloki-Assanga, S.B., Lewis-Luján, L.M., Lara-Espinoza, C.L., Gil-Salido, A.A., Fernandez-Angulo, D., Rubio-Pino, J.L. and Haines, D.D. 2015. Solvent effects on phytochemical constituent profiles and antioxidant activities, using four different extraction formulations for analysis of Bucida buceras L. and Phoradendron californicum Complementary and Alternative Medicine. BMC Res. Notes 8: 1–14. Doi: 10.1186/s13104-015-1388-1.

Jaffri, S.B. and Ahmad, K.S. 2019. Foliar-mediated Ag:ZnO nanophotocatalysts: Green synthesis, characterization, pollutants degradation, and in vitro biocidal activity. Green Process Synth. 8: 172–182. Doi: 10.1515/gps-2018-0058.

Janakiraman, V., Govindarajan, K. and Magesh, C.R. 2019. Biosynthesis of silver nanoparticles from endophytic fungi, and its cytotoxic activity. Bionanoscience 9: 573–579. Doi: 10.1007/s12668-019-00631-1.

Joshi, C.G., Danagoudar, A., Poyya, J., Kudva, A.K. and BL, D. 2017. Biogenic synthesis of gold nanoparticles by marine endophytic fungus-Cladosporium cladosporioides isolated from seaweed and evaluation of their antioxidant and antimicrobial properties. Process Biochem. 63: 137–144. Doi: 10.1016/j.procbio.2017.09.008.

Jutz, G. and Böker, A. 2011. Bionanoparticles as functional macromolecular building blocks—A new class of nanomaterials. Polymer (Guildf) 52: 211–232. Doi: 10.1016/j.polymer.2010.11.047.

Kalia, A. and Kaur, G. 2018. Biosynthesis of nanoparticles using mushrooms. pp. 351–360. *In:* Singh, B. and Lallawmsanga, P.A. (eds.). Biology of Macrofungi, Fungal Biology. Springer, Cham.

Kalia, A., Manchanda, P., Bhardwaj, S. and Singh, G. 2020. Biosynthesized silver nanoparticles from aqueous extracts of sweet lime fruit and callus tissues possess variable antioxidant and antimicrobial potentials. Inorg. Nano-Metal Chem. 50: 1053–1062. Doi: 10.1080/24701556.2020.1735420.

Kashyap, P.L., Kumar, S., Srivastava, A.K. and Sharma, A.K. 2013. Myconanotechnology in agriculture: A perspective. World J. Microbiol. Biotechnol. 29: 191–207.

Kaur, G., Kalia, A. and Sodhi, H.S. 2020. Size controlled, time-efficient biosynthesis of silver nanoparticles from Pleurotus florida using ultra-violet, visible range, and microwave radiations. Inorg. Nano-Metal Chem. 50: 35–41. Doi: 10.1080/24701556.2019.1661466.

Kedare, S.B. and Singh, R.P. 2011. Genesis and development of DPPH method of antioxidant assay. J. Food Sci. Technol. 48: 412–422. Doi: 10.1007/s13197-011-0251-1.

Khan, A., Rashid, A., Younas, R. and Chong, R. 2015. A chemical reduction approach to the synthesis of copper nanoparticles. Int. Nano Lett. 6: 21–26. Doi: 10.1007/s40089-015-0163-6.

Khan, S.A. and Ahmad, A. 2013. Fungus mediated synthesis of biomedically important cerium oxide nanoparticles. Mater. Res. Bull. 48: 4134–4138. Doi: 10.1016/j.materresbull.2013.06.038.

Khandel, P. and Shahi, S.K. 2018a. Mycogenic nanoparticles and their bio-prospective applications: Current status and future challenges. J. Nanostructure Chem. 8: 369–391. Doi: 10.1007/s40097-018-0285-2.

Khandel, P. and Shahi, S.K. 2018b. Mycogenic nanoparticles and their bio-prospective applications: Current status and future challenges. J. Nanostructure Chem. 8: 369–391. Doi: 10.1007/s40097-018-0285-2.

Kitching, M., Ramani, M. and Marsili, E. 2015. Fungal biosynthesis of gold nanoparticles: Mechanism and scale up. Microb. Biotechnol. 8: 904–917.

Konappa, N., Udayashankar, A.C., Dhamodaran, N., Krishnamurthy, S., Jagannath, S., Uzma, F., Pradeep, C.K., De Britto, S., Chowdappa, S. and Jogaiah, S. 2021. Ameliorated antibacterial and antioxidant properties by trichoderma harzianum mediated green synthesis of silver nanoparticles. Biomolecules 11. Doi: 10.3390/biom11040535.

Kumar, H., Bhardwaj, K., Kuča, K., Kalia, A., Nepovimova, E., Verma, R. and Kumar, D. 2020. Flower-based green synthesis of metallic nanoparticles: Applications beyond fragrance. Nanomaterials 10: 766. Doi: 10.3390/nano10040766.

Kumar, S.A., Abyaneh, M.K., Gosavi, S.W., Kulkarni, S.K., Pasricha, R., Ahmad, A. and Khan, M.I. 2007. Nitrate reductase-mediated synthesis of silver nanoparticles from AgNO3. Biotechnol. Lett. 29: 439–445. Doi: 10.1007/s10529-006-9256-7.

Lal, S., Jana, U., Manna, P.K., Mohanta, G.P., Manavalan, R. and Pal, S.L. 2011. Nanoparticle: An overview of preparation and characterization. J. Appl. Pharm. Sci. 2011: 228–234.

Lee, S.H. and Jun, B.H. 2019. Silver nanoparticles: Synthesis and application for nanomedicine. Int. J. Mol. Sci. 20.

Li, X., Xu, H., Chen, Z.S. and Chen, G. 2011. Biosynthesis of nanoparticles by microorganisms and their applications. J. Nanomater. 2011: 1–16. Doi: 10.1155/2011/270974.

Manjunath Hulikere, M. and Joshi, C.G. 2019. Characterization, antioxidant and antimicrobial activity of silver nanoparticles synthesized using marine endophytic fungus—Cladosporium cladosporioides. Process Biochem. 82: 199–204. Doi: 10.1016/j.procbio.2019.04.011.

Mistry, H., Thakor, R., Patil, C., Trivedi, J. and Bariya, H. 2021. Biogenically proficient synthesis and characterization of silver nanoparticles employing marine procured fungi Aspergillus brunneoviolaceus along with their antibacterial and antioxidative potency. Doi: 10.1007/s10529-020-03008-7.

Mohammed, A.E., Bin Baz, F.F. and Albrahim, J.S. 2018. Calligonum comosum and Fusarium sp. extracts as bio-mediator in silver nanoparticles formation: characterization, antioxidant and antibacterial capability. 3 Biotech 8: 1–8. Doi: 10.1007/s13205-017-1046-5.

Molnár, Z., Bódai, V., Szakacs, G., Erdélyi, B., Fogarassy, Z., Sáfrán, G., Varga, T., Kónya, Z., Tóth-Szeles, E., Szucs, R. and Lagzi, I. 2018. Green synthesis of gold nanoparticles by thermophilic filamentous fungi. Sci. Rep. 8. Doi: 10.1038/s41598-018-22112-3.

Moon, J.K. and Shibamoto, T. 2009. Antioxidant assays for plant and food components. J. Agric. Food Chem. 57: 1655–1666.

Mousa, S.A., El-Sayed, E-S.R., Mohamed, S.S., Abo El-Seoud, M.A., Elmehlawy, A.A. and Abdou, D.A.M. 2021. Novel mycosynthesis of Co3O4, CuO, Fe3O4, NiO, and ZnO nanoparticles by the endophytic Aspergillus terreus and evaluation of their antioxidant and antimicrobial activities. Appl. Microbiol. Biotechnol. 105: 741–753. Doi: 10.1007/s00253-020-11046-4.

Munusamy, S., Bhakyaraj, K., Vijayalakshmi, L., Stephen, A. and Narayanan, V. 2014. Synthesis and characterization of cerium oxide nanoparticles using Curvularia lunata and their antibacterial properties. Int. J. Innov. Res. Sci. Eng. 12: 1401–1413.

Nagajyothi, P.C., Sreekanth, T.V.M., Lee, J.Il and Lee, K.D. 2014. Mycosynthesis: Antibacterial, antioxidant and antiproliferative activities of silver nanoparticles synthesized from Inonotus obliquus (Chaga mushroom) extract. J. Photochem. Photobiol. B Biol. 130: 299–304. Doi: 10.1016/j.jphotobiol.2013.11.022.

Netala, V.R., Bethu, M.S., Pushpalatha, B., Baki, V.B., Aishwarya, S. and Rao, J.V. 2016. Biogenesis of silver nanoparticles using endophytic fungus Pestalotiopsis microspora and evaluation of their antioxidant and anticancer activities. Int. J. Nanomedicine 11: 5683–5696. Doi: 10.2147/IJN. S112857.

Owaid, M.N. and Ibraheem, I.J. 2017. Mycosynthesis of nanoparticles using edible and medicinal mushrooms. Eur. J. Nanomedicine 9: 5–23. Doi: 10.1515/ejnm-2016-0016.

Pandey, G. and Jain, P. 2020. Assessing the nanotechnology on the grounds of costs, benefits, and risks. Beni-Suef Univ. J. Basic Appl. Sci. 9. Doi: 10.1186/s43088-020-00085-5.

Patra, J.K. and Baek, K.H. 2016. Biosynthesis of silver nanoparticles using aqueous extract of silky hairs of corn and investigation of its antibacterial and anticandidal synergistic activity and antioxidant potential. IET Nanobiotechnology 10: 326–333. Doi: 10.1049/iet-nbt.2015.0102.

Prior, R.L., Wu, X. and Schaich, K. 2005. Standardized methods for the determination of antioxidant capacity and phenolics in foods and dietary supplements. J. Agric. Food Chem. 53: 4290–4302. Doi: 10.1021/jf0502698.

Pulido, R., Bravo, L. and Saura-Calixto, F. 2000. Antioxidant activity of dietary polyphenols as determined by a modified ferric reducing/antioxidant power assay. J. Agric. Food Chem. 48: 3396–3402. Doi: 10.1021/jf9913458.

Qidwai, A., Pandey, A., Kumar, R., Shukla, S.K. and Dikshit, A. 2018. Advances in biogenic nanoparticles and the mechanisms of antimicrobial effects. Indian J. Pharm. Sci. 80: 592–603.

Ragunath, A., Parveen Nisha, A. and Kumuthakalavalli, R. 2017. Mycosynthesis of silver nanoparticles: Characterization, antioxidant and anti-inflammatory activity from Pleurotus Florida (MONT) singer: A macro fungi. Asian J. Pharm. Clin. Res. 10: 186–191. Doi: 10.22159/ajpcr.2017.v10i9.19090.

Rai, M., Bonde, S., Golinska, P., Trzcińska-Wencel, J., Gade, A., Abd-Elsalam, K., Shende, S., Gaikwad, S. and Ingle, A.P. 2021. Fusarium as a novel fungus for the synthesis of nanoparticles: Mechanism and applications. J. Fungi. 7: 1–24.

Rajeshkumar, S. and Naik, P. 2018. Synthesis and biomedical applications of cerium oxide nanoparticles—A Review. Biotechnol. Reports 17: 1–5. Doi: 10.1016/j.btre.2017.11.008.

Rajput, S., Werezuk, R., Lange, R.M. and Mcdermott, M.T. 2016. Fungal isolate optimized for biogenesis of silver nanoparticles with enhanced colloidal stability. Langmuir 32: 8688–8697. Doi: 10.1021/acs.langmuir.6b01813.

Ram Kumar Pandian, S., Deepak, V., Kalishwaralal, K., Muniyandi, J., Rameshkumar, N. and Gurunathan, S. 2009. Synthesis of PHB nanoparticles from optimized medium utilizing dairy industrial waste using Brevibacterium casei SRKP2: A green chemistry approach. Colloids Surfaces B Biointerfaces 74: 266–273. Doi: 10.1016/j.colsurfb.2009.07.029.

Rautaray, D., Sanyal, A., Adyanthaya, S.D., Ahmad, A. and Sastry, M. 2004. Biological synthesis of strontium carbonate crystals using the fungus Fusarium oxysporum. Langmuir 20: 6827–6833. Doi: 10.1021/la049244d.

Reddy, P.V.L., Hernandez-Viezcas, J.A., Peralta-Videa, J.R. and Gardea-Torresdey, J.L. 2016. Lessons learned: Are engineered nanomaterials toxic to terrestrial plants? Sci. Total Environ. 568: 470–479. Doi: 10.1016/j.scitotenv.2016.06.042.

Sabri, M.A., Umer, A., Awan, G.H., Hassan, M.F. and Hasnain, A. 2016. Selection of suitable biological method for the synthesis of silver nanoparticles. Nanomater. Nanotechnol. 6. Doi: 10.5772/62644.

Saravanakumar, K., Chelliah, R., MubarakAli, D., Jeevithan, E., Oh, D.H., Kathiresan, K. and Wang, M.H. 2018. Fungal enzyme-mediated synthesis of chitosan nanoparticles and its biocompatibility, antioxidant and bactericidal properties. Int. J. Biol. Macromol. 118: 1542–1549. Doi: 10.1016/j.ijbiomac.2018.06.198.

Saxena, J., Sharma, P.K., Sharma, M.M. and Singh, A. 2016. Process optimization for green synthesis of silver nanoparticles by Sclerotinia sclerotiorum MTCC 8785 and evaluation of its antibacterial properties. Springerplus 5. Doi: 10.1186/s40064-016-2558-x.

Shah, M., Fawcett, D., Sharma, S., Tripathy, S.K. and Poinern, G.E.J. 2015. Green synthesis of metallic nanoparticles via biological entities. Materials (Basel). 8: 7278–7308.

Siddiqi, K.S. and Husen, A. 2016. Fabrication of metal nanoparticles from fungi and metal salts: Scope and application. Nanoscale Res. Lett. 11: 1–15. Doi: 10.1186/s11671-016-1311-2.

Singh, A.V., Laux, P., Luch, A., Sudrik, C., Wiehr, S., Wild, A.M., Santomauro, G., Bill, J. and Sitti, M. 2019. Review of emerging concepts in nanotoxicology: Opportunities and challenges for safer nanomaterial design. Toxicol. Mech. Methods 29: 378–387. Doi: 10.1080/15376516.2019.1566425.

Singh, P., Kim, Y.J., Zhang, D. and Yang, D.C. 2016. Biological synthesis of nanoparticles from plants and microorganisms. Trends Biotechnol. 34: 588–599. Doi: 10.1016/j.tibtech.2016.02.006.

Syed, A. and Ahmad, A. 2013. Extracellular biosynthesis of CdTe quantum dots by the fungus Fusarium oxysporum and their anti-bacterial activity. Spectrochim. Acta—Part A Mol. Biomol. Spectrosc. 106: 41–47. Doi: 10.1016/j.saa.2013.01.002.

Taha, Z.K., Hawar, S.N. and Sulaiman, G.M. 2019. Extracellular biosynthesis of silver nanoparticles from Penicillium italicum and its antioxidant, antimicrobial and cytotoxicity activities. Biotechnol. Lett. 41: 899–914. Doi: 10.1007/s10529-019-02699-x.

Thakkar, K.N., Mhatre, S.S. and Parikh, R.Y. 2010. Biological synthesis of metallic nanoparticles. Nanomedicine Nanotechnology Biol. Med. 6: 257–262.

Tiquia-Arashiro, S. and Rodrigues, D. 2016. Nanoparticles synthesized by microorganisms. pp. 1–51. *In*: Extremophiles: Applications in Nanotechnology. Springer International Publishing.

Venkatesh, K.S., Gopinath, K., Palani, N.S., Arumugam, A., Jose, S.P., Bahadur, S.A. and Ilangovan, R. 2016. Plant pathogenic fungus *F. solani* mediated biosynthesis of nanoceria: Antibacterial and antibiofilm activity. RSC Adv. 6: 42720–42729. Doi: 10.1039/C6RA05003D.

Weaver, J.D. and Stabler, C.L. 2015. Antioxidant cerium oxide nanoparticle hydrogels for cellular encapsulation. Acta Biomater. 16: 136–144. Doi: 10.1016/j.actbio.2015.01.017.

Yahyaei, B. and Pourali, P. 2019. One step conjugation of some chemotherapeutic drugs to the biologically produced gold nanoparticles and assessment of their anticancer effects. Sci. Rep. 9. Doi: 10.1038/s41598-019-46602-0.

Yan, S., He, W., Sun, C., Zhang, X., Zhao, H., Li, Z., Zhou, W., Tian, X., Sun, X. and Han, X. 2009. The biomimetic synthesis of zinc phosphate nanoparticles. Dye Pigment 80: 254–258. Doi: 10.1016/j.dyepig.2008.06.010.

Yang, H., Dong, Y., Du, H., Shi, H., Peng, Y. and Li, X. 2011. Antioxidant compounds from propolis collected in Anhui, China. Molecules 16: 3444–3455. Doi: 10.3390/molecules16043444.

Zhang, W., Zhang, J., Ding, D., Zhang, L., Muehlmann, L.A., Deng, S.e., Wang, X., Li, W. and Zhang, W. 2018. Synthesis and antioxidant properties of Lycium barbarum polysaccharides capped selenium nanoparticles using tea extract. Artif. Cells Nanomedicine Biotechnol. 46: 1463–1470. Doi: 10.1080/21691401.2017.1373657.

Zomorodian, K., Pourshahid, S., Sadatsharifi, A., Mehryar, P., Pakshir, K., Rahimi, M.J. and Arabi Monfared, A. 2016. Biosynthesis and characterization of silver nanoparticles by aspergillus species. Biomed. Res. Int. 2016. Doi: 10.1155/2016/5435397.

7

Fungal Biofilms in Nanotechnology Era
Diagnosis, Drug Delivery, and Treatment Strategies

Mohammadhassan Gholami-Shabani,[1]
Masoomeh Shams-Ghahfarokhi,[2] *Fatemehsadat Jamzivar*[1]
and *Mehdi Razzaghi-Abyaneh*[1,*]

Introduction

Recent advances in new microscopy approaches and behavioral genetic research have provided evidence that shows the most widespread mode of growth of microorganisms in nature is biofilms formation (Blanco-Cabra et al. 2021). Biofilms are structures formed by microorganisms that protect them from environmental stresses (Yin et al. 2019). In the biofilm, microbes are surrounded by a self-produced matrix (generally containing polysaccharides, proteins, lipids and nucleic acids) that has several functions including protection and communication between cells inside the biofilms, as well as with the outside world (Flemming and Wingender 2010). This growth form presumably allows microbial cells to survive in hostile environments, enhances their resistance to physicochemical pressures, and promotes metabolic cooperation (Martinez and Fries 2010). In cases such as the waste water industry and the removal of oil pollution, microbial biofilms can be useful (Asri et al. 2018). However, they are mostly detrimental to health, social manufacture and living activities (Muhammad et al. 2020). Under diverse environmental conditions, microbes make biofilms at

---
[1] Department of Mycology, Pasteur Institute of Iran, Tehran 1316943551, Iran.
[2] Department of Mycology, Faculty of Medical Sciences, Tarbiat Modares University, Tehran 14115-331, Iran.
* Corresponding author: mrab442@yahoo.com, mrab442@pasteur.ac.ir

different levels, including the surfaces of natural water systems, water pipes, living tissues, teeth, and medical devices (Muhammad et al. 2020). Most microorganisms (bacteria, fungi, yeast) on earth live in various aggregates biofilms (Guzmán-Soto et al. 2021). Biofilm formation involves the following steps (Choudhary et al. 2020) as shown in Figure 1. Initial biofilm reversible attachment (initial step of biofilm formation), biofilm irreversible attachment, biofilm proliferation, biofilm growth or maturation (equilibrium between the accumulation and detachment) and biofilm dispersion (detachment and reversion to planktonic growth). Because of the complex matrix structure, biofilms are more tolerant to antimicrobials than planktonic cells (Sharma et al. 2019) and frequently develop antibiotic resistance, thus being more difficult to control (Dincer et al. 2020). As a result, conventional approaches (e.g., ultraviolet radiation, mechanical cleaning, traditional chemical antimicrobials) were employed against biofilms (Zea et al. 2020) and some novel methods have been investigating potential strategies for microbial biofilms control, such as quorum sensing inhibitors, antimicrobial peptides, enzymes, biomimetic surfaces, nanomaterials (NMs) (Zhou et al. 2020). However, many of these methods have confirmed a modest antimicrobial efficacy and still have barriers in successfully controlling microbial biofilms (Flemming et al. 2009). Among the above methods of controlling biofilms, nanotechnology holds great promise and has received enormous attention towards effective antimicrobial capacity (Mba and Nweze 2021). Nanomaterials (NMs) are particles possessing grain sizes of the order of 1–100 nm (Khan et al. 2019). NMs are more reactive and effective due to their very large surface

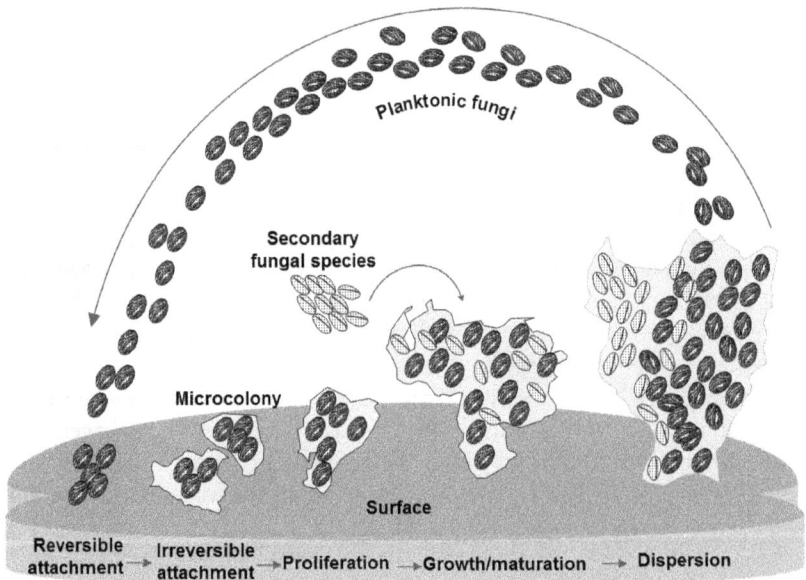

Figure 1. The fungal biofilm life cycle. Fungal biofilm formation is a multistage process that involves reversible attachment of cells to a surface, secretion of adhesins and extracellular polymeric substances that result in irreversible attachment of the biofilm and cell start to proliferate, growth/maturation of the biofilm and microcolony formation, and cell death in microcolonies and dispersal of single cells that return to a planktonic phase, completing the biofilm life cycle.

area compared to their size and are preferred over other previous antimicrobials. NMs have been proven to have antimicrobial and antibiotic properties against various bacterial, fungal and other microorganisms (Wang et al. 2017). Therefore, nanomaterials can be used as an alternative to inactivate pathogenic microorganisms. Unlike traditional chemical antimicrobials, antimicrobials based on NMs are cost-effective, easy to use, relatively ineffective in water and are not strong oxidizing agents (Wang et al. 2017). This chapter describes the potential of fungal biofilms in the production of NMs as well as the application of various NMs as antimicrobial agents to control microbial biofilms. The mechanisms of antimicrobial action and the principal applications of NPs are presented. Limitations for large-scale production of NMs and solutions to overcome them are also discussed. We also look at the application of nanomaterials on human health, as well as any usage interference with treatment processes, and provide guidance for future investigation.

Fungal Biofilms: Formation, Establishment and Dissemination

Varied fungi have shown the ability to colonize surfaces and organize biofilms (Martinez and Fries 2010). Most studies on fungal biofilms have been performed on *Candida albicans*. However, in recent years, several researchers have reported the involvement of yeast and Candida species, as well as filamentous fungi, in biofilm formation, such as *Aspergillus* species, *Blastoschizomyces*, *Cryptococcus gattii*, *Cryptococcus neoformans*, *Coccidioides immitis*, *Fusarium* species, *Histoplasma capsulatum*, *Malassezia pachydermatis*, *Mucorales*, *Paracoccidioides brasiliensis*, *Pneumocystis* species, *Rhodotorula* species, *Saccharomyces cerevisiae* and *Trichosporon asahii* (Kernien et al. 2018). There is a present interest in describing the specific properties of the biofilm creation using these fungi. One major concern is the control of biofilms, demanding knowledge of biofilm mechanisms (Coenye et al. 2020). However, our knowledge and information of these microbial communities are limited as a result of the complexity of these systems and the metabolic interactions which remain unknown (Coenye et al. 2020). Infections related to biofilm formation are known as a significant and growing clinical concern; therefore, mycology research has been progressively focused on biofilm genotyping (Lebeaux et al. 2014). Recent developments in molecular methods and confocal scanning laser microscopy have displayed that biofilm formation is the normal and preferred form of fungal growth and a major reason for persistent human infections (Blanco-Cabra et al. 2021). Biofilm microorganisms develop in multicellular communities and generate an extracellular matrix that makes available protection against host defense mechanisms and antifungal medications (Gebreyohannes et al. 2019). Table 1 summarizes the most common fungi that cause fungal biofilms infections in humans.

Clinical Relevance of Fungal Biofilms

The use of wide-range antibiotic and corticosteroid drugs, invasive medical surgical procedures, COVID-19, and the AIDS epidemic, are linked to a significantly increased prevalence of invasive fungal diseases that are basically challenging to treat

Table 1. Fungi that cause fungal biofilms infections in humans.

Fungi	Disease	Distribution	Treatment	References
Aspergillus spp.	Aspergillosis	Worldwide	Itraconazole Voriconazole Lipid Amphotericin Posaconazole Isavuconazole	(Morelli et al. 2021)
Blastomyces dermatitidis	Blastomycosis	United States Canada Africa India	Amphotericin B Itraconazole	(Baumgardner 2016)
Candida spp.	Candidiasis	Worldwide	Caspofungin Micafungin Anidulafungin Fluconazole Amphotericin B	(Nobile and Johnson 2015, Cavalheiro and Teixeira 2018)
Coccidioides immitis	Coccidioidomycosis	United States Mexico Central and South America	Fluconazole	(Fanning and Mitchell 2012)
Cryptococcus neoformans	Cryptococcosis	Worldwide	Fluconazole Amphotericin B corticosteroids	(Fanning and Mitchell 2012)
Trichophyton spp.	Dermatophytosis	Worldwide	Clotrimazole Miconazole Fluconazole	(Danielli et al. 2017, Dalla Lana et al. 2019)
Microsporum spp.				
Histoplasma capsulatum	Histoplasmosis	North and Central America	Itraconazole	(Gonçalves et al. 2020)
Mucormycetes	Mucormycosis	United States	amphotericin B Posaconazole Isavuconazole	(Firacative 2020)
Paracoccidioides brasiliensis	Paracoccidioidomycosis	Central and South America	Itraconazole Amphotericin B trimethoprim/ sulfamethoxazole	(Sardi et al. 2015)
Pneumocystis carinii	Pneumocystosis	United States Canada	trimethoprim/ sulfamethoxazole	(Cushion et al. 2009)

(Ghosh et al. 2021). Fungal infectious biofilms can develop on a variety of surfaces including host tissues and implanted biomaterials including vascular catheters (Tsui et al. 2016). This makes fungal biofilms a major clinical and economic problem. The frequency of fungal infections by biofilm is increasing due to immunodeficiency viruses and immunosuppressant drugs. The extent to which fungal biofilms have an impact on the human host is extensive, and usually, *Candida* species are predominant (Tsui et al. 2016).

C. albicans is the most common human pathogenic fungus causing illnesses ranging from mucous membrane infections to systemic infections (Nobile and Johnson 2015, Cavalheiro and Teixeira 2018). However, any disturbance to the host environment or under immune dysfunction conditions, *C. albicans* may possibly proliferate and invade almost any host site. The number of *Candida* species (non-*albicans*) that show biofilm formation and cause infections is constantly increasing and therefore causes great concern. *Candida* spp. that cause nosocomial infections include *C. glabrata*, *C. parapsilosis*, *C. krusei*, and *C. tropicalis* (Jahagirdar et al. 2018). Figure 2 depicts some of the major body sites colonized with fungal biofilms, which will now be examined in detail.

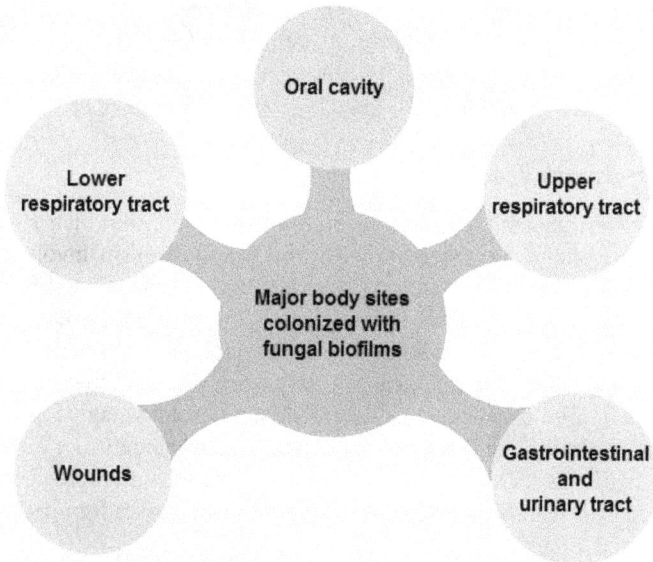

Figure 2. Major body sites colonized with fungal biofilms.

Oral Cavity

The oral cavity is one of the main entrance portals for microorganisms and is a substantial site for the presence of various microbial biofilms (Deo and Deshmukh 2019). Microbial biofilms are involved in cavities, periodontal diseases, endodontic and mucous membrane infections (Deo and Deshmukh 2019). Among the fungi related to the oral cavity, *Candida* species are usually the most plenty and important (Vila et al. 2020). The oral cavity provides an optimum environment for the growth of biofilms by various fungal species (Deo and Deshmukh 2019). Apart from *Candida* species, other fungi have been isolated from patients such as *Cladosporium*, *Saccharomycetales*, *Aureobasidium*, *Aspergillus*, *Cryptococcus* and *Fusarium* (Limon et al. 2017). Oral candidiasis is one of the most highly defined fungal biofilm infections of soft and hard tissues, creating complex biofilms in relation to host components and bacteria. *Candida* species have been isolated from periodontal pouches, orthodontic appliances, enamel, dentures and mucous membranes of the patients (Vila et al. 2020).

Upper Respiratory Tracts

Sinusitis (also known as rhinosinusitis) is demarcated as an inflammation of the mucosa that covers the paranasal sinuses. It is classified as acute or chronic rhinosinusitis classified as acute recurrent, sub-acute, and acute chronic exacerbation. All these sinusitis types show similar symptoms and are therefore often difficult to distinguish. Most adults have had sinusitis at some time in their lives, and this disease has a major impact on their quality of life. There is increasing recognition that chronic rhinosinusitis is characterized by the growth of fungal biofilm (Fastenberg et al. 2016). Although there is considerable evidence of the role of bacterial biofilms in sinus infection, there is disagreement with fungi (Foreman et al. 2012). Considering the anatomical location of the sinuses and the lack of appropriate diagnosis, it is difficult to determine whether they form defined biofilms (Foreman et al. 2012, Fastenberg et al. 2016). However, clinically paranasal sinus fungal balls have been reported which have been likened to fungal biofilms (Fastenberg et al. 2016).

Lower Respiratory Tracts

The airways are frequently exposed to biofilm formation, of which *Pseudomonas aeruginosa* is undoubtedly the most important microorganism involved (Grainha et al. 2020). However, it is increasingly acknowledged that fungal biofilms may persist in the lungs and contribute to infection. Filamentous molds like *A. fumigatus* may cause a spectrum of respiratory diseases, including aspergilloma, invasive aspergillosis and allergic bronchopulmonary aspergillosis. Aspergilloma has the most obvious biofilm characteristics, which in some people may be asymptomatic, while others typically experience chronic cough and hemoptysis (Williams et al. 2016). *Aspergillus* bronchitis is also a disease associated with problematic biofilm, characterized by bronchial plasters containing compact masses forming mycelium.

Gastrointestinal and Urinary Tracts

Fungal biofilms are unusual causes of urinary tract infection in healthy persons but common in hospitals or in patients with predisposing diseases and structural anomalies of the kidney and collector system (Fisher et al. 2011). The urinary system can be invaded by either the bloodstream or the urethra and bladder. *Candida* spp. use a range of pathogens, including phenotypic alteration, dimorphism, galvano and tigmotropism, and hydrolytic enzymes for colonization and invasion of the urinary tract. The antegrade disease occurs primarily in people with a predisposition to candidemia (Fisher et al. 2011). Fungal biofilm infection of the gastrointestinal tract may result from a proliferation of yeast in the intestine, exposure to contaminated water and food or invasive fungal infections that are disseminated from other sites. The vastness of the infection is dependent upon the major risk factors, for example, diabetes or immunosuppression, and ranges from colonization, localized infection, or fungemia, to aggressive life-threatening gastrointestinal tract infections. *Candida* spp. are the commonest cause of mucosal infection, although mould infections are increasingly reported (Schelenz 2017).

Wounds

Wound infection is the invasion of a wound through the proliferation of microorganisms at a level that invokes a local and/or systemic response in the host. The presence of microbes within the wound results in damage to local tissues and prevents wound healing (Kalan and Grice 2018). Intervention is usually required to help the host's defenses destroy invasive microorganisms. Studies published in recent years have confirmed that biofilms form in wounds (James et al. 2008, Bjarnsholt et al. 2008, Davis et al. 2008). Scanning electron microscopy revealed that more than sixty percent (60%) of chronic wounds contained biofilm (James et al. 2008, Bjarnsholt et al. 2008, Davis et al. 2008). Since then, a fast-growing body of scientific work has reported the impact of microbial biofilm on a wound (Kalan and Grice 2018). Despite the rapid development in laboratory biofilm research and the development of biofilm-based wound care principles, the link between wound biofilm and clinical outcomes remains ambiguous (Cavalheiro and Teixeira 2018). The increasing understanding and acceptance of the role of microbial biofilm in wound infection has led to an evolution in the clinical management of chronic, non-healing wound that seeks to address the potential presence of fungal biofilm (Kalan and Grice 2018).

Fungal Biofilm-Derived Nanomaterials as Novel Antimicrobials and Antibiofilms

Nanotechnology is the study and production of materials on a scale of 1 to 100 nm (Gholami-Shabani et al. 2018). Due to the variation of physicochemical properties, materials at the nano-scale have diverse properties as compared to their original characteristics. Having various materials with different size/shape-dependent characteristics, presents unique occasions to researchers to improve the novel form of material with activity in broad spectrum fields of knowledge and technology. Nanoparticles are a vast part of nanomaterials with novel structures and properties with wide application in various aspects of our life, for example, agriculture, medicine, industry, and the environment (Gholami-Shabani et al. 2021a, b). The variety of different physicochemical properties has drawn attention to the production of nanoparticles more than other nanomaterial sectors. Currently, some nanomaterials have been produced on an industrial scale and more and more have developed their applications. Nanoparticles are a promising part of nanomaterials that have reached industrial manufacture due to their importance and efficiency in many aspects of our lives (Njuguna et al. 2021).

In many instances, nature is a guide to the advancement of human technology. Looking for a more environmentally friendly approach to synthesizing nanomaterials, scientists were inspired by how microorganisms interact with metals in the natural environment. Fungi have a long history of absorbing and accumulating inorganic metallic ions from their environment. Different fungi were exploited for the green synthesis of metallic nanomaterials. Some typical metallic nanoparticles synthesized by a biofilm of fungi are summarized in Table 2.

Table 2. Typical nanoparticles synthesized by fungi and their antimicrobial and antibiofilm activity.

Fungi	NPs	Size (nm)	Shape	Antimicrobial and Antibiofilm Activity	References
Alternaria alternata	Silver	20–60	Spherical	*Phoma glomerata Trichoderma* sp. *Candida albicans*	(Gajbhiye et al. 2009)
	Iron	9	cubic	*Bacillus subtilis Escherichia coli Staphylococcus aureus Pseudomonas aeruginosa*	(Mohamed et al. 2015)
Aspergillus clavatus	Silver	10–25	Spherical Hexagonal	*Candida albicans Pseudomonas fluorescens Escherichia coli*	(Verma et al. 2010)
Aspergillus flavus	Silver	33.5	Spherical	*Proteus vulgaris Escherichia coli Pseudomonas aeruginosa Staphylococcus aureus Aspergillus niger Penicillium chrysogenum Alternaria alternata Fusarium culmorum*	(Sulaiman et al. 2015)
	Selenium	51.5	Spherical	*Aspergillus* spp. *Candida* spp.	(Bafghi et al. 2021)
Aspergillus fumigatus	Zinc oxide	60–80	Spherical	*Klebsiella pneumoniae Pseudomonas aeruginosa Escherichia coli Staphylococcus aureus Bacillus subtilis*	(Rajan et al. 2016)
	Silver	ND	ND	*Escherichia coli Klebsiella pneumoniae Enterococcus* sp. *Staphylococcus albus*	(Bala and Arya 2013)
Aspergillus niger	Cerium oxide	5–20	cubic	*Streptococcus pneumoniae Bacillus subtilis Proteus vulgaris Escherichia coli*	(Gopinath et al. 2015)
	Magnesium Oxide	40–95	Spherical	*Staphylococcus aureus Pseudomonas aeruginosa*	(Ibrahem et al. 2017)
	Silver	10–100	ND	*Fusarium oxysporum Aspergillus flavus Penicillium digitatum*	(Al-Zubaidi et al. 2019)

Table 2 contd. ...

...Table 2 contd.

Fungi	NPs	Size (nm)	Shape	Antimicrobial and Antibiofilm Activity	References
Aspergillus oryzae	Silver	3–105	Spherical	*Escherichia coli Staphylococcus aureus Bacillus subtilis Klebseilla pneumoniae*	(Phanjom and Ahmed 2017)
	Selenium	55	Sphere isotropic	*Acinetobacter calcoaceticus Staphylococcus aureus Candida albicans Aspergillus flavus*	(Mosallam et al. 2018)
	Iron	10–25	Spherical	ND	(Tarafdar and Raliya 2013)
Aspergillus sydowii	Silver	1–24	Spherical	*Aspergillus* spp. *Fusarium* spp. *Candida* spp. *Cryptococcus neoformans Sporothrix schenckii*	Wang et al. 2021)
Aspergillus terreus	Silver	10–18	Spherical	*Aspergillus ochraceus Aspergillus niger Aspergillus parasiticus*	(Ammar and El-Desouky 2016)
Aspergillus tubingensis	Silver	25–45	ND	*Escherichia coli Staphylococcus aureus Bacillus subtilis Candida* spp.	(Rodrigues et al. 2021)
Aspergillus kambarensis	Silver	> 50	Different	*Escherichia coli Bacillus subtilis Proteus vulgaris Staphylococcus aureus Candida albicans Candida tropicalis Fusarium oxysporum Aspergillus niger*	(Gholami-Shabani et al. 2021a)
	Gold				
	Copper				
	Zinc oxide				
Aureobasidium pullulans	Silver	2–53	Spherical	*Escherichia coli Salmonella typhi Staphylococcus aureus Pseudomonas* sp.	(Rahi et al. 2018)
Bipolaris nodulosa	Silver	10–60	Spherical Semipentagonal hexahedral	*Bacillus subtilis Bacillus cereus Pseudomonas aeruginosa Proteus vulgarius Escherichia coli Micrococcus luteus*	(Saha et al. 2010)

Table 2 contd. ...

...Table 2 contd.

Fungi	NPs	Size (nm)	Shape	Antimicrobial and Antibiofilm Activity	References
Candida albicans	Zinc oxide	15–25	Quasi-spherical	ND	(Mashrai et al. 2017)
	Silver	61–66	Spherical	*Staphylococcus aureus* *Escherichia coli* *Bacillus cereus* *Vibrio cholerae* *Proteus vulgaris*	(Bhat et al. 2015)
Candida glabrata	Silver	2–15	Spherical	*Candida albicans*	(Jalal et al. 2018)
Cladosporium cladosporioides	Silver	5–50	Different	*Trichophyton* spp. *Candida* spp. *Rhodotorula* spp. *Penicllium* spp. *Mucor* spp. *Syncephalastrum* spp. *Aspergillus niger*	(Lafta et al. 2019)
	Gold	40	Irregular	*Escherichia coli* *Staphylococcus aureus* *Bacillus subtilis* *Pseudomonas aeruginosa* *Aspergillus niger*	(Joshi et al 2017)
Cochliobolus lunatus	Silver	3–21	Spherical	ND	(Salunkhe et al. 2011)
Colletotrichum sp.	Sulphur oxide	50	Spherical	*Salmonella typhi* *Chromobacterium violaceum* *Listeria monocytogenes* *Aspergillus flavus* *Fusarium oxysporum*	(Suryavanshi et al. 2017)
	Aluminium oxide	30			
Coriolus versicolor	Silver	15–30	Spherical	ND	(Deniz et al. 2019)
Cylindrocladium floridanum	Gold	14–19	Spherical	ND	(Narayanan and Sakthivel 2013)
Epicoccum nigrum	Silver	37	Spherical	*Alternaria solani*	(Abdel-Hafez et al. 2017)
Fusarium oxysporum	Silver	14–25	Spherical	*Aspergillus fumigatus* *Alternaria alternata* *Trichoderma parceramosum* *Penicillium citrinum* *Paecilomyces variotii* *Candida albicans* *Candida glabrata* *Trichophyton mentagrophytes* *Microsporum gypseum*	(Gholami-Shabani et al. 2014)

Table 2 contd. ...

...Table 2 contd.

Fungi	NPs	Size (nm)	Shape	Antimicrobial and Antibiofilm Activity	References
	Vanadium Oxide	10–20	Spherical	*Fusarium oxysporum* *Fusarium graminearum* *Aspergillus fumigatus* *Aspergillus niger* *Aspergillus flavus* *Alternaria alternata* *Penicillium citrinum*	(Gholami-Shabani et al. 2021b)
	Platinum	25	Cubical Spherical Truncated triangular	*Staphylococcus aureus* *Pseudomonas aeruginosa* *Klebsiella pneumoniae* *Escherichia coli* *Aspergillus niger*	(Gupta and Chundawat 2019)
Fusarium semitectum	Silver	1–50	Spherical Ellipsoid	*Pseudomonas aeruginosa* *Klebsiella pneumoniae*	(Shelar and Chavan 2014)
	Selenium	32–103	Spherical	*Staphylococcus aureus* *Pseudomonas aeruginosa* *Klebsiella pneumoniae* *Acinetobacter baumanni* *Escherichia coli* *Proteus vulgaris*	(Abbas and Abou Baker 2020)
Guingnardia mangiferae	Silver	5–30	Spherical	*Aspergillus niger* *Colletotrichum* sp. *Fusarium* sp. *Rhizoctonia solani* *Curvularia lunata*	(Balakumaran et al. 2015)
Hansenula anomala	Gold	14	Different	*Bacillus cereus* *Pseudomonas putida*	(Arumugam and Berchmans 2011)
Hormoconis resinae	Silver	20–80	Different	ND	(Varshney et al. 2009)
Hypocrea lixii	Copper	24.5	Spherical	ND	(Salvadori et al. 2013)
Macrophomina phaseolina	Silver	5–40	Spherical	*Escherichia coli* *Agrobacterium tumefaciens*	(Chowdhury et al. 2014)
Monascus purpureus	Silver	1–7	Spherical Cuboids	*Staphylococcus aureus* *Staphylococcus epidermidis* *Streptococcus spyrogenes* *Escherichia coli* *Pseudomonas aeruginosa* *Salmonella typhimurium* *Candida albicans* *Candida tropicalis* *Candida glabrata*	(El-Baz et al. 2016)

Table 2 contd. ...

...Table 2 contd.

Fungi	NPs	Size (nm)	Shape	Antimicrobial and Antibiofilm Activity	References
	Cobalt ferrite	6.5	Spherical	*Staphylococcus aureus Escherichia coli Klebsiella pneumonia Pseudomonas aeruginosa Aspergillus niger Alternaria solani Fusarium oxysporum Candida albicans*	(El-Sayed et al. 2020)
Neurospora intermedia	Silver	19–84	Spherical	*Escherichia coli*	(Hamedi et al. 2014)
Pleurotus citrinopileatus	Silver	> 40	Spherical	*Bacillus subtilis Bacillus cereus Staphylococcus aureus Escherichia coli Pseudomonas aeruginosa*	(Al-Bahrani et al. 2017)
Pleurotus djamor	Silver	5–50	Spherical	ND	(Raman et al. 2015)
Penicillium nalgiovense	Silver	25	Spherical	ND	(Maliszewska et al. 2014)
Pleurotus ostreatus	Silver	4–15	Spherical	*Candida albicans Candida glabrata Candida parapsilosis Candida krusei*	(Yehia and Al-Sheikh 2014)
Penicillium brevicompactum	Silver	30–50	Spherical	*Staphylococcus epidermidis Staphylococcus aureus Vibro cholerae Proteus vulgaris Escherichia coli*	(Majeed et al. 2016)
Penicillium chrysogenum	Magnesium oxide	5–13	Rounded	*Enterococcus faecalis Klebsiella pneumoniae Candida albicans*	(El-Sayyad et al. 2018)
Pestalotia sp.	Silver	10–40	Spherical	*Staphylococcus aureus Salmonella typhi*	(Raheman et al. 2011)
Phanerochaete chrysosporium	Cadmium sulfide	1.5–2	Spherical	ND	(Chen et al. 2014)
Phoma glomerata	Silver	60–0	Spherical	*Escherichia coli Pseudomonas aeruginosa Staphylococcus aureus*	(Birla et al. 2009)
Phoma sorghina	Silver	20–25	Spherical	*Staphylococcus aureus Brevibacillus borstelensis*	(Sonar et al. 2017)

Table 2 contd. ...

...Table 2 contd.

Fungi	NPs	Size (nm)	Shape	Antimicrobial and Antibiofilm Activity	References
Phytopthora infestans	Silver	ND	ND	*Shiegella dysentriae* *Escherichia coli* *Salmonella typhi* *Klebsiella pneumoniae* *Proteus vulgaris* *Bacillus subtilis* *Staphylococcus aureus*	(Thirumurugan et al. 2009)
Pleurotus florida	Silver	5–40	Spherical	*Staphylococcus* sp. *Bacillus* sp.	(Kaur et al. 2018)
Pleurotus sajor caju	Silver	3–33	Spherical	*Candida albicans*	(Musa et al. 2018)
Pycnoporus sanguineus	Silver	39–87	Spherical	*Escherichia coli* *Staphylococcus aureus* *Staphylococcus epidermidis* *Candida albicans* *Aspergillus niger*	(Chan and Don 2012)
Rhizopus oryzae	Magnesium oxide	10–30	Spherical	*Staphylococcus aureus* *Bacillus subtilis* *Pseudomonas aeruginosa* *Escherichia coli* *Candida albicans*	(Hassan et al. 2021)
Rhizopus stolonifer	Silver	5–30	Spherical	*Candida glabrata* *Candida parapsilosis* *Candida krusei*	(Rathod et al. 2012)
Saccharomyces boulardii	Silver	60–100	Spherical	*Acinetobacter baumannii* *Cryptococcus neoformans* *Escherichia coli* *Klebsiella pneumonia* *Pantoea agglomerans* *Pseudomonas aeruginosa* *Staphylococcus aureus* *Saccharomyces boulardii* *Streptococcus pyogenes* *Salmonella Typhi*	(Sahib et al. 2017)
Saccharomyces cerevisiae	Selinium	30–100	Spherical	*Escherichia coli* *Pseudomonas aeruginosa* *Klebsiella pneumoniae* *Salmonella typhimurium* *Staphylococcus aureus* *Bacillus subtilis*	(Hariharan et al. 2012)

Table 2 contd. ...

...Table 2 contd.

Fungi	NPs	Size (nm)	Shape	Antimicrobial and Antibiofilm Activity	References
Tricholoma crassum	Silver	5–50	Spherical, hexagonal	*Escherichia coli Agrobacterium tumifaciens Magnaporthe oryzae*	(Chowdhury et al. 2011)
Trichoderma harzianum	Silver	20–30	Spherical	*Trichoderma harzianum Sclerotinia sclerotiorum Escherichia coli Staphylococcus aureus Candida albicans*	(Guilger et al. 2017)
Trichoderma koningii	Silver	8–24	Spherical	*Salmonella typhimurium*	(Tripathi et al. 2013)
Trichoderma reesei	Silver	1–25	Spherical	ND	(Gemishev et al. 2019)
Trichoderma viride	Silver	1–50	Globular	*Shigella boydii Acinetobacter baumannii Salmonella typhimurium Shigella sonnei*	(Elgorban et al. 2016)
Trichoderma asperellum	Chitosan	150–350	Spherical	*Fusarium oxysporum Sclerotium rolfsii Rhizoctonia solani*	(Boruah and Dutta 2021)
Thraustochytrium kinnei	Gold	10–85	cubical	*Proteus mirabilis Staphylococcus aureus Streptococcus pyogens Vibrio cholerae Vibrio parahaemolyticus Escherichia coli Klebsiella pneumoniae Klebsiella oxytoca Salmonella typhi Salmonella paratyphi*	(Kalidasan et al. 2021)
	Silver	5–90			
Yarrowia lipolytica	Silver	5–60	Spherical	*Staphylococcus aureus Escherichia coli Enterococcus faecalis Proteus vulgaris Streptococcus pyogenes Pseudomonas aeruginosa*	(Bolbanabad et al. 2020)

NPs = nanoparticles; ND = not determined

Nanomaterials as Diagnostic Agents for Fungal biofilms

Early detection is critical to the effective treatment of potentially lethal infections caused by pathogenic fungi because the delayed diagnosis of systemic infection is almost always an adverse prognosis (Hussain et al. 2020). The field of fungal diagnosis includes some tests that are relatively simple, fast to perform, and potentially appropriate at the point of care. However, there are also more complex high-tech methods that offer new chances regarding the scale and precision of fungal infection diagnosis, but are more limited in their transportability and affordability. Future advances in nanotechnology can overcome these problems (Prattes et al. 2016, Hussain et al. 2020).

Nanomaterials have some unique physicochemical and biological properties which are fundamentally different from those of the corresponding bulk material and could be widely used in medical testing (Alharbi and Al-Sheikh 2014, Khan et al. 2019). One of the possible applications of bionanotechnology is the timely diagnosis of various diseases (Figure 3). Many infectious agents of fungi which cause Aspergillosis, Candidiasis, Blastomycosis, Dermatophytosis, Coccidioidomycosis, Cryptococcosis, Histoplasmosis, Mucormycosis, Paracoccidioidomycosis, Pneumocystosis, etc., spread out very fast and cause epidemic outbreaks with very high morbidity and mortality (Kernien et al. 2018). Therefore, there is great potential in using nanotechnology for diagnosing infectious agents.

Integrating molecular diagnostics and nanotechnologies is a promising technology for the fast and accurate identification of fungal pathogens (Campos et al. 2020). Presently, several nanodevices and nanosystems have been used in diagnostics as well as sequenced DNA molecules (Mukhtar et al. 2021). Assays which use nano-size devices to investigate DNA sequences and diagnose the disease are becoming faster, more flexible and more sensitive (Mukhtar et al. 2021). It should be mentioned here that newly developed nanomaterials with special nanometric characteristics offer an enormous breakthrough in the technology of detection and diagnosis of fungal pathogens. Nanotechnology is also driving the development of lab-on-chip systems for detecting pathogens, toxicity in water, observing nutrients in irrigation water, and controlling the quality of food products (Alharbi and Al-Sheikh 2014).

Metallic nanoparticles were applied to biosensors as markers to replace enzymes as a label (Malekzad et al. 2017). Quantum dots (QD) are a class of luminescent semiconductor nanocrystals that emit light of specific wavelengths, in which the size of the nanoparticle determines the wavelength. The larger the size, the higher the wavelength of the infrared light emitted (Gaikwad et al. 2019). They have several advantages in comparison with organic dyes, based on wide excitation spectra. Quantum dots have narrow adjustable emission peak, longer fluorescence lifetime, photobleaching resistance and a molar extinction coefficient 10 to 100 times higher. These quantum dot properties allow multicolored quantum dots to produce brighter probes than conventional fluorophores (which are excited from a source by ordinary fluorescent dyes without emitting signal overlap).

Gaikwad et al. (2019) demonstrated carbon dots (CDs)-based thin film as a sensor for detection of fungal spores from the environment. The detection procedure is based on fluorescence, observed in the film of carbon points placed on quartz plates using the Blodgett method (Gaikwad et al. 2019). It is observed that the CDs film shows quenching in the fluorescence intensity by the substrate, namely, fungal spores' (*Aspergillus niger, Penicillium chrysogenum, Alternaria alternata*).

Rispail et al. (2014) showed the QD composed of a cadmium selenide (CdSe) core and a zinc sulphide (ZnS) shell, rapidly interacted with the fungal hypha, labeling the presence of the pathogenic fungus *F. oxysporum* (Rispail et al. 2014). Kattke et al. (2011) showed fluorescence resonance energy transfer (FRET)-Based QD immunoassay for rapid and sensitive detection of *Aspergillus amstelodami* (Kattke et al. 2011). Safarpour et al. (2012) reported a novel quantum dots FRET-based biosensor for efficient detection of *Polymyxa betae* (Safarpour et al. 2012). Yu et al. (2019) reported a fluorimetric detection of *C. albicans* using cornstalk n-carbon quantum dots modified with amphotericin B (Yu et al. 2019). Chen et al. (2020) reported red-emitting carbon dots for bioimaging of fungal cells and detecting Hg^{2+} and ziram in an aqueous solution (Chen et al. 2020).

Nanomaterials as Drug Delivery Systems

Over the last few decades, the application of nanotechnology in medicine has been extensively explored in many medical areas, especially in drug delivery (Figure 3). By loading drugs into nanomaterials through physical or chemical conjugation, encapsulation, adsorption, etc., the clinical pharmacokinetics (such as time course of drug absorption, distribution, metabolism, and excretion) and therapeutic index of the drugs can be noticeably improved in contrast to the free drug similitudes (Vargason et al. 2021). Many advantages of nanomaterials-based drug delivery methods have been known, including enhancing serum solubility of the drugs, extending the systemic circulation lifetime, releasing drugs in a sustained and controlled way, especially delivering drugs to the tissues and cells of interest, and simultaneously delivering multiple therapeutic agents to the same cells for combination treatment (Vargason et al. 2021). In addition, drug-laden nanomaterials may enter host cells using endocytosis, and then release the drug payloads to cure and remove microbe-induced intracellular infections. Consequently, a number of nanomaterials-based drug delivery methods have been approved for medical applications in the treatment of a variety of diseases and numerous other formulations of therapeutic nanomaterials are currently at different stages of clinical testing. Table 3 provides an overview of new antifungal drug delivery methods. Therefore, antifungal/antibiofilm drug delivery methods have allowed the development of numerous pharmaceutical products that improve patient health using increased delivery of a therapeutic to its target spot, reducing off-target accumulation and helping patient compliance. As green therapeutic approaches developed, targeted drug delivery systems were adapted to address the novel challenges that emerged (Vargason et al. 2021).

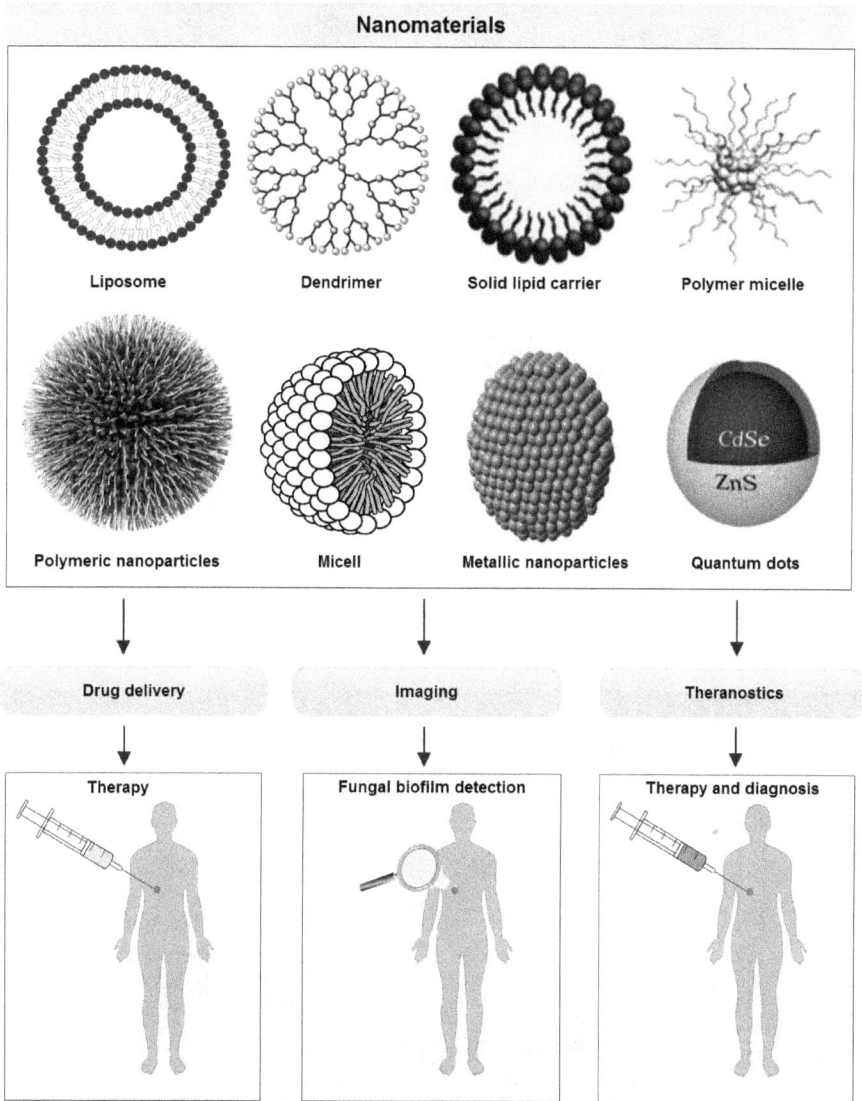

Figure 3. Smart nanomaterials for diagnostic, drug delivery and treatment of fungal disease.

Nanomaterials as Therapeutic Agents for Fungal Biofilms

Several studies have confirmed the effectiveness of nanomaterials in preventing and eliminating fungal biofilms. These nanomaterials mainly include gold, silver, zinc, copper, magnesium, iron, etc.

Silver Nanoparticles (AgNPs)

Silver nanoparticles have attracted a lot of attention as a result of their unique physical, chemical, biological properties and also for their vast application such as

Table 3. Some important drug delivery systems to develop antifungal drugs.

Common Antifungal Drugs	Drug Delivery Systems	Routes of Medication Administration	Treatment Disease or Fungi	References
Amphotericin B	Liposomes	Intravenous	Aspergillosis, Candidiasis, Blastomycosis, Paracoccidioidomycosis, Coccidioidomycosis, Cryptococcosis, Histoplasmosis, Mucormycosis, Hyalohyphomycosis, Phaeohyphomycosis, Dermatophytosis, Dermatomycosis, Sporotrichosis, Talaromycosis (Formerly Penicilliosis), and Trichosporonosis	(Jain and Kumar 2010, Saldanha et al. 2016, Stone et al. 2016, Sosa et al. 2017, Cuddihy et al. 2019, Faustino and Pinheiro 2020, Saqib et al. 2020, Fernández-García et al 2020, Nimtrakul et al. 2020)
	SLN/NLC	Oral		
		Topical		
	Magnetic nanoparticles	Nasal instillation		
	Nanoemulsion	Topical		
	Polymeric nanoparticles	Intravenous		
		Oral		
	Transfersomes	Topical		
	Copolymeric micelles	Oral		
Butenafine	Microemulsion	Topical	Tinea (pityriasis) versicolor due to *Malassezia furfur*, as well as athlete's foot (Tinea pedis), ringworm (Tinea corporis) and jock itch (Tinea cruris) due to *Epidermophyton floccosum*, *Trichophyton mentagrophytes*, *Trichophyton rubrum*, and *Trichophyton tonsurans*. It also displays superior activity against *Candida albicans* than terbinafine and naftifine. Butenafine demonstrates low minimum inhibitory concentrations against *Cryptococcus* and *Aspergillus*	(Pillai et al. 2015)

Drug	Formulation	Route	Description	References
Clotrimazole	Liposomes	Topical	Treatment of vaginal and skin infections due to yeasts and dermatophytes. *In vitro*, it is most active against *Candida* spp., *Trichophyton* spp., *Microsporum* spp. and *Malassezia furfur* (*Pityrosporon orbiculare*)	(Sawyer et al. 1975, Ning et al. 2005, Souto and Müller 2006, Alam et al. 2017, Makhmalzade and Chavoshy 2018, Osanloo et al. 2018, Bolla et al. 2019, Madhul.atha et al. 2020, Fathalla et al. 2020, Gosecka and Gosecki 2021)
	Nanosponges	Topical		
	Ethosomes	Topical		
	Niosomes	Topical		
	Ufosomes	Topical		
	Polymeric emulgel	Topical		
	Polymeric micelles	Topical		
	SLN/NLC	Topical		
	Microemulsion	Topical		
Ciclopirox	Liposomes	Topical	Dermatologic treatment of superficial mycoses. It is most useful against Tinea versicolor	(Asadi et al. 2021)
Econazole	Microemulsion	Percutaneous	Tinea corporis, Tinea pedis, Tinea cruris, Tinea versicolor, Yeast infection of the skin (cutaneous candidiasis).	(Sanna et al. 2007, Sharma and Pathak 2011, Geet al. 2014, Verma and Utreja 2018)
	SLN/NLC	Topical		
	Transethosomes	Transdermal		
	Nanosponges	Topical		
Fluconazole	Microemulsion	Topical	Dermatophytosis, Vaginal candidiasis, Cryptococcosis and Coccidioidomycosis, *Malassezia* infections	(Bhalaria et al. 2009, Fetih 2016, El-Housiny et al. 2018, Sakita et al. 2019, Soliman et al. 2019, Fernandes et al. 2020, Mahant et al. 2020, Bafrui et al. 2020, Bezerra et al. 2020, Albayaty et al. 2021, Vlaia et al. 2021)
	Niosomes	Topical		
	Liposomes	Topical		
	SLN/NLC	Topical		
	Microsponges	Topical		
	Ethosomes	Topical		
	Polymeric micelles	Topical		

Table 3 contd. ...

...Table 3 contd.

Common Antifungal Drugs	Drug Delivery Systems	Routes of Medication Administration	Treatment Disease or Fungi	References
Itraconazole	Transfersomes	Pulmonary	Dermatophytes (tinea infections) Yeasts such as candida and *Malassezia* infections Systemic fungal infections such as *Histoplasma, Aspergillus,* coccidiodomycosis, chromoblastomycosis	(Rhee et al. 2007, Alam et al. 2009, Alomrani et al. 2015, Leal et al. 2015, Aghdam et al. 2016, Ling et al. 2016, Zeb et al. 2017, Ho et al. 2020)
	SLN/NLC	Oral and Parenteral		
	Miosomes	Topical		
	Microemulsion	Transdermal		
		Parenteral		
	Liposomes	Topical		
	Polymeric Nanoparticle	Transdermal		
		Intravenous		
Ketoconazole	SLN/NLC	Topical	Dermatophytosis, Candidiasis, Onychomycosis, Blastomycosis, Coccidioidomycosis, Histoplasmosis, Chromomycosis, and Paracoccidioidomycosis	(Souto et al. 2005, Dave et al. 2017, Kaur et al. 2018, Amra and Momin 2019, Mlynarczyk et al. 2021)
	Liposomes	Topical		
	Dendrimers	Topical		
	Microemulsion	Topical		
	Niosomes	Topical		
Miconazole	Liposomes	Topical	Candidiasis, Dermatophytosis, Cryptococcosis, treat ring worm, Pityriasis versicolor, and yeast infections of the vagina, mouth and skin	(Firthouse et al. 2011, Ofokansi et al. 2012, Aljaeid and Hosny 2016, Firooz et al. 2016, Sadique et al. 2018, Fernandes et al. 2019, Tejada et al. 2020)
	SLN/NLC	Topical		
		Oral		
	Niosomes	Transdermal		
	Polymeric Nanoparticle	Topical		
	Microemulsion	Topical		

Drug	Nanoformulation	Route	Uses	References
Nystatin	SLN/NLC	Topical	Treat *Candida* infections of the skin including diaper rash, thrush, esophageal candidiasis, and vaginal yeast infections, Aspergillosis, Cryptococcosis	(Carrillo-Munoz et al. 1999, Offner et al. 2004, El-Ridy et al. 2011, Campos et al. 2012, Khalil et al. 2013)
	Nanoemulsion	Topical		
	Liposomes	Intravenous		
	Niosomes	Parenteral		
Terbinafine	Polymeric Nanoparticle	Topical	Onychomycosis, Dermatophytosis Pheohyphomycosis, Hyalohyphomycosis, Aspergillosis, Candidiasis	(Koutsoulas et al. 2014, Gaba et al. 2015, Elmataeeshy et al. 2018, Zhang et al. 2020, Mirshekari et al. 2021)
	SLN/NLC	Topical		
	Ethosomes	Dermal		
	Liposomes	Topical		
	Nanoemulsion	Transdermal		
Voriconazole	Nanoemulsion	Transcorneal	Candidiasis, Aspergillosis, Coccidioidomycosis, Histoplasmosis, Penicilliosis, and infections by *Scedosporium* or *Fusarium*	(Song et al. 2014, Veloso et al. 2018, Rasoanirina et al. 2020, Mumtaz et al. 2021)
	SLN/NLC	Topical		
	Liposomes	Intravenous		
	Polymeric Nanoparticle	Topical		

Solid Lipid Nanoparticles (SLN), Nanostructured Lipid Carriers (NLC) antibacterial, antiviral, antifungal, anti-inflammatory, and, etc. (Galatage et al. 2021). Silver has a promising role in the treatment of fungal infections related to wounds and burns (Khansa et al. 2019). Different methods have now been developed and introduced to synthesize AgNPs including physicochemical, and biological methods, among which the biological methods have been preferred for being more simple, cost-effective, less toxic, and for having a high yield, and environmentally friendly procedures (Gajbhiye et al. 2009, Sulaiman et al. 2015, Galatage et al. 2021). To rule out the disadvantages of chemical and physical methods for synthesizing AgNPs, Gholami-Shabani et al. (2014) applied a green method using fungal enzymes of *Fusarium oxysporum*. Antibiofilm and antimicrobials of AgNPs compared with standard and traditional antimicrobials were investigated against a series of fungi and bacteria. AgNPs have shown growth inhibition of many fungi like *A. fumigates*, *A. niger*, *A. flavus*, *T. rubrum*, *C. albicans*, and *Penicillium* species. Currently, AgNPs work by producing reactive oxygen species and free radicals and motive denaturation/damage to protein, nucleic acid and proton pump, lipid peroxidation, and cell wall damage. Consequently, they alter the cell membrane penetrability, causing cell death.

Several surveys have confirmed the effectiveness of AgNPs in preventing fungal biofilm-related infections. Lara et al. (2015) studied the effect of AgNPs on *C. albicans* biofilms. Their nanoparticles were manufactured using microwave-assisted methods. A phenotypic analyze of *C. albicans* displayed a potent dose-dependent inhibitory effect of AgNPs on biofilm formation, with an IC_{50} of 0.089 ppm. Also, AgNPs displayed efficacy when tested against preformed *C. albicans* biofilms resulting in an IC_{50} of 0.48 ppm. The cytotoxicity analyze resulted in a CC_{50} of 7.03 ppm. The ultrastructural differences visualized under SEM with AgNPs treatment were changes in the surface appearance of the yeast from smooth to rough thus showing outer cell wall damage. On the fungal preformed biofilm, the true hyphae were mostly absent because filamentation was inhibited. TEM study of the cell-wall width of *C. albicans* after treatment resulted in substantial enlargement (206 ± 11 nm) demonstrating membrane permeabilization (Lara et al. 2015).

Zhou et al. (2021) reported the antifungal activity of AgNPs against *C. albicans* and its mechanisms were tested *in vitro*. Moreover, AgNPs were loaded in chitosan (CS) composite dressing and applied to skin wound healing in mice. As a result, AgNPs showed excellent antifungal activity with minimum inhibitory concentrations (MIC) of 1.25, 2.5, and 5 μg/mL at *C. albicans* concentrations of 1×105, 1×106, and 1×107 CFU/mL, respectively. The MIC value was still kept at 2.5 μg/mL against *C. albicans* (105 CFU/mL) after 15 regeneration, showing less induction of drug resistance to the pathogenic fungus. The antifungal mechanisms of AgNPs against *C. albicans* have been recognized as the proliferation of membrane permeability, damage of cell membrane integrity, and leakage of cellular protein and nucleic acids. In wound healing application, iturin-AgNP-CS composite dressing significantly accelerated the healing of *C. albicans* infected skin wounds in the early 10 days (Zhou et al. 2021).

Kamikawa et al. (2014) showed *C. glabrata* and *C. albicans* increasingly adhered to the resin surface of the control piece over time, but the adhesion AgNP of both *Candida* species to the AgNP-coated surface was meaningfully inhibited (Kamikawa et al. 2014).

Gold Nanoparticles (AuNPs)

AuNPs with different sizes, shapes, and aggregation ability, display tunable and unique surface plasmon resonance (SPR). It has been reported that AuNPs exhibit antifungal, antibacterial and antibiofilm activity (Narayanan and Sakthivel 2013, Joshi et al. 2017, Gholami-Shabani et al. 2015, 2021a).

de Alteriis et al. (2018) showed that treatment with the AuNPs significantly inhibits the capacity of *C. albicans* to form biofilms and impairs preformed mature biofilms. Treatment with AuNPs–indolicidin results in an increase in the kinetics of Rhodamine 6G efflux and a reduction in the expression of biofilm-related genes (de Alteriis et al. 2018).

Judan Cruz et al. (2021) showed the green method to synthesis gold nanoparticles and their effect to control the biofilm of *C. albicans* (Judan Cruz et al. 2021).

Sherwani et al. (2015) showed that AuNPs in combination with photodynamic therapy effectively killed both *C. albicans* planktonic cells and biofilm populating hyphal forms. The mixture of AuNPs conjugated to dual diverse photosensitizer meaningfully depleted the hyphal *C. albicans* burden against superficial skin and oral *C. albicans* infection in mice (Sherwani et al. 2015).

Piktel et al. (2020) showed Rod-shaped AuNPs exert potent candidacidal activity and reduction in the adhesion of fungal cells (Piktel et al. 2020).

Yadav et al. (2021) showed that tyrosol functionalized chitosan AuNPs (Chi-TYAuNPs) were prepared as an alternative treatment strategy to combat fungal infections. The antifungal and antibiofilm activities of Chi-TY-AuNPs were evaluated against *C. albicans* and *C. glabrata*. Chi-TY-AuNPs showed a strong fungicidal effect against both sessile and planktonic cells of *Candida* spp. Furthermore, Chi-TYAuNPs completely eradicated (100%) the mature biofilms of both the *Candida* spp. (Yadav et al. 2021).

Zinc Oxide Nanoparticles (ZnONPs)

ZnO nanoparticles have received considerable attention due to their superior antifungal and antibiofilm properties (Gholami-Shabani et al. 2021a).

Hosseini et al. (2018) showed the influence of ZnONPs on *C. albicans* isolates biofilm formed on the urinary catheter. Out of 133 *C. albicans* isolates, 20 (15%) fluconazole-resistant and 113 (85%) susceptible isolates were determined using the disk diffusion technique. Results showed that both isolates adhered to biofilm formation on the catheter surfaces. A significantly ($P < 0.05$) higher number of CFUs was evident in fluconazole-resistant biofilms compared to those formed by susceptible isolates. ZnO-np reduced biofilm biomass and CFUs of dual isolate biofilms ($P < 0.05$). ZnO nanoparticles had a significantly ($P < 0.05$) greater effect on reducing fluconazole-resistant *C. albicans* biofilm biomass compared to susceptible isolates (Hosseini et al. 2018).

Jalal et al. (2018) synthesized ZnO NPs by using leaf extract of *Crinum latifolium* and its effect on *Candida* species. Their nanoparticles showed potent inhibitory effects on principal virulence factors of *C. albicans* and non-albicans such as germ tube formation, secretion of hydrolytic phospholipases and proteinases and biofilm formation has been investigated. ZnONPs remarkably reduced the germ tube formation of *C. albicans* at 1 (86.4%), 0.5 (75.0%), 0.25 (61.4%), 0.125 (34.1%) and 0.062 mg/ml (11.4%). ZnO NPs significantly lowered the phospholipase and proteinase secretion by 58.8% and 95.2% at 0.25 mg/ml, respectively. Confocal laser scanning microscopy (CLSM) results showed that ZnONPs suppressed biofilm formation up to 85% at 0.25 mg/ml. SEM and TEM micrographs showed that ZnONPs penetrated inside the cell and caused extensive damage in the cell wall and cell membrane (Jalal et al. 2018).

Lakshmeesha et al. (2019) reported the green synthesis of zinc oxide nanoparticles from *Syzygium aromaticum* flower bud extract by combustion method and investigated their application for controlling of growth and mycotoxins of *Fusarium graminearum*. The synthesized SaZnO NPs reduced the growth and production of deoxynivalenol and zearalenone of *F. graminearum* in broth culture. Further analysis revealed that treatment of mycelia with ZnONPs resulted in the accumulation of ROS in a dose-dependent manner. Also, ZnONPs treatment enhanced lipid peroxidation, depleted ergosterol content, and caused detrimental damage to the membrane integrity of fungi. Moreover, SEM observations revealed the presence of diverged micro-morphology (wrinkled, rough and shrank surface) in the macroconidia treated with ZnONPs (Lakshmeesha et al. 2019).

Ficociello et al. (2018) studied the antifungal activity of graphene nanoplateles decorated with zinc oxide nanorods against the human pathogen *C. albicans*. They observed that ZnONPs were able to induce significant mortality in fungal cells, as well as to affect the main virulence factors of this fungus or rather the hyphal development and biofilm formation (Ficociello et al. 2018).

Copper Oxide Nanoparticles (CuONPs)

CuONPs have received considerable attention due to their superior antifungal and antibiofilm properties (Gholami-Shabani et al. 2021a).

Muñoz-Escobar and Reyes-López (2020) studied the antifungal susceptibility of *Candida* species to copper oxide nanoparticles on polycaprolactone fibers (PCL-CuONPs). The antimycotic applicability of the composite was evaluated to determine the antifungal activity in three species of the genus Candida (*C. albicans*, *C. glabrata* and *C. tropicalis*). PCL-CuONPs exhibit considerable antifungal effect on all species tested. The preparation of PCL-CuONPs was simple, fast and low-cost for practical application as an antifungal dressing (Muñoz-Escobar and Reyes-López 2020).

Mudiar and Kelkar-Mane (2020) reported 92–100% cytotoxicity to the fluconazole-resistant *Candida* species with CuONPs as well as tamrajal during 24 hr time-killing assay. The study also confirmed complete germ tube inhibition by copper in both its forms in addition to the reduction in biofilm production (Mudiar and Kelkar-Mane 2020).

Chalandar et al. (2017) reported successful synthesis of copper and copper oxide nanoparticles by chemical reduction and precipitation methods. Antifungal properties of synthesized nanoparticles were tested against *Penicillium* on Orange fruit by Disc-Diffusion method. The results showed that antifungal properties were elevated by increasing the concentration of nanoparticles to 15% (Chalandar et al. 2017).

Magnesium Oxide Nanoparticles (MgONPs)

Among different mineral nanoparticles, MgONPs are promising antimicrobial agents that have the benefit of being non-toxic and easy to obtain. Antifungal and inhibitory effects of MgONPs have been investigated against various fungi (Abdel-Aziz et al. 2020). Abdel-Aziz et al. (2020) synthesized magnesium oxide nanoparticles (MgO NPs) using the cell filtrate of the endobacterium Burkholderia rinojensis. Their MgONPs showed considerable antifungal and antibiofilm activities against *F. oxysporum* f. sp. lycopersici. At the concentration of 15.36 μg/ml, the MgONPs completely inhibited the mycelial growth of the fungus. The biofilm formation of the pathogen was completely suppressed by MgONPs at 1.92 μg/ml. The MgONPs caused severe morphological changes in the hyphal morphology and biofilm formation of the fungus with significant damage to the fungal membrane integrity (Abdel-Aziz et al. 2020).

Kong et al. (2020) studied the effects of various concentrations of magnesium MgO-MNPs on the growth and key virulence factors of *C. albicans*. The minimum inhibitory concentration (MIC) of MgOMNPs against *C. albicans* was determined by the micro-broth dilution method. A time-killing curve of MgOMNPs and *C. albicans* was established to investigate the aging effect of MgOMNPs on *C. albicans*. Crystal violet staining, the MTT assay, and invert fluorescence microscopy were employed to determine the effects of MgOMNPs on *C. albicans* adhesion, two-phase morphological transformation, biofilm biomass, and metabolic activity. The time-killing curve showed that MgOMNPs had fungicidal activity against *C. albicans* in a time and concentration dependent manner (Kong et al. 2020).

Sidhu et al. (2020) studied the surface and electric properties of nontoxic sepiolite that have been blended with the antifungal properties of metabolizable MgONPs as a greener alternative to preparing their nanocomposites. They compared a sepiolite-MgO (SE-MgO) nanocomposite with MgO MNPs in an aqua dispersed form (aqMgO-MNPs) for their antifungal evaluation against different phytopathogenic fungi of rice. The SE-MgO nanocomposite was more potent in comparison to aqMgO-MNPs with ED90 > 230 and 249 μg/mL, respectively, against the test fungi better than standard fungicides. Ultramicroscopic studies revealed hyphal distortion and spore collapse as the cause of antimycotic activity. The *in vitro* seed treatment revealed 100% hyphal reduction with SE-MgO at 250 μg/mL of MgO as an active ingredient (Sidhu et al. 2020).

Iron Oxide Nanoparticles

Iron oxide MNPs are known in bioprocesses applications due to unique magnetic properties, superior biocompatibility, and high surface/volume ratio (Chifiriuc et al. 2013, Niemirowicz et al. 2016, Arias et al. 2018, Golipour et al. 2019).

Golipour et al. (2019) studied the effects of superparamagnetic Iron Oxide MNPs (SPIO-MNPs) on *C. albicans* biofilm formation. Their MNPs was about 70 nm. Antifungal analysis of SPIO-MNPs against *C. albicans* was 100 ppm and 200 ppm which reduced biofilm formation by 87.2% and 100%, respectively. SPIO-MNPs showed significant inhibitory effects on *C. albicans* growth and biofilm formation (Golipour et al. 2019).

Chifiriuc et al. (2013) investigated the influence of magnetic MNPs of Fe_3O_4/ oleic acid-core/shell on the adherence to cellular and inert substrata of clinical fungal using direct and inverted optic microscopy and Confocal Laser Scanning Microscopy. The biofilm development capability is diverse with the tested species (*C. albicans* > *C. tropicalis* > *C. glabrata* > *S. cerevisiae* > *C. krusei* > *C. famata*), its intensity being correlated with early germ tubes production. The magnetite NPs strongly inhibited the mature *Candida* biofilms development (of 48 and 72 hours) as well as the *Candida* adherence to the cellular substratum, suggesting that magnetic MNPs could be used for obtaining novel materials using anti-adherence properties (Chifiriuc et al. 2013).

Niemirowicz et al. (2016) studied the antifungal/antibiofilm and hemolytic properties of two polyene antifungals, nystatin (NYS) and amphotericin B (AMB), attached to the surface of magnetic MNPs against clinical isolates of *Candida* species and human red blood cells. The developed nanosystems, MNP-AMB and MNPs-NYS, exposed stronger fungicidal activity than free AMB or NYS. Synergistic activity was detected using a combination of polyenes and MNPs against all tested *Candida* strains. Nanosystems were more effective than free antifungal agents when tested against *Candida* strains in the presence of infection and as agents capable to prevent *Candida* biofilm formation. The observed inactivation of catalase Cat1 in *Candida* cells during treatment with the nanosystems recommends that interruption of the oxidation-reduction balance is a mechanism leading to inhibition of *Candida* yeast growth. The substantial reduction of polyenes lytic activity against *candida* host cells after their attachment to MNPs surface displays improvement in their biocompatibility (Niemirowicz et al. 2016).

Arias et al. (2018) studied the impact of antifungal of a novel miconazole-carrier nanosystem (NS) on dual yeast species biofilms of *C. albicans* and *C. glabrata*. The NS was manufactured by loading the chitosan-coated MNPs with miconazole. NS showed an antifungal effect against dual-species *Candida* biofilms, that stimulates the development of alternative treatments against recalcitrant fungal infections (Arias et al. 2018).

Conclusions

The global problem of drug-resistant fungi needs a modern solution. Antifungal properties of nanostructures depend on their physicochemical properties, as well as the structure of the fungus. As highlighted in this chapter, almost unlimited possibilities of nanobiotechnology products allow them to be used not only as active antifungal agents but also as nanoplatforms for the diagnosis and delivery of drugs or other compounds. However, it has to be kept in mind that the use of nanomaterials instead of or in combination with antifungal therapy should be strictly monitored. Unfortunately, there are still no appropriate technically advanced analytical tools, which could clearly confirm or exclude the possible toxicity. Public concerns about the widespread use of nanomaterials are related to the safe use of controversial elements, such as heavy metals, and their ability to penetrate and accumulate in the body. Crucial issues regarding the use of nanomaterials as antifungal agents and drug carriers remain to be further investigated.

Acknowledgments

Research reported in this publication was supported by Elite Researcher Grant Committee under award numbers [958634 and 963646] from the National Institute for Medical Research Development (NIMAD), Tehran, Iran to MRA.

References

Abbas, H. and Abou Baker, D. 2020. Biological evaluation of selenium nanoparticles biosynthesized by *Fusarium semitectum* as antimicrobial and anticancer agents. Egyptian J. Chem. 63: 1119–1133. https://doi.org/10.21608/ejchem.2019.15618.1945.

Abdel-Aziz, M.M., Emam, T.M. and Elsherbiny, E.A. 2020. Bioactivity of magnesium oxide nanoparticles synthesized from cell filtrate of endobacterium *Burkholderia rinojensis* against *Fusarium oxysporum*. Mat. Sci. Eng.: C 109: 110617. https://doi.org/10.1016/j.msec.2019.110617.

Abdel-Hafez, S.I., Nafady, N.A., Abdel-Rahim, I.R., Shaltout, A.M., Daròs, J.A. and Mohamed, M.A. 2017. Biosynthesis of silver nanoparticles using the compound curvularin isolated from the endophytic fungus *Epicoccum nigrum*: Characterization and antifungal activity. J. Pharm. Appl. Chem. 3: 135–146. http://doi.org/10.18576/jpac/030207.

Aghdam, M.H., Ghanbarzadeh, S., Javadzadeh, Y. and Hamishehkar, H. 2016. Aggregated nanotransfersomal dry powder inhalation of itraconazole for pulmonary drug delivery. Adv. Pharmaceut. Bulletin 6: 57–64. http://doi.org/10.15171/apb.2016.009.

Alam, M.A., Al-Janoobi, F.I., Alzahrani, K.A., Al-Agamy, M.H., Abdelgalil, A.A. and Al-Mohizea, A.M. 2017. *In-vitro* efficacies of topical microemulsions of clotrimazole and ketoconazole; and *in-vivo* performance of clotrimazole microemulsion. J. Drug Deliv. Sci. Technol. 39: 408–416. https://doi.org/10.1016/j.jddst.2017.04.025.

Alam, S., Iqbal, Z., Ali, A., Khar, R.K., Ahmad, F.J., Akhter, S. and Talegaonkar, S. 2009. Microemulsion as a potential transdermal carrier for poorly water soluble antifungal drug itraconazole. J. Disper. Sci. Technol. 31: 84–94. https://doi.org/10.1080/01932690903107265.

Al-Bahrani, R., Raman, J., Lakshmanan, H., Hassan, A.A. and Sabaratnam, V. 2017. Green synthesis of silver nanoparticles using tree oyster mushroom *Pleurotus ostreatus* and its inhibitory activity against pathogenic bacteria. Mat. Lett. 186: 21–25. http://dx.doi.org/10.1016/j.matlet.2016.09.069.

Albayaty, Y.N., Thomas, N., Ramírez-García, P.D., Davis, T.P., Quinn, J.F., Whittaker, M.R. and Prestidge, C.A. 2021. Polymeric micelles with anti-virulence activity against *Candida albicans* in a single- and dual-species biofilm. Drug Deliv. Translat. Res. 13: 1–2. https://doi.org/10.1007/s13346-021-00943-4.

Alharbi, K.K. and Al-Sheikh, Y.A. 2014. Role and implications of nanodiagnostics in the changing trends of clinical diagnosis. Saudi J. Biologic. Sci. 21: 109–117. https://doi.org/10.1016/j.sjbs.2013.11.001.

Aljaeid, B.M. and Hosny, K.M. 2016. Miconazole-loaded solid lipid nanoparticles: Formulation and evaluation of a novel formula with high bioavailability and antifungal activity. Int. J. Nanomed. 11: 441. https://doi.org/10.2147/IJN.S100625.

Alomrani, A.H., Al-Agamy, M.H. and Badran, M.M. 2015. *In vitro* skin penetration and antimycotic activity of itraconazole loaded niosomes: Various non-ionic surfactants. J. Drug Deliv. Sci. Technol. 28: 37–45. https://doi.org/10.1016/j.jddst.2015.04.009.

Al-Zubaidi, S., Al-Ayafi, A. and Abdelkader, H. 2019. Biosynthesis, characterization and antifungal activity of silver nanoparticles by *Aspergillus niger* isolate. J. Nanotechnol. Res. 1: 022–035.

Ammar, H.A. and El-Desouky, T.A. 2016. Green synthesis of nanosilver particles by *Aspergillus terreus* HA 1N and *Penicillium expansum* HA 2N and its antifungal activity against mycotoxigenic fungi. J. Appl. Microbiol. 121: 89–100. https://doi.org/10.1111/jam.13140.

Amra, K. and Momin, M. 2019. Formulation evaluation of ketoconazole microemulsion-loaded hydrogel with nigella oil as a penetration enhancer. J. Cosmetic Dermatol. 18: 1742–1750. https://doi.org/10.1111/jocd.12945.

Arias, L.S., Pelim Pessan, J., Botazzo Delbem, A.C., Facholi Afanaci, L., De Souza Neto, F.N., Rodrigues de Camargo, E. and Monteiro, D.R. 2018. Antifungal activity of a miconazole-carrier magnetic nanosystem on dual species *Candida* biofilms. Archives Health Invest. 7: 02–04.

Arumugam, P. and Berchmans, S. 2011. Synthesis of gold nanoparticles: An ecofriendly approach using *Hansenula anomala*. ACS Appl. Mat. Interfac. 3: 1418–1425. https://doi.org/10.1021/am200443.

Asadi, P., Mehravaran, A., Soltanloo, N., Abastabar, M. and Akhtari, J. 2021. Nanoliposome-loaded antifungal drugs for dermal administration: A review. Curr. Med. Mycol. 7: 71–78. https://dx.doi.org/10.18502%2Fcmm.7.1.6247.

Asri, M., Elabed, S., Koraichi, SI. and El Ghachtouli, N. 2018. Biofilm-based systems for industrial wastewater treatment. Handbook of Environmental Materials Management 2018: 1–21. https://doi.org/10.1007/978-3-319-58538-3_137-1.

Bafghi, M.H., Darroudi, M., Zargar, M., Zarrinfar, H. and Nazari, R. 2021. Biosynthesis of selenium nanoparticles by *Aspergillus flavus* and *Candida albicans* for antifungal applications. Micro Nano Lett. 1–14. https://doi.org/10.1049/mna2.12096.

Bafrui, N.M., Hazaveh, S.J. and Bayat, M. 2020. *In-vitro* activity of nano fluconazole and conventional fluconazole against clinically important dermatophytes. Iranian J. Pub. Health. 49: 1970. https://doi.org/10.18502/ijph.v49i10.4701.

Bala, M. and Arya, V. 2013. Biological synthesis of silver nanoparticles from aqueous extract of endophytic fungus *Aspergillus fumigatus* and its antibacterial action. Int. J. Nanomat. Biostruct. 3: 37–41.

Balakumaran, M.D., Ramachandran, R. and Kalaichelvan, P.T. 2015. Exploitation of endophytic fungus, *Guignardia mangiferae* for extracellular synthesis of silver nanoparticles and their *in vitro* biological activities. Microbiologic. Res. 178: 9–17. https://doi.org/10.1016/j.micres.2015.05.009.

Baumgardner, D.J. 2016. Disease-causing fungi in homes and yards in the Midwestern United States. J. Patient-Centered Res. Rev. 3: 99–110. https://doi.org/10.17294/2330-0698.1053.

Bezerra, C.F., de Alencar Júnior, J.G., de Lima Honorato, R., Dos Santos, A.T., da Silva, J.C., da Silva, T.G., Alves Borges Leal, A.L., Esmeraldo Rocha, J., Sampaio de Freitas, T., Tavares Vieira, T.A., Fonseca Bezerra, M.C., Lima Sales, D., Kerntopf, M.R., de Araujo Delmondes, G., Barbosa Filho, J.M., Rangel Peixoto, L., Pontes Pinheiro, A., Ribeiro-Filho, J., Melo Coutinho, H.D., Bezerra Morais-Braga, M.F. and Gonçalves da Silva, T. 2020. Antifungal activity of farnesol incorporated in liposomes and associated with fluconazole. Chem. Physic Lipid 233: 104987. https://doi.org/10.1016/j.chemphyslip.2020.104987.

Bhalaria, M.K., Naik, S. and Misra, A.N. 2009. Ethosomes: A novel delivery system for antifungal drugs in the treatment of topical fungal diseases. Indian J. Exp. Biol. 47: 368–375.

Bhat, M.A., Nayak, B.K. and Nanda, A. 2015. Evaluation of bactericidal activity of biologically synthesised silver nanoparticles from *Candida albicans* in combination with ciprofloxacin. Mat. Today: Proceed. 2: 4395–4401. https://doi.org/10.1016/j.matpr.2015.10.036.

Birla, S.S., Tiwari, V.V., Gade, A.K., Ingle, A.P., Yadav, A.P. and Rai, M.K. 2009. Fabrication of silver nanoparticles by *Phoma glomerata* and its combined effect against *Escherichia coli, Pseudomonas aeruginosa* and *Staphylococcus aureus*. Lett. Appl. Microbiol. 48: 173–179. https://doi.org/10.1111/j.1472-765X.2008.02510.x.

Bjarnsholt, T., Kirketerp-Møller, K., Jensen, P.Ø., Madsen, K.G., Phipps, R., Krogfelt, K., Alves Borges Leal, A.L., Esmeraldo Rocha, J., Sampaio de Freitas, T., Tavares Vieira, T.A., Fonseca Bezerra, M.C., Sales, D.L., Kerntopf, M.R., de Araujo Delmondes, G., José Maria Barbosa Filho, J.M., Peixoto, L.R., Pinheiro, A.P., Ribeiro-Filho, J., Melo Coutinho, H.D., Morais-Braga, M.F.B. and

Gonçalves da Silva, T. 2008. Why chronic wounds will not heal: a novel hypothesis. Wound Repair Regenerat. 16: 2–10. https://doi.org/10.1111/j.1524-475X.2007.00283.x.

Blanco-Cabra, N., López-Martínez, M.J., Arévalo-Jaimes, B.V., Martin-Gómez, M.T., Samitier, J. and Torrents, E. 2021. A new BiofilmChip device for testing biofilm formation and antibiotic susceptibility. NPJ Biofilms and Microbiomes 7: Article number: 62. 1–9. https://doi.org/10.1038/s41522-021-00236-1.

Bolbanabad, E.M., Ashengroph, M. and Darvishi, F. 2020. Development and evaluation of different strategies for the clean synthesis of silver nanoparticles using *Yarrowia lipolytica* and their antibacterial activity. Proc. Biochem. 94: 319–328. https://doi.org/10.1016/j.procbio.2020.03.024.

Bolla, P.K., Meraz, C.A., Rodriguez, V.A., Deaguero, I., Singh, M., Yellepeddi, V.K. and Renukuntla, J. 2019. Clotrimazole loaded ufosomes for topical delivery: Formulation development and *in-vitro* studies. Molecules 24: 3139. https://doi.org/10.3390/molecules24173139.

Boruah, S. and Dutta, P. 2021. Fungus mediated biogenic synthesis and characterization of chitosan nanoparticles and its combine effect with *Trichoderma asperellum* against *Fusarium oxysporum*, *Sclerotium rolfsii* and *Rhizoctonia solani*. Indian Phytopathol. 74: 81–93. https://doi.org/10.1007/s42360-020-00289-w.

Campos, E.V., de Oliveira, J.L., Abrantes, D.C., Rogério, C.B., Bueno, C., Miranda, V.R., Monteiro, R.A. and Fraceto, L.F. 2020. Recent developments in nanotechnology for detection and control of aedes aegypti-borne diseases. Front. Bioeng. Biotechnol. 8: 102. https://doi.org/10.3389/fbioe.2020.00102.

Campos, F.F., Campmany, A.C., Delgado, G.R., Serrano, O.L. and Naveros, B.C. 2012. Development and characterization of a novel nystatin-loaded nanoemulsion for the buccal treatment of candidosis: Ultrastructural effects and release studies. J. Pharmaceut. Sci. 2012 Oct 1; 101(10): 3739–52. https://doi.org/10.1002/jps.23249.

Carrillo-Munoz, A.J., Quindos, G., Tur, C., Ruesga, M.T., Miranda, Y., Valle, O.D., Cossum, P.A. and Wallace, T.L. 1999 *In-vitro* antifungal activity of liposomal nystatin in comparison with nystatin, amphotericin B cholesteryl sulphate, liposomal amphotericin B, amphotericin B lipid complex, amphotericin B desoxycholate, fluconazole and itraconazole. J. Antimicrobial Chemotherapy 44: 397–401. https://doi.org/10.1093/jac/44.3.397.

Cavalheiro, M. and Teixeira, M.C. 2018. *Candida* biofilms: Threats, challenges, and promising strategies. Front. Med. 5: 28. https://doi.org/10.3389/fmed.2018.00028.

Chalandar, H.E., Ghorbani, H.R., Attar, H. and Alavi, S.A. 2017. Antifungal effect of copper and copper oxide nanoparticles against *Penicillium* on orange fruit. Biosci. Biotechnol. Res. Asia 14: 279–284. http://dx.doi.org/10.13005/bbra/2445.

Chan, S. and Don, M. 2012. Characterization of Ag nanoparticles produced by white-rot fungi and its *in vitro* antimicrobial activities. Int. Arabic J. Antimicrobial Agents 2: 1–8. https://doi.org/10.3823/717.

Chen, G., Yi, B., Zeng, G., Niu, Q., Yan, M., Chen, A., Du, J., Huang, J. and Zhang, Q. 2014. Facile green extracellular biosynthesis of CdS quantum dots by white rot fungus *Phanerochaete chrysosporium*. Colloids and Surfaces B: Biointerfaces 117: 199–205. https://doi.org/10.1016/j.colsurfb.2014.02.027.

Chen, Y., Sun, X., Wang, X., Pan, W., Yu, G. and Wang, J. 2020. Carbon dots with red emission for bioimaging of fungal cells and detecting Hg^{2+} and ziram in aqueous solution. Spectrochim Acta Part A: Molecul. Biomolecul. Spectroscopy 233: 118230. https://doi.org/10.1016/j.saa.2020.118230.

Chifiriuc, M.C., Grumezescu, A.M., Saviuc, C., Hristu, R., Grumezescu, V., Bleotu, C., Stanciu, G., Mihaiescu, D.E., Andronescu, E., Lazar, V. and Radulescu, R. 2013. Magnetic nanoparticles for controlling *in vitro* fungal biofilms. Curr. Organic Chem. 17: 1023–1028. https://doi.org/10.2174/1385272811317100004.

Choudhary, P., Singh, S. and Agarwal, V. 2020. Microbial biofilms. *In*: Dincer, S., Özdenefe, M.S. and Arkut, A. (eds.). Bacterial Biofilms. London, England: IntechOpen. https://doi.org/10.5772/intechopen.90790.

Chowdhury, S., Basu, A. and Kundu, S. 2011. Extracellular biosynthesis of silver nanoparticles using the mycorrhizal mushroom *Tricholoma crassum* (berk.) sacc.: Its antimicrobial activity against pathogenic bacteria and fungus, including multidrug resistant plant and human bacteria. Digest. J. Nanomat. Biostruct. 6: 1289–1299.

Chowdhury, S., Basu, A. and Kundu, S. 2014. Green synthesis of protein capped silver nanoparticles from phytopathogenic fungus *Macrophomina phaseolina* (Tassi) Goid with antimicrobial properties against multidrug-resistant bacteria. Nanoscale Res. Lett. 9: 365. https://doi.org/10.1186/1556-276X-9-365.

Coenye, T., Kjellerup, B., Stoodley, P. and Bjarnsholt, T. 2020. The future of biofilm research–Report on the '2019 Biofilm Bash'. Biofilm 2: 100012. https://doi.org/10.1016/j.bioflm.2019.100012.

Cuddihy, G., Wasan, E.K., Di, Y. and Wasan, K.M. 2019. The development of oral amphotericin B to treat systemic fungal and parasitic infections: Has the myth been finally realized? Pharmaceutics 11: 99. http://doi.org/10.3390/pharmaceutics11030099.

Cushion, M.T., Collins, M.S. and Linke, M.J. 2009. Biofilm formation by *Pneumocystis* spp. Eukaryotic Cell 8: 197–206. https://doi.org/10.1128%2FEC.00202-08.

Dalla Lana, D.F., Reginatto, P., Lopes, W., Vainstein, M.H. and Fuentefria, A.M. 2019. Invasion of human nails by *Microsporum canis*. J. Infectiol. 2: 36–38. https://doi.org/10.29245/2689-9981/2019/4.1151.

Danielli, L.J., Lopes, W., Vainstein, M.H., Fuentefria, A.M. and Apel, M.A. 2017. Biofilm formation by *Microsporum canis*. Clin. Microbiol. Infect. 23: 941–942. http://doi.org/10.1016/j.cmi.2017.06.006.

Dave, V., Sharma, S., Yadav, R.B. and Agarwal, U. 2017. Herbal liposome for the topical delivery of ketoconazole for the effective treatment of seborrheic dermatitis. Appl. Nanosci. 7: 973–987. https://doi.org/10.1007/s13204-017-0634-3.

Davis, S.C., Ricotti, C., Cazzaniga, A., Welsh, E., Eaglstein, W.H. and Mertz, P.M. 2008. Microscopic and physiologic evidence for biofilm-associated wound colonization *in vivo*. Wound Repair Regenerat. 16: 23–29. https://doi.org/10.1111/j.1524-475X.2007.00303.x.

de Alteriis, E., Maselli, V., Falanga, A., Galdiero, S., Di Lella, F.M., Gesuele, R., Guida, M. and Galdiero, E. 2018. Efficiency of gold nanoparticles coated with the antimicrobial peptide indolicidin against biofilm formation and development of *Candida* spp. clinical isolates. Infect. Drug Resist. 11: 915–925. https://doi.org/10.2147%2FIDR.S164262.

Deniz, F., Adigüzel, A.O. and Mazmanci, M.A. 2019. The biosynthesis of silver nanoparticles by cytoplasmic fluid of *Coriolus versicolor*. Turkish J. Eng. 3: 92–96. https://doi.org/10.31127/tuje.429072.

Deo, P.N. and Deshmukh, R. 2019. Oral microbiome: Unveiling the fundamentals. J. Oral Maxillofacial Pathol: JOMFP 23: 122–128. https://doi.org/10.4103%2Fjomfp.JOMFP_304_18.

Dincer, S., Masume Uslu, F. and Delik, A. 2020. Antibiotic resistance in biofilm. *In*: Dincer Sadik, Özdenefe, M.S. and Arkut, A. (eds.). Bacterial Biofilms. London, England: IntechOpen. https://doi.org/10.5772/intechopen.92388.

El-Baz, A.F., El-Batal, A.I., Abomosalam, F.M., Tayel, A.A., Shetaia, Y.M. and Yang, S.T. 2016. Extracellular biosynthesis of anti-*Candida* silver nanoparticles using *Monascus purpureus*. J. Basic Microbiol. 56: 531–540. https://doi.org/10.1002/jobm.201500503.

Elgorban, A.M., Al-Rahmah, A.N., Sayed, S.R., Hirad, A., Mostafa, A.A. and Bahkali, A.H. 2016. Antimicrobial activity and green synthesis of silver nanoparticles using *Trichoderma viride*. Biotechnol. Biotechnol. Equip. 30: 299–304. https://doi.org/10.1080/13102818.2015.1133255.

El-Housiny, S., Shams Eldeen, M.A., El-Attar, Y.A., Salem, H.A., Attia, D., Bendas, E.R. and El-Nabarawi, M.A. 2018. Fluconazole-loaded solid lipid nanoparticles topical gel for treatment of *Pityriasis versicolor*: Formulation and clinical study. Drug Delivery 25: 78–90. https://doi.org/10.1080/10717544.2017.1413444.

Elmataeeshy, M.E., Sokar, M.S., Bahey-El-Din, M. and Shaker, D.S. 2018. Enhanced transdermal permeability of Terbinafine through novel nanoemulgel formulation; Development, *in vitro* and *in vivo* characterization. Future J. Pharmaceut. Sci. 4: 18–28. https://doi.org/10.1016/j.fjps.2017.07.003.

El-Ridy, M.S., Abdelbary, A., Essam, T., Abd EL-Salam, R.M. and Aly Kassem, A.A. 2011. Niosomes as a potential drug delivery system for increasing the efficacy and safety of nystatin. Drug Develop Indust. Pharma. 37: 1491–1508. https://doi.org/10.3109/03639045.2011.587431.

El-Sayed, E.S., Abdelhakim, H.K. and Zakaria, Z. 2020. Extracellular biosynthesis of cobalt ferrite nanoparticles by *Monascus purpureus* and their antioxidant, anticancer and antimicrobial activities: Yield enhancement by gamma irradiation. Mat. Sci. Eng.: C 107: 110318. https://doi.org/10.1016/j.msec.2019.110318.

El-Sayyad, G.S., Mosallam, F.M. and El-Batal, A.I. 2018. One-pot green synthesis of magnesium oxide nanoparticles using *Penicillium chrysogenum* melanin pigment and gamma rays with antimicrobial activity against multidrug-resistant microbes. Advanced Powder Technology 29: 2616–2625. https://doi.org/10.1016/j.apt.2018.07.009.

Fanning, S. and Mitchell, A.P. 2012. Fungal biofilms. PLoS Pathogens 8: e1002585. https://doi. org/10.1371/journal.ppat.1002585.

Fastenberg, J.H., Hsueh, W.D., Mustafa, A., Akbar, N.A. and Abuzeid, W.M. 2016. Biofilms in chronic rhinosinusitis: Pathophysiology and therapeutic strategies. World J. Otorhinolaryngol-head Neck Surg. 2: 219–229. https://doi.org/10.1016%2Fj.wjorl.2016.03.002.

Fathalla, D., Youssef, E.M. and Soliman, G.M. 2020. Liposomal and ethosomal gels for the topical delivery of anthralin: Preparation, comparative evaluation and clinical assessment in psoriatic patients. Pharmaceutics 12: 446. http://doi.org/10.3390/pharmaceutics12050446.

Faustino, C. and Pinheiro, L. 2020. Lipid systems for the delivery of amphotericin B in antifungal therapy. Pharmaceutics 12: 29. http://doi.org/10.3390/pharmaceutics12010029.

Fernandes Costa, A., Evangelista Araujo, D., Santos Cabral, M., Teles Brito, I., Borges de Menezes Leite, L., Pereira, M. and Amaral, A.C. 2019. Development, characterization, and *in vitro-in vivo* evaluation of polymeric nanoparticles containing miconazole and farnesol for treatment of vulvovaginal candidiasis. Medical Mycology 57: 52–62. https://doi.org/10.1093/mmy/myx155.

Fernandes, A.V., Pydi, C.R., Verma, R., Jose, J. and Kumar, L. 2020. Design, preparation and *in vitro* characterizations of fluconazole loaded nanostructured lipid carriers. Brazilian J. Pharmaceut. Sci. 56. https://doi.org/10.1590/s2175-97902019000318069.

Fernández-García, R., Sttats, L., Jesus, J.A., Dea-Ayuela, M.A., Bolás-Fernández, F., Ballesteros, M.P., Laurenti, M.D.,Passero, L.F.D., Lalatsa, A. and Serrano, D.R. 2020. Topical delivery of amphotericin b utilizing transferosomes for the treatment of cutaneous Leishmaniasis. In Multidisciplinary Digital Publishing Institute Proceedings 78: 26. https://doi.org/10.3390/IECP2020-08669.

Fetih, G. 2016. Fluconazole-loaded niosomal gels as a topical ocular drug delivery system for corneal fungal infections. J. Drug Delive. Sci. Technol. 35: 8–15. https://doi.org/10.1016/j.jddst.2016.06.002.

Ficociello, G., De Caris, M.G., Trillò, G., Cavallini, D., Sarto, M.S., Uccelletti, D. and Mancini, P. 2018. Anti-candidal activity and *in vitro* cytotoxicity assessment of graphene nanoplatelets decorated with zinc oxide nanorods. Nanomaterials 8: 752. https://doi.org/10.3390/nano8100752.

Firacative, C. 2020. Invasive fungal disease in humans: Are we aware of the real impact? Memórias Instituto Oswaldo Cruz 115: e200430. https://doi.org/10.1590/0074-02760200430.

Firooz, A., Namdar, R., Nafisi, S. and Maibach, H. 2016 Nano-sized technologies for miconazole skin delivery. Curr. Pharmaceut. Biotechnol. 17: 524–531. https://doi.org/10.2174/1389201017666160 301102459.

Firthouse, P.M., Halith, S.M., Wahab, S.U., Sirajudeen, M. and Mohideen, S.K. 2011. Formulation and evaluation of miconazole niosomes. Int. J. Pharm. Tech. Res. 3: 1019–1022.

Fisher, J.F., Kavanagh, K., Sobel, J.D., Kauffman, C.A. and Newman, C.A. 2011. *Candida* urinary tract infection: Pathogenesis. Clin. Infect. Dis. 52: S437–451. https://doi.org/10.1093/cid/cir110.

Flemming, H.C. and Ridgway, H. 2009. Biofilm control: Conventional and alternative approaches. *In*: Flemming, H.C., Murthy, P.S., Venkatesan, R. and Cooksey, K. (eds.). Marine and Industrial Biofouling. Springer Series on Biofilms, vol 4. Springer, Berlin, Heidelberg. https://doi. org/10.1007/978-3-540-69796-1_5.

Flemming, H.C. and Wingender, J. 2010. The biofilm matrix. Nat. Rev. Microbiol. 8: 623–633. https:// doi.org/10.1038/nrmicro2415.

Foreman, A., Boase, S., Psaltis, A. and Wormald, P.J. 2012. Role of bacterial and fungal biofilms in chronic rhinosinusitis. Curr. Allerg. Asthma Rep. 12: 127–135. https://doi.org/10.1007/s11882-012-0246-7.

Gaba, B., Fazil, M., Khan, S., Ali, A., Baboota, S. and Ali, J. 2015. Nanostructured lipid carrier system for topical delivery of terbinafine hydrochloride. Bulletin Faculty Pharma, Cairo Univ. 53: 147–159. https://doi.org/10.1016/j.bfopcu.2015.10.001.

Gaikwad, A., Joshi, M., Patil, K., Sathaye, S. and Rode, C. 2019. Fluorescent carbon-dots thin film for fungal detection and bio-labeling applications. ACS Appl. Bio. Mat. 2: 5829–5840. https://doi. org/10.1021/acsabm.9b00795.

Gajbhiye, M., Kesharwani, J., Ingle, A., Gade, A. and Rai, M. 2009. Fungus-mediated synthesis of silver nanoparticles and their activity against pathogenic fungi in combination with fluconazole. Nanomed.: Nanotechnol. Biol. Med. 5: 382–386. https://doi.org/10.1016/j.nano.2009.06.005.

Galatage, S.T., Hebalkar, A.S., Dhobale, S.V., Mali, O.R., Kumbhar, P.S., Nikade, S.V. and Killedar, S.G. 2021. Silver nanoparticles: Properties, synthesis, characterization, applications and future trends. In

Silver Micro-Nanoparticles-Properties, Synthesis, Characterization, and Applications. IntechOpen. https://doi.org/10.5772/intechopen.99173.

Ge, S., Lin, Y., Lu, H., Li, Q., He, J., Chen, B., Wu, C. and Xu, Y. 2014. Percutaneous delivery of econazole using microemulsion as vehicle: Formulation, evaluation and vesicle-skin interaction. Inte. J. Pharmaceut. 465: 120–131. https://doi.org/10.1016/j.ijpharm.2014.02.012.

Gebreyohannes, G., Nyerere, A., Bii, C. and Sbhatu, D.B. 2019. Challenges of intervention, treatment, and antibiotic resistance of biofilm-forming microorganisms. Heliyon 5: e02192. https://doi.org/10.1016/j.heliyon.2019.e02192.

Gemishev, O.T., Panayotova, M.I., Mintcheva, N.N., Djerahov, L.P., Tyuliev, G.T. and Gicheva, GD. 2019. A green approach for silver nanoparticles preparation by cell-free extract from *Trichoderma reesei* fungi and their characterization. Mat. Res. Exp. 6: 095040. https://doi.org/10.1088/2053-1591/ab2e6a.

Gholami-Shabani, M., Akbarzadeh, A., Norouzian, D., Amini, A., Gholami-Shabani, Z., Imani, A., Chiani, M., Riazi, G., Shams-Ghahfarokhi, M. and Razzaghi-Abyaneh, M. 2014. Antimicrobial activity and physical characterization of silver nanoparticles green synthesized using nitrate reductase from *Fusarium oxysporum*. Appl. Biochem. Biotechnol. 172: 4084–4098. https://doi.org/10.1007/s12010-014-0809-2.

Gholami-Shabani, M., Gholami-Shabani, Z., Shams-Ghahfarokhi, M. and Razzaghi-Abyaneh, M. 2018. Application of nanotechnology in mycoremediation: Current status and future prospects. *In*: Prasad, R., Kumar, V., Kumar, M. and Wang, S. (eds.). Fungal Nanobionics: Principles and Applications. Springer, Singapore. https://doi.org/10.1007/978-981-10-8666-3_4.

Gholami-Shabani, M., Shams-Ghahfarokhi, M., Gholami-Shabani, Z., Akbarzadeh, A., Riazi, G., Ajdari, S., Amani, A. and Razzaghi-Abyaneh, M. 2015. Enzymatic synthesis of gold nanoparticles using sulfite reductase purified from *Escherichia coli*: A green eco-friendly approach. Proc. Biochem. 50: 1076–1085. https://doi.org/10.1016/j.procbio.2015.04.004.

Gholami-Shabani, M., Sotoodehnejadnematalahi, F., Shams-Ghahfarokhi, M., Eslamifar, A. and Razzaghi-Abyaneh, M. 2021a. Physicochemical properties, anticancer and antimicrobial activities of metallic nanoparticles green synthesized by *Aspergillus kambarensis*. IET Nanobiotechnol. https://doi.org/10.1049/nbt2.12070.

Gholami-Shabani, M., Sotoodehnejadnematalahi, F., Shams-Ghahfarokhi, M., Eslamifar, A. and Razzaghi-Abyaneh, M. 2021b. Mycosynthesis and physicochemical characterization of vanadium oxide nanoparticles using the cell-free filtrate of *Fusarium oxysporum* and evaluation of their cytotoxic and antifungal activities. J. Nanomat. 2021: Article ID 7532660. https://doi.org/10.1049/nbt2.12070.

Ghosh, A., Sarkar, A., Paul, P. and Patel, P. 2021. The rise in cases of mucormycosis, candidiasis and aspergillosis amidst COVID19. Fungal Biol. Rev. 38: 67–91. https://doi.org/10.1016%2Fj.fbr.2021.09.003.

Golipour, F., Habibipour, R. and Moradihaghgou, L. 2019. Investigating effects of superparamagnetic iron oxide nanoparticles on *Candida albicans* biofilm formation. Med. Lab. J. 13: 44–50. https://doi.org/10.29252/mlj.13.6.44.

Gonçalves, L.N., Costa-Orlandi, C.B., Bila, N.M., Vaso, C.O., Da Silva, R.A., Mendes-Giannini, M.J., Taylor, M.L. and Fusco-Almeida, A.M. 2020. Biofilm formation by *Histoplasma capsulatum* in different culture media and oxygen atmospheres. Fron. Microbiol. 11: 1455. https://doi.org/10.3389/fmicb.2020.01455.

Gopinath, K., Karthika, V., Sundaravadivelan, C., Gowri, S. and Arumugam, A. 2015. Mycogenesis of cerium oxide nanoparticles using *Aspergillus nig*er culture filtrate and their applications for antibacterial and larvicidal activities. J. Nanostruct. Chem. 5: 295–303. https://doi.org/10.1007/s40097-015-0161-2.

Gosecka, M. and Gosecki, M. 2021. Antimicrobial polymer-based hydrogels for the intravaginal therapies—engineering considerations. Pharmaceutics 1: 1393. https://doi.org/10.3390/pharmaceutics13091393.

Grainha, T., Jorge, P., Alves, D., Lopes, S.P. and Pereira, M.O. 2020. Unraveling *Pseudomonas aeruginosa* and *Candida albicans* communication in coinfection scenarios: Insights through network analysis. Front. Cellul. Infect. Microbiol. 10: 550505. https://doi.org/10.3389/fcimb.2020.550505.

Guilger, M., Pasquoto-Stigliani, T., Bilesky-Jose, N., Grillo, R., Abhilash, P.C., Fraceto, L.F. and Lima, R.D. 2017. Biogenic silver nanoparticles based on *Trichoderma harzianum*: Synthesis, characterization, toxicity evaluation and biological activity. Sci. Rep. 7: 44421. https://doi.org/10.1038/srep44421.

Gupta, K. and Chundawat, T.S. 2019. Bio-inspired synthesis of platinum nanoparticles from fungus *Fusarium oxysporum*: Its characteristics, potential antimicrobial, antioxidant and photocatalytic activities. Mat. Res. Exp. 6: 1050d6. https://doi.org/10.1088/2053-1591/ab4219.

Guzmán-Soto, I., McTiernan, C., Gonzalez, M., Ross, A., Gupta, K., Suuronen, E.J., Mah, T-F., Griffith, M. and Alarcon, E.I. 2021. Mimicking biofilm formation and development: Recent progress in *in vitro* and *in vivo* biofilm models. iScience 24: 102443. https://doi.org/10.1016/j.isci.2021.102443.

Hamedi, S., Shojaosadati, S.A., Shokrollahzadeh, S. and Hashemi-Najafabadi, S. 2014. Extracellular biosynthesis of silver nanoparticles using a novel and non-pathogenic fungus, *Neurospora intermedia*: Controlled synthesis and antibacterial activity. World J. Microbiol. Biotechnol. 30: 693–704. https://doi.org/10.1007/s11274-013-1417-y.

Hariharan, H., Al-Harbi, N., Karuppiah, P. and Rajaram, S. 2012. Microbial synthesis of selenium nanocomposite using *Saccharomyces cerevisiae* and its antimicrobial activity against pathogens causing nosocomial infection. Chalcogenide Lett. 9: 509–515.

Hassan, S.E., Fouda, A., Saied, E., Farag, M., Eid, A.M., Barghoth, M.G., Awad, M.A., Hamza, M.F. and Awad, M.F. 2021. *Rhizopus Oryzae*-mediated green synthesis of magnesium oxide nanoparticles (MgO-NPs): A promising tool for antimicrobial, mosquitocidal action, and tanning effluent treatment. J. Fungi. 7: 372. https://doi.org/10.3390/jof7050372.

Ho, H.N., Le, T.G., Dao, T.T., Le, T.H., Dinh, T.T., Nguyen, D.H., Tran, T.C. and Nguyan, C.N. 2020. Development of itraconazole-loaded polymeric nanoparticle dermal gel for enhanced antifungal efficacy. J. Nanomat. 2020. https://doi.org/10.1155/2020/8894541.

Hosseini, S.S., Ghaemi, E. and Koohsar, F. 2018. Influence of ZnO nanoparticles on *Candida albicans* isolates biofilm formed on the urinary catheter. Iranian J. Microbiol. 10: 424–432.

Hussain, K., Malavia, D., Johnson, E., Littlechild, J., Winlove, C.P., Vollmer, F. and Gow, N.A. 2020. Biosensors and diagnostics for fungal detection. J. Fungi. 6: 349. https://doi.org/10.3390%2Fjof6040349.

Ibrahem, E.J., Thalij, K.M. and Badawy, A.S. 2017. Antibacterial potential of magnesium oxide nanoparticles synthesized by *Aspergillus niger*. Biotechnol. J. Int. 18: 1–7. https://doi.org/10.9734/BJI/2017/29534.

Jahagirdar, V.L., Davane, M.S., Aradhye, S.C. and Nagoba, B.S. 2018. *Candida* species as potential nosocomial pathogens—A review. Electronic J. General Med. 15: Article No: em05. https://doi.org/10.29333/ejgm/82346.

Jain, J.P. and Kumar, N. 2010. Development of amphotericin B loaded polymersomes based on (PEG) 3-PLA co-polymers: Factors affecting size and *in vitro* evaluation. European J. Pharmaceut. Sci. 40: 456–465. https://doi.org/10.1016/j.ejps.2010.05.005.

Jalal, M., Ansari, M.A., Ali, S.G., Khan, H.M. and Rehman, S. 2018. Anticandidal activity of bioinspired ZnO NPs: Effect on growth, cell morphology and key virulence attributes of *Candida* species. Artificial Cells, Nanomed. Biotechnol. 46: 912–925. https://doi.org/10.1080/21691401.2018.1439837.

Jalal, M., Ansari, M.A., Alzohairy, M.A., Ali, S.G., Khan, H.M., Almatroudi, A. and Raees, K. 2018. Biosynthesis of silver nanoparticles from oropharyngeal *Candida glabrata* isolates and their antimicrobial activity against clinical strains of bacteria and fungi. Nanomaterials 8: 586. https://doi.org/10.3390/nano8080586.

James, G.A., Swogger, E., Wolcott, R., Pulcini, E.D., Secor, P., Sestrich, J., Costerton, J.W. and Stewart, P.S. 2008. Biofilms in chronic wounds. Wound Repair Regenerat. 16: 37–44. https://doi.org/10.1111/j.1524-475X.2007.00321.x.

Joshi, C.G., Danagoudar, A., Poyya, J., Kudva, A.K. and Dhananjaya, B.L. 2017. Biogenic synthesis of gold nanoparticles by marine endophytic fungus—*Cladosporium cladosporioides* isolated from seaweed and evaluation of their antioxidant and antimicrobial properties. Process Biochemistry 63: 137–144. http://doi.org/10.1016/j.procbio.2017.09.008.

Judan Cruz, K.G., Alfonso, E.D., Fernando, S.I. and Watanabe, K. 2021. *Candida albicans* biofilm inhibition by ethnobotanicals and ethnobotanically-synthesized gold nanoparticles. Front. Microbiol. 12: 1175. https://doi.org/10.3389/fmicb.2021.665113.

Kalan, L. and Grice, E.A. 2018. Fungi in the wound microbiome. Adv. Wound Care 7: 247–255. https://doi.org/10.1089%2Fwound.2017.0756.

Kalidasan, K., Asmathunisha, N., Gomathi, V., Dufossé, L. and Kathiresan, K. 2021. Isolation and optimization of culture conditions of *Thraustochytrium kinnei* for biomass production, nanoparticle synthesis, antioxidant and antimicrobial activities. J. Marine Sci. Eng. 9: 678. https://doi.org/10.3390/jmse9060678.

Kamikawa, Y., Hirabayashi, D., Nagayama, T., Fujisaki, J., Hamada, T., Sakamoto, R., Kamikawa, Y. and Sugihara, K. 2014. *In vitro* antifungal activity against oral *Candida* species using a denture base coated with silver nanoparticles. J. Nanomat. 2014: Article ID 780410. https://doi.org/10.1155/2014/780410.

Kattke, M.D., Gao, E.J., Sapsford, K.E., Stephenson, L.D. and Kumar, A. 2011. FRET-based quantum dot immunoassay for rapid and sensitive detection of *Aspergillus amstelodami*. Sensors 11: 6396–6410. https://doi.org/10.3390/s110606396.

Kaur, P., Dua, J.S. and Prasad, D.N. 2018. Formulation and evaluation of ketoconazole niosomal gel. Asian J. Pharmaceut. Res. Developm. 6: 71–75. https://doi.org/10.22270/ajprd.v6i5.424.

Kaur, T., Kapoor, S. and Kalia, A. 2018. Synthesis of silver nanoparticles from *Pleurotus florida*, characterization and analysis of their antimicrobial activity. Int. J. Curr. Microbiol. Appl. Sci. 7: 4085–4095. https://doi.org/10.20546/ijcmas.2018.707.475.

Kernien, J.F., Snarr, B.D., Sheppard, D.C. and Nett, J.E. 2018. The interface between fungal biofilms and innate immunity. Front. Immun. 8: 1968. https://doi.org/10.3389/fimmu.2017.01968.

Khalil, R., Kassem, M., Elbary, A.A., El Ridi, M. and AbouSamra, M. 2013. Preparation and characterization of nystatin-loaded solid lipid nanoparticles for topical delivery. Int. Journal Pharmaceut. Sci. Res. 4: 2292. http://doi.org/10.13040/IJPSR.0975-8232.4(6).2292-00.

Khan, I., Saeed, K. and Khan, I. 2019. Nanoparticles: Properties, applications and toxicities. Arabian J. Chemist. 12: 908–931. https://doi.org/10.1016/j.arabjc.2017.05.011.

Khansa, I., Schoenbrunner, A.R., Kraft, C.T. and Janis, J.E. 2019. Silver in wound care—friend or foe? A comprehensive review. Plastic Reconstruct. Surg. Glob. Open. 7: e2390. https://doi.org/10.1097%2FGOX.0000000000002390.

Kong, F., Wang, J., Han, R., Ji, S., Yue, J., Wang, Y. and Ma, L. 2020. Antifungal activity of magnesium oxide nanoparticles: Effect on the growth and key virulence factors of *Candida albicans*. Mycopathologia 185: 485–494. https://doi.org/10.1007/s11046-020-00446-9.

Koutsoulas, C., Pippa, N., Demetzos, C. and Zabka, M. 2014. Preparation of liposomal nanoparticles incorporating terbinafine *in vitro* drug release studies. J. Nanosci. Nanotechnol. 14: 4529–4533. https://doi.org/10.1166/jnn.2014.9026.

Lafta, A.K., Ajah, H.A., Dakhil, O.A. and AL-Wattar, W.M. 2019. Biosynthesis of silver nanoparticles using biomass of *cladosporium cladosporioides* and antifungal activity against pathogenic fungi causing onychomycosis. Plant Archives 19: 4391–4396.

Lakshmeesha, T.R., Kalagatur, N.K., Mudili, V., Mohan, C.D., Rangappa, S., Prasad, B.D., Ashwini, B.S., Hashem, Alqarawi, A.A., Malik, J.A., Abd_Allah, E.F., Gupta, V.K., Siddaiah, C.N. and Niranjana, S.R. 2019. Biofabrication of zinc oxide nanoparticles with *Syzygium aromaticum* flower buds extract and finding its novel application in controlling the growth and mycotoxins of *Fusarium graminearum*. Front. Microbiol. 10: 1244. https://doi.org/10.3389/fmicb.2019.01244.

Lara, H.H., Romero-Urbina, D.G., Pierce, C., Lopez-Ribot, J.L., Arellano-Jiménez, M.J. and Jose-Yacaman, M. 2015. Effect of silver nanoparticles on *Candida albicans* biofilms: An ultrastructural study. J. Nanobiotechnol. 13: 91. https://doi.org/10.1186/s12951-015-0147-8.

Leal, A.F., Leite, M.C., Medeiros, C.S., Cavalcanti, I.M., Wanderley, A.G., Magalhães, N.S. and Neves, R.P. 2015. Antifungal activity of a liposomal itraconazole formulation in experimental *Aspergillus flavus* keratitis with endophthalmitis. Mycopathologia 179: 225–229. https://doi.org/10.1007/s11046-014-9837-2.

Lebeaux, D., Ghigo, J.M. and Beloin, C. 2014. Biofilm-related infections: Bridging the gap between clinical management and fundamental aspects of recalcitrance toward antibiotics. Microbiol. Molecul. Biol. Rev. 78: 510–543. https://doi.org/10.1128/MMBR.00013-14.

Limon, J.J., Skalski, J.H. and Underhill, D.M. 2017. Commensal fungi in health and disease. Cell host Microbe. 22: 156–165. https://doi.org/10.1016%2Fj.chom.2017.07.002.

Ling, X., Huang, Z., Wang, J., Xie, J., Feng, M., Chen, Y., Abbas, F., Tu, J., Wu, J. and Sun, C. 2016. Development of an itraconazole encapsulated polymeric nanoparticle platform for effective antifungal therapy. J. Mat. Chem. B 4: 1787–1796. https://doi.org/10.1039/C5TB02453F.

MadhuLatha, A.V., Sojana, N., Mounika, N., Priyanka, G., Venkatesh, A. and Kumar, J.S. 2020. Design and optimization of clotrimazole emulgel by using various polymers. World J. Adv. Res. Rev. 7: 188–199. https://doi.org/10.30574/wjarr.2020.7.2.0281.

Mahant, S., Kumar, S., Nanda, S. and Rao, R. 2020. Microsponges for dermatological applications: Perspectives and challenges. Asian J. Pharmaceut. Sci. 15: 273–291. https://doi.org/10.1016/j.ajps.2019.05.004.

Majeed, S., Abdullah, M.S., Nanda, A. and Ansari, M.T. 2016. *In vitro* study of the antibacterial and anticancer activities of silver nanoparticles synthesized from *Penicillium brevicompactum* (MTCC-1999). J. Taibah. Univ. Sci. 10: 614–620. https://doi.org/10.1016/j.jtusci.2016.02.010.

Makhmalzade, B.S. and Chavoshy, F. 2018. Polymeric micelles as cutaneous drug delivery system in normal skin and dermatological disorders. J. Adv. Pharmaceut. Technol. Res. 9: 2–8. https://doi.org/10.4103/japtr.JAPTR_314_17.

Malekzad, H., Zangabad, P.S., Mirshekari, H., Karimi, M. and Hamblin, M.R. 2017. Noble metal nanoparticles in biosensors: Recent studies and applications. Nanotechnol. Rev. 6: 301–329. https://doi.org/10.1515/ntrev-2016-0014.

Maliszewska, I., Juraszek, A. and Bielska, K. 2014. Green synthesis and characterization of silver nanoparticles using ascomycota fungi *Penicillium nalgiovense* AJ12. J. Clust. Sci. 25: 989–1004. https://doi.org/10.1007/s10876-013-0683-z.

Martinez, L.R. and Fries, B.C. 2010. Fungal biofilms: Relevance in the setting of human disease. Curr. Fungal Infect. Rep. 4: 266–275. https://doi.org/10.1007/s12281-010-0035-5.

Mashrai, A., Khanam, H. and Aljawfi, R.N. 2017. Biological synthesis of ZnO nanoparticles using *C. albicans* and studying their catalytic performance in the synthesis of steroidal pyrazolines. Arabian J. Chem. 10: S1530–1536. https://doi.org/10.1016/j.arabjc.2013.05.004.

Mba, I.E. and Nweze, E.I. 2021. Nanoparticles as therapeutic options for treating multidrug-resistant bacteria: Research progress, challenges, and prospects. World J. Microbiol. Biotechnol. 37: 108. https://doi.org/10.1007/s11274-021-03070-x.

Mirshekari, M., Ghomi, A.B. and Mehravaran, A. 2021. Smart terbinafine recent nano-advances in delivery of terbinafine. Nanomed. J. 8: 1590. https://doi.org/10.22038/NMJ.2021.57263.1590.

Mlynarczyk, D.T., Dlugaszewska, J., Kaluzna-Mlynarczyk, A. and Goslinski, T. 2021. Dendrimers against fungi—A state of the art review. J. Control Release 330: 599–617. https://doi.org/10.1016/j.jconrel.2020.12.021.

Mohamed, Y.M., Azzam, A.M., Amin, B.H. and Safwat, N.A. 2015. Mycosynthesis of iron nanoparticles by *Alternaria alternata* and its antibacterial activity. African J. Biotechnol. 14: 1234–1241. https://doi.org/10.5897/AJB2014.14286.

Morelli, K.A., Kerkaert, J.D. and Cramer, R.A. 2021. *Aspergillus fumigatus* biofilms: Toward understanding how growth as a multicellular network increases antifungal resistance and disease progression. PLoS Pathogens 17: e1009794. https://doi.org/10.1371/journal.ppat.1009794.

Mosallam, F.M., El-Sayyad, G.S., Fathy, R.M. and El-Batal, A.I. 2018. Biomolecules-mediated synthesis of selenium nanoparticles using *Aspergillus oryzae* fermented Lupin extract and gamma radiation for hindering the growth of some multidrug-resistant bacteria and pathogenic fungi. Microbial Pathogenesis 122: 108–116. https://doi.org/10.1016/j.micpath.2018.06.013.

Mudiar, R. and Kelkar-Mane, V. 2020. Original research article (experimental): Targeting fungal menace through copper nanoparticles and Tamrajal. J. Ayurveda Integrative Med. 11: 316–321. https://doi.org/10.1016/j.jaim.2018.02.134.

Muhammad, M.H., Idris, A.L., Fan, X., Guo, Y., Yu, Y., Jin, X., Qiu, J., Guan, X. and Huang, T. 2020. Beyond risk: Bacterial biofilms and their regulating approaches. Fron. Microbiol. 2020; 11: 928. https://doi.org/10.3389/fmicb.2020.00928.

Mukhtar, M., Sargazi, S., Barani, M., Madry, H., Rahdar, A. and Cucchiarini, M. 2021. Application of nanotechnology for sensitive detection of low-abundance single-nucleotide variations in genomic DNA: A review. Nanomaterials 11: 1384. https://doi.org/10.3390/nano11061384.

Mumtaz, T., Ahmed, N., ul Hassan, N., Badshah, M. and Khan, S. 2021. Voriconazole nanoparticles-based film forming spray: An efficient approach for potential treatment of topical fungal infections. J. Drug Deliv. Sci. Technol. 18: 102973. https://doi.org/10.1016/j.jddst.2021.102973.

Muñoz-Escobar, A. and Reyes-López, S.Y. 2020. Antifungal susceptibility of *Candida* species to copper oxide nanoparticles on polycaprolactone fibers (PCL-CuONPs). PLoS One 15: e0228864. https://doi.org/10.1371/journal.pone.0228864.

Musa, S.F., Yeat, T.S., Kamal, L.Z., Tabana, Y.M., Ahmed, M.A., El Ouweini, A., Lim, V., Keong, L.C. and Sandai, D. 2018. *Pleurotus sajor-caju* can be used to synthesize silver nanoparticles with antifungal activity against *Candida albicans*. J. Sci. Food Agric. 98: 1197–1207. https://doi.org/10.1002/jsfa.8573.

Narayanan, K.B. and Sakthivel, N. 2013. Mycocrystallization of gold ions by the fungus *Cylindrocladium floridanum*. World J. Microbiol. Biotechnol. 29: 2207–2211. https://doi.org/10.1007/s11274-013-1379-0.

Niemirowicz, K., Durnaś, B., Tokajuk, G., Głuszek, K., Wilczewska, A.Z., Misztalewska, I., Mystkowska, J., Michalak, G., Sodo, A., Wątek, M., Kiziewicz, B., Góźdź, S., Głuszek, S. and Bucki, R. 2016. Magnetic nanoparticles as a drug delivery system that enhance fungicidal activity of polyene antibiotics. Nanomed.: Nanotechnol. Biol. Med. 12: 2395–4204. https://doi.org/10.1016/j.nano.2016.07.006.

Nimtrakul, P., Williams, D.B., Tiyaboonchai, W. and Prestidge, C.A. 2020. Copolymeric micelles overcome the oral delivery challenges of amphotericin B. Pharmaceuticals 13: 121. https://doi.org/10.3390%2Fph13060121.

Ning, M., Guo, Y., Pan, H., Chen, X. and Gu, Z. 2005. Preparation, *in vitro* and *in vivo* evaluation of liposomal/niosomal gel delivery systems for clotrimazole. Drug Develop. Indust. Pharma. 31: 375–383. https://doi.org/10.1081/ddc-54315.

Njuguna, J., Ansari, F., Sachse, S., Rodriguez, V.M., Siqqique, S. and Zhu, H. 2021. Nanomaterials, nanofillers, and nanocomposites: Types and properties. In Health and Environmental Safety of Nanomaterials. Woodhead Publishing, pp. 3–37. https://doi.org/10.1016/B978-0-12-820505-1.00011-0.

Nobile, C.J. and Johnson, A.D. 2015. *Candida albicans* biofilms and human disease. Annu. Rev. Microbiol. 69: 71–92. https://doi.org/10.1146/annurev-micro-091014-104330.

Offner, F., Krcmery, V., Boogaerts, M., Doyen, C., Engelhard, D., Ribaud, P., Cordonnier, C., de Pauw, B., Durrant, S., Marie, J-P., Moreau, P., Guiot, H., Samonis, G., Sylvester, R. and Herbrecht, R. 2004. Liposomal nystatin in patients with invasive aspergillosis refractory to or intolerant of amphotericin B. Antomicrob. Agents Chemother. 48: 4808–4812. https://doi.org/10.1128/aac.48.12.4808-4812.2004.

Ofokansi, K.C., Kenechukwu, F.C., Charles, L. and Attama, A.A. 2012. Topical delivery of miconazole-loaded microemulsion: Formulation design and evaluation. J. Pharmaceutical Allied Sci. 9: 1458–1471.

Osanloo, M., Assadpour, S., Mehravaran, A., Abastabar, M. and Akhtari, J. 2018. Niosome-loaded antifungal drugs as an effective nanocarrier system: A mini review. Curr. Medi. Mycol. 4: 31. http://doi.org/10.18502/cmm.4.4.384.

Phanjom, P. and Ahmed, G. 2017. Effect of different physicochemical conditions on the synthesis of silver nanoparticles using fungal cell filtrate of *Aspergillus oryzae* (MTCC No. 1846) and their antibacterial effect. Adv. Nat. Sci.: Nanosci. Nanotechnol. 8: 045016. https://doi.org/10.1088/2043-6254/aa92bc.

Piktel, E., Suprewicz, Ł., Depciuch, J., Cieśluk, M., Chmielewska, S., Durnaś, B., Król, G., Wollny, T., Deptuła, P., Kochanowicz, J., Kułakowska, A., Fiedoruk, K., Maximenko, A., Parlińska-Wojtan, M. and Bucki, R. 2020. Rod-shaped gold nanoparticles exert potent candidacidal activity and decrease the adhesion of fungal cells. Nanomedicine 15: 2733–2752. https://doi.org/10.2217/nnm-2020-0324.

Pillai, A.B., Nair, J.V., Gupta, N.K. and Gupta, S. 2015. Microemulsion-loaded hydrogel formulation of butenafine hydrochloride for improved topical delivery. Archiv. Dermatol. Res. 307: 625–633. https://doi.org/10.1007/s00403-015-1573-z.

Prattes, J., Heldt, S., Eigl, S. and Hoenigl, M. 2016. Point of care testing for the diagnosis of fungal infections: Are we there yet? Curr. Fungal Infect. Rep. 10: 43–50. https://doi.org/10.1007/s12281-016-0254-5.

Raheman, F., Deshmukh, S., Ingle, A., Gade, A. and Rai, M. 2011. Silver nanoparticles: Novel antimicrobial agent synthesized from an endophytic fungus *Pestalotia* sp. isolated from leaves of *Syzygium cumini* (L). Nano Biomed. Eng. 3: 174–178. https://doi.org/10.5101/nbe.v3i3.p174-178.

Rahi, D.K., Manhas, P.L., Kaur, M., Malik, D. and Rahi, S. 2018. Extracellular synthesis of silver nanoparticles by an indigenous yeast *Aureobasidium Pullulans* Rylf 10: Characterization and evaluation of antibacterial potential. Int. J. Phar. Biol. Sci. 8: 312.

Rajan, A., Cherian, E. and Baskar, G. 2016. Biosynthesis of zinc oxide nanoparticles using *Aspergillus fumigatus* JCF and its antibacterial activity. Int. J. Mod. Sci. Technol. 1: 52–57.

Raman, J., Reddy, G.R., Lakshmanan, H., Selvaraj, V., Gajendran, B., Nanjian, R., Chinnasamy, A. and Sabaratnamb, V. 2015. Mycosynthesis and characterization of silver nanoparticles from *Pleurotus djamor* var. roseus and their *in vitro* cytotoxicity effect on PC3 cells. Process Biochemistry 50: 140–147. https://doi.org/10.1016/j.procbio.2014.11.003.

Rasoanirina, B.N., Lassoued, M.A., Miladi, K., Razafindrakoto, Z., Chaâbane-Banaoues, R., Ramanitrahasimbola, D., Cornet, M. and Sfar, S. 2020. Self-nanoemulsifying drug delivery system to improve transcorneal permeability of voriconazole: *in-vivo* studies. J. Pharma Pharmacol. 72: 889–896. https://doi.org/10.1111/jphp.13265.

Rathod, V., Banu, A. and Ranganath, E. 2012. Biosynthesis of highly stabilized silver nanoparticles by *Rhizopus stolonifer* and their anti-fungal efficacy. Int. J. Cur. Biomed. Phar. Res. 2: 241–245.

Rhee, Y.S., Park, C.W., Nam, T.Y., Shin, Y.S., Chi, S.C. and Park, E.S. 2007. Formulation of parenteral microemulsion containing itraconazole. Archiv. Pharma Res. 30: 114–123.

Rispail, N., De Matteis, L., Santos, R., Miguel, A.S., Custardoy, L., Testillano, P.S., Risueño, M.C., Pérez-de-Luque, A., Maycock, C., Fevereiro, P., Oliva, A., Fernández-Pacheco, R., Ricardo Ibarra, M., de la Fuente, J.M., Marquina, K., Rubiales, D. and Prats, E. 2014. Quantum dot and superparamagnetic nanoparticle interaction with pathogenic fungi: Internalization and toxicity profile. ACS Appl. Mat. Interface 6: 9100–9110. https://doi.org/10.1021/am501029g.

Rodrigues, A.G., Ruiz, R.D., Selari, P.J., Araújo, W.L. and Souza, A.O. 2021. Anti-biofilm action of biological silver nanoparticles produced by *Aspergillus tubingensis* and antimicrobial activity of fabrics carrying it. Biointerface Res. Appl. Chem. 11: 14764–14774. https://doi.org/10.33263/BRIAC116.1476414774.

Sadique, A., Khalid, S.H., Asghar, S. and Irfan, M. 2018. Miconazole nitrate microemulsion: Preparation, characterization and evaluation for enhancement of antifungal activity. Latin American J. Pharma. 37: 1578–1586.

Safarpour, H., Safarnejad, M.R., Tabatabaei, M., Mohsenifar, A., Rad, F., Basirat, M., Shahryari, F. and Hasanzadeh, F. 2012. Development of a quantum dots FRET-based biosensor for efficient detection of *Polymyxa betae*. Canadian J. Plant Pathol. 34: 507–515. https://doi.org/10.1080/07060661.2012.709885.

Saha, S., Sarkar, J., Chattopadhyay, D., Patra, S., Chakraborty, A. and Acharya, K. 2010. Production of silver nanoparticles by a phytopathogenic fungus *Bipolaris nodulosa* and its antimicrobial activity. Dig. J. Nanomater. Biostruct. 5: 887–895.

Sahib, F.H., Aldujaili, N.H. and Alrufae, M.M. 2017. Biosynthesis of silver nanoparticles using *Saccharomyces boulardii* and study their biological activities. European J. Pharmaceut. Med. Res. 4: 65–74.

Sakita, K.M., Conrado, P.C., Faria, D.R., Arita, G.S., Capoci, I.R., Rodrigues-Vendramini, F.A., Pieralisi, N., Cesar, G.B., Gonçalves, R.S., Caetano, W., Hioka, N., Kioshima, E.S., Ie Svidzinski, T., Patricia, S. and Bonfim-Mendonça, P.S. 2019. Copolymeric micelles as efficient inert nanocarrier for hypericin in the photodynamic inactivation of *Candida* species. Future Microbiol. 14: 519–531. https://doi.org/10.2217/fmb-2018-0304.

Saldanha, C.A., Garcia, M.P., Iocca, D.C., Rebelo, L.G., Souza, A.C., Bocca, A.L., Menezes Almeida Santos, M.F., Morais, P.C. and Azevedo, R.B. 2016. Antifungal activity of amphotericin B conjugated to nanosized magnetite in the treatment of paracoccidioidomycosis. PLoS Negl. Trop. Dis. 10: e0004754. https://dx.doi.org/10.1371%2Fjournal.pntd.0004754.

Salunkhe, R.B., Patil, S.V., Patil, C.D. and Salunke, B.K. 2011. Larvicidal potential of silver nanoparticles synthesized using fungus *Cochliobolus lunatus* against Aedes aegypti (Linnaeus, 1762) and Anopheles stephensi Liston (*Diptera; Culicidae*). Parasitol. Res. 109: 823–831. https://doi.org/10.1007/s00436-011-2328-1.

Salvadori, M.R., Lepre, L.F., Ando, R.A., Oller do Nascimento, C.A. and Corrêa, B. 2013. Biosynthesis and uptake of copper nanoparticles by dead biomass of *Hypocrea lixii* isolated from the metal mine in the Brazilian Amazon region. PLoS One 8: e80519. https://doi.org/10.1371/journal.pone.0080519.

Sanna, V., Gavini, E., Cossu, M., Rassu, G. and Giunchedi, P. 2007. Solid lipid nanoparticles (SLN) as carriers for the topical delivery of econazole nitrate: *in-vitro* characterization, *ex-vivo* and *in-vivo* studies. J. Pharma Pharmacol. 59: 1057–1064. https://doi.org/10.1211/jpp.59.8.0002.

Saqib, M., Ali Bhatti, A.S., Ahmad, N.M., Ahmed, N., Shahnaz, G., Lebaz, N. and Elaissari, A. 2020. Amphotericin B loaded polymeric nanoparticles for treatment of leishmania infections. Nanomaterials 10: 1152. https://doi.org/10.3390/nano10061152.

Sardi, J.D., Pitangui, N.D., Voltan, A.R., Braz, J.D., Machado, M.P., Fusco Almeida, A.M. and Mendes Jiannini, M.J.S. 2015. *In vitro Paracoccidioides brasiliensis* biofilm and gene expression of adhesins and hydrolytic enzymes. Virulence 6: 642–651. https://doi.org/10.1080/21505594.2015.1031437.

Sawyer, P.R., Brogden, R.N., Pinder, K.M., Speight, T.M. and Avery, G.S. 1975. Clotrimazole: A review of its antifungal activity and therapeutic efficacy. Drugs 9: 424–447. https://doi.org/10.2165/00003495-197509060-00003.

Schelenz, S. 2017. Fungal diseases of the gastrointestinal tract. Oxford Textbook of Medical Mycology. https://doi.org/10.1093/med/9780198755388.003.0026.

Sharma, D., Misba, L. and Khan, A.U. 2019. Antibiotics versus biofilm: An emerging battleground in microbial communities. Antimicrob. Resist. Infect. Cont. 8: 76. https://doi.org/10.1186/s13756-019-0533-3.

Sharma, R. and Pathak, K. 2011. Polymeric nanosponges as an alternative carrier for improved retention of econazole nitrate onto the skin through topical hydrogel formulation. Pharmaceut. Develop. Technol. 16: 367–376. https://doi.org/10.3109/10837451003739289.

Shelar, G.B. and Chavan, A.M. 2014. *Fusarium semitectum* mediated extracellular synthesis of silver nanoparticles and their antibacterial activity. Int. J. Biomed. Adv. Res. 5: 20–24. https://doi.org/10.7439/ijbar.

Sherwani, M.A., Tufail, S., Khan, A.A. and Owais, M. 2015. Gold nanoparticle-photosensitizer conjugate based photodynamic inactivation of biofilm producing cells: Potential for treatment of *C. albicans* infection in BALB/c mice. PLoS One 10: e0131684. https://doi.org/10.1371/journal.pone.0131684.

Sidhu, A., Bala, A., Singh, H., Ahuja, R. and Kumar, A. 2020. Development of MgO-sepoilite nanocomposites against phytopathogenic fungi of rice (*Oryzae sativa*): A green approach. ACS Omega 5: 13557–13565. https://doi.org/10.1021/acsomega.0c00008.

Soliman, O.A., Mohamed, E.A. and Khatera, N.A. 2019. Enhanced ocular bioavailability of fluconazole from niosomal gels and microemulsions: Formulation, optimization, and *in vitro–in vivo* evaluation. Pharmaceuti. Develop. Technol. 24: 48–62. https://doi.org/10.1080/10837450.2017.1413658.

Sonar, H., Nagaonkar, D., Ingle, A.P. and Rai, M. 2017. Mycosynthesized silver nanoparticles as potent growth inhibitory agents against selected waterborne human pathogens. CLEAN–Soil, Air, Water 45: 1600247. https://doi.org/10.1002/clen.201600247.

Song, S.H., Lee, K.M., Kang, J.B., Lee, S.G., Kang, M.J. and Choi, Y.W. 2014. Improved skin delivery of voriconazole with a nanostructured lipid carrier-based hydrogel formulation. Chem. Pharmaceut. Bulletin 62: 793–798. https://doi.org/10.1248/cpb.c14-00202.

Sosa, L., Clares, B., Alvarado, H.L., Bozal, N., Domenech, O. and Calpena, A.C. 2017. Amphotericin B releasing topical nanoemulsion for the treatment of candidiasis and aspergillosis. Nanomed.: Nanotechnol. Biol. Med. 13: 2303–2312. https://doi.org/10.1016/j.nano.2017.06.021.

Souto, E.B. and Müller, R.H. 2005. SLN and NLC for topical delivery of ketoconazole. J. Microencapsulat. 22: 501–510. https://doi.org/10.1080/02652040500162436.

Souto, E.B. and Müller, R.H. 2006. Investigation of the factors influencing the incorporation of clotrimazole in SLN and NLC prepared by hot high-pressure homogenization. J. Microencapsulat. 23: 377–388. https://doi.org/10.1080/02652040500435295.

Stone, N.R., Bicanic, T., Salim, R. and Hope, W. 2016. Liposomal amphotericin B (AmBisome®): A review of the pharmacokinetics, pharmacodynamics, clinical experience and future directions. Drugs 76: 485–500. https://doi.org/10.1007/s40265-016-0538-7.

Sulaiman, G.M.T., Hussien, H.I. and Saleem, M.M. 2015. Biosynthesis of silver nanoparticles synthesized by *Aspergillus flavus* and their antioxidant, antimicrobial and cytotoxicity properties. Bullet. Mate. Sci. 38: 639–644. https://doi.org/10.1007/s12034-015-0905-0.

Suryavanshi, P., Pandit, R., Gade, A., Derita, M., Zachino, S. and Rai, M. 2017. *Colletotrichum* sp.-mediated synthesis of sulphur and aluminium oxide nanoparticles and its *in vitro* activity against selected food-borne pathogens. LWT-Food Sci. Technol. 81: 188–194. http://doi.org/10.1016/j.lwt.2017.03.038.

Tarafdar, J.C. and Raliya, R. 2013. Rapid, low-cost, and ecofriendly approach for iron nanoparticle synthesis using *Aspergillus oryzae* TFR9. J. Nanopart. 2013: Article ID 141274. https://doi.org/10.1155/2013/141274.

Tejada, G., Barrera, M.G., García, P., Sortino, M., Lamas, M.C., Lassalle, V., Alvarez, V. and Leonardi, D. 2020. Nanoparticulated systems based on natural polymers loaded with miconazole nitrate and lidocaine for the treatment of topical Candidiasis. AAPS Pharm. Sci. Tech. 21: 1–3. https://doi.org/10.1208/s12249-020-01826-6.

Thirumurugan, G., Shaheedha, S.M. and Dhanaraju, M.D. 2009. *In vitro* evaluation of antibacterial activity of silver nanoparticles synthesised by using *Phytophthora infestans*. Int. J. Chem. Tech. Res. 1: 714–716.

Tripathi, R.M., Gupta, R.K., Shrivastav, A., Singh, M.P., Shrivastav, B.R. and Singh, P. 2013. *Trichoderma koningii* assisted biogenic synthesis of silver nanoparticles and evaluation of their antibacterial activity. Adv. Nat. Sci.: Nanoscie. Nanotechnol. 4: 035005. https://doi.org/10.1088/2043-6262/4/3/035005.

Tsui, C., Kong, E.F. and Jabra-Rizk, M.A. 2016. Pathogenesis of *Candida albicans* biofilm. Pathog. Dis. 74: ftw018. https://doi.org/10.1093%2Ffemspd%2Fftw018.

Vargason, A.M., Anselmo, A.C. and Mitragotri, S. 2021. The evolution of commercial drug delivery technologies. Nat. Biomed. Eng. 1: 1–7. https://doi.org/10.1038/s41551-021-00698-w.

Varshney, R., Mishra, A.N., Bhadauria, S. and Gaura, M.S. 2009. Novel microbial route to synthesize silver nanoparticles using fungus *Hormoconis resinae*. Digest. J. Nanomat. Biostruct. (DJNB) 4: 349–355.

Veloso, D.F., Benedetti, N.I., Ávila, R.I., Bastos, T.S., Silva, T.C., Silva, M.R., Batista, A.C., Valadares, M.C. and Lima, E.M. 2018. Intravenous delivery of a liposomal formulation of voriconazole improves drug pharmacokinetics, tissue distribution, and enhances antifungal activity. Drug Delivery 25: 1585–1594. https://doi.org/10.1080/10717544.2018.1492046.

Verma, S. and Utreja, P. 2018. Transethosomes of econazole nitrate for transdermal delivery: Development, *in-vitro* characterization, and *ex-vivo* assessment. Pharmaceut. Nanotechnol. 6: 171–179. https://doi.org/10.2174/2211738506666180813122102.

Verma, V.C., Kharwar, R.N. and Gange, A.C. 2010, Biosynthesis of antimicrobial silver nanoparticles by the endophytic fungus *Aspergillus clavatus*. Nanomedicine 5: 33–40. https://doi.org/10.2217/nnm.09.77.

Vila, T., Sultan, A.S., Montelongo-Jauregui, D. and Jabra-Rizk, M.A. 2020. Oral candidiasis: A disease of opportunity. J. Fungi. 6: 15. https://doi.org/10.3390%2Fjof6010015.

Vlaia, L., Coneac, G., Muţ, A.M., Olariu, I., Vlaia, V., Anghel, D.F., Maxim, M.E., Dobrescu, A., Hîrjău, M. and Lupuleasa, D. 2021. Topical biocompatible fluconazole-loaded microemulsions based on essential oils and sucrose esters: Formulation design based on pseudo-ternary phase diagrams and physicochemical characterization. Processes 9: 144. https://doi.org/10.3390/pr9010144.

Wang, D., Xue, B., Wang, L., Zhang, Y., Liu, L. and Zhou, Y. 2021. Fungus-mediated green synthesis of nano-silver using *Aspergillus sydowii* and its antifungal/antiproliferative activities. Scientific Reports 11: 1–9. https://doi.org/10.1038/s41598-021-89854-5.

Wang, L., Hu, C. and Shao, L. 2017. The antimicrobial activity of nanoparticles: Present situation and prospects for the future. Int. J. Nanomed. 12: 1227. https://doi.org/10.2147/IJN.S121956.

Williams, C., Rajendran, R. and Ramage, G. 2016. *Aspergillus* biofilms in human disease. Fungal Biofilm Relat. Infect. 931: 1–11. https://doi.org/10.1007/5584_2016_4.

Yadav, T.C., Gupta, P., Saini, S., Pruthi, V. and Prasad, R. 2021. Plausible mechanistic insights in biofilm eradication potential of against *Candida* spp. using *in situ* synthesized tyrosol functionalized chitosan gold nanoparticles as a versatile antifouling coating on implant surfaces. 2021: 1–40. https://doi.org/10.1101/2021.09.30.462644.

Yehia, R.S. and Al-Sheikh, H. 2014. Biosynthesis and characterization of silver nanoparticles produced by *Pleurotus ostreatus* and their anticandidal and anticancer activities. World J. Microbiol. Biotechnol. 30: 2797–2803. https://doi.org/10.1007/s11274-014-1703-3.

Yin, W., Wang, Y., Liu, L. and He, J. 2019. Biofilms: The microbial "protective clothing" in extreme environments. Int. J. Molecul. Sci. 20: 3423. https://doi.org/10.3390/ijms20143423.

Yu, D., Wang, L., Zhou, H., Zhang, X., Wang, L. and Qiao, N. 2019. Fluorimetric detection of *Candida albicans* using cornstalk N-carbon quantum dots modified with amphotericin B. Bioconjugate Chem. 30: 966–973. https://doi.org/10.1021/acs.bioconjchem.9b00131.

Zea, L., McLean, R.J., Rook, T.A., Angle, G., Carter, D.L., Delegard, A., Denvir, A., Gerlachf, R., Gortid, S., McIlwaine, D., Nur, M., Peyton, B.M., Stewart, P.S., Sturman, P., Ann, Y. and Justiniano, V. 2020. Potential biofilm control strategies for extended spaceflight missions. Biofilm 2: 100026. https://doi.org/10.1016/j.bioflm.2020.100026.

Zeb, A., Qureshi, O.S., Kim, H.S., Kim, M.S., Kang, J.H., Park, J.S. and Kim, J.K. 2017. High payload itraconazole-incorporated lipid nanoparticles with modulated release property for oral and parenteral administration. J. Pharma. Pharma. 69: 955–966. https://doi.org/10.1111/jphp.12727.

Zhang, L., Li, X., Zhu, S., Zhang, T., Maimaiti, A., Ding, M. and Shi, S. 2020. Dermal targeting delivery of terbinafine hydrochloride using novel multi-ethosomes: A new approach to fungal infection treatment. Coatings 10: 304. https://doi.org/10.3390/coatings10040304.

Zhou, L., Zhang, Y., Ge, Y., Zhu, X. and Pan, J. 2020. Regulatory mechanisms and promising applications of quorum sensing-inhibiting agents in control of bacterial biofilm formation. Front. Microbiol. 11: 2558. https://doi.org/10.3389/fmicb.2020.589640.

Zhou, L., Zhao, X., Li, M., Lu, Y., Ai, C., Jiang, C., Liu, Y., Pan, Z. and Shi, J. 2021. Antifungal activity of silver nanoparticles synthesized by iturin against *Candida albicans in vitro* and *in vivo*. Appl. Microbiol. Biotechnol. 105: 3759–3770. https://doi.org/10.1007/s00253-021-11296-w.

Section III

Applications in Food, Agriculture and Veterinary

8

Strategic Role of Myconanotechnology in Agriculture for Control of Fungal Pathogens

Graciela Dolores Avila-Quezada,[1,*] *Mahendra Rai,*[2,3]
Nuvia Orduño-Cruz,[1] *Denisse Yatzely Mercado-Meza*[1]
and *Hilda Karina Sáenz-Hidalgo*[1,4]

Introduction

During the last 50 years, the use of technologies such as monocultures has made it possible to optimize the soil for food production. However, this has caused the depletion of soil, turning it into soil with a low microbial population causing less interaction, less microbial competition, an increase in the population of pathogens, resulting in losses in food production (Madrid-Delgado et al. 2021, González-Escobedo et al. 2022).

Recently an increase in the number of diseases caused by phytopathogens has been observed. According to the phytosanitary conditions related to the disease tetrahedron, the severity in crops can be devastating (Guarnaccia and Crous 2017, Avila-Quezada et al. 2018).

[1] Facultad de Ciencias Agrotecnológicas, Universidad Autónoma de Chihuahua, Escorza 900, Chihuahua 31000, México.
[2] Nanobiotechnology Lab., Department of Biotechnology, Sant Gadge Baba Amravati University, Amravati, Maharashtra 444601, India.
[3] Department of Microbiology, Nicolaus Copernicus University, Toruń 87-100, Poland.
[4] Centro de Investigación en Alimentación y Desarrollo AC. Av. 4 Sur 3828, Pablo Gómez, Delicias, Chihuahua 33088, México.
* Corresponding author: gdavila@uach.mx

Plant pathogens and pests are responsible for up to 40% of the production losses of crops such as corn, potatoes, rice, soybeans, and wheat worldwide (Savary et al. 2019). Diseases cost the world economy $220 billion dollars annually (Savary et al. 2019).

Pathogen infections occur during agricultural production (Nazarov et al. 2020). Even infected seeds can be vehicles to disseminate the pathogen to the seedlings (Avila-Quezada and Rai 2022). Therefore, the combat of these microorganisms is of the utmost importance. Generally, to keep the pathogens in low populations, it is carried out with synthetic fungicides, which have caused the horizontal transfer of resistance genes, and its use is associated with serious environmental and human health risks (Soanes and Richards 2014, Avila-Quezada et al. 2018).

When pathogens are devastating economically important crops, measures such as epidemiological surveillance programs and quarantines are taken, in which the movement of diseased plant material between areas is restricted (Avila-Quezada et al. 2016). For this, various international agencies related to plant health publish lists of quarantine pests to try to exclude them (Iftikhar and Sajid 2020). However, these efforts are never sufficient for the total control of pathogens.

In addition to these phytosanitary problems, post-harvest losses are added. For this reason, it is projected that food production will increase by up to 60% by 2050, in order to supply the world population (Beddington 2009, FAO ITU 2017). To achieve this, it is necessary to apply technological innovations in the agricultural sector, which will allow the control of pathogens and the reduction of high amounts of inputs (Rao et al. 2017). An alternative to the application of fungicides is a more environment friendly approach that involves biological control that helps in the management of diseases with a beneficial effect on crop production (Singh et al. 2018, Hernandez-Montiel et al. 2021, Savín-Molina et al. 2021).

A recent strategic technology is nanotechnology, which has aroused the interest of the scientific and business community worldwide. Among the most promising applications of nanotechnology in agriculture, detection and control of pests and pathogens is an area of interest. Thus, one of the technological innovations is nanomaterials (NMs) (Avila-Quezada et al. 2021). Its use in agriculture and plant biotechnology occupies an outstanding place in the transformation of agriculture and food production (Deka et al. 2018). NMs have been used as fertilizers, pesticides, herbicides, fungicides, insecticides, and for seed germination (Priyanka et al. 2020).

Regarding their antimicrobial effect, it should be noted that nanoparticles (NPs) have antifungal activity. NPs affect the integrity of the membrane, causing the formation of transmembrane pores that allow the exit of cellular components until cell death (Abdel-Aziz et al. 2018). Other NMs that cause myotoxicity are metal and metal oxides, carbon-based, silica-based, polymeric, and hybrid. Considering the large number of agrochemicals that are discharged into the environment, the small amount that is really useful, and the high economic cost that they represent, is convenient to explore strategies such as the potential of myconanotechnology to control fungal diseases of crops.

Mycosynthesis of Nanoparticles: Principles of Bioreduction of the Metal Ions

The term myconanotechnology coined by Rai et al. in 2009 is used to refer to the manufacture of NPs by fungi. Fungi use their antioxidant systems to detoxify metal ions to protect themselves from metals (Jha and Prasad 2016). In addition, they act differently against each metal due to their metabolism. Under this principle, fungi or fungal biomass is used for the detoxification of industrial effluents that contain metallic radionuclides.

Microorganisms are a very important tool for the production of NPs because they have potential to synthesize molecules (Vinod et al. 2011). That is because fungi are ideal candidates for nanofabrication, since they are a renewable source, easy to preserve and reproduce in the laboratory. In addition, production is easy to scale due to the mechanical resistance of the fungal cell wall, which does not suffer alterations during agitation in the bioreactor (Rao et al. 2017).

The biosynthesis of NPs does not require high energy supplies for its production nor does it generate toxicity, therefore, its use has potential (Youssef et al. 2017). To date, NPs of silver, gold, copper, cadmium, platinum, silica, palladium, among other metals, have been synthesized with fungi. Various genera of fungi have been successfully exploited for the biological synthesis of NPs. Among the most frequent are different species of *Verticillium*, *Aspergillus* (Iranmanesh et al. 2020), and *Fusarium* (Rai et al. 2021a).

It is known that the synthesis of NMs is divided into two categories: the top-down method where NMs are obtained from a bulk structure, and the bottom-up method where the atoms, molecules or groups assemble to form NPs (Ghadam et al. 2021). Similarly, obtaining NPs in fungi has a bottom-up approach and is generated intracellularly or extracellularly. Here the NM is obtained through the addition of atom by atom, molecule by molecule or group by group. Precision is required in the entire process, in physical factors, kinetics, and proportion of all ingredients to obtain homogeneous NPs (Boddula et al. 2018).

For example, the synthesis of AgNPs by means of fungi requires precursors such as $AgNO_3$ or $AgClO_4$. It also requires reducers like $NaHB_4$ or dextrose or ethylene glycol; sodium hydroxide as accelerators; and starch, surfactin or polyvinylpyrrolidone (PVP) as stabilizers (Tai et al. 2008, Sharma and Kumar 2021). NP mycosynthesis is a three-step process involving metal ion capture, bioreduction, and finally synthesis. Myconanofabrication is the result of the reaction between fungal biomass and saline solution, which occurs extracellularly or intracellularly, although the former predominates (Rai et al. 2011).

The process occurs when metal ions are captured on the surface of the fungal cell, probably due to electrostatic interaction in the enzymes present in the fungal cell wall. Subsequently, the enzymes reduce the metal ions by the use of the enzyme nitrate reductase or by the entrapment of ions in the fungal cell wall. Finally, the aggregation of metal ions and the synthesis of NPs are channeled (Sastry et al. 2003, Youssef et al. 2017).

According to Gade et al. (2022) three steps are involved in NPs synthesis: *nucleation* where proteins are the capping agent; *elongation*, here the photosensitized anthraquinone derivative takes the electron donated by the inorganic nitrate and transfers it to silver ions, and thus it reduces them to form silver particles (Ag^0); and *termination* which will occur when the anthraquinone molecule involved in the synthesis is recruited by another nucleation center for elongation or is unable to release electrons.

Intracellular Synthesis

In this biological synthesis, NP formation occurs within fungal cells. The microorganism comes into direct contact with silver ions that it absorbs and metabolizes (Figure 1).

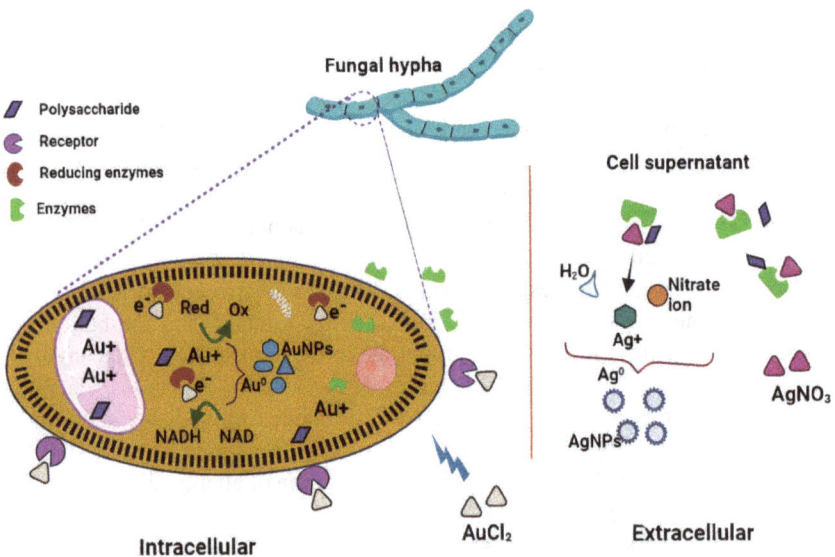

Figure 1. Principle of fungi-mediated synthesis (reduction) of gold nanoparticles (AuNPs) in an intracellular way and silver nanoparticles (AgNPs) in an extracellular way.

Fungal cells protect themselves from metals and use their antioxidant systems to detoxify metal ions (Jha and Prasad 2016). The synthesis of NP in fungi cells is due to the action of cellular ATPases and hydrogenases. For instance, *Fusarium oxysporum* produces AuNPs intracellularly in cytoplasmic vacuoles, with plasma membrane ATPase, 3-glucan binding enzyme, and glyceraldehyde-3-phosphate dehydrogenase (Vahabi and Dorcheh 2014). Hydrogenases generate cytoplasmic hydrogen to precipitate metallic NPs (Whiteley et al. 2011).

Cells use intracellular glutathione and metal-binding proteins in detoxification. This is how the biological reduction of metal ions occurs. The fungi that can carry out this intracellular synthesis usually generate small and stable NPs. Although it requires the separation of the NPs from the cellular structure (de Souza and Rodriguez 2015, Crisan et al. 2021).

The synthesis of NPs by fungi depends on the microorganism involved. According to the species, the culture medium and technique must be selected (Subashini and Bhuvaneswari 2018). The intracellular synthesis of AgNPs has been successfully developed in *Verticillium* spp. (Senapati et al. 2004) and for AuNPs synthesis, *Penicillium* spp. and *Verticillium* spp. have been used (Mukherjee et al. 2001, Du et al. 2011). Intracellular synthesis with *Verticillium luteoalbum* yielded spherical, triangular, hexagons, and rod AuNPs of size 10–100 nm in media of pH 5 (Gericke and Pinches 2006).

Factors such as pH, temperature, metal ions, incubation time, and nature of the compound, influence the NPs synthesis, which also influence the fungal growth. Therefore, the optimal point will generate enough biomass to improve the yield of the final product (Subashini and Bhuvaneswari 2018).

Extracellular Synthesis

In extracellular synthesis with fungal biomass, silver ions interact with proteins on the outer surface of the fungal cell wall. The cell wall is negatively charged by carboxylic and amine groups. Then the resulting AgNPs adhere to the surface of the fungi (Rai et al. 2021b). Moreover, when aqueous extracts derived from fungal biomass are used, the action of the biomolecules present in these extracts occurs, therefore, NPs are formed in a "clean" medium free of mycelium (Bansod et al. 2013) (Figure 2).

The AgNPs extracellular synthesis technique has been standardized by various authors with very particular features in each case. AbdelRahim et al. (2015) used *Rhizopus* aqueous mycelium extract and silver ions at 40°C at 180 rpm for 3 days, obtaining 10 nm spherical AgNPs. On the other hand, cell filtering of *Aspergillus fumigatus* (72 h) with 1 mM silver ions at 24 hr of incubation has also been used to obtain AgNPs of the size range of 5 to 25 nm (Bhainsa and D'Souza 2006). Moreover, Basavaraja et al. (2008) used a cell filtrate of *Fusarium semitectum* and added 1 mM of silver ions, stirring and keeping at 27°C. They obtained AgNPs in an average size range of 10 to 60 nm in a spherical shape.

Mukherjee et al. (2002) obtained AuNPs by extracellular synthesis using *Fusarium oxysporum*, while Nachiyar et al. (2015) synthesized AuNPs from three different isolates of endophytic fungi obtaining NPs in the ranges of 15 to 35 nm. Species of *Alternaria* such *A. alternata* (Sarkar et al. 2011) and species of *Trichoderma* such *T. longibrachiatum* (Elamawi et al. 2018) have been used to produce AgNPs and for the ZnONP synthesis, *Aspergillus fumigatus* has been used (Raliya et al. 2016). Extracellular synthesis with *Macrophomina phaseolina* produced spherical AgNPs in the size range 5–30 nm (Spagnoletti et al. 2019). The extracellular synthesis of AgNPs by a filtrate of the fungus *Phoma glomerata*, resulted in rapid synthesis of AgNPs with a protein layer, which increases the stability of NPs (Gade et al. 2014).

Losses of agricultural crops by pathogens are high. In the case of phytopathogenic fungi and oomycetes, yield losses and even postharvest losses can reach 80% and can even cause total losses (Sánchez-Chávez et al. 2017, Avila-Quezada and Rai 2022). The strategies for the control of phytopathogens depend on the agrochemicals. These

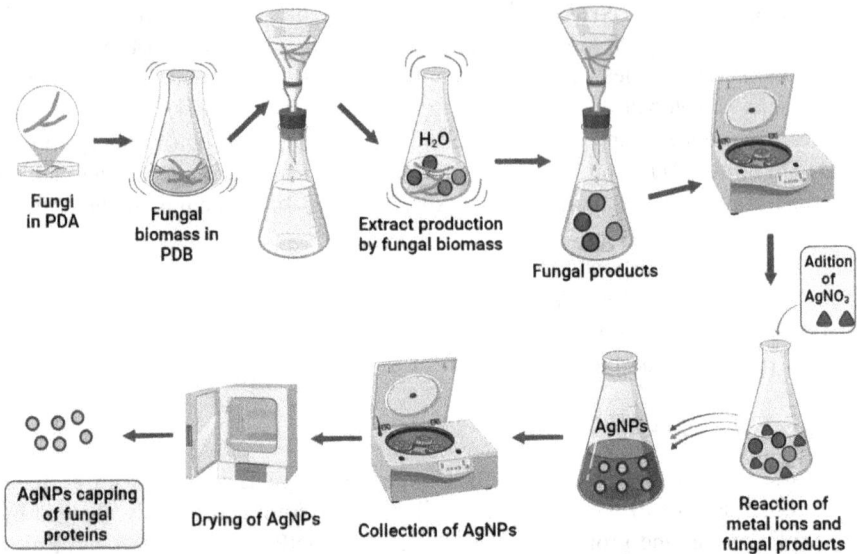

Figure 2. Schematic representation of extracellular biosynthesis of silver nanoparticles (AgNPs) with the cell-free fungal extract. AgNPs capping of proteins is in the final product. The reduction of silver for the synthesis of AgNPs requires fungal products and release of nicotinamide adenina dinucleótida (NADH)-dependent reductases. PDA= potato dextrose agar. PDB= potato dextrose broth. Fungal products are polysaccharides, proteins, flavonoids, terpenoids, phenolic acid, organic acids, alkaloids, and more.

generate negative impacts on the environment and on animal and human health. Nanotechnology in a sustainable and ecological way that can help reduce these negative effects. For instance, it is possible to improve the solubility of fungicides, increase the useful life of fruits and vegetables, and reduce microbial contamination. In addition, by replacing conventional fungicides, due to their high volume used, we will reduce toxicity.

Despite the advantages that nanoparticles offer, there are few commercial agricultural products based on nanoparticles. This may be related to the low number of field experiments.

Applications of Nanomaterials in Foliar Diseases

Conventional agriculture relies on the use of synthetic products to control pathogens (Kumar et al. 2021). Unfortunately, more than 90% of the applied pesticides are not deposited in the target sites but are lost by volatilization or leaching, negatively affecting the health of living beings, ecosystems and causing resistance in pathogens (Guha et al. 2020).

Therefore, foliar application of NP has been highly studied. It offers advantages such as reducing the production of reactive oxygen species (ROS), improving absorption, and increasing the useful life of nanofungicides (Hong et al. 2021). The advantages offered by these delivery systems are regulation for time-controlled or self-regulated release to overcome biological barriers. Considering that solar radiation, pH, and other factors can affect the effectiveness of NP or modify them chemically, nanoencapsulation or nanocarriers are being used. In this way the degradation of AI in the carrier before its release at the target site is avoided; penetration and solubility of AI are ensured, and AI degradation can be monitored (Kumar et al. 2021).

The success of NM is related to the targeted application of AI, which can be achieved through nanocarriers, which in addition to allowing slow release of AI, improves dispersion, wettability, and provides greater protection for AI without risk of runoff and loss (Zhao et al. 2017). Other notable characteristics of pesticide nanocomposites are thermal stability, specificity, and biodegradable nature. Only a small fraction of NPs enter the plant when applied in the foliar área. In addition, a part of the applied product remains in the epidermis (Su et al. 2019). In relation to the host plant, the entry of NPs depends on the properties of the cuticle, the size of the stomata or other natural openings and the abundance of these (Avila-Quezada et al. 2021).

The fate of NPs absorbed by plant cells depends on various parameters related to NPs such as size, charge, or colloidal stability (Oh et al. 2011). Once NPs enter, they are added to the surface of the leaf. A study with the CuONPs of 24–37 nm demonstrated that after 4 hrs of application on lettuce leaves, NP formed aggregates of 230–400 nm (Keller et al. 2018).

NPs can also be used in organic production. For instance, *Artemisia arborescens* L. essential oil which has antifungal properties was incorporated in solid lipid NPs. The ecological formulation reduced the evaporation of the oil compared to other emulsified products (Lai et al. 2006). Por otra parte, chemical and biosynthesized NPs act differrent, besides, they are more effective when mixed with secondary metabolites (Kumari et al. 2019).

Table 1 shows various NPs that have been investigated against pathogenic fungi. Even when only a small fraction of NP enters the leaf in the foliar application, this fraction can be effective in inhibiting plant pathogens (Su et al. 2019).

Applications of Nanomaterials in Root Diseases

Some of the genera of the fungi that cause root diseases are primarily *Heterobasidion*, *Rosellinia*, *Rhizoctonia*, *Fusarium*, and *Phoma* (Bodah 2017). Nowadays the use of nanotechnology for fungal control has exponentially increased due to the potential to provide solutions for disease management. Some NMs are metal and metal oxides, carbon-based, silica-based, polymeric, and hybrid.

NPs of metal or oxide metal have been used to protect plants from pathogens. The simplest way to apply them is directly to the plant or seeds (Khan and Tanveer 2014). Most of the NPs used for root diseases are made of metals such as silver (Ag), copper (Cu), zinc (Zn), and titanium dioxide (Hamza et al. 2016, Karimi and Sadeghi 2019). Besides, new materials have been developed based on metallic, polymeric, and inorganic NPs to improve nanosystems capable of gradual release of active ingredients in the soil, and nanosensors (Koli et al. 2015, Dubey and Mailapalli 2016).

Cu-based NM can be an effective strategy in the management of crop diseases because they are effective and because of the low amounts of copper that would be used. The study of Borgatta et al. (2018) in a greenhouse with $Cu_3(PO_4)_2 \cdot 3H_2O$

Table 1. Antifungal activity of nanoparticles/nanomaterials.

Nanoparticles	Size (nm)	Target Pathogen	References
AgNPs	35	*Macrophomina phaseolina, Rhizoctonia solani, Botrytis cinerea, Curvularia lunata, Alternaria alternata, Sclerotinia sclerotiorum*	Krishnaraj et al. 2012
AgNPs		*Bipolaris sorokiniana, Magnapothe grisea*	Jo et al. 2009
AgNPs + fluconazole	20–60	*Phoma glomerata, Trichoderma* sp., *Candida albicans*	Gajbhiye et al. 2009
AgNPs	5–24	*Colletotrichum gloesporioides*	Aguilar-Méndez et al. 2011
AgNPs	7–25	*Sclerotium cepivorum*	Jung et al. 2010
AgNPs	~ 8–22	*Bipolaris sorokiniana*	Mishra et al. 2014
AgNPs	10	*Fusarium solani*	Karimi and Sadeghi 2019
AgNPs	–	*Rhizoctonia solani, Fusarium oxysporum, F. redolens*	Aleksandrowicz-Trzcińska et al. 2018
AuNPs–chitosan	80	*Fusarium oxysporum*	Lipsa et al. 2020
TiO$_2$ NPs	20	*Botrytis cinerea*	Hao et al. 2017
TiO$_2$ NPs	30	*Podosphaera pannosa*	Hao et al. 2019
CeO$_2$ NPs		*Fusarium oxysporum*	Adisa et al. 2018
Cu NP		*Fusarium* spp., *Verticillium* spp.	Elmer and White 2016
CuNPs	–	*Rhizoctonia solani, Fusarium oxysporum, F. redolens*	Aleksandrowicz-Trzcińska et al. 2018
CuONP	30	*F. oxysporum* f. sp. *niveum*	Elmer et al. 2018
Copper oxide (CuO)	20–40	*Podosphaera pannosa*	Hao et al. 2019
CuNPs	50 nm	*Botrytis cinerea, B. fabae, Colletotrichum acutatum, Fusarium oxysporum* f. sp. *ciceris, F. o. f. sp. lycopersici, F. o. f. sp. melonis, Verticillium dahliae, Verticillium albo-atrum Alternaria alternata*	Banik and Luque 2017
MgONPs	50 ± 10 nm	*Alternaria alternata, Fusarium oxysporum, Rhizopus stolonifera, Mucor plumbeus*	Wani and Shah 2012
MnO	40	*F. oxysporum* f. sp. *niveum*	Elmer et al. 2018
SiO	20–30	*F. oxysporum* f. sp. *niveum*	Elmer et al. 2018
SeNPs	60.48 to 123.16	*Pyricularia grisea Colletotrichum capsici Alternaria solani*	Joshi et al. 2019
TiO$_2$	–	*Rhizoctonia solani*	Hamza et al. 2016
TiO$_2$	30	*F. oxysporum* f. sp. *niveum*	Elmer et al. 2018

Table 1 contd. ...

...Table 1 contd.

Nanoparticles	Size (nm)	Target Pathogen	References
TiO$_2$ NPs	20	*Botrytis cinerea*	Hao et al. 2017
Zn NPs	–	*Cercospora beticola*	Derbalah et al. 2013
Zn NPs	20	*Botrytis cinerea, Penicillium expansum*	He et al. 2011
ZnO NPs	–	*Sphaerotheca fuliginea*	Hamza et al. 2015
ZnO NPs	~ 30 ± 10 nm	*Alternaria alternata Fusarium oxysporum, Rhizopus stolonifera, Mucor plumbeus*	Wani and Shah 2012
ZnO	10–30	*F. oxysporum* f. sp. *niveum*	Elmer et al. 2018
Fullerene (C$_{60}$)	50	*Botrytis cinerea*	Hao et al. 2017
Carbon nanoparticles	119	*Fusarium oxysporum*	Lipsa et al. 2020
Single-walled carbon nanotubes (SWCNTs)	128	*Fusarium graminearum* *Fusarium poae*	Wang et al. 2014
Multi-walled carbon nanotubes (MWCNTs)	20–30	*Botrytis cinerea*	Hao et al. 2017
Multi-walled carbon nanotubes (MWCNTs)	78.8	*Fusarium graminearum* *Fusarium poae*	Wang et al. 2014
Multi-wall carbon nanotubes (MWCNTs)	20–30	*Podosphaera pannosa*	Hao et al. 2019
Graphene oxide (GO)	68.06	*Fusarium graminearum Fusarium poae*	Wang et al. 2014
Reduced graphene oxide (rGO)	105.7	*Fusarium graminearum Fusarium poae*	Wang et al. 2014
Reduced graphene oxide (rGO)	500 nm. Thickness of a single layer 0.55 to 3.74 nm	*Botrytis cinerea*	Hao et al. 2017
Reduced graphene oxide (rGO)	500 nm. Single-layer thickness 0.55 to 3.74 nm	*Podosphaera pannosa*	Hao et al. 2019
MgO NPs	100	*Thielaviopsis basicola*	Chen et al. 2020

nanosheets at 10 mg/L, repressed the disease caused by *Fusarium oxysporum* f. sp. *niveum* in watermelon (*Citrullus lanatus*) seedlings where a 58% reduction of the disease was observed. In addition, they tested CuONP where they found that the disease was reduced by 50.6% at a dose of 1000 mg/L. The same authors in field studies observed suppression of diseases similar to that of the greenhouse, with the nanosheets of $Cu_3(PO_4)_2 \cdot 3H_2O$.

A strategy used by various researchers includes immersing the root ball in NM. This technique is effective due to the homogeneity of the coverage throughout the root, as shown by Borgatta et al. (2018). ZnONPs reduced *Fusarium graminearum* growth *in vitro* (Dimkpa et al. 2013) and CuNPs have shown good activity against *Alternaria alternata, Curvularia lunata,* and *Fusarium oxysporum* (Kanhed et al. 2014). Thus, NP can be used to form nanocomposites with antimicrobial activity. Nanocomposites based on copper oxide and graphene oxide had a lethal effect on spores of *Fusarium oxysporum* in root rot affected tomato and pepper plants (El-Abeid et al. 2020).

Chitosan NPs have low toxicity and antimicrobial activity, such as controlling Fusarium crown (Kashyap et al. 2015). Chitosan polymer (poly b-(1,4) N-acetyl-D-glucosamine) is produced by chemical deacetylation of chitin found in arthropod exoskeletons. Chitosan is used for its film-forming ability, and for its antimicrobial properties (Gómez-Estaca et al. 2011).

The study of Boruah and Dutta (2021) showed that Chitosan NPs combined with *Trichoderma asperellum* suppressed the mycelial growth of *Fusarium oxysporum, Rhizoctonia solani,* and *Sclerotium rolfsii* in comparison with agrochemical control. Nanotechnology can also be used for biological control, through the release of extracts of microorganisms, an *in vitro* study has been carried out to control *Phytophthora* sp. with *Chaetomium brasiliense* NP. The results showed highly significant inhibition of the root rot disease at 40% (Tongon et al. 2018).

Carbon nanotubes (CNT) which are engineered from graphene oxide sheets, exhibit strong antifungal activity against *Fusarium graminearum* and *F. poae* (Wang et al. 2014). Because the nanotubes were deposited on the surface of the spores inhibiting water uptake, CNT blockage of water channels is possible (Figure 3). The results also showed evidence of plasmolysis.

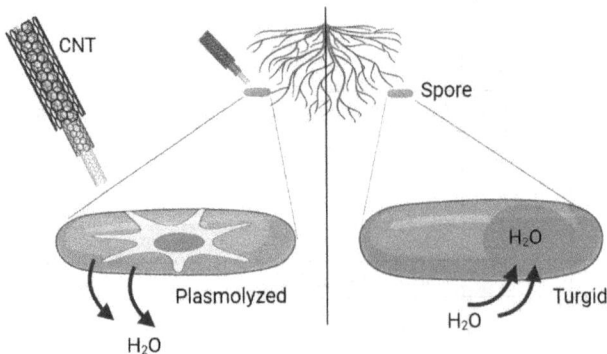

Figure 3. Carbon nanotubes (CNT) around the surface of the fungal spore. Possibly CNT block water channels.

The study of Ming et al. (2018) showed that the structure of the fungus *Phanerochaete chrysosporium* was severely altered when the fungus was exposed to pristine multi-walled carbon nanotubes (p-MWCNT). These nanotubes were able to penetrate the cell plasma and inhibit the activity of the laccase enzyme. These nanotubes could be used successfully for fungi that affect roots. Unfortunately, in this study, degradation of the wood exposed to p-MWCNT was observed, thus, its use would put the environment at risk.

Polymeric NMs as dendrimers are tree-shaped NPs. It is possible that in the future dendrimers will be applied as chemical delivery agents to control root diseases or delivery of biomolecules into plant cells (Samuel et al. 2014). Many authors thought that their utilization in plant pathology will be successful (Elmer and White 2018).

Hybrid NMs include nanobiosensors. Another application of nanotechnology in plant health is the use of nanosensors. Its application has been studied in monitoring plant growth by utilizing the cross-talk between roots and rhizosphere (Dimri et al. 2020). A nanobiosensor is a device that can detect chemical and biological agents in real-time, with high sensitivity and selectivity. Therefore, it is composed of a biological receptor (proteins, DNA) that specifically detects a substance, and a transducer or sensor, which interprets the biological recognition reaction and translates it into a quantifiable signal (Farooq et al. 2021). Biosensors are novel tools that can be used for early detection for pathogens. For instance, they are successfully used for the detection of *Agrobacterium tumefaciens* in *in vitro* assays (Choi et al. 2019).

Mechanism of Action of Biogenic Nanoparticles on Fungal Pathogens

Although the biosynthesis of AgNPs has been successfully achieved (Gade et al. 2022) several mechanisms are still unknown. Through strategies to improve the antifungal effect of AgNPs, such as the modification of the charge and the potential of the surface, the shape, the modification of the size, and the manipulation of the surface coating (Kumari et al. 2017), we now know that the main damage occurs in the fungal cell membrane (Figure 4) (Kim et al. 2009), and in ROS-dependent fungal cell death.

Some experiments lead us to understand that biosynthesized NPs have better properties than chemical NPs. In the study of Kumari et al. (2019), chemically synthesized AgNPs were not able to suppress the growth of spores, whilst, biosynthesized AgNPs added with *Trichoderma viride* metabolites, completely inhibited the spore germination of *Alternaria brassicicola* and *Fusarium oxysporum*. Biosynthesized NPs are promising for the control of phytopathogens for having synergistic effects of amylolytic and proteolytic activities, as well as metal chelation of cell extracts as reported for *T. viride* by Kumari et al. (2017).

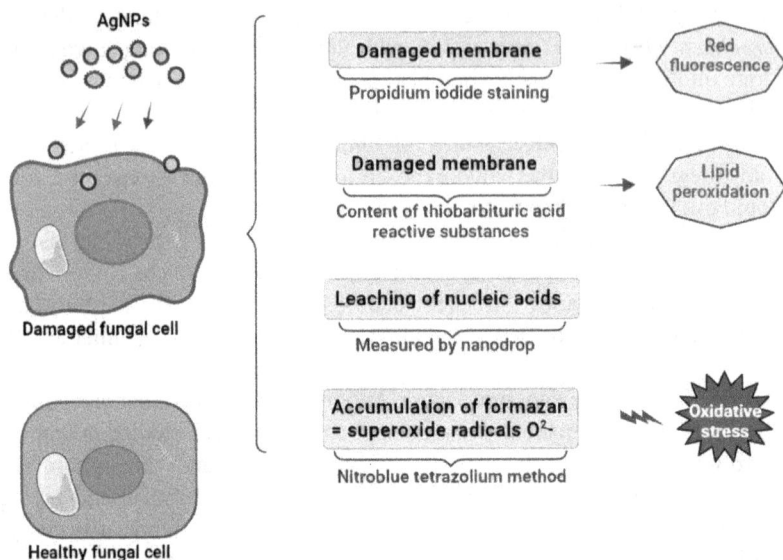

Figure 4. AgNPs deform the structure of the mycelium, affecting directly the cell membrane, triggering fungal cell toxicity. Biosynthestized AgNPs appear to cause higher cell damage of fungal pathogen than chemically synthesized.

Principles of the Metal Ions Affecting Fungi

The toxicity of metals in the metal-fungus interaction is the key to its function and is the basis of many fungicidal products (Baldrian 2003). Hence, the mechanisms of metal toxicity represent a useful way to design the control of phytopathogenic fungi.

The toxicity effect will depend on the organism, the type of metal, and its concentration, among other factors. Metals are toxic to almost all metabolic processes. The principle of toxicity is based on the effect on the ion and nutrient transport system, on the substitution of essential metal ions, conformational modification, and enzymes denaturation, besides the damage to the membrane of cells and organelles (Gadd 1994).

Furthermore, fungal cell-NP interactions induce pore formation by increasing the internalization of NPs through the cell wall. After crossing the cell wall, NPs reach the plasma membrane. NPs can also enter cells through endocytosis (Nomura et al. 2016). The internalization of NPs to the cell is mediated by receptors where the acting force is the main driver of endocytosis (Xu et al. 2020).

Once NPs cross cell membranes via carrier proteins or ion channels, within the cell, they can bind to different types of organelles and compartments, to interfere with metabolic processes (Rai et al. 2021b). Examples are the swelling of the endoplasmic reticulum, changes in the vacuole and phagosomes that were observed by Jia et al. (2005) in macrophage cells exposed to high doses of single-walled nanotubes. Furthermore, the nucleus of macrophage cells degenerated at high concentrations of multi-walled nanotubes (Jia et al. 2005).

By damaging the mitochondrial membrane, NPs disturb the activity of the respiratory chain and ATP synthesis, generating intracellular reactive oxygen species (ROS). ROS generation results from NPs-cell interactions, leading to mitochondrial dysfunction and oxidative stress (Rai et al. 2021a). However, not all NPs induce oxidative stress. The generation of ROS will depend on the physical and chemical properties, such as the crystalline phase, the adsorption capacity and the solubility of the NPs (Horie and Tabei 2021).

NPs are subject to environmental transformation processes. For instance, AgNPs and silver ions can interact with several chemical groups, such as sulfide and chloride (Ghobashy et al. 2021, Prasher et al. 2022). This issue must be thoroughly studied in each pathosystem and its environment, to ensure its effectiveness and possible toxicity towards non-target organisms.

Related to proteins, thiol molecules are conjugated to cell membrane proteins and mitochondria and can serve as targets for NPs or metal ions. The AgNP when joining thiol groups such as NADH dehydrogenase triggers ROS (Rai et al. 2016, Rai et al. 2021b). The effect of NPs on the nucleic acids of the fungal cell is that AgNPs smaller than 10 nm enter the nucleus, causing DNA damage and chromosomal alterations (Rai et al. 2021a).

It has been discovered that the toxicity of NPs depends on the specific surface of NPs. In addition, very small NPs show greater toxicity to the fungal cell. Hund-Rinke and Simon (2006) showed the concentration-effect relationship of TiO_2 in the alga *Desmodesmus subspicatus*. The small NPs were toxic and as the NPs were larger they caused less toxicity. This could be due to possible direct damage to DNA with NP of size approximately 10 nm.

An interesting finding of Horky et al. (2018) is that they eliminated mycotoxins with NPs. They concluded that NMs have interesting adsorption properties, which makes them promising for the removal of mycotoxins in food. Moreover, plant pathogenic fungi can sometimes survive the toxic effect of metal ions, especially those with pigmented cell walls, or when the extracellular polysaccharides or metabolites they produce can detoxify metals (Figure 5).

The success of the metal toxicity towards the fungus will also depend on the physical-chemical properties of the environment. For instance, the presence of other anions and cations that will result in changes in pH and availability of ligands (Fomina and Gadd 2014). In addition, these conditions also determine the development of fungi. It is documented that fungi themselves have the ability to modulate the pH of their environment for their benefit (Guetsky et al. 2005, Miranda-Gómez et al. 2014). The main advantage of the use of NPs for managing plant diseases is the low proportion of metals that are applied compared to conventional metal fungicides, in which huge amounts are used to reduce populations of phytopathogens.

Challenges

The basis for the effectiveness of NPs is their toxicity to fungal cells. These same properties cause concern in the community due to their possible effect on animal and human health. Knowing the NPs can be applied in our favor, as is the case of the damage caused by the AgNP in human lung fibroblast cells and human glioblastoma

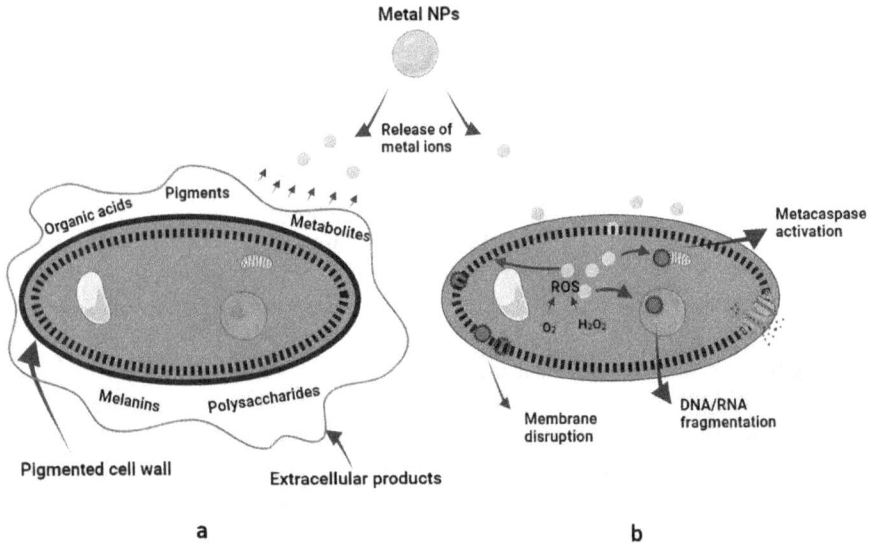

Figure 5. Fungi can survive the toxic effect of metal ions when they have pigmented cell walls, or when the extracellular polysaccharides or metabolites they produce can detoxify metals (a). On the other hand, the passive mechanisms of metal ions intervene in fungi, inhibiting the electron transport chain, displacing metal ions, denaturing proteins, and directly damaging the wall, membrane, and organelles (b).

cells (Kim and Ryu 2013). The covers used for NPs offer many alternatives according to needs. Interestingly capped AgNPs are more effective than antibiotics (Abbas et al. 2021).

Other studies show that AgNP coated with chitosan based crosslinked gelatin/ polyvinyl pyrrolidone, enhances the effectiveness of NP even with antibiotic-resistant microorganisms (El-Aassar et al. 2021). Therefore, the study of covers for NPs offers the possibility of their use by integrating effectiveness and food safety. Although silver-based NM has even been used in medicine, debate continues about potential risks (Tortella et al. 2020).

Possible disadvantages are related to toxicity due to the shape and size of the NPs, in addition to the concentration and the plant organ to which they will be applied. Applying NPs to edible fruits will require much research due to the threat of oral ingestion of NPs. A possibility is an application in the foliar area of ornamental plants. In the case of spray application, the use of conventional protective equipment for the worker will be required, while the possible risks due to entry through the respiratory tract are investigated. In addition, research will be required on the transport of NPs from the aerial part towards the root and then to the rhizosphere. It is also essential to know more about the transformation of NPs in the environment and whether it could cause risks to the various members of an ecosystem.

Another interesting aspect that should be investigated is the possible harmful effect of NPs on endophytic microorganisms and endomycorrhizal fungi. While more knowledge is generated, as far as possible, copper hydroxide, copper oxide, and copper oxychloride can be used in a minimal amount to avoid or reduce accumulation in the soil.

The need for regulations in our opinion needs time. First, more research needs to be developed to give answers to our questions.

Conclusions

Myconanotechnology plays an important strategic role in controlling plant pathogens. Research results show that NMs show promise for owing an inhibitory effect against phytopathogenic fungi.

However, to guarantee the successful control, features such as the NPs morphology, the surface and NPs size must be studied in each pathosystem. Therefore, much research is still required on the subject before launching fungicidal agents of the new generation, based on nanotechnology on the market.

To optimize NMs application for root diseases, the most effective way is to immerse the seedling's root ball in the NMs solution. This method is effective due to the homogeneity of the coverage throughout the root. NP protein capping is the key to successful results since they are a biological matrix that interacts with the environment, avoiding direct damage and increasing effects.

Acknowledgements

All figures were created by Biorender. Mahendra Rai is thankful to the Polish National Agency for Academic Exchange (NAWA) for financial support (Project No. PPN/ULM/2019/1/00117/A/DRAFT/00001) to visit the Department of Microbiology, Nicolaus Copernicus University, Toruń, Poland.

References

Abbas, M., Atiq, A., Xing, R. and Yan, X. 2021. Silver-incorporating peptide and protein supramolecular nanomaterials for biomedical applications. J. Mater. Chem. B 9: 4444–4458. https://doi.org/10.1039/D1TB00025J

Abdel-Aziz, S.M., Prasad, R., Hamed, A.A. and Abdelraof, M. 2018. Fungal nanoparticles: A novel tool for a green biotechnology? *In*: Prasad, R., Kumar, V., Kumar, M. and Wang, S. (eds.). Fungal Nanobionics: Principles and Applications. Springer, Singapur. https://doi.org/10.1007/978-981-10-8666-3_3.

AbdelRahim, K., Mahmoud, S.Y. and Ali, A.M. 2015. Extracellular biosynthesis of silver nanoparticles using *Rhizopus stolonifer*. Saudi J. Biol. Sci. 24: 208–216. https://doi.org/10.1016/j.sjbs.2016.02.025.

Adisa, I.O., Pullagurala, V.L., Rawat, S., Hernandez Viezcas, J.A., Dimkpa, C., Elmer, W.H. and Gardea Torresdey, J.L. 2018. Role of cerium compounds in Fusarium wilt suppression and growth enhancement in tomato (*Solanum lycopersicum*). J. Agric. Food Chem. 66: 5959–5970. https://doi.org/10.1021/acs.jafc.8b01345.

Aguilar-Méndez, M.A., San Martín-Martínez, E., Ortega-Arroyo, L., Cobián-Portillo, G. and Sánchez-Espíndola, E. 2011. Synthesis and characterization of silver nanoparticles: Effect on phytopathogen *Colletotrichum gloesporioides*. J. Nanopart. Res. 13: 2525–532. https://doi.org/10.1007/s11051-010-0145-6.

Aleksandrowicz-Trzcińska, M., Szaniawski, A., Olchowik, J. and Drozdowski, S. 2018. Effects of copper and silver nanoparticles on growth of selected species of pathogenic and wood-decay fungi *in vitro*. For. Chron. 94: 109–116. https://doi.org/10.5558/tfc2018-017.

Avila-Quezada, G., Silva-Rojas, H.V., Sanchez-Chávez, E., Leyva-Mir, G., Martínez-Bolaños, L., Guerrero-Prieto, V., Garcia-Avila, C., Gardea-Béjar, A. and Muñoz-Castellanos, L.N. 2016. Seguridad alimentaria: la continua lucha contra las enfermedades de los cultivos. TECNOCIENCIA Chihuahua 10: 133–142. https://vocero.uach.mx/index.php/tecnociencia/article/view/176/139.

Avila-Quezada, G.D. and Rai, M. 2022. Diseases of fruits, tubers, and seeds caused by *Phoma* sensu lato species complex. In Phoma: Diversity, Taxonomy, Bioactivities, and Nanotechnology (pp. 57–64). Springer, Cham. https://doi.org/10.1007/978-3-030-81218-8_4.

Avila-Quezada, G.D., Esquivel, J.F., Silva-Rojas, H.V., Leyva-Mir, S.G., de Jesus Garcia-Avila, C., Noriega-Orozco, L., Rivas-Valencia, P., Ojeda-Barrios, D. and Melgoza-Castillo, A. 2018. Emerging plant diseases under a changing climate scenario: Threats to our global food supply. Emir. J. Food Agricul. 30: 1–10. https://doi.org/10.9755/ejfa.2018.v30.i6.1715.

Avila-Quezada, G.D., Golinska, P. and Rai, M. 2021. Engineered nanomaterials in plant diseases: Can we combat phytopathogens? Appl. Microbiol. Biotechnol. 1–13. https://doi.org/10.1007/s00253-021-11725-w.

Baldrian, P. 2003. Interactions of heavy metals with white-rot fungi. Enzyme Microb. Technol. 32: 78–91. https://doi.org/10.1016/S0141-0229(02)00245-4.

Banik, S. and Luque, A.P. 2017. *In vitro* effects of copper nanoparticles on plant pathogens, beneficial microbes and crop plants. Span. J. Agric. Res. 15: 1–15. https://doi.org/10.5424/sjar/2017152-10305.

Bansod, S., Bonde, S., Tiwari, V., Bawaskar, M., Deshmukh, S., Gaikwad, S., Gade, A. and Rai, M. 2013. Bioconjugation of gold and silver nanoparticles synthesized by *Fusarium oxysporum* and their use in rapid identification of *Candida* species by using bioconjugate-nano-polymerase chain reaction. J. Biomed. Nanotechnol. 9: 1962–1971. https://doi.org/10.1166/jbn.2013.1727.

Basavaraja, S., Balaji, S.D., Lagashetty, A., Rajasab, A.H. and Venkataraman, A. 2008. Extracellular biosynthesis of silver nanoparticles using the fungus *Fusarium semitectum*. Mater. Res. Bull. 43: 1164–1170. https://doi.org/10.1016/j.materresbull.2007.06.020.

Beddington, J. 2009. Food, Energy, Water, and the Climate: A Perfect Storm of Global Events? Paper Presented at the Sustainable Development UK Annual Conference, QEII Conference Centre, London, 19 March 2009, Government Office for Science, London.

Bhainsa, K.C. and D'Souza, S.F. 2006. Extracellular biosynthesis of silver nanoparticles using the fungus *Aspergillus fumigatus*. Colloids Surf. B: Biointerfaces 47: 160–164. https://doi.org/10.1016/j.colsurfb.2005.11.026.

Bodah, T.E. 2017. Root rot diseases in plants: A review of common causal agents and management strategies. Agri. Res. Tech.: Open Access J. 5: 555661. https://doi.org/10.19080/ARTOAJ.2017.04.555661.

Boddula, R., Dubey, P., Gautam, S., Pothu, R. and Saran, A. 2018. Nanofabrication of myconanoparticles: A future prospect. *In*: Prasad, R., Kumar, V., Kumar, M. and Wang, S. (eds.). Fungal Nanobionics: Principles and Applications. Springer, Singapur. https://doi.org/10.1007/978-981-10-8666-3_8.

Borgatta, J., Ma, C., Hudson-Smith, N., Elmer, W., Plaza Pérez, C.D., De La Torre-Roche, R., Zuverza-Mena, N., Haynes, C.L., White, J.C. and Hamers, R.J. 2018. Copper based nanomaterials suppress root fungal disease in watermelon (*Citrullus lanatus*): Role of particle morphology, composition and dissolution behavior. ACS Sustain. Chem. Eng. 6: 14847–14856. https://doi:10.1021/acssuschemeng.8b03379.

Boruah, S. and Dutta, P. 2021. Fungus mediated biogenic synthesis and characterization of chitosan nanoparticles and its combine effect with *Trichoderma asperellum* against *Fusarium oxysporum, Sclerotium rolfsii* and *Rhizoctonia solani*. Indian Phytopathol. 74: 81–93. https://doi.org/10.1007/s42360-020-00289-w.

Chen, J., Wu, L., Lu, M., Lu, S., Li, Z. and Ding, W. 2020. Comparative study on the fungicidal activity of metallic MgO nanoparticles and macroscale MgO against soilborne fungal phytopathogens. Front. Microbiol. 11: 365. https://doi.org/10.3389/fmicb.2020.00365.

Choi, O., Bae, J., Kang, B., Lee, Y., Kim, S., Fuqua, C. and Kim, J. 2019. Simple and economical biosensors for distinguishing Agrobacterium-mediated plant galls from nematode-mediated root knots. Sci. Rep. 9: 1–9. https://doi.org/10.1038/s41598-019-54568-2.

Crisan, C.M., Mocan, T., Manolea, M., Lasca, L.I., Tăbăran, F.-A. and Mocan, L. 2021. Review on silver nanoparticles as a novel class of antibacterial solutions. Appl. Sci. 11: 1120. https://doi.org/10.3390/app11031120.

de Souza, A.O. and Rodrigues, A.G. 2015. Biosynthesis of silver nanoparticles by fungi. pp. 115–135. *In*: Gupta, V.K., Mach, R.L. and Sreenivasaprasad, S. (eds.). Fungal Biomolecules. John Wiley and Sons, Inc.: Hoboken, NJ, USA, 2015.

Deka, D., Rabha, J. and Jha, D.K. 2018. Application of myconanotechnology in the sustainable management of crop production system. *In*: Prasad, R. (ed.). Mycoremediation and Environmental Sustainability. Fungal Biology. Springer, Cham. https://doi.org/10.1007/978-3-319-77386-5_11.

Derbalah, A.S., El-Moghazy, S.M. and Godah, M.I. 2013. Alternative control methods of sugar-beet leaf spot disease caused by the fungus *Cercospora beticola* (Sacc). Egypt. J. Biolog. Pest Control. 23: 247–254.

Dimkpa, C., McLean, J.E., Britt, D.W. and Anderson, A.J. 2013. Antifungal activity of ZnO nanoparticles and their interactive effect with a biocontrol bacterium on growth antagonism of the plant pathogen *Fusarium graminearum*. Biometals 26: 913–924. https://doi.org/10.1007/s10534-013-9667-6.

Dimri, A., Pathak, N. and Sharma, S. 2020. Nanosensors for root zone parameters influencing plant growth. In Nanomaterials for Agriculture and Forestry Applications, 387–406. Elsevier. https://doi.org/10.1016/B978-0-12-817852-2.00015-9.

Du, L., Xian, L. and Feng, J.X. 2011. Rapid extra-/intracellular biosynthesis of gold nanoparticles by the fungus *Penicillium* sp. J. Nanopart. Res. 13: 921–930. https://doi.org/10.1007/s11051-010-0165-2.

Dubey, A. and Mailapalli, D.R. 2016. Nanofertilisers, nanopesticides, nanosensors of pest and nanotoxicity in agriculture. In Sustainable Agriculture Reviews, 307–330. Springer, Cham. https://doi.org/10.1007/978-3-319-26777-7_7.

El-Aassar, M.R., Ibrahim, O.M., Fouda, M.M., Fakhry, H., Ajarem, J., Maodaa, S.N., Allam, A.A. and Hafez, E.E. 2021. Wound dressing of chitosan-based-crosslinked gelatin/polyvinyl pyrrolidone embedded silver nanoparticles, for targeting multidrug resistance microbes. Carbohydr. Polym. 255: 117484. https://doi.org/10.1016/j.carbpol.2020.117484.

El-Abeid, S., Ahmed, Y., Daròs, J.A. and Mohamed, M.A. 2020. Reduced graphene oxide nanosheet-decorated copper oxide nanoparticles: A potent antifungal nanocomposite against fusarium root rot and wilt diseases of tomato and pepper plants. Nanomater. 10: 1001. https://doi.org/10.3390/nano10051001.

Elamawi, R.M., Al-Harbi, R.E. and Hendi, A.A. 2018. Biosynthesis and characterization of silver nanoparticles using *Trichoderma longibrachiatum* and their effect on phytopathogenic fungi. Egypt. J. Biol. Pest Control. 28: 1–11. https://doi.org/10.1186/s41938-018-0028-1.

Elmer, W. and White, J.C. 2018. The future of nanotechnology in plant pathology. Annu. Rev. Phytopathol. 56: 111–133. https://doi.org/10.1146/annurev-phyto-080417-050108.

Elmer, W., De La Torre-Roche, R., Pagano, L., Majumdar, S., Zuverza-Mena, N., Dimkpa, C., Gardea-Torresdey, J. and White, J.C. 2018. Effect of metalloid and metal oxide nanoparticles on Fusarium wilt of watermelon. Plant Dis. 102: 1394–1401. https://doi.org/10.1094/PDIS-10-17-1621-RE.

Elmer, W.H. and White, J. 2016. Nanoparticles of CuO improves growth of eggplant and tomato in disease infested soils. Environ. Sci. Nano. 3: 1072–1079.

FAO, ITU. 2017. Food and Agriculture Organization (FAO) & The International Telecommunication Union (ITU). E-agriculture in Action. http://www.fao.org/3/a-i6972e.pdf.

Farooq, A., Bhat, K.A., Mir, R.A., Mahajan, R., Nazir, M., Sharma, V. and Zargar, S.M. 2021. Emerging trends in developing biosensor techniques to undertake plant phosphoproteomic analysis. J. Proteomics 104458. https://doi.org/10.1016/j.jprot.2021.104458.

Fomina, M. and Gadd, G.M. 2014. Biosorption: current perspectives on concept, definition and application. Bioresour. Technol. 160: 3–14. https://doi.org/10.1016/j.biortech.2013.12.102

Gadd, G.M. 1994. Interactions of fungi with toxic metals. The Genus Aspergillus, 361–374.

Gade, A., Gaikwad, S., Duran, N. and Rai, M. 2014. Green synthesis of silver nanoparticles by *Phoma glomerata*. Micron. 59: 52–59. https://doi.org/10.1016/j.micron.2013.12.005.

Gade, A., Shende, S. and Rai, M. 2022. Potential role of Phoma spp. for mycogenic synthesis of silver nanoparticles. In Phoma: Diversity, Taxonomy, Bioactivities, and Nanotechnology (pp. 313–325). Springer, Cham. https://doi.org/10.1007/978-3-030-81218-8_17.

Gajbhiye, M., Kesharwani, J., Ingle, A., Gade, A. and Rai, M. 2009. Fungus-mediated synthesis of silver nanoparticles and their activity against pathogenic fungi in combination with fluconazole. Nanomed.: Nanotechnol. Biol. Med. 5: 382–386. https://doi.org/10.1016/j.nano.2009.06.005.

Gericke, M. and Pinches, A. 2006. Biological synthesis of metal nanoparticles. Hydrometallurgy 83: 132–140. https://doi.org/10.1016/j.hydromet.2006.03.019.

Ghadam, P., Mohammadi, P. and Ali, A.A. 2021. Silver-based nanoantimicrobials: Mechanisms, ecosafety, and future perspectives. In Silver Nanomaterials for Agri-Food Applications (pp. 67–99). Elsevier. https://doi.org/10.1016/B978-0-12-823528-7.00012-3.

Ghobashy, M.M., Abd Elkodous, M., Shabaka, S.H., Younis, S.A., Alshangiti, D.M., Madani, M., Al-Gahtany, S.A., Elkhatib, W.F., Noreddin, A.M., Nady, N. and El-Sayyad, G.S. 2021. An overview of methods for production and detection of silver nanoparticles, with emphasis on their fate and toxicological effects on human, soil, and aquatic environment. Nanotechnol. Rev. 10: 954–977. https://doi.org/10.1515/ntrev-2021-0066.

Gómez-Estaca, J., Gómez-Guillén, M.C., Fernández-Martín, F. and Montero, P. 2011. Effects of gelatin origin, bovine-hide and tuna-skin, on the properties of compound gelatin–chitosan films. Food Hydrocoll. 25: 1461–1469. https://doi.org/10.1016/j.foodhyd.2011.01.007.

González-Escobedo, R., Muñoz-Castellanos, L.N., Muñoz-Ramirez, Z.Y., López, C.G. and Avila-Quezada, G.D. 2022. Microbial community analysis of rhizosphere of healthy and wilted pepper (*Capsicum annuum* L.) in an organic farming system. Terra Latinoam.

Guarnaccia, V. and Crous, P.W. 2017. Emerging citrus diseases in Europe caused by species of *Diaporthe*. IMA Fungus 8: 317–334. https://doi.org/10.5598/imafungus.2017.08.02.07.

Guetsky, R., Kobiler, I., Wang, X., Perlman, N., Gollop, N., Avila-Quezada, G., Hadar, I. and Prusky, D. 2005. Metabolism of the flavonoid epicatechin by laccase of *Colletotrichum gloeosporioides* and its effect on pathogenicity on avocado fruits. Phytopathology 95: 1341–1348. https://doi.org/10.1094/PHYTO-95-1341.

Guha, T., Gopal Geetha, Kundu, R. and Mukherjee, A. 2020. Nanocomposites for delivering agrochemicals: A comprehensive review. J. Agricul. Food Chem. 68: 3691–3702. https://doi.org/10.1021/acs.jafc.9b06982.

Hamza, A., Mohamed, A. and Derbalah, A. 2016. Unconventional alternatives for control of tomato root rot caused by *Rhizoctonia solani* under greenhouse conditions. J. Plant Protect. Res. 56: 298–305. https://doi.org/10.1515/jppr-2016-0046.

Hamza, A.M., Essa, T.A., Derbalah, A.S. and Mohamed, A.A. 2015. Performance of some fungicide alternatives for controlling powdery mildew on cucumber under greenhouse conditions. Egypt. J. Biol. Pest Control. 25: 631–637.

Hao, Y., Cao, X., Ma, C., Zhang, Z., Zhao, N., Ali, A., Hou, T., Xiang, Z., Zhuang, J., Wu, S. and Xing, B. 2017. Potential applications and antifungal activities of engineered nanomaterials against gray mold disease agent *Botrytis cinerea* on rose petals. Front. Plant Sci. 8: 1332. https://doi.org/10.3389/fpls.2017.01332.

Hao, Y., Fang, P., Ma, C., White, J.C., Xiang, Z., Wang, H., Zhang, Z., Rui, Y. and Xing, B. 2019. Engineered nanomaterials inhibit *Podosphaera pannosa* infection on rose leaves by regulating phytohormones. Environ. Res. 170: 1–6. https://doi.org/10.1016/j.envres.2018.12.008.

He, L., Liu, Y., Mustapha, A. and Lin, M. 2011. Antifungal activity of zinc oxide nanoparticles against *Botrytis cinerea* and *Penicillium expansum*. Microbiol. Res. 166: 207–215. https://doi.org/10.1016/j.micres.2010.03.003.

Hernandez-Montiel, L.G., Droby, S., Preciado-Rangel, P., Rivas-García, T., González-Estrada, R.R., Gutiérrez-Martínez, P. and Ávila-Quezada, G.D. 2021. A sustainable alternative for postharvest disease management and phytopathogens biocontrol in fruit: Antagonistic yeasts. Plants 10: 2641. https://doi.org/10.3390/plants10122641.

Hong, J., Wang, C., Wagner, D.C., Gardea-Torresdey, J.L., He, F. and Rico, C.M. 2021. Foliar application of nanoparticles: Mechanisms of absorption, transfer, and multiple impacts. Environ. Sci.: Nano. 8: 1196–1210. https://doi.org/10.1039/d0en01129k.

Horie, M. and Tabei, Y. 2021. Role of oxidative stress in nanoparticles toxicity. Free Radic. Res. 55: 331–342. https://doi.org/10.1080/10715762.2020.1859108.

Horky, P., Skalickova, S., Baholet, D. and Skladanka, J. 2018. Nanoparticles as a solution for eliminating the risk of mycotoxins. Nanomater. 8: 727. https://doi.org/10.3390/nano8090727.

Hua, X., Chi, W., Su, L., Li, J., Zhang, Z. and Yuan, X. 2017. ROS-induced oxidative injury involved in pathogenesis of fungal keratitis via p38 MAPK activation. Scientific Reports 7: 1–12. https://doi.org/10.1038/s41598-017-09636-w.

Hund-Rinke, K. and Simon, M. 2006. Ecotoxic effect of photocatalytic active nanoparticles (TiO$_2$) on algae and daphnids (8 pp). Environ. Sci. Pollut. Res. 13: 225–232. https://doi.org/10.1065/espr2006.06.311.

Iftikhar, Y. and Sajid, A. 2020. Quarantine and regulations. In Plant Disease Management Strategies for Sustainable Agriculture through Traditional and Modern Approaches (pp. 279–293). Springer, Cham. https://doi.org/10.1007/978-3-030-35955-3_14.

Iranmanesh, S., Bonjar, G.H.S. and Baghizadeh, A. 2020. Study of the biosynthesis of gold nanoparticles by using several saprophytic fungi. SN Appl. Sci. 2: 1–11. https://doi.org/10.1007/s42452-020-03704-z.

Jha, A.K. and Prasad, K. 2016. Understanding mechanism of fungus mediated nanosynthesis: A molecular approach. In Advances and Applications Through Fungal Nanobiotechnology (pp. 1–23). Springer, Cham. https://doi.org/10.1007/978-3-319-42990-8_1.

Jia, G., Wang, H., Yan, L., Wang, X., Pei, R., Yan, T., Zhao, Y. and Guo, X. 2005. Cytotoxicity of carbon nanomaterials: Single-wall nanotube, multi-wall nanotube, and fullerene. Environ. Sci. Technol. 39: 1378–1383. https://doi.org/10.1021/es0487291.

Jo, Y.K., Kim, B.H. and Jung, G. 2009. Antifungal activity of silver ions and nanoparticles on phytopathogenic fungi. Plant Dis. 93: 1037–1043. https://doi.org/10.1094/PDIS-93-10-1037.

Joshi, S.M., De Britto, S., Jogaiah, S. and Ito, S.I. 2019. Mycogenic selenium nanoparticles as potential new generation broad spectrum antifungal molecules. Biomol. 9: 419. https://doi.org/10.3390/biom9090419.

Jung, J.H., Kim, S.W., Min, J.S., Kim, Y.J., Lamsal, K., Kim, K.S. and Lee, Y.S. 2010. The effect of nano-silver liquid against the white rot of the green onion caused by *Sclerotium cepivorum*. Mycobiol. 38: 39–45. https://doi.org/10.4489/MYCO.2010.38.1.039.

Kanhed, P., Birla, S., Gaikwad, S., Gade, A., Seabra, A.B., Rubilar, O. and Rai, M. 2014. *In vitro* antifungal efficacy of copper nanoparticles against selected crop pathogenic fungi. Mater. Lett. 115: 13–17. https://doi.org/10.1016/j.matlet.2013.10.011.

Karimi, E. and Sadeghi, A. 2019. Toxicity effect of silver nanoparticles on two plant growth promoting *Streptomyces* spp. strains, phytopathogenic fungi *Fusarium Solani* and phytopathogenic oomycetes *Pythium aphanidermatum* and *Pythium ultimum*. Modares. J. Biotechnol. 10: 23–27. https://biot.modares.ac.ir/browse.php?a_id=16269&sid=22&slc_lang=en.

Kashyap, P.L., Xiang, X. and Heiden, P. 2015. Chitosan nanoparticle based delivery systems for sustainable agriculture. Int. J. Biol. Macromol. 77: 36–51. https://doi.org/10.1016/j.ijbiomac.2015.02.039.

Keller, A.A., Huang, Y. and Nelson, J. 2018. Detection of nanoparticles in edible plant tissues exposed to nano-copper using single-particle ICP-MS. J. Nanopart. Res. 20: 1–13. https://doi.org/10.1007/s11051-018-4192-8.

Khan, M.R. and Tanveer, R. 2014. Nanotechnology: scope and application in plant disease management. Plant Pathol. J. 13: 214–231. https://doi.org/10.3923/ppj.2014.214.231.

Kim, K.J., Sung, W.S., Suh, B.K., Moon, S.K., Choi, J.S., Kim, J.G. and Lee, D.G. 2009. Antifungal activity and mode of action of silver nano-particles on *Candida albicans*. Biometals 22: 235–242. https://doi.org/10.1007/s10534-008-9159-2.

Kim, S. and Ryu, D.Y. 2013. Silver nanoparticle-induced oxidative stress, genotoxicity and apoptosis in cultured cells and animal tissues. J. Appl. Toxicol. 33: 78–89. https://doi.org/10.1002/jat.2792.

Koli, P., Singh, B.B., Shakil, N.A., Kumar, J. and Kamil, D. 2015. Development of controlled release nanoformulations of carbendazim employing amphiphilic polymers and their bioefficacy evaluation against *Rhizoctonia solani*. J. Environ. Sci. Health. Part B 50: 674–681. https://doi.org/10.1080/03601234.2015.1038961.

Krishnaraj, C., Ramachandran, R., Mohan, K. and Kalaichelvan, P. 2012. Optimization for rapid synthesis of silver nanoparticles and its effect on phytopathogenic fungi. Spectrochim. Acta Part A Mol. Biomol. Spectrosc. 93: 95–99. https://doi.org/10.1016/j.saa.2012.03.002.

Kumar, A., Choudhary, A., Kaur, H., Mehta, S. and Husen, A. 2021. Smart nanomaterial and nanocomposite with advanced agrochemical activities. Nanoscale Res. Lett. 16: 1–26. https://doi.org/10.1186/s11671-021-03612-0.

Kumari, M., Giri, V.P., Pandey, S., Kumar, M., Katiyar, R., Nautiyal, C.S. and Mishra, A. 2019. An insight into the mechanism of antifungal activity of biogenic nanoparticles than their chemical counterparts. Pesticide Biochem. Physiol. 157: 45–52. https://doi.org/10.1016/j.pestbp.2019.03.005.

Kumari, M., Shukla, S., Pandey, S., Giri, V.P., Bhatia, A., Tripathi, T., Kakkar, P., Nautiyal, C.S. and Mishra, A. 2017. Enhanced cellular internalization: A bactericidal mechanism more relative to biogenic nanoparticles than chemical counterparts. ACS Appl. Mater. Interfaces 9: 4519–4533. https://doi.org/10.1021/acsami.6b15473.

Lai, F., Wissing, S.A., Müller, R.H. and Fadda, A.M. 2006. *Artemisia arborescens* L. essential oil-loaded solid lipid nanoparticles for potential agricultural application: Preparation and characterization. Aaps. Pharm. Sci. Tech. 7: E10–E18. https://doi.org/10.1208/pt070102.

Lipsa, F.D., Ursu, E.L., Ursu, C., Ulea, E. and Cazacu, A. 2020. Evaluation of the antifungal activity of gold–chitosan and carbon nanoparticles on *Fusarium oxysporum*. Agronomy 10: 1143. https://doi.org/10.3390/agronomy10081143.

Madrid-Delgado, G., Orozco-Miranda, M., Cruz-Osorio, M., Hernández-Rodríguez, O.A., Rodríguez-Heredia, R., Roa-Huerta, M. and Avila-Quezada, G.D. 2021. Pathways of phosphorus absorption and early signaling between the mycorrhizal fungi and plants. Phyton. 90: 1321. https://doi.org/10.32604/phyton.2021.016174.

Ming, Z., Feng, S., Yilihamu, A., Yang, S., Ma, Q., Yang, H., Bai, Y. and Yang, S.T. 2018. Toxicity of carbon nanotubes to white rot fungus *Phanerochaete chrysosporium*. Ecotoxicol. Environ. Saf. 162: 225–234. https://doi.org/10.1016/j.ecoenv.2018.07.011.

Miranda-Gómez, B., García-Hernández, A., Muñoz-Castellanos, L., Ojeda-Barrios, D.L. and Avila-Quezada, G.D. 2014. Pectate lyase production at high and low pH by *Colletotrichum gloeosporioides* and *Colletotrichum acutatum*. Afr. J. Micr. Res. 8: 948–1954. https://doi.org/10.5897/AJMR2014.6765.

Mishra, S., Singh, B.R., Singh, A., Keswani, C., Naqvi, A.H. and Singh, H.B. 2014. Biofabricated silver nanoparticles act as a strong fungicide against *Bipolaris sorokiniana* causing spot blotch disease in wheat. PLOS ONE 9: e97881. https://doi.org/10.1371/journal.pone.0097881.

Mukherjee, P., Ahmad, A., Mandal, D., Senapati, S., Sainkar, S.R., Khan, M.I., Ramani, R., Parischa, R., Ajayakumar, P.V., Alam, M. and Sastry, M. 2001. Bioreduction of $AuCl_4^-$ ions by the fungus *Verticillium* sp. and surface trapping of the gold nanoparticles formed. Angew. Chem. Int. Edition. 40: 3585–3588. https://doi.org/10.1002/1521-3773(20011001)40:19<3585::AID-ANIE3585>3.0.CO;2-K.

Mukherjee, P., Senapati, S., Mandal, D., Ahmad, A., Khan, M.I., Kumar, R. and Sastry, M. 2002. Extracellular synthesis of gold nanoparticles by the fungus *Fusarium oxysporum*. Chembiochem. 3: 461–463. https://doi.org/10.1002/1439-7633(20020503)3:5<461::AID-CBIC461>3.0.CO;2-X.

Nachiyar, V., Sunkar, S. and Prakash, P. 2015. Biological synthesis of gold nanoparticles using endophytic fungi. Der. Pharma. Chem. 7: 31–38.

Nazarov, P.A., Baleev, D.N., Ivanova, M.I., Sokolova, L.M. and Karakozova, M.V. 2020. Infectious plant diseases: Etiology, current status, problems and prospects in plant protection. Acta Naturae 12: 46–59. https://doi.org/10.32607/actanaturae.11026.

Nomura, T., Tani, S., Yamamoto, M., Nakagawa, T., Toyoda, S., Fujisawa, E., Yasui, A. and Konishi, Y. 2016. Cytotoxicity and colloidal behavior of polystyrene latex nanoparticles toward filamentous fungi in isotonic solutions. Chemosphere 149: 84–90. https://doi.org/10.1016/j.chemosphere.2016.01.091.

Oh, E., Delehanty, J.B., Sapsford, K.E., Susumu, K., Goswami, R., Blanco-Canosa, J.B., Dawson, P.E., Granek, J., Shoff, M., Zhang, Q., Goering, P.L., Huston, A. and Medintz, I.L. 2011. Cellular uptake and fate of PEGylated gold nanoparticles is dependent on both cell-penetration peptides and particle size. ACS Nano 5: 6434–6448. https://doi.org/10.1021/nn201624c.

Prasher, P., Sharma, M., Mudila, H., Verma, A. and Bhatt, P. 2022. Silver nanoparticles in natural ecosystems: Fate, transport, and toxicity. In Green Synthesis of Silver Nanomaterials (pp. 649–668). Elsevier. https://doi.org/10.1016/B978-0-12-824508-8.00004-6.

Priyanka, P., Kumar, D., Yadav, A. and Yadav, K. 2020. Nanobiotechnology and its application in agriculture and food production. *In*: Thangadurai, D., Sangeetha, J. and Prasad, R. (eds.). Nanotechnology for Food, Agriculture, and Environment. Nanotechnology in the Life Sciences. Springer, Cham. https://doi.org/10.1007/978-3-030-31938-0_6.

Rai, M., Bonde, S., Golinska, P., Trzcińska-Wencel, J., Gade, A., Abd-Elsalam, K., Shende, S., Gaikwad, S. and Ingle, A. 2021a. *Fusarium* as a novel fungus for the synthesis of nanoparticles: mechanism and applications. J. Fungi. 7: 139. https://doi.org/10.3390/jof7020139.

Rai, M., Ingle, A.P., Trzcińska-Wencel, J., Wypij, M., Bonde, S., Yadav, A., Kratošová, G. and Golińska, P. 2021b. Biogenic silver nanoparticles: What we know and what do we need to know? Nanomaterials 11: 2901. https://doi.org/10.3390/nano11112901.

Rai, M., Gade, A. and Yadav, A. 2011. Biogenic nanoparticles: An introduction to what they are, how they are synthesized and their applications. *In*: Rai, M. and Durán, N. (eds.). Metal Nanoparticles Microbiology. Springer, Berlin. https://doi.org/10.1007/978-3-642-18312-6_1.

Rai, M., Ingle, A.P., Birla, S., Yadav, A. and Santos, C.A.D. 2016. Strategic role of selected noble metal nanoparticles in medicine. Crit. Rev. Microbiol. 42: 696–719. https://doi.org/10.3109/104084 1X.2015.1018131.

Rai, M., Yadav, A., Bridge, P. and Gade, A. 2009. Myconanotechnology: A new and emerging science. Appl. Mycol. 258–267. https://doi.org/10.1079/9781845935344.0258.

Raliya, R., Franke, C., Chavalmane, S., Nair, R., Reed, N. and Biswas, P. 2016. Quantitative understanding of nanoparticles uptake in watermelon plants. Front. Plant Sci. 7: 1288. https://doi.org/10.3389/fpls.2016.01288.

Rao, M., Jha, B., Jha, A.K. and Prasad, K. 2017. Fungal nanotechnology: A pandora to agricultural science and engineering. *In*: Prasad, R. (ed.). Fungal Nanotechnology. Fungal Biology. Springer, Cham. https://doi.org/10.1007/978-3-319-68424-6_1.

Samuel, J.P., Samboju, N.C., Yau, K.Y., Webb, S.R. and Burroughs, F. 2014. Use of dendrimer nanotechnology for delivery of biomolecules into plant cells. U.S. Patent No. 8,686,222. Washington, DC: U.S. Patent and Trademark Office. https://patents.google.com/patent/US8686222B2/en.

Sánchez-Chávez, E., Silva-Rojas, H.V., Leyva-Mir, G., Villarreal-Guerrero, F., Jiménez-Castro, J.A., Molina-Gayosso, E., Gardea-Bejar, A. and Ávila-Quezada, G.D. 2017. An effective strategy to reduce the incidence of Phytophthora root and crown rot in bell pepper. Interciencia 42: 229–235. https://www.redalyc.org/journal/339/33950546006/html/.

Sarkar, J., Chattopadhyay, D., Patra, S., Singh Deo, S., Sinha, S., Ghosh, M., Mukherjee, A. and Acharya, K. 2011. *Alternaria alternata* mediated synthesis of protein capped silver nanoparticles and their genotoxic activity. Dig. J. Nanomater. Biostructures 6: 563–573. https://citeseerx.ist.psu.edu/viewdoc/download?doi=10.1.1.1071.538&rep=rep1&type=pdf.

Sastry, M., Ahmad, A., Khan, M.I. and Kumar, R. 2003. Biosynthesis of metal nanoparticles using fungi and actinomycetes. Curr. Sci. 85: 162–170. http://www.jstor.org/stable/24108579.

Savary, S., Willocquet, L., Pethybridge, S.J., Esker, P., McRoberts, N. and Nelson, A. 2019. The global burden of pathogens and pests on major food crops. Nat. Ecol. Evol. 3: 430–439. https://doi.org/10.1038/s41559-018-0793-y.

Savín-Molina, J., Hernández-Montiel, L.G., Ceiro-Catasú, W., Ávila-Quezada, G.D., Palacios-Espinosa, A., Ruiz-Espinoza, F.H. and Romero-Bastidas, M. 2021. Morphological characterization and biocontrol potential of *Trichoderma* species isolated from semi-arid soils. Revista Mex Fitopatología 39: 435–451. https://doi.org/10.18781/r.mex.fit.2106-7.

Senapati, S., Mandal, D., Ahmad, A., Khan, M.I., Sastry, M. and Kumar, R. 2004. Fungus mediated synthesis of silver nanoparticles: A novel biological approach. Indian J. Phys. 78: 101–105.

Sharma, A. and Kumar, S. 2021. Synthesis and Green Synthesis of Silver Nanoparticles. Polymer Nanocomposites Based on Silver Nanoparticles: Synthesis, Characterization and Applications, 25–64. Springer Nature.

Singh, S., Bhatnagar, S., Choudhary, S., Nirwan, B. and Sharma, K. 2018. Fungi as biocontrol agent: An alternate to chemicals. *In*: Gehlot, P. and Singh, J. (eds.). Fungi and their Role in Sustainable Development: Current Perspectives. Springer, Singapore. https://doi.org/10.1007/978-981-13-0393-7_2.

Soanes, D. and Richards, T.A. 2014. Horizontal gene transfer in eukaryotic plant pathogens. Annu. Rev. Phytopathol. 52: 583–614. https://doi.org/10.1146/annurev-phyto-102313-050127. PMID: 25090479.

Spagnoletti, F.N., Spedalieri, C., Kronberg, F. and Giacometti, R. 2019. Extracellular biosynthesis of bactericidal Ag/AgCl nanoparticles for crop protection using the fungus *Macrophomina phaseolina*. J. Environ. Manag. 231: 457–466. https://doi.org/10.1016/j.jenvman.2018.10.081.

Su, Y., Ashworth, V., Kim, C., Adeleye, A.S., Rolshausen, P., Roper, C., White, J. and Jassby, D. 2019. Delivery, uptake, fate, and transport of engineered nanoparticles in plants: A critical review and data analysis. Environ. Sci.: Nano. 6: 2311–2331. https://doi.org/10.1039/C9EN00461K.

Subashini, G. and Bhuvaneswari, S. 2018. Nanoparticles from fungi (myconanoparticles). *In*: Gehlot, P. and Singh, J. (eds.). Fungi and their Role in Sustainable Development: Current Perspectives. Springer, Singapore. https://doi.org/10.1007/978-981-13-0393-7_39.

Tai, C.Y., Wang, Y.H. and Liu, H.S. 2008. A green process for preparing silver nanoparticles using spinning disk reactor. AIChE J. 54: 445–452. https://doi.org/10.1002/aic.11396.

Tongon, R., Soytong, K., Kanokmedhakul, S. and Kanokmedhakul, K. 2018. Nano-particles from *Chaetomium brasiliense* to control *Phytophthora palmivora* caused root rot disease in durian var Montong. Int. J. Agric. Technol. 14(7 Special Issue): 2163–2170. https://www.thaiscience.info/Journals/Article/IJAT/10992547.pdf.

Tortella, G.R., Rubilar, O., Durán, N., Diez, M.C., Martínez, M., Parada, J. and Seabra, A.B. 2020. Silver nanoparticles: Toxicity in model organisms as an overview of its hazard for human health and the environment. J. Hazard. Mater. 390: 121974. https://doi.org/10.1016/j.jhazmat.2019.121974.

Vahabi, K. and Dorcheh, S.K. 2014. Biosynthesis of silver nano-particles by Trichoderma and its medical applications. In Biotechnology and biology of *Trichoderma* (pp. 393–404). Elsevier. https://doi.org/10.1016/B978-0-444-59576-8.00029-1.

Vinod, V.T.P., Saravanan, P., Sreedhar, B., Devi, D.K. and Sashidhar, R.B. 2011. A facile synthesis and characterization of Ag, Au and Pt nanoparticles using a natural hydrocolloid gum kondagogu (*Cochlospermum gossypium*). Colloids Surf. B 83: 291–298. https://doi.org/10.1016/j.colsurfb.2010.11.035.

Wang, X., Liu, X., Chen, J., Han, H. and Yuan, Z. 2014. Evaluation and mechanism of antifungal effects of carbon nanomaterials in controlling plant fungal pathogen. Carbon 68: 798–80. https://doi.org/10.1016/j.carbon.2013.11.072.

Wani, A.H. and Shah, M.A. 2012. A unique and profound effect of MgO and ZnO nanoparticles on some plant pathogenic fungi. J. Appl. Pharm. Sci. 2: 40–44. https://www.japsonline.com/admin/php/uploads/394_pdf.pdf.

Whiteley, C., Govender, Y., Riddin, T. and Rai, M. 2011. Enzymatic synthesis of platinum nanoparticles: Prokaryote and eukaryote systems. In Metal Nanoparticles in Microbiology (pp. 103–134). Springer, Berlin, Heidelberg. https://doi.org/10.1007/978-3-642-18312-6_5.

Xu, W., Jia, K. and Liu, X. 2020. Receptor-mediated endocytosis of nanoparticle based on the co-rotational grid method. Phys. Scr. 96: 015009. https://doi.org/10.1088/1402-4896/abc9f0.

Youssef, K., Hashim, A.F., Hussien, A. and Abd-Elsalam, K.A. 2017. Fungi as ecosynthesizers for nanoparticles and their application in agriculture. pp. 55–75. *In*: Prasad, R. (eds.). Fungal Nanotechnology. Springer, Cham. https://doi.org/10.1007/978-3-319-68424-6_3.

Zhao, X., Cui, H., Wang, Y., Sun, C., Cui, B. and Zeng, Z. 2017. Development strategies and prospects of nano-based smart pesticide formulation. J. Agric. Food Chem. 66: 6504–6512. https://doi.org/10.1021/acs.jafc.7b02004.

9

Biosynthesis of Nanoparticles by Fungi and their Applications as Nanofertilizers

Sarika Bhalerao[1,*] and *Mahendra Rai*[2,3]

Introduction

Sustainable and high productive farming is important in reducing the perils of hunger and increasing food security. Food production and distribution are under increasing and ongoing global pressure due to climate change, population growth, and declining fertile land and water resources (Usman et al. 2020). This challenge can be solved with technological advancement coupled with significant changes in existing global food production systems (Dwivedi et al. 2016, Shang et al. 2019). Currently, modern agriculture is heavily supported by the use of high levels of agrochemicals. Synthetic chemical fertilizers are used for growth and productivity of plants, but so far, accepted farming methods have not been particularly effective in simultaneously improving plant absorption, nutrient utilization (NUE), and crop production (Adnan et al. 2020, Seleiman et al. 2020a). In most cases, organic fertilizers used in agriculture have low NUE value (Guo et al. 2018), indicating that more than half of the fertilizer distributed in the fields is lost and does not reach the intended sites for various reasons such as photolysis, hydrolysis, leaching, and microbial immobilization and degradation (Green and Beestman 2007). Lower NUE can lead to deeper use of synthetic fertilizers to increase crop production (Guo et al. 2018). Over time, however, these vigorous fertilizers can lead to serious environmental hazards such as air pollution, soil degradation, water eutrophication and groundwater pollution (Czymmek et al. 2020, Eid et al. 2020, Seleiman et al.

[1] Department of Plant Biotechnology, Vilasrao Deshmukh College of Agricultural Biotechnology, V.N.M.K.V., Latur, Maharashtra, India.
[2] Nanobiotechnology Lab., Department of Biotechnology, SGB Amravati University, Amravati-444 602, Maharashtra, India.
[3] Department of Microbiology, Nicolaus Copernicus University, 87-100 Torun, Poland.
* Corresponding author: sarikasshende@gmail.com

2020a). In addition, overuse of synthetic fertilizers increases the cost of their production and reduces the profit margins of farmers (Diatta et al. 2020, Seleiman et al. 2020a). Therefore, agricultural sustainability can be achieved through the use and implementation of new strategies (Shang et al. 2019) that can improve global food production while also protecting natural and environmental resources (Arora 2018). Recent research has suggested that nanotechnology may have the potential to transform the current synthetic framework used in modern agricultural systems (Prasad et al. 2017) by increasing the efficiency of novel agrochemicals (Kerry et al. 2017) and providing solutions for environmental and agricultural problems (Usman et al. 2020).

A nanoparticle is defined as one having a dimension of 100 nm or less in size (Ghorbani et al. 2011). Recent research has shown that microorganisms, plant and fungi can produce nanoparticles through biological methods (Sastry et al. 2003, Abou El-Nour et al. 2010, Ghorbani et al. 2011). They have received much attention worldwide due to their rapid formation and eco-friendly nature compared to physical and chemical methods. With the advent of modern nanotechnology in the 1980s, mold has always been important in providing an alternative to the chemically produced nanoparticles. The synthesis of nanoparticles using a fungus is called mycosynthesis (Ingle et al. 2008, 2009, Rai et al. 2009a, Yadav et al. 2015) and is treated under myconanotechnology (Rai et al. 2009b). Gade et al. (2010) reported the benefits of using filamentous fungi over other biological agents to synthesize nanoparticles. This mainly includes high tolerance for heavy metals, easy to culture fungi at mass level, the extracellular synthesis of nanoparticles that reduces cost of down streaming processing, etc.

In addition, fungi are preferred over other biological systems due to their widespread distribution in the environment. Therefore, many fungi have been explored to produce different metal nanoparticles of various shapes and sizes. The most common nanoparticles are silver and gold. However, fungi have been used to synthesize other types of nanoparticles including ferrous, zinc oxide, platinum, magnetite, zirconia, silica, titanium, cadmium sulfide, quantum selenide and cadmium dots. The fungal system has been found to be a dynamic biological system capable of synthesizing metal nanoparticles inside (intracellular) and outside cells (extracellular). The synthesis of silver nanoparticles has been investigated using many ubiquitous fungal species including *Trichoderma* (Basavaraja et al. 2008, Vahabi et al. 2011), *Fusarium* (Durán et al. 2005, Gaikwad et al. 2013, Rai et al. 2021), *Penicillium* (Naveen et al. 2010), *Rhizoctonia* (Potbhare et al. 2020), *Pleurotus* and *Aspergillus* (Bhainsa and D'Sousa, 2006, Rai et al. 2022). Extracellular synthesis has been demonstrated by *Trichoderma viride, T. reesei, Fusarium oxysporum, F. semitectum, F. solani, Aspergillus niger, A. flavus* (Ingle et al. 2009, Jain et al. 2011), *A. fumigatus, A. clavatus, Pleurotus ostreatus, Cladosporium cladosporioides* (Vahabi et al. 2011), *Penicillium brevicompactum, P. fellutanum*, endophytic *Rhizoctonia* sp., *Epicoccum nigrum, Chrysosporium tropicum*, and *Phoma glomerata* (Gade et al. 2013), while intracellular synthesis was shown to occur in a *Verticillium* (Mukherjee et al. 2001) species, and in *Neurospora crassa*. Several other types of fungi can be used in the nanoparticle synthesis as described by Mahmoud et al. (2013), Elamawi et al. (2018), Noor et al. (2020) and much more.

To date, several fungi have been successfully exploited for biosynthesis of nanoparticles, but different species of *Fusarium* are the prime choice of scientific community. Different species of *Fusarium* such as *F. oxysporum, F. semitectum, F. acuminatum, F. solani, F. culmorum*, etc., and their various strains (Ahmad et al. 2003, Bansal et al. 2004, Basavaraja et al. 2008, Reyes et al. 2009, Gaikwad et al. 2013, Siddiqi and Husen 2016, Khan et al. 2017) have been used in the synthesis of nanoparticles such as silver, gold, platinum, silica, palladium, etc.

Unzipping the Mechanism of Nanoparticle Synthesis from Fungi

Apart from the various fungi used in the synthesis of nanoparticles, *Fusarium* spp., have been the preferred choice of many researchers (Ahmad et al. 2003, Yadav et al. 2015, Rai et al. 2021). Members of the genus *Fusarium* can synthesize metal nanoparticles both intracellularly and extracellularly. In terms of the extracellular mycosynthesis pathway, it is proposed that metabolites such as enzymes, proteins, polysaccharides, flavonoids, alkaloids, phenolic and organic acids, etc., produced by the fungi for their survival when exposed to the different environmental stresses are mostly responsible for the reduction of metal ions to metal nanoparticles by catalytic effect (Srivastava et al. 2019). In addition, the same metabolites act as reducing and stabilizing agents and are responsible for the growth and stability of biogenic metal nanoparticles (Mahmoud et al. 2013, Yadav et al. 2015). The general mechanism involved in the synthesis, growth, and stabilization of metal nanoparticles employing fungi like *Fusarium* is schematically illustrated in Figure 1.

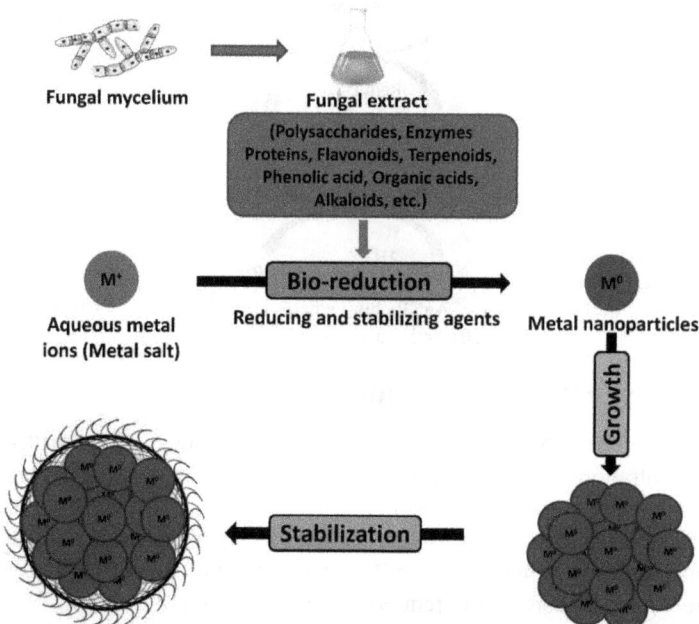

Figure 1. General mechanism involved in the synthesis, growth, and stabilization of metal nanoparticles using fungi (Adapted from Rai et al. (2021) an open-access article).

In addition, various other studies conducted on mycosynthesis have suggested the role of other different enzymes and proteins. However, among all these extracellular mycosynthesis methods, a hypothetical mechanism involving the role of NADH-dependent nitrate reductase enzyme has been widely accepted (Figure 2). Nitrate-dependent reductases and extracellular shuttle quinone, are involved in the silver nanoparticle synthesis of *Fusarium oxysporum* (Duran et al. 2005). Jain et al. (2011) showed that the silver nanoparticle synthesis of *A. flavus* occurs initially with the protein "33kDa" followed by a protein (cysteine and free amine groups) electrostatic attraction that stabilizes the nanoparticle by forming a capping agent. Intracellular silver nanoparticle synthesis is not fully understood but fungal cell wall surface electrostatic attraction, reduction, and accumulation is proposed (Ahmad et al. 2002). Extracellular gold nanoparticle synthesis by *P. chrysosporium* was attributed to laccase, whereas intracellular gold nanoparticle synthesis was attributed to ligninase (Ahmad et al. 2002).

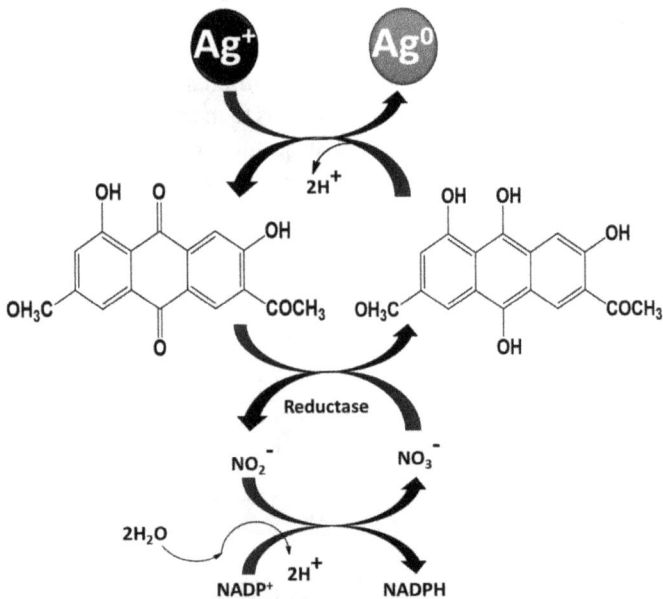

Figure 2. Hypothetical mechanism of AgNPs biosynthesis from *F. oxysporum* (Adapted from Duran et al. (2005), an open-access article).

Culture Techniques and Conditions

Fungal cell wall has a high binding potential with metal ions and can withstand high metal concentrations. The production of NPs using the fungus is much more efficient and less expensive than bacteria, as the mold has a higher tendency to accumulate metals. In addition, biomass treatment and downstream processing of NPs is easy on fungus-based biosynthesis for NPs. Therefore, the fungus has been studied extensively for synthesis of different NPs, such as silver, gold, etc.

Cultural techniques and media vary depending on the fungal isolate involved. However, in the general process fungal hyphae are usually placed in liquid growth media and placed in shake culture until the fungal culture has increased in biomass. The fungal hyphae is removed from the growth media, rinsed with distilled water to remove growth media, immersed in distilled water and incubated on shake culture for 24 to 48 hours. The fungal hyphae are separated from the supernatant, and an aliquot of the supernatant is added to the ion solution of 1.0 mM. The ion solution was then monitored for 2 to 3 days to form nanoparticles. Another traditional method is to add the washed hyphae directly to the 1.0 mM ion solution instead of using the fungal filtrate.

Factors such as moderate pH, reaction time and ionic concentration have a significant impact on yield and NP size. Therefore, many studies have been conducted to study the effect of these parameters on the biofabrication of NPs. Gold nanoparticles can vary in shape and size depending on the pH of the ion solution. For example, Gericke and Pinches (2006) reported that spherical gold nanoparticles are formed at pH 3 in *V. luteoalbum* small (cc.10 nm), and larger (spherical, triangular, hexagon and rods) gold nanoparticles are formed at pH 5, and at pH 7 to pH 9. Large nanoparticles often do not have a defined shape. The temperature interactions of both silver and gold nanoparticles were similar. Lower temperatures produces larger nanoparticles while higher temperatures resulted in smaller nanoparticles (Gericke and Pinches 2006).

Bhargava et al. (2016) investigated the effect of ionic concentration and pH on the biosynthesis of AuNPs using the fungi *Cladosporium oxysporum* and reported high yields of AuNPs at pH 7.0, 1 mM ionic concentration and 1:5 biomass to water ratio. Similarly, Rajput et al. (2016) also studied the effect of pH, temperature and isolate selections on AgNP biosynthesis using 12 different isolate of the fungi *Fusarium oxysporum*. The different medium pH (3, 5, 7 and 9) was found to affect AgNP formation using *Fusarium oxysporum* 405. At pH 3.0, AgNPs of triangular, spherical, rod and others abnormal shapes are produced, while at pH 5 and 7, monodisperse and mostly spherical NP are formed (Rajput et al. 2016). Changes in the pH of the medium have led to the formation of different sizes of NPs because change in pH affect the basic or acidic environment of the amino acids, which are involved in the formation of NPs (Gericke and Pinches 2006, Rajput et al. 2016). Similarly, in studies of the effect of different temperatures on the amount of AgNPs produced from *F. oxysporum* 405, Rajput et al. (2016) found that maximum NPs were formed at temperatures between 50 and 70°C, while minimum AgNPs formation was observed at 25°C. Birla et al. (2013) studied the effect of temperature on AgNPs biosynthesis using *Fusarium oxysporum* and reported that temperatures 40 to 60°C were optimal for the synthesis of AgNPs. The incubation temperatures also affect the size of newly formed AgNPs. In another study, El Domany et al. (2018) demonstrated extracellular biofabrication of 10–30 nm size AuNP using the edible fungi *Pleurotus ostreatus* and reported that AuNP formation was significantly affected by salt concentration, temperature, pH and incubation time. In this study, it was reported that the biofabrication rate of AuNPs was directly proportional to the incubation period, salt concentration and temperature. As the incubation period, salt concentration and temperature were increased to 12–48 hours, 1–5 mM and 30–40°C, respectively, the rate of AuNP biofabrication also increased. However, higher AuNPs were produced at pH 3.0.

Application of Nanofertilizers NFs in Sustainable Agriculture

Nanotechnology may have the potential to transform the current synthetic framework used in modern agricultural systems (Prasad et al. 2017, Avila-Quezada et al. 2021) by increasing the efficiency of agrochemicals (Kerry et al. 2017) and providing solutions to environmental and agricultural problems (Usman et al. 2020). Therefore, research on the use of nanoparticles (NPs) has gained attention among agricultural researchers in recent years (Kerry et al. 2017, Kah et al. 2019, Seleiman et al. 2020c). From a sustainable agricultural perspective, nanotechnology has the potential to develop new types of fertilizers such as nanofertilizers (NFs) to increase global food production to feed the world's population (Feregrino-Pérez et al. 2018, Diatta et al. 2020, Seleiman et al. 2020c).

The term nanofertilizer indicates nanomaterial, which can be a plant nutrient itself (micro or macro-nutrients) or a carrier of plant nutrients. Nutrients that are encapsulated or coated with nanomaterials are also called NFs (DeRosa et al. 2010). NFs can be developed from synthetic materials (i.e., modified forms of synthetic fertilizers) or green synthesized from different parts of plants using various chemicals, mechanical, or biological methods using nanotechnology (Singh and Kumar 2017). NFs are used to increase soil fertility, bioavailability of plant nutrients (Chhipa 2017, Singh and Kumar 2017) and product quality (Brunner et al. 2006). Based on plant nutrient requirements, NFs are generally categorized into: macro NFs, micro NF, and nanoparticulate fertilizers (Chippa 2017). NFs have larger surface areas and characteristics, slow and steady nutrient release, both of which make them suitable for modern agricultural use (Prasad et al. 2017, Pitambara and Shukla 2019, Seleiman et al. 2020c). Only small amounts of fertilizers reach the target areas of the plants, leading to low nutrient availability at plant level (Rafiullah et al. 2020). Therefore, farmers use high concentrations of synthetic fertilizers to obtain high yields, which increase salinity and affect the natural balance of soil nutrients, ultimately adversely affecting crop production. This depends on the climatic zones and the type of soil and vegetation. Therefore, it is of paramount importance to develop a novel fertilizer that can release its nutrients slowly and consistently to increase crop yield, improve quality, and improve the overall sustainability of agricultural systems.

In conclusion, nanotechnology can increase agricultural production, which includes nano-formulations of agrochemicals to be used as pesticides and plant development fertilizers, nano-biosensors in plant protection to diagnose diseases and residues of agrochemicals, and nano devices for the genetic manipulation of plants, etc.

Potential Advantages of Nanofertilizers Over Synthetic Fertilizers and their Use for Sustainable Agriculture

Depending on the type of composition, nanofertilizers are divided into three categories (Mastronardi et al. 2015):

1. Nanoscale fertilizer, which corresponds to a standard fertilizer reduced in size, usually in the form of nanoparticles

2. Nanoscale additive fertilizer, a traditional fertilizer containing additional nanomaterials and

3. Nanoscale coating fertilizer, refers to nutrients encapsulated with nanofilms or intercalated into nanoscale pores of host material.

Nano-fertilizers are developed using both mechanical and biochemical processes, i.e., materials are grounded to obtain nano-sized particles using mechanical methods and biochemical techniques used to achieve effective nanoscale formulation. NF-coated or nanomaterials regulate nutrient release depending on plant requirements, and this results in an increase in NUE values of plant (Qureshi et al. 2018). Remarkably, NFs can release their nutrients in 40–50 days, while synthetic fertilizers do the same in 4–10 days. Recently, N was modified in the NF form by covering urea with hydroxyapatite NPs, which led to a slower release of N to plants (Kottegoda et al. 2017). Similarly, studies have shown that nanohybrid of urea (i.e., a modified form of hydroxyapatite) can release N about 12 times slower than synthetic urea in rice fields (*Oryza sativa* L.), and can increase grain yield by only 50% the rate used with normal urea (Kottegoda et al. 2017). Similarly, synthetic P fertilizers have a low absorption rate and high levels of fixation in soil (Shenoy and Kalagudi 2005), whereas the nano-formulations of P can reduce nutrient loss by direct internalization of plant (Dwivedi et al. 2016). For example, the use of porous nanomaterials, such as chitosan and zeolite, has been found to significantly improve the efficiency by controlling demand based release and reduce N loss (Millán et al. 2008, Abdel-Aziz et al. 2016). The use of P-enriched hydroxyapatite NPs was found to significantly increase plant length, shoot growth, and grain yield (18%) of soybeans (*Glycine max* L.) compared to plant grown with synthetic P fertilizers (Liu and Lal 2014). Similarly, carbon-based nanomaterials (e.g., graphene oxide films) have the potential to extend the process of potassium nitrate release, thereby reducing leaching losses (Shalaby et al. 2016).

Nanoparticles also have an effect on other plant metabolic processes that influence the ability to bind nutrients like P to plants (Zahra et al. 2015). For example, to increase the absorption efficiency of synthetic P, zinc NPs are used for mobilization (Seleiman and Abdelaal 2018). In addition, biosensors can be linked to NFs to regulate nutrient release and their bioavailability depending on the growth phase of the plant (León-Silva et al. 2018), a technology that does not apply to synthetic fertilizers. Finally, the prices and costs of NF applications are generally lower than the synthetic fertilizer, as NF is required in smaller quantities (León-Silva et al. 2018).

Salient Characteristics of NFs for Plant Performance

NFs have distinctive characteristics (Figure 3) that make them more valuable than synthetic fertilizers (Singh and Kumar 2017, Pitambara and Shukla 2019). One of the most important characteristics of NF is its ability to penetrate plants when used as foliar or soil supplements due to the small particle size (< 100 nm) (Liscano et al. 2000). NFs have high surface areas, and this may provide maximum reactivity, higher efficiency and increase both the availability of nutrients and plants NUE (Liscano

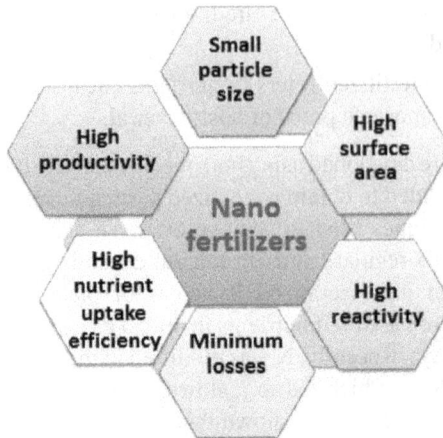

Figure 3. The most important advantages of NFs.

et al. 2000, Siddiqi and Husen 2017). In addition, NFs are water soluble and can increase the diffusion of nutrients in the soil and increase their availability to plants. Besides this, the slow and targeted release of nutrients (Singh and Kumar 2017) by NFs (Siddiqi and Husen 2017) reduces their toxicity to plants (Sohair et al. 2018) and reduces N loss through volatilization, leaching, fixation, and denitrification, and the accumulation of salt in the soil.

Mechanisms of NFs uptake by Plants

The NF composition, size of NPs, plant physiology, and the pore diameter (5–20 nm) of the cell wall (Rico et al. 2011, Pitambara and Shukla 2019) affect the transport and accumulation of nutrients released from NFs in plants (Corredor et al. 2009). If NFs choose to move through xylem, then the best possible application of NFs through an irrigation system, but if NFs transport through phloem, then an external application is recommended and appropriate (Pitambara and Shukla 2019).

Foliar Exposure and Uptake of NFs

Several studies have reported that NPs larger than 5.0 nm can enter plants by foliar application (Lv et al. 2019). In foliar applications, NPs have to pass a cuticular barrier before entering into plant tissues (Pollard et al. 2008). The cuticle layer is a waxy covering on leaves that has two entry points, namely, the lipophilic or cuticular pathway and the hydrophilic or stomatal pathway. Polar nanoparticles, on the other hand, can enter through the hydrophilic or stomatal pathway (Eichert et al. 2008). However, differences in leaf morphology and the number and size of stomata across plant species may affect the uptake of foliar NPs (Wiesner et al. 2009). However, due to the physiological function and the unique geometric structure of the stomata, the exact limitation of the size of the stomatal hole for NP diffusion is not yet clear (Seleiman et al. 2021). Following the stomatal pathway, nanoparticles can travel long distances through the vascular system of a plant after entering the leaf apoplast

(Lv et al. 2019). As the vascular system of plants is unidirectional, the nutrients or photosynthates moving towards shoots (xylem) or roots (phloem) do not return to their original sites (Lough and Lucas 2006). Therefore, foliar based nanoparticles have only the phloem system option for uptake and transport from the leaves to the roots.

Root Exposure and Uptake of NFs in Plants

Many factors such as plant growth, growth stage, exposure conditions, particle size, and rhizosphere processes affect the acquisition of NPs by roots. Surface charge also influences the uptake and translocation of NPs in plants (Lv et al. 2019). Avellan et al. (2017) noted that the roots of *Arabidopsis thaliana* produced a mucilage that facilitated the acquisition of positively charged gold NPs (12 nm) by the roots, while the same size (12 nm) of negatively charged gold NPs did not penetrate the root tissue. Nanoparticles applied to soil first adsorbed over the roots and then crossed several barriers to reach the plant's vascular system (Lv et al. 2019). The first barrier is the root cuticle layer, which has a similar composition to that of the leaf cuticle layer. Nanoparticles cross the root cuticle and reach the root epidermis. When NPs reach the root epidermis, they may follow apoplastic or symplastic pathways. The main obstacle to the apoplastic pathway is the Casparian strip around the vascular system, which prevents direct entry of NPs into the vascular cylinder (Roppolo et al. 2011), although studies have shown that ZnO NPs (30 nm) can penetrate from the lateral root junction of maize to the vascular system (Lv et al. 2015). Another potential pathway is the symplastic pathway, in which NPs move from cell to cell *via* plasmodesmata (Rico et al. 2011, Lv et al. 2019). When nanoparticles reach the central cylinder, they can move the upper parts of the plant *via* transpiration stream through xylem (LaRue et al. 2012).

Macronutrient Nanofertilizers (npk) and their Effects on Plants

Macronutrients such as N, P, and K are needed for plants in large quantities (Battaglia et al. 2018, Kumar et al. 2019, Adeyemi et al. 2020, Czymmek et al. 2020). Since most of these nutrients cannot be efficiently absorbed by plants, farmers often use high doses of fertilizer to partially adjust their NUE levels, resulting in a potentially harmful effect on soil, water and the environment (Seleiman et al. 2013, Chhipa 2017, Czymmek et al. 2020, Seleiman et al. 2020a). The use of NFs can increase NUE of fertilizer, improve crop yields and crop quality, and reduce the negative effects of organic fertilizers in the context of more sustainable farming (Liu and Lal 2015, Battaglia et al. 2018, Seleiman et al. 2020c). NFs or nanoenabled fertilizers accurately release nutrients from the root zone of plants by preventing rapid changes in soil chemical composition, which in turn reduces loss of nutrients. Different types of NFs are produced depending on the material or carrier present, e.g., hydroxyapatite nanoparticles, zeolite, mesoporous silica nanoparticle, nitrogen, copper, zinc, silica, carbon, and polymeric nanoparticles (Liscano et al. 2000, Mikhak et al. 2017, Guo et al. 2018). In Table 1, we present the type and range doses of NF types based

Table 1. The effects of different NF and/or NP types and their dose ranges on different crops. (Adopted from Seleiman et al. (2021) An open-access article).

NFs/NPs	Range of Doses	Plant/Crop	Effects	References
Zn NFs	5–20 mg/L	*Allium cepa* L.	Reduced root growth	Ahmed et al. 2018
Zn NFs	100–500 ppm	*Capsicum annuum* L.	Increased seed germination	Tantawy et al. 2015
Zn NFs	500 mg/kg	*Pisum sativum* L.	Reduced H_2O_2 and chlorophyll molecules	Nair and Chung 2014
Zn NFs	1000 mg/kg	*Cucumis sativus* L.	Inhibited root growth	Zhao et al. 2014
ZnO NPs	20 mg/L	*Triticum aestivum* L.	Increased biological and grain yield	Du et al. 2019
ZnO NPs	10 mg/L	*Cyamopsis tetragonoloba* (L.) Taub.	Increased growth, biological yield, and nutrient contents	Raliya and Tarafdar 2013
ZnO NPs	10 mg/L	*Zea mays* L.	NPs Increased root shoot length, plant height, leaf area, chlorophyll content, and grain quality	Raliya et al. 2016
ZnO NPs	5–20 mg/L	*Solanum melongena* L.	Reduced germination, root length, and leaf area under culture media but increased these parameters under soil conditions	Thunugunta et al. 2018
Cu NPs	20–80 mg/kg	*Coriandrum sativum* L.	Decreased germination and shoot growth	Zuverza-Mena et al. 2015
Cu NPs	50–500 mg/L	*Solanum lycopersicum* L.	Increased antioxidant contents and fruit firmness	Ahmed et al. 2018
Cu NPs	10–20 mg/L	*Lactuca sativa* L.	Decreased seedling growth and dry weight of seedlings; affected water relationships and nutrient contents	Trujillo-Reyes et al. 2014
Cu NPs	130–660 mg/kg	*Lactuca sativa* L.	Increased shoot/root length ratio	Hong et al. 2015
Cu NPs	200 mg/kg	*Spinacia oleracea* L.	Increased fresh biomass and photosynthesis rate	Wang et al. 2019
CuO NPs	500 mg/kg	*Triticum aestivum* L.	Increased biological yield	Dimkpa et al. 2012
Fe-based NFs	30–60 ppm	*Pisum sativum* L.	Increased chlorophyll contents and seed weight	Giorgetti et al. 2019
Fe-based NFs	10–20 mg/L	*Lactuca sativa* L.	Increased antioxidants and enzymatic activities but decreased overall growth	Trujillo-Reyes et al. 2014

Table 1 contd. ...

...Table 1 contd.

NFs/NPs	Range of Doses	Plant/Crop	Effects	References
Nano-iron oxide (Fe)	500–1000 mg/L	*Cuminum cyminum* L.	Increased stem length, yield (130%), and Fe concentration in plant (110%)	Sabet and Mortazaeinezhad 2018
FeO	1–50 ppm	*Lactuca sativa* L.	Germination was maximum at 1 ppm of FeO but a high root length was noted at 10 ppm	Liu et al. 2016
FeS2	80–100 g/mL	*Cicer arietinum* L.	High germination rate and crop yield	Das et al. 2016
Nano-nitrogen (N)	25–100%	*Oryza sativa* L.	Increased tillers per plant, height, and dry weight	Rathnayaka et al. 2018
Nano-apatite (P)	100 mg/L	*Glycine max* (L.) Merr.	Increased biological yield (18.2%) and root length	Liu and Lal 2015
Hydroxyapatite (P)	200 mg P/kg	*Lactuca sativa* L.	Increased P content and dry weight	Taskın et al. 2018
Nano-potash (K)	1500–2500 mg/L	*Arachis hypogaea* L.	Increased shoot length, stem diameter, biological yield, and number of flowers per plant	Asgari et al. 2018
Chitosan-NPK	(500, 60, and 400 ppm; respectively) 10%, 25%, and 100%	*Triticum aestivum* L.	Increased P and K contents but decreased protein content	Abdel-Aziz et al. 2018
MgO	7–10 g/mL	*Solanum lycopersicum* L.	Decreased bacterial wilt disease caused by *Ralstonia solanacearum* L.	Imada et al. 2016
MnO	0.25–50 ppm	*Lactuca sativa* L.	No effect on germination but increased root length	Liu et al. 2016
Nano-silica (SiO$_2$)	30–60 mg/L	*Triticum aestivum* L.	Increased relative water content (84%) and final yield (18–25%)	Behboudi et al. 2018
SiO$_2$ NPs	15 kg/ha	*Zea mays* L.	Improved growth parameters	Suriyaprabha et al. 2012
Sulfur NPs	500–4000 ppm	*Vigna radiata* L.	Increased dry weight	Patra et al. 2013

on different plant types research, as well as their effects on plant growth and their physiological, biochemical, and productivity characteristics.

Nitrogen NFs

Nitrogen (N), considered to be the most important mineral component in plants, is a basic component of several amino acids, proteins, DNA (deoxyribonucleic acid), ATP (adenine triphosphate), chlorophyll, and cell structure units. Most of the metabolic functions and regulatory pathways in plants depend on a sufficient amount of N. Plants uptake N in the form of NO_3 and NH_4^+ (Preetha and Balakrishnan 2017, Seleiman et al. 2020a).

One of the main constraints of synthetic N-fertilizer is the high volatilization and leaching rates that occur during and after their application in the field. To minimize these losses, N-based NFs may be used for continuous N supply with a slower release rate. Urea-modified zeolites have been found to increase the seed yield of soybeans (*Glycine max* L.) over synthetic fertilizers (Liu and Lal 2015). N-NF formed by coating urea in nanofilm has been successfully used in *Brassica napus* L. (DeRosa et al. 2010). Similarly, both nano-N and chelated nano-N were effective in increasing the yield of potato plant (*Solanum tuberosum* L.) and in reducing nitrate leaching (Zareabyaneh and Bayatvarkeshi 2015). Recently, Ha et al. (2018) used NPK-coated NFs in coffee plants grown under greenhouse conditions. The authors stated that such an application of NPK NF increased the nutrient uptake and growth of coffee plants by increasing the number of leaves and the photosynthetic plants area, compared to controls (zero NF).

In conclusion, nitrogen in the form of NF is highly recommended because it can cause slow N release, reduce volatilization and leaching rates, lead to higher nutrient uptake, and improve plant growth and productivity.

Phosphorus NFs

After N, phosphorus (P) is considered to be the second most important nutrient for plant growth, as it is an important component of energy transfer molecules, ATP (adenine triphosphate), ADP, phospholipids, and sugar phosphate, and it has vital role in processes such as photosynthesis, respiration, and biosynthesis of DNA (Soliman et al. 2016). Different parameters of plant productivity such as root and shoot length, plant vigor, disease resistance, number of reproductive buds, yield, and quality are strongly influenced by P availability (Preetha and Balakrishnan 2017).

Recent studies have shown that NFs can gradually deliver P for 40–50 days following their use, whereas a typical P fertilizer delivers all nutrients within 8–10 days after application (Liu and Lal 2015). Therefore, it has been suggested that the use of NF or slow release substances such as zeolites may have the potential to increase NUE of P for several field plants (Liu and Lal 2015). Nano-fertilizer, the source of P, has been found to significantly increase fresh and dry biomass, increase fruit yield, and improve quality several times, in addition to leading to higher NUE (Patra et al. 2013).

Potassium NFs

Potassium (K) is the third most vital macronutrient after N and P, and plays an important regulatory role in all physiochemical activities of plants to maintain normal growth and development. Among other processes, K is involved in plant stomatal opening, photosynthesis, photosynthates transfer, protein synthesis, ionic balance, water relationships, and activation of more than 60 enzymes (Preetha and Balakrishnan 2017).

Plants with adequate K value have been shown to be highly resistant to abiotic stresses such as water pressure and high/low temperatures (Wang et al. 2013, Sohair et al. 2018, Taha et al. 2020). On the other hand, K deficiency adversely affects root growth, seed content within fruit, size, shape, color, taste, and final crop yield (Preetha and Balakrishnan 2017). Li et al. (2010) found that zeolites loaded with K increase yield, harvest index, K concentration, and chlorophyll content in hot peppers (*Capsicum annuum* L.). Similarly, nano-K fertilizer using foliar application significantly improved the growth, biomass, and quality of *Cucurbita pepo* (Gerdini 2016). Therefore, using K NFs can protect soil health and improve water quality by reducing K losses in soil and can enhance physiological and yield traits.

Micronutrient Nanofertilizers Zn, Fe, Mn, Cu, and Si and their Effects on Plants

Micronutrients are also important in increasing plant production and quality and rising plant tolerance against multiple stresses (Seleiman et al. 2012, 2013, 2020a). The synthesis of micronutrients by nanosized structures may increase their solubility and bioavailability, facilitate the acquisition of the uniform dispersion of these nutrients in the soil, and reduce the adsorption and fixation of micronutrients in soil colloid (Seleiman et al. 2021).

Zinc NFs

Plant growth is highly dependent on Zn nutrition because Zn is part of a structural component co-factor that combines various proteins and enzymes. Zinc also plays a role in regulating auxins, protein metabolism, biosynthesis of carbohydrates, and plant protection against bacteria and environmental stresses (Broadley et al. 2007).

Zinc NF-based ZnO is commonly used in modern agriculture (Seleiman et al. 2020c) as it is more efficient and cost-effective than synthetic Zn fertilizers (Khanm et al. 2018, Seleiman et al. 2020c) and can be used for soil mixing, seed priming (Sharifi et al. 2016) and foliar spray (Seleiman et al. 2020c). Studies have shown that the use of Zn NFs can increase germination, seedling growth, yield, and crop quality (Seleiman et al. 2020c). Similarly, the use of Z NFs increased the growth of shoots, leaf area, dry weight, final yield and protein content of sunflower (*Helianthus annuus* L.), pearl millet (*Pennisetum americanum* L.), rice, corn, sugarcane (*Saccharum officinarum* L.) and potatoes (Tarafdar et al. 2014, Monreal et al. 2016, Moghaddasi et al. 2017, Seleiman et al. 2020c).

Iron NFs

Iron (Fe) is an important component involved in the synthesis of chlorophyll, DNA, chloroplast formation, respiration, and several metabolic pathways. Although plants need Fe in small amounts to grow, its deficiency or excess has harmful effects on the physiological and metabolic functions of plants, thereby reducing their yield (Palmqvist et al. 2017).

Various studies have shown that Fe NFs increased germination and improved growth of different plants compared to control and/or synthetic Fe sources. Srivastava et al. (2014) reported that iron pyrite NPs increased spinach growth (*Spinacia oleracea* L.). Rui et al. (2016) documented better root growth in peanut (*Arachis hypogaea* L.) plants treated with Fe NPs compared with non-treated plants under field conditions. Raju et al. (2016) observed higher radical lengths during germination in green gram (*Vigna radiata* L.) and higher biomass with application of Fe NPs (2–6 nm) compared with controls (ferrous sulphate; $FeSO_4$). In conclusion, Fe NFs can be an optimal alternative source, especially in soils suffering from Fe deficiency.

Manganese NFs

Manganese (Mn) is a vital micronutrient involved in N metabolism, photosynthesis and biosynthesis of fatty acids, ATP, and proteins (Palmqvist et al. 2017). Manganese also helps plants to deal with different stresses.

Studies have shown that Mn applications significantly improve the growth and yield of wheat, maize, sugarcane. Mn treatments also improve yield of egg plant (*Solanum melongena* L.) by 22% (Elmer and White 2016) and significantly increase root lengths of lettuce (*Lactuca sativa* L.) soybeans, and common beans compared to controls, i.e., Mn ions (Fageria 2001, Dimkpa and Bindraban 2016). However, there was no effect of Mn NPs on the length of white mustard root (*Sinapis alba*) (Landa et al. 2016), lettuce seed germination (Liu et al. 2016), or watermelon (*Citrullus lanatus*) yield.

At the physiological level, Mn NPs adhere to the chlorophyll binding protein (CP43) of photosystem II, and this leads to an increase in the activity of the electron transport chain, thus, the complete efficiency of the photosynthesis process. As a result, Mn NPs fertilized plants showed higher levels of nitrogen assimilation and metabolism compared to their conventional bulk counterparts (Pradhan et al. 2013).

Copper NFs

Copper (Cu) is a component of regulatory proteins that contribute to photosynthesis and plant respiration and is a cofactor of antioxidants such as superoxide dismutase and ascorbate oxidase. Lack of copper leads to various disorders such as necrosis, stunted growth, low seeds set, grains, and fruits; and ultimately lower yield (Rai et al. 2018).

In a recent study, field use of the CuO NPs nanofertilizer improved germination and growth of soybean roots and chickpeas (*Cicer arietinum* L.) (Adhikari et al. 2012). Similarly, soybean seeds treated with nanocrystalline powder Cu, Co, and Fr (40–60 nm) had germination rates of 65%, 80%, and 80%, respectively, which were higher than the germination rate of 55% in control sample (zero NF) (Ngo et al. 2014). Similarly, different concentrations of Cu NPs increased growth and yield of wheat due to improvements in leaf area, chlorophyll content, grain content per spike, and grain weight. In addition, there has been an improvement in flavonoid content, sulfur assimilation, and biosynthesis of proline and glutathione in *Arabidopsis thaliana* after the application of Cu NPs in a dose of 5 mg L^{-1} (Nair and Chung 2014). On the contrary, Cu NPs application negatively affected the growth of water lettuce (*Pistia stratiotes* L.) (Olkhovych et al. 2016) and decreased fruit firmness in cucumber plants (Hong et al. 2016).

Silicon NFs

Silicon (Si) is not essential to complete the plant life cycle. However, it does provide some benefits to plants under normal and stressful conditions. Si absorption from soil by plants occurs only in the form of mono-silicic acid. It has been suggested that Si can play a significant role in improving plant tolerance against heavy metal toxicity, as well as heat, water, and salinity stresses (Rastogi et al. 2019, Seleiman et al. 2019). In addition, the use of SiO_2 in organic fertilizers has the potential to improve overall crop production (Janmohammadi et al. 2016, Seleiman et al. 2020b). The mesoporous structure of the Si NP enables them to be suitable nanocarriers of various molecules that are beneficial to agricultural systems. Therefore, Si NPs are successfully used in agriculture to monitor soil moisture and improve soil water retention and have the potential to be used as fertilizers for certain plants that cannot survive without the appropriate quantity of silicon or as nano-carriers to improve agriculture sustainability.

Boron NFs

Boron (B) is an important micronutrient that plays a key role in the elongation of pollen grains and tubes, the formation of cellular walls, the transfer of photosynthetes from leaves to active areas, and the increase in flower and fruit yields (Davarpanah et al. 2016).

Studies have shown that B NFs or NPs can improve plant growth and yield. Genaidy et al. (2020) sprayed nano-boron at 20ppm and nano-zinc at 200 ppm on olive trees, and the plants produced a maximum number fruits with high seed oil content. Similarly, Davarpanah et al. (2016) reported higher number of fruits and yields in pomegranate (*Punica granatum*) after the application of B nanofertilizer (34 mg B per plant). Taherian et al. (2019) applied a B nanofertilizer to the alfalfa plant (*Medicago sativa*) planted in calcareous soil and harvested maximum yields with good fodder quality. In conclusion, the B application of NF and/and NPs can improve crop quality and yield.

Nanofertilizers for Abiotic and Biotic Stress Tolerance

Abiotic and biotic stresses are major constraints to plant productivity that adversely affect plant growth and production, and are a major threat to food security worldwide (Seleiman and Kheir 2018, Battaglia et al. 2019, Seleiman et al. 2020b). Among the abiotic stresses, droughts, floods, heat, hail, salinity, heavy metals, and mineral deficiencies are considered to be major stresses affecting growth, yield, and crop quality (Battaglia et al. 2019, Seleiman et al. 2020a, 2020b, 2020c). On the other hand, different types of pests and diseases are biotic stresses that reduce crop yields. According to the Food and Agriculture Organization (FAO), the biggest challenge for agricultural scientists is to increase crop production by 70% by 2050 (Food 2009). Therefore, clear identification and appropriate use of new technologies or approaches to overcome current yield limiting factors and increase efficiency of resource utilization is essential. In light of this concern, many studies have recently confirmed the importance of NFs in sustainable agriculture.

Drought Stress

Drought is a major abiotic stress that significantly reduces agricultural production. In addition to the cultivation of drought resilient crops, the use of stress reducing agents such as NF has great potential to decrease the negative effects of drought stress on plants (Singh and Husen 2019) by increasing water holding soil capacity. Under stress conditions, the increased production of reactive oxygen species (ROS) causes lipid peroxidation, damages cell membranes, and leads to leakage of solutes in cells and causes cell death. Studies have shown that NPs can increase the content of antioxidants and proline, thereby reducing the production of H_2O_2 and malondialdehyde (Janmohammadi et al. 2016).

Foliar spray application of Fe NPs was found to mitigate the effects of water stress and increase yields (Table 2) and the oil percentage in safflower (*Carthamus tinctorius* L.) (Davar et al. 2014). In addition, a foliar spray of 0.02% TiO$_2$ NPs was found to increase the tiller number, grain weight, final grain yield, and harvest index of a wheat crop under water stress (Jaberzadeh et al. 2013). Silver NPs ameliorated the effects of drought stress, improved lentil (*Lens culinaris* Medik) seed germination, and improved dry root weight (Hojjat and Ganjali 2016). Mahmoud and Swaefy (2020) reported that the use of nano-NPK and nano-zeolite loaded N reduced the effects of water stress and increased the growth of sage (*Salvia officinalis*).

Salinity Stress

Excessive accumulation of Na^+, Cl^-, and SO_4^{2-} ions in the root zone of plants reduces osmotic potential, decreases water uptake, and inhibits plant growth, thereby causing plant mortality in some cases (Rasool et al. 2013). Salt affected soils have a low osmotic potential, creating imbalances in plant nutrients and increasing certain ionic toxicity.

In this context, the use of NF can be a positive approach to overcome the rising global soil salinity problem. The use of SiO$_2$ NPs has been found to increase leaf dry weight, chlorophyll, proline, and antioxidant content under salinity stress (Kalteh et al. 2018). Savvas et al. (2009) reported that SiO$_2$ NPs reduced Na^+ ion toxicity and increased plant growth under salt stress. Tantawy et al. (2014) applied nano-calcium to *Solanum lycopersicum* grown under salt stress, and reported that plants fertilized with nano-calcium showed more fruit per plant and had a higher yield (76%) than those grown with synthetic monophosphate. El-Hefnawy (2020) applied nano-NPK (50–100 ppm) for foliar spray on pea plants under salinity stress, nano-NPK reduced the severe effects of salt and increased growth and productivity.

Table 2. The effects of different methods of application of NF and/or NP types on different crops grown under different environmental stresses. (Adopted from Seleiman et al. (2021) An open-access article).

NFs/NPs	Method of Application	Stress	Plant/Crop	Effects	References
Fe	Foliar	Drought	*Carthamus tinctorius* L.	Reduced effect drought and increased yield	Davar et al. 2014
Fe		Drought	*Fragaria ananassa* Duch.	Increased drought resistance in the field	Mozafari et al. 2018
ZnO		Drought	*Glycine max* L.	Increased germination	Sedghi et al. 2013
SiO$_2$		Drought	*Crataegus* sp.	Increased photosynthesis by improving stomatal conductance and increased yield	Ashkavand et al. 2015
Na$_2$SeO$_4$		Heat	*Lycopersicon esculentum* L.	Improved water relationships of plants and increased chlorophyll contents	Haghighi et al. 2014
Se		Heat	*Lycopersicon esculentum* L.	Increased growth and yield	Djanaguiraman et al. 2018
CuO 20–2000 _g/mL	Pre-sowing	Oxidative stress	*Allium cepa* L.	Increased antioxidant activities	Ahmed et al. 2018
SiO$_2$ 1.5–7.5 g/L	Pre-sowing	Salinity	*Cucurbita pepo* L.	Increased germination, photosynthesis, and antioxidants; decreased production of H$_2$O$_2$	Ashkavand et al. 2018
Nano-urea Hydroxyapatite 25–100%	Pre-sowing	Salinity	*Prunus dulcis* L.	Increased germination plant height, and secondary roots/plants, yield	Badran and Savin 2018
Chitosan-Cu 10 mg	Post-transplanting	Salinity	*Solanum lycopersicum* L.	Increased plant growth and gene expression for jasmonic acid	(Haghighi and Pessarakli 2013)
Si 1–5 mg/L	Post-transplanting	Salinity	*Capsicum frutescens* L.	Increased salt tolerance	Tantawy et al. 2015

Table 2 contd. ...

...Table 2 contd.

NFs/NPs	Method of Application	Stress	Plant/Crop	Effects	References
Nano-Ca 0.5–1 g/L	Post-transplanting	Salinity	*Solanum lycopersicum* L.	Increased flowers/plants, yield and improved stem diameter	(Younes and Nassef 2016)
Nano-silicon 1–2 mM	Foliar application	Salinity	*Jatropha* sp.	Integerrima enhanced vegetative parameters and chemical constituents	Ashour and Mahmoud 2017
Na_2SiO_3 10 µM P.	Post-transplanting	Heavy metal	*Pisum sativum* L.	Decreased uptake of heavy metal and increased antioxidants activities	Delfani et al. 2014
SiO_2		Salinity	*Ocimum basilicum* L.	Increased fresh and dry weights and chlorophyll and proline contents	Rasool et al. 2013
SiO_2		Salinity	*Glycine max* L.	Increased antioxidant enzymes and decreased oxidative stress	Farhangi-Abriz and Torabian 2018
ZnO		Salinity	*Helianthus annuus* L.	Increased CO_2 assimilation and photosynthesis rate; reduced Na content in leaves	Torabian et al. 2015
SiO_2		Mineral nutrient	*Carthamus tinctorius* L.	Increased yield	Janmohammadi et al. 2016
Zn		Mineral nutrient	*Pennisetum americanum* L.	Improved leaf area, chlorophyll content, and enzyme activities	Tarafdar et al. 2014

Temperature and Stress

Extremely high temperatures create oxidative stress and adversely affect the net photosynthesis rates, chlorophyll content, and plant growth (Kai and Iba 2014). Low doses of Se NPs have been found to significantly reduce the effects of heat stress by improving water relations, chlorophyll content, and the activities of plant antioxidants (Haghighi et al. 2014). Under high temperatures or heat stress conditions, plants produce heat shock proteins that can reduce the effects of heat or temperature stress (Wahid et al. 2007, Tantawy et al. 2014).

Studies have shown that multiwalled carbon nanotubes can help to induce gene expression for heat-shock proteins. Wheat plants under conditions of heat stress treated with foliar spray of Ag NPs (50–75 mg L^{-1}), showed improved root lengths by 5.0% and 5.4%, shoot lengths by 22.2% and 26.1%, root numbers by 6.6% and 7.5%, fresh weights by 1.3% and 2.0%, and dry weights by 0.36% and 0.60%, respectively (Iqbal et al. 2019). Foliar application of 10 mg L^{-1} of Se NPs under high temperature stress during booting stage of sorghum (*Sorghum bicolor* L. Moench) increased pollen germination, enhanced the system of antioxidant defenses, and thus increased the seeds yield of plant compared to those obtained from control treatment (Djanaguiraman et al. 2018).

Biotic Stress

Globally, annual crop losses due to disease and pest infestations are estimated to be between 20% and 40% (Pestovsky and Martínez-Antonio 2017). To reduce the negative impact of pests on plant productivity, farmers around the world use millions of metric tons of pesticides each year, increasing environmental pollution, disruption of the ecosystem, residual toxicity in food and feed, reduced soil fertility, and insects pest resistance (Chowdappa and Gowda 2013).

Various studies have shown that the use of NPs or NFs has the potential to demolish the population of various deleterious soils and plant micro-organisms, as they can easily invade and disrupt bacterial or fungal cells (Sotelo-Boyás et al. 2016). Nano-Cu has been found to effectively control bacterial infections (*Xanthomonas campestris* pv. *phaseoli*) in mung plants and bacterial blight of rice (*Xanthomonas oryzae* pv. *oryzae*) (Vigneshwaran et al. 2006). Chitosan NPs can control fungal infections, bacterial, and even viruses because chitosan NP binds to microbial cell wall, disrupts cells, alters membrane stability, or attaches to DNA and stops replication. Nawaz et al. (2011) reported that the application of ZnO NPs was effective against the *E. coli*, *Clostridium perfringens*, and *Bacillus subtilis*. Cu-based NPs can also be used to kill fungi and bacterial species that affect agricultural crops, as accomplished by Ramyadevi et al. (2012) who reported antimicrobial potential of Cu NPs against various fungi (*Aspergillus niger*, *Aspergillus flavus*, and *Candida albicans*) and bacterial species (*Escherichia coli*, *Pseudomonas aeruginosa*, *Staphylococcus aureus*, *Klebsiella pneumoniae* and *Micrococcus luteus*). Kale et al. (2017) and Divte et al. (2019) demonstrated significant antifungal activity of copper nanoparticles against fungal pathogen *Sclerotium rolfsii* and *Colletotrichum capsici* at a concentration of 200 ppm in field conditions and effectively reduced stem rot lesions in tomatoes and anthracnose disease in chilli plants without noticeable phytotoxicity. In conclusion, the use of NPs and/or NFs on infected plants has great potential to reduce the different noxious soil and plant micro-organisms because they can easily penetrate and disrupt bacterial or fungal cells.

Nanobiofertilizers

Nanobiofertilizers can be defined as an amalgamation of biofertilizers with nanostructures or nanoparticles to enhance plant growth (Simarmata et al. 2016). Some of the most important aspects in the development of nanobiofertilizers are the interactions between nanoparticles and microorganisms, the shelf life of biofertilizers and their delivery. The interaction between gold nanoparticles and plant growth promoting rhizobacteria has been shown to have positive effects (Malusá et al. 2012, Shukla et al. 2015). In contrast, silver nanoparticles cannot be used with biofertilizers because they cause adverse effects on biological processes in microorganisms, such as alterations in cell membrane structure and functions (Duhan et al. 2017). Use of nanoformulations can be helpful to enhance the stability of biofertilizers with respect to desiccation, heat, and UV inactivation. For example, polymeric nanoparticle coatings can be used to develop formulations resistant to desiccation and thereby improve the useful life of these products (Simarmata et al. 2016, Jampílek and Kráľová 2017). In addition, nanomaterials can be used to improve the delivery of biofertilizers to soil and plants.

Effects of NFs on Yield and Quality of Plants

Numerous field and greenhouse studies have reported yield benefits following the use of different NFs and NPs. Foliar application of NPK NFs has been found to improve yields and yield parameters for chickpeas (Drostkar et al. 2016). Tarafdar et al. (2014) reported that zinc nano-fertilizer increased pearl millet (*Pennisetum americanum* L.) grain yield by 37.7%. Singh and Kumar (2017) observed a greater achene yield of sunflower fertilized with ZnO than those with other treatments. Foliar application of ZnO, MgO, and CuO NPs increased cotton yields between 18% and 23% (Nagesh 2019).

The use of NF can increase the quality of agricultural products. For example, Afshar and Rahimi Haghighi (2014) reported high Zn content and seed proteins without yield penalties following Zn NF application. In cowpeas (*Vigna unguiculata*), the use of nano-Fe increased the protein content of the seed by 2% compared to Fe from synthetic fertilizers (Delfani et al. 2014). In fodder maize, Sharifi (2016) found that Zn and Fe NFs foliar application significantly improve crude protein, P, and carbohydrate content, as well as biological yields, compared to crops grown with synthetic fertilizers. In sunflower, Sham (2017) documented that ZnO NPs used as a foliar spray increased both the achene carbohydrate and oil content as compared to other treatments.

Limitation of Nanofertilizers

Although NF and NP technologies have the clear potential to transform the agricultural sector and its production, some of these benefits may come at a higher cost. Other studies have reported the phytotoxic effect of nanoparticles (Ebbs et al. 2016), and the detection, uptake, translocation, transformation and accumulation (phytotoxicity) of NPs in plants depend on species, dose and application method as well as type of NPs (composition, size, shape, surface properties) (Ebbs et al. 2016). Importantly, nanomaterials are highly effective due to their minute size with an enhanced surface area (Konate et al. 2018). Reactivity and variability of these materials are also an apprehension. This raises safety concerns for farm workers who may be exposed to xenobiotics during their application (Nair 2018).

NPs also cause imbalances in nutrients and induce molecular changes in plants, e.g., CuO NPs have been found to affect hormone (e.g., indole-3-acetic acid and abscisic acid) levels in plants (Le Van et al. 2016). Iron based nanomaterials (nFeOx) affect hydraulic conductivity of roots due to the particle aggregation on the root surface, leading to low absorption of water and nutrients such as Ca, K, Mg, and S (Martínez-Fernández et al. 2016).

Lack of production and availability of nano fertilizers at the required price. This limits the acceptance of a wide range of nano-fertilizers as a source of plant nutrients. The high cost of nano fertilizer limits its use in agricultural sector. Nanofertilizers have constraints concerning research gaps, meagerness of recognized formulations and standardization of products, and lack of scrupulous monitoring and risk associated management hampers the development and adoption of nanoparticles as nanofertilizers (Remedios et al. 2012, Iqbal 2019). Deliberate introduction of nanoparticles such as nanofertilizers in agricultural activities can lead to many unintended irretrievable consequences (Kah 2015). Phytotoxicity of nanomaterials is an important problem as different plants react differently to different doses of nanomaterials (Ashkavand et al. 2018). Elongation of roots in cucumber, soybeans, cabbage, maize and carrots is inhibited due to the phytotoxicity of the uncoated nano-Al_2O_3 particles (Yang and Watts 2005).

NFs for Developing Smart Agriculture

In the coming decades, the agricultural sector will face increasing pressures to provide food security to the world's fastest growing population without expanding its overall environmental footprint. Another option to achieve higher biomass and grain yields would be modification of current fertilization techniques. Nutrients such as N, P, K, Ca, Mg, Cu, and Zn are essential for plant growth and reproduction. Although crop yields have grown significantly since the use of chemical fertilizers in the early 1960s, NUE values for these fertilizers are low. The use of smart fertilizers such as NFs has been proposed as a way to increase overall NUE values of fertilizers through more controlled, slower nutrient release that better suits needs of crops across time (DeRosa et al. 2010, Bley et al. 2017). Consistent and slow release of nutrients for a long time can be achieved through semipermeable coatings, which control the dissolution of fertilizers in water or soil solutions in or around the surfaces of fertilizers (Naz and Sulaiman 2016). This will lead to a new fertilizer framework that will deliver the right amount of nutrients at the right time, as well as a dramatic reduction in nutrient loss in the environment. In addition, nanosensors can be linked to NF or NPs to deliver specific nutrients to target areas within living systems. Nanosensors are a promising and powerful tool for agriculture and food production. Nanosensors provide real-time information about field conditions, plant growth, pesticides, and plant diseases, and can help predict environmental stresses (Chen and Yada 2011). Yao et al. (2009) used fluorescent nanoprobes of silica NPs to detect local bacterial infection in Solanaceous plants caused by *Xanthomonas axonopodis*. Similarly, Sharon and Sharon (2008) synthesized carbon nanomaterial-based chemical sensors to detect pesticide residues in plants. Prasad et al. (2014) report that nanobiosensors can monitor glyphosate and glufosinate herbicides in soil using a nanofilm-modified pencil graphite electrode.

The nano-sensing system is transforming conventional agriculture into precision farming system, and real-time monitoring has reduced the excessive use of fertilizers and pesticides, which is beneficial in protecting the environment from pollution (Chhipa and Joshi 2016).

Limitation of using NFs in Terms of Ethical and Safety Issues

There are various safety and ethical issues related to the use of NF or NPs. In this context, unintended health and environmental issues may limit the use of NPs or NFs in agricultural crop production (Diatta et al. 2020). Furthermore, NPs and NFs can enter the food chain, thereby increasing their distribution in non-targeted living organisms. Studies have shown that NPs can alter genetic expression in animals because of their size, allowing them to enter various animal tissues, cells, and organelles and interact with DNA (Xia et al. 2009). NPs affect organisms in different ways, e.g., carbon-based NPs alter DNA structure and gene expression levels in plant tissue (Lahiani et al. 2015, Zuverza-Mena et al. 2017).

NFs such as macronutrient and micronutrient fertilizers are currently used in agriculture (Chhipa 2017), but the supra-optimum application rates may lead to the deposition of nano-based macro and micronutrients and result in nanotoxicity and reduction in water quality (Chhipa 2017). Nanomaterials interact with microorganisms in the soil and alter nutrient absorption in plants. ZnO NPs affect the symbiotic relationships in legumes and delay the nitrogen fixation process (Huang et al. 2014). Mycorrhizal colonization in *Helianthus annuus* L. decreased due to Ag nanoparticles (Dubchak et al. 2010).

Particle size (Yilma et al. 2013), particle shape (Yin et al. 2013), particle surface properties (Braydich-Stolle et al. 2005), biological fluid properties, and formation affects corona formation and thus adverse effects on human health and the environment (Lesniak et al. 2012, Navya and Daima 2016). Nanoparticles by damage to the mitochondrial membrane disrupt the function of the respiratory chain and ATP synthesis, which may produce ROS, leading to oxidative stress and ultimately apoptosis (Asharani et al. 2009, Gurunathan et al. 2015). Many authors have reported that silver ions and AgNPs can interact with various chemical groups, including sulfide and chloride (Andersson 1972, Li and Lenhart 2012). Thiol molecules are found to be attached to several membrane proteins in the cell membrane, cytoplasm and mitochondria, which may serve as targets of AgNPs or Ag^+ ions (Almofti et al. 2003). AgNPs also bind to thiol groups of enzymes, such as NADH dehydrogenase, and disrupt the respiratory chain, eventually producing ROS. AgNPs, especially those smaller than 10 nm, have been shown to diffuse through the nuclear pores into the nucleus, causing DNA damage, chromosomal aberrations, and cell cycle arrest, leading to genotoxicity in human cell lines (e.g., genes. fibroblasts and glioblastoma cells) (Andersson 1972, Li and Lenhart 2012).

Conclusion and Future Perspectives

World's population is estimated to be over 9.7 billion by 2050 (FAO 2018). Therefore, it is expected that current crop production needs to be increased to 70% to meet future food needs (Hunter et al. 2017). This major challenge will require concerted efforts to conserve natural resources to support sustainable agriculture while limiting the negative impact on the environment (Lee et al. 2006, Hunter et al. 2017, Xie et al. 2019). Excessive use of mineral fertilizers and organic amendments has adversely affected soil and water quality worldwide (Bashir et al. 2020). In the last few decades, nanotechnology has been considered a projecting technology with plentiful applications (Marchiol et al. 2020). The application of nanofertilizers in agriculture may serve as an opportunity to achieve sustainable global food production. In the context of sustainable agriculture, the use of nanotechnology for the development of new fertilizers is considered one of the most promising options for significantly improving global crop production to meet the growing food needs of people with the added benefits of sustainability under the current scenario of climate change (Raliya et al. 2017, Feregrino-Pérez et al. 2018). NFs have potential as part of smart crop production systems under a sustainable agricultural framework because they can characteristically release nutrients at a slow and steady pace. These promising features make them ideal for use in modern agriculture. The use of NFs can increase agricultural productivity and resistance to biotic and abiotic stresses. Therefore, the use of NFs in the agricultural sector will not be ignored. The use of NFs may help to reduce the amount of fertilizer by smart delivery of active ingredients, increase the absorption of nutrients and NUE values, and reduce the loss of fertilizer from volatilization, leaching, runoff, and energy used during production.

Furthermore, the use of seed coatings by NF and nanosensors may reduce the cost of agricultural production and environmental issues. In addition, bio-synthesized NFs or nano-biofertilizers should be explored in order to further increase yields in sustainable agriculture.

They also require the perfect timing for the application, as climatic conditions affect efficiency. The cost of many applications can be too high to be profitable, the suspension of nanoformulations, and the lack of uniform size of nanoparticles (Iqbal 2019), and improving foliar use of nanofertilizers are challenges that need to be addressed in future research. More information is needed to understand that nanofertilizers are fully converted to ionic forms in a plant and later incorporated into various proteins and metabolites, or if some of them remain intact and reach consumers through food chains (Iqbal 2019).

Acknowledgements

MR is thankful to the Polish National Agency for Academic Exchange (NAWA) for financial support (Project No. PPN/ULM/2019/1/00117/A/DRAFT/00001) to visit the Department of Microbiology, Nicolaus Copernicus University, Toruń, Poland.

References

Abdel-Aziz, H.M.M., Hasaneen, M.N.A. and Omer, A.M. 2016. Nano chitosan-NPK fertilizer enhances the growth and productivity of wheat plants grown in sandy soil. Span. J. Agric. Res. 14(1): e0902.

Abdel-Aziz, H.M., Hasaneen, M.N. and Omar, A. 2018. Effect of foliar application of nano chitosan NPK fertilizer on the chemical composition of wheat grains. Egypt. J. Bot. 58: 87–95.

Abou El-Nour, K.M.M., Eftaiha, A., Al-Warthan, A. and Ammar, R.A.A. 2010. Synthesis and applications of silver nanoparticles. Arab. J. Chem. 3(3): 135–140. Doi: 10.1016/j.arabjc.2010.04.008.

Adeyemi, O., Keshavarz-Afshar, R., Jahanzad, E., Battaglia, M.L., Luo, Y. and Sadeghpour, A. 2020. Effect of wheat cover crop and split nitrogen application on corn yield and nitrogen use efficiency. Agronomy 10: 1081.

Adhikari, T., Kundu, S., Biswas, A.K., Tarafdar, J.K. and Rao, A.S. 2012. Effect of copper oxide nano particle on seed germination of selected crops. J. Agric. Sci. Technol. 2: 815–823.

Adnan, M., Fahad, S., Zamin, M., Shah, S., Mian, I.A., Danish, S., Zafar-ul-Hye, M., Battaglia, M.L., Naz, R.M.M., Saeed, B., Saud, S., Ahmad, I., Yue, Z., Brtnicky, M., Holatko, J. and Datta, R. 2020. Coupling phosphate-solubilizing bacteria with phosphorus supplements improve maize phosphorus acquisition and growth under lime induced salinity stress. Plants 9: 900.

Afshar, I. and Rahimi Haghighi, A. 2014. Comparison the effects of spraying different amounts of nano zincoxide and zinc oxide on wheat. Int. J. Adv. Biol. Biomed. Res. 2: 318–325.

Ahmad, A., Mukherjee, P., Mandal, D., Senapati, S., Khan, M.I., Kumar, R. and Sastry, M. 2002. Enzyme mediated extracellular synthesis of cds nanoparticles by the fungus, *Fusarium oxysporum*. J. Am. Chem. Soc. 124(41): 12108–12109.

Ahmad, A., Mukherjee, P., Senapati, S., Mandal, D., Khan, M.I., Kumar, R. and Sastry, M. 2003. Extracellular biosynthesis of silver nanoparticles using the fungus *Fusarium oxysporum*. Colloids Surf. B: Biointerfaces 28(4): 313–318.

Ahmed, B., Shahid, M., Khan, M.S. and Musarrat, J. 2018. Chromosomal aberrations, cell suppression and oxidative stress generation induced by metal oxide nanoparticles in onion (*Allium cepa*) bulb, Metallomics 10: 1315–1327.

Almofti, M.R., Ichikawa, T., Yamashita, K., Terada, H. and Shinohara, Y. 2003. Silver ion induces a cyclosporine a-insensitive permeability transition in rat liver mitochondria and release of apoptogenic cytochrome C. J. Biochem. 134(1): 43–49.

Andersson, L.O. 1972. Study of some silver-thiol complexes and polymers: Stoichiometry and optical effects. J. Polym. Sci. A Polym. Chem. 10: 1963–1973.

Arora, N.K. 2018. Agricultural sustainability and food security. Environ. Sustain. 1: 217–219.

Asgari, S., Moradi, H. and Afshari, H. 2018. Evaluation of some physiological and morphological characteristics of narcissus tazatta under BA treatment and nano-potassium fertilizer. J. Chem. Health Risks. 4.

AshaRani, P.V., Mun, G.L.K., Hande, M.P. and Valiyaveettil, S. 2009. Cytotoxicity and genotoxicity of silver nanoparticles in human cells. ACS Nano. 3(2): 279–290.

Ashkavand, P., Kouchaksaraei, M.T., Zarafshar, M., Tomaskova, I. and Struve, D. 2015. Effect of SiO_2 nanoparticles on drought resistance in hawthorn seedlings. For. Res. Pap. 76: 350–359.

Ashkavand, P., Zarafshar, M., Tabari, M., Mirzaie, J., Nikpour, A., Bordbar, S.K., Struve, D. and Striker, G.G. 2018. Application of SiO_2 nanoparticles as pretreatment alleviates the impact of drought on the physiological performance of *Prunus mahaleb* (Rosaceae). Boletín de la Sociedad Argentina de Botánica 53(2): 207–219.

Ashour, H.A. and Mahmoud, A.W.M. 2017. Response of *Jatropha integerrima* plants irrigated with different levels of saline water to nano silicon and gypsum. J. Agric. Stud. 5: 136–160.

Avellan, A., Schwab, F., Masion, A., Chaurand, P., Borschneck, D., Vidal, V., Rose, J., Santaella, C. and Levard, C. 2017. Nanoparticle uptake in plants: Gold nanomaterial localized in roots of *Arabidopsis thaliana* by x-ray computed nanotomography and hyperspectral imaging. Environ. Sci. Technol. 51: 8682–8691.

Avila-Quezada, G.D., Golinska, P. and Rai, M. 2021. Engineered nanomaterials in plant diseases: Can we combat phytopathogens? Appl. Microbiol. Biotechnol. 10.1007/s00253-021-11725-w. Advance online publication. https://doi.org/10.1007/s00253-021-11725-w.

Badran, A. and Savin, I. 2018. Effect of nano-fertilizer on seed germination and first stages of bitter almond seedlings' growth under saline conditions. BioNanoScience 8: 742–751.

Bansal, V., Rautaray, D., Ahmad, A. and Sastry, M. 2004. Biosynthesis of zirconia nanoparticles using the fungus *Fusarium oxysporum*. J. Mater. Chem. 14: 3303–3305.

Basavaraja, S., Balaji, S.D., Lagashetty, A., Rajasab, A.H. and Venkataraman, A. 2008. Extracellular biosynthesis of silver nanoparticles using the fungus *Fusarium semitectum*. Mater. Res. Bull. 43: 1164–1170. Doi: 10.1016/j.materresbull.2007.06.020.

Bashir, I., Lone, F.A., Bhat, R.A., Mir, S.A., Dar, Z.A. and Dar, S.A. 2020. Concerns and threats of contamination on aquatic ecosystems. In Bioremediation and Biotechnology, Sustainable Approaches to Pollution Degradation, Berlin, Germany: Springer. 1–26. Doi: 10.1007/978-3-030-35691-0_1.

Battaglia, M.L., Groover, G. and Thomason, W.E. 2018. Harvesting and nutrient replacement costs associated with corn stover removal in virginia; Virginia Cooperative Extension Publication: Ettrick, VA, USA. CSES-229NP.

Battaglia, M., Lee, C., Thomason, W., Fike, J. and Sadeghpour, A. 2019. Hail damage impacts on corn productivity: A review. Crop. Sci. 59: 1–14.

Behboudi, F., Sarvestani, T., Kassaee, M.Z., Modares Sanavi, S.A.M. and Sorooshzadeh, A. 2018. Improving growth and yield of wheat under drought stress via application of SiO$_2$ nanoparticles. J. Agric. Sci. Technol. 20: 1479–1492.

Bhainsa, K.C. and D'Souza, S.F. 2006. Extracellular biosynthesis of silver nanoparticles using the fungus *Aspergillus fumigatus*. Colloids Surf. B: Biointerfaces 47(2): 160–164. Doi: 10.1016/j.colsurfb.2005.11.026. PMID 16420977.

Bhargava, A., Jain, N., Khan, M.A., Pareek, V., Dilip, R.V. and Panwar, J. 2016. Utilizing metal tolerance potential of soil fungus for efficient synthesis of gold nanoparticles with superior catalytic activity for degradation of rhodamine B. J. Environ. Manag. 183: 22–32.

Birla, S.S., Gaikwad, S.C., Gade, A.K. and Rai, M.K. 2013. Rapid synthesis of silver nanoparticles from *Fusarium oxysporum* by optimizing physicocultural conditions. Sci. World J. 1–12. Article ID 796018.

Bley, H., Gianello, C., Santos, L. da S. and Selau, L.P.R. 2017. Nutrient release, plant nutrition, and potassium leaching from polymer-coated fertilizer. Rev. Brasil Ciênc Solo 41: e0160142.

Braydich-Stolle, L., Hussain, S., Schlager, J.J. and Hofmann, M.C. 2005. *In vitro* cytotoxicity of nanoparticles in mammalian germline stem cells. Toxicol. Sci. 88(2): 412–419.

Broadley, M.R., White, P.J., Hammond, J.P., Zelko, I. and Lux, A. 2007. Zinc in plants. New Phytol. 173(4): 677–702.

Brunner, T.J., Wick, P., Manser, P., Spohn, P., Grass, R.N., Limbach, L.K., Bruinink, A. and Stark, W.J. 2006. *In vitro* cytotoxicity of oxide nanoparticles: Comparison to asbestos, silica, and the effect of particle solubility. Environ. Sci. Technol. 40(14): 4374–4381.

Chen, H. and Yada, R. 2011. Nanotechnologies in agriculture: New tools for sustainable development. Trends Food Sci. Technol. 22(11): 585–594.

Chhipa, H. 2017. Nanofertilizers and nanopesticides for agriculture. Environ. Chem. Lett. 15: 15–22.

Chhipa, H. and Joshi, P. 2016. Nanofertilisers, nanopesticides and nanosensors in agriculture. pp. 247–282. *In*: Ranjan, S., Dasgupta, N. and Lichtfouse, E. (eds.). Nanoscience in Food and Agriculture, vol 1. Sustainable Agriculture Reviews. Springer, Cham.

Chowdappa, P. and Gowda, S. 2013. Nanotechnology in crop protection: Status and scope. Pest. Manag. Hortic. Ecosyst. 19: 131–151.

Corredor, E., Testillano, P.S., Coronado, M.-J., González-Melendi, P., Fernández-Pacheco, R., Marquina, C., Ibarra, M.R., de la Fuente, J.M., Rubiales, D., Pérez-de-Luque, A. and Risueño, M.C. 2009. Nanoparticle penetration and transport in living pumpkin plants: *In situ* subcellular identification. BMC Plant Biol. 9: 45. Doi: 10.1186/1471-2229-9-45.

Czymmek, K., Ketterings, Q., Ros, M., Battaglia, M., Cela, S., Crittenden, S., Gates, D., Walter, T., Latessa, S. and Klaiber, L. 2020. The New York Phosphorus Index 2.0. Agronomy Fact Sheet Series. Fact Sheet #110; Cornell University Cooperative Extension: New York, NY, USA.

Das, C.K., Srivastava, G., Dubey, A., Roy, M., Jain, S., Sethy, N.K., Saxena, M., Harke, S., Sarkar, S. and Misra, K. 2016. Nano-iron pyrite seed dressing: A sustainable intervention to reduce fertilizer consumption in vegetable (beetroot, carrot), spice (fenugreek), fodder (alfalfa), and oilseed (mustard, sesamum) crops. Nanotechnol. Environ. Eng. 1: 1–12.

Davar, Z.F., Roozbahani, A. and Hosnamidi, A. 2014. Evaluation the effect of water stress and foliar application of fe nanoparticles on yield, yield components and oil percentage of safflower (*Carthamus tinctorious* L.). Int. J. Adv. Biol. Biomed. Res. 2: 1150–1159.

Davarpanah, S., Tehranifar, A., Davarynejad, G., Abadía, J. and Khorasani, R. 2016. Effects of foliar applications of zinc and boron nano-fertilizers on pomegranate (*Punica granatum* cv. Ardestani) fruit yield and quality. Sci. Hortic. 210: 57–64.

Delfani, M., Baradarn Firouzabadi, M., Farrokhi, N. and Makarian, H. 2014. Some physiological responses of black-eyed pea to iron and magnesium nanofertilizers. Commun. Soil Sci. Plant. Anal. 45(4): 530–540.

DeRosa, M.C., Monreal, C., Schnitzer, M., Walsh, R. and Sultan, Y. 2010. Nanotechnology in fertilizers. Nat. Nanotechnol. 5(2): 91–91. Doi.org/10.1038/nnano.2010.2.

Diatta, A.A., Thomason, W.E., Abaye, O., Thompson, T.L., Battaglia, M.L., Vaughan, L.J., Lo, M. and Filho, J.F.D.C.L. 2020. Assessment of nitrogen fixation by mungbean genotypes in different soil textures using 15n natural abundance method. J. Soil Sci. Plant Nutr. 20: 2230–2240.

Dimkpa, C.O., McLean, J.E., Latta, D.E., Manangón, E., Britt, D.W., Johnson, W.P., Boyanov, M.I. and Anderson, A.J. 2012. CuO and ZnO nanoparticles: Phytotoxicity, metal speciation, and induction of oxidative stress in sand-grown wheat. J. Nanoparticle Res. 14: 1–15.

Dimkpa, C.O. and Bindraban, P.S. 2016. Fortification of micronutrients for efficient agronomic production: A review. Agron. Sustain. Dev. 36(1). Doi.org/10.1007/s13593-015-0346-6.

Divte, P.R., Shende, S.S., Limbalkar, O.M. and Kale, R.A. 2019. Characterization of biosynthesised copper nanoparticle from *Citrus sinesis* and *in-vitro* evaluation against fungal pathogen *Colletotrichum capsici*. Int. J. Chem. Stud. 7(5): 325–330.

Djanaguiraman, M., Belliraj, N., Bossmann, S.H. and Prasad, P.V.V. 2018. High-temperature stress alleviation by selenium nanoparticle treatment in grain sorghum. ACS Omega 3(3): 2479–2491. Doi.org/10.1021/acsomega.7b01934.

Drostkar, E., Talebi, R. and Kanouni, H. 2016. Foliar application of Fe, Zn and NPK NF on seed yield and morphological traits in chickpea under rainfed condition. J. Resour. Ecol. 4: 221–228.

Du, W., Yang, J., Peng, Q., Liang, X. and Mao, H. 2019. Comparison study of zinc nanoparticles and zinc sulphate on wheat growth: From toxicity and zinc biofortification. Chemosphere 227: 109–116.

Dubchak, S., Ogar, A., Mietelski, J.W. and Turnau, K. 2010. Influence of silver and titanium nanoparticles on arbuscular mycorrhiza colonization and accumulation of radiocaesium in *Helianthus annuus*. Span. J. Agric. Res. 8(S1): S103–S108.

Duhan, J.S., Kumar, R., Kumar, N., Kaur, P., Nehra, K. and Duhan, S. 2017. Nanotechnology: The new perspective in precision agriculture. Biotechnol. Rep. 15: 11–23.

Durán, N., Marcato, P.D., Alves, O.L., de Souza, G.I. and Esposito, E. 2005. Mechanistic aspects of biosynthesis of silver nanoparticles by several *Fusarium oxysporum* strains. J. Nanobiotechnol. 3(1): 8. Doi: 10.1186/1477-3155-3-8.

Dwivedi, S., Saquib, Q., Al-Khedhairy, A.A. and Musarrat, J. 2016. Understanding the role of nanomaterials in agriculture. *In*: Microbial Inoculants in Sustainable Agricultural Productivity. New Delhi: Springer India, 271–288.

Ebbs, S.D., Bradfield, S.J., Kumar, P., White, J.C., Musante, C. and Ma, X. 2016. Accumulation of zinc, copper, or cerium in carrot (*Daucus carota*) exposed to metal oxide nanoparticles and metal ions. Environ. Sci.: Nano. 3(1): 114–126. Doi: 10.1039/C5EN00161G.

Eichert, T., Kurtz, A., Steiner, U. and Goldbach, H.E. 2008. Size exclusion limits and lateral heterogeneity of the stomatal foliar uptake pathway for aqueous solutes and water-suspended nanoparticles. Physiol. Plant 134(1): 151–160.

Eid, M.A.M., Abdel-Salam, A.A., Salem, H.M., Mahrous, S.E., Seleiman, M.F., Alsadon, A.A., Solieman, T.H.I. and Ibrahim, A.A. 2020. Interaction effects of nitrogen source and irrigation regime on tuber quality, yield, and water use efficiency of *Solanum tuberosum* L. Plants 9: 110. Doi.org/10.3390/plants9010110.

Elamawi, R.M., Al-Harbi, R.E. and Hendi, A.A. 2018. Biosynthesis and characterization of silver nanoparticles using *Trichoderma longibrachiatum* and their effect on phytopathogenic fungi. Egypt. J. Biol. Pest Control. 28(28): 1–11. Doi.org/10.1186/s41938-018-0028-1.

El Domany, E.B., Essam, T.M., Ahmed, A.E. and Farghali, A.A. 2018. Biosynthesis physico-chemical optimization of gold nanoparticles as anti-cancer and synergetic antimicrobial activity using *Pleurotus ostreatus* fungus. J. Appl. Pharm. Sci. 8(5): 119–128.

El-Hefnawy, S.F.M. 2020. Nano NPK and growth regulator promoting changes in growth and mitotic index of pea plants under salinity stress. J. Agric. Chem. Biotechnol. 11: 263–269.

Elmer, W.H. and White, J.C. 2016. The use of metallic oxide nanoparticles to enhance growth of tomatoes and eggplants in disease infested soil or soil less medium. Environ. Sci.: Nano. 3(5): 1072–1079. DOI: 10.1039/c6en00146g.

Fageria, V.D. 2001. Nutrient interactions in crop plants. J. Plant Nutr. 24(8): 1269–1290. Doi.org/10.1081/PLN-100106981.

FAO. 2018. The future of food and agriculture – Alternative pathways to 2050. Summary version. Rome, Italy, 60 Licence: CC BY-NC-SA 3.0 IGO. Available at: http://www.fao.org/3/CA1553EN/ca1553en.pdf.

Farhangi-Abriz, S. and Torabian, S. 2018. Nano-silicon alters antioxidant activities of soybean seedlings under salt toxicity. Protoplasma 255: 953–962.

Feregrino-Pérez, A.A., Magaña-López, E., Guzman, C. and Esquivel, K. 2018. A general overview of the benefits and possible negative effects of the nanotechnology in horticulture. Sci. Hortic. 238: 126–137.

Food, F. 2009. Agriculture Organization of the United Nations (2009) How to feed the world 2050; Forum: 12–13 October, Report; FAO: Rome, Italy.

Gade, A., Ingle, A., Whiteley, C. and Rai, M. 2010. Mycogenic metal nanoparticles: Progress and applications. Biotechnol. Lett. 32(5): 593–600.

Gade, A.K., Gaikwad, S.C., Duran, N. and Rai. M.K. 2013. Green synthesis of silver nanoparticles by *Phoma glomerata*. Micron. 59: 52–59.

Gaikwad, S., Birla, S.S., Ingle, A.P., Gade, A.K., Marcato, P.D., Rai, M. and Duran, N. 2013. Screening of different *Fusarium* species to select potential species for the synthesis of silver nanoparticles. J. Braz. Chem. Soc. 24(12): 1974–1982.

Genaidy, E.A.E., Abd-Alhamid, N., Hassan, H.S.A., Hassan, A.M. and Hagagg, L.F. 2020. Effect of foliar application of boron trioxide and zinc oxide nanoparticles on leaves chemical composition, yield and fruit quality of *Olea europaea* L. cv. *Picual*. Bull. Natl. Res. Cent. 44: 106. Doi.org/10.1186/s42269-020-00335-7.

Gerdini, F. 2016. Effect of nano potassium fertilizer on some parchment pumpkin (Cucurbita pepo) morphological and physiological characteristics under drought conditions. Intl. J. Farm Alli Sci. 5: 367–371.

Gericke, M. and Pinches, A. 2006. Microbial production of gold nanoparticles. Gold Bull. 39: 22–28. Doi.org/10.1007/BF03215529.

Ghorbani, H.R., Safekordi, A.A., Attar, H. and Rezayat Sorkhabadi, S.M. 2011. Biological and non-biological methods for silver nanoparticles synthesis. Chem. Biochem. Eng. Q 25(3): 317–326.

Giorgetti, L., Spanò, C., Muccifora, S., Bellani, L., Tassi, E., Bottega, S., Di Gregorio, S., Siracusa, G., Sanità di Toppi, L. and Ruffini Castiglione, M. 2019. An integrated approach to highlight biological responses of *Pisum sativum* root to nano-TiO_2 exposure in a biosolid-amended agricultural soil. Sci. Total Environ. 650: 2705–2716.

Green, J.M. and Beestman, G.B. 2007. Recently patented and commercialized formulation and adjuvant technology. Crop Prot. 26: 320–327.

Guo, H., White, J.C., Wang, Z. and Xing, B. 2018. Nano-enabled fertilizers to control the release and use efficiency of nutrients. Curr. Opin. Environ. Sci. Health 6: 77–83.

Gurunathan, S., Park, J.H., Han, J.W. and Kim, J.H. 2015. Comparative assessment of the apoptotic potential of silver nanoparticles synthesized by *Bacillus tequilensis* and *Calocybe indica* in MDA-MB-231 human breast cancer cells: Targeting p53 for anticancer therapy. Int. J. Nanomed. 10: 4203–4223.

Ha, N.M.C., Nguyen, T.H., Wang, S.L. and Nguyen, A.D. 2018. Preparation of NPK nanofertilizer based on chitosan nanoparticles and its effect on biophysical characteristics and growth of coffee in green house. Res. Chem. Intermed. 45(1): 51–63.

Haghighi, M. and Pessarakli, M. 2013. Influence of silicon and nano-silicon on salinity tolerance of cherry tomatoes (*Solanum lycopersicum* L.) at early growth stage. Sci. Hortic. 161: 111–117.

Haghighi, M., Abolghasemi, R. and da Silva, J.A.T. 2014. Low and high temperature stress affect the growth characteristics of tomato in hydroponic culture with Se and nano-Se amendment. Sci. Hortic. 178: 231–240.

Hojjat, S.S. and Ganjali, A. 2016. The effect of silver nanoparticle on lentil seed germination under drought stress. Int. J. Farm. Allied Sci. 5: 208–212.

Hong, J., Wang, L., Sun, Y., Zhao, L., Niu, G., Tan, W., Rico, C.M., Peralta-Videa, J.R. and Gardea-Torresdey, J.L. 2016. Foliar applied nanoscale and microscale CeO_2 and CuO alter cucumber (*Cucumis sativus*) fruit quality. Sci. Total Environ. 563-564: 904–911. Doi: 10.1016/j.scitotenv.2015.08.029.

Hong, J., Rico, C.M., Zhao, L., Adeleye, A.S., Keller, A.A., Peralta-Videa, J.R. and Gardea-Torresdey, J.L. 2015. Toxic effects of copper-based nanoparticles or compounds to lettuce (*Lactuca sativa*) and alfalfa (*Medicago sativa*). Environ. Sci. Process Impacts 17: 177–185.

Huang, Y.C., Fan, R., Grusak, M.A., Sherrier, J.D. and Huang, C.P. 2014. Effects of nano-ZnO on the agronomically relevant Rhizobium-legume symbiosis. Sci. Total Environ. 497-498: 78–90. Doi: 10.1016/j.scitotenv.2014.07.100.

Hunter, M.C., Smith, R.G., Schipanski, M.E., Atwood, L.W. and Mortensen, D.A. 2017. Agriculture in 2050: Recalibrating targets for sustainable intensification. BioScience 67: 386–391. Doi: 10.1093/biosci/bix010.

Imada, K., Sakai, S., Kajihara, H., Tanaka, S. and Ito, S.I. 2016. Magnesium oxide nanoparticles induce systemic resistance in tomato against bacterial wilt disease. Plant Pathol. 65: 551–560.

Ingle, A., Gade, A., Pierrat, S., Sonnichsen, C. and Rai, M. 2008. Mycosynthesis of silver nanoparticles using the fungus *Fusarium acuminatum* and its activity against some human pathogenic bacteria. Curr. Nanosci. 4(2): 141–144.

Ingle, A., Rai, M., Gade, A and Bawaskar, M. 2009. *Fusarium solani*: A novel biological agent for the extracellular synthesis of silver nanoparticles. J. Nanoparticle Res. 11(8): 2079–2085.

Iqbal, M.A. 2019. Nano-fertilizers for sustainable crop production under changing climate: A global perspective. *In*: Sustainable Crop Production. IntechOpen, 1–12. Doi: 10.5772/intechopen.89089.

Iqbal, M., Raja, N.I., Mashwani, Z.U.R., Hussain, M., Ejaz, M. and Yasmeen, F. 2019. Effect of silver nanoparticles on growth of wheat under heat stress. Iran. J. Sci. Technol. Trans. A Sci. 43: 387–395.

Jaberzadeh, A., Moaveni, P., Tohidi Moghadam, H.R. and Zahedi, H. 2013. Influence of bulk and nanoparticles titanium foliar application on some agronomic traits, seed gluten and starch contents of wheat subjected to water deficit stress. Not. Bot. Horti Agrobot. Cluj-Napoca. 41(1): 201–207.

Jain, N., Bhargava, A., Majumdar, S., Tarafdar, J.C. and Panwar, J. 2011. Extracellular biosynthesis and characterization of silver nanoparticles using *Aspergillus flavus* NJP08: A mechanism perspective. Nanoscale 3(2): 635–641. Doi: 10.1039/c0nr00656d. PMID 21088776.

Jampílek, J. and Králová, K. 2017. Nanomaterials for delivery of nutrients and growth-promoting compounds to plants. pp. 177–226. *In*: Prasad, R., Kumar, M. and Kumar, V. (eds.). Nanotechnology: An Agricultural Paradigm, Springer, Singapore.

Janmohammadi, M., Amanzadeh, T., Sabaghnia, N. and Ion, V. 2016. Effect of nano-silicon foliar application on safflower growth under organic and inorganic fertilizer regimes. Bot. Lith. 22(1): 53–64. Doi: 10.1515/botlit-2016-0005.

Kah, M. 2015. Nanopesticides and nanofertilizers: Emerging contaminants or opportunities for risk mitigation? Front. Chem. 3: 64.

Kah, M., Tufenkji, N. and White, J.C. 2019. Nano-enabled strategies to enhance crop nutrition and protection. Nat. Nanotechnol. 14(6): 532–540. Doi: 10.1038/s41565-019-0439-5.

Kai, H. and Iba, K. 2014. Temperature stress in plants. *In*: eLS. John Wiley & Sons, Ltd: Chichester, Hoboken, NJ, USA. Doi: 10.1002/9780470015902.a0001320.pub2.

Kale, R.A., Shende, S.S., Deshmukh, A.M. and Wghmare, D.H. 2017. Green synthesis of copper nanoparticles by *Citrus sinensis* and evaluation of its role in management of stem rot of tomato. Trends in Biosciences 10(23): 4708–4715.

Kalteh, M., Alipour, Z.T., Ashraf, S., MarashiAliabadi, M. and Falah Nosratabadi, A. 2018. Effect of silica nanoparticles on basil (*Ocimum basilicum*) under salinity stress. Journal J. Chem. Health Risks 4(3): 49–55.

Kerry, R.G., Gouda, S., Das, G., Vishnuprasad, C.N. and Patra, J.K. 2017. Agricultural nanotechnologies: Current applications and future prospects. pp. 3–28. *In*: Jayanta Kumar Patra, Chethala N. Vishnuprasad and Gitishree Das (eds.). Vol. 1 Microbial Biotechnology.

Khan, N.T., Jameel, M. and Jameel, J. 2017. Silver nanoparticles biosynthesis by *Fusarium oxysporum* and determination of its antimicrobial potency. J. Nanomed. Biotherapeutic. Discov. 7 (1): 1–3.

Khanm, H., Vaishnavi, B.A. and Shankar, A.G. 2018. Raise of nano-fertilizer era: Effect of nano scale zinc oxide particles on the germination, growth and yield of tomato (*Solanum lycopersicum*). Int. J. Curr. Microbiol. Appl. Sci. 7(05): 1861–1871. Doi: https://doi.org/10.20546/ijcmas.2018.705.219.

Konate, A., Wang, Y., He, X., Adeel, M., Zhang, P., Ma, Y., Ding, Y., Zhang, J., Yang, J., Kizito, S., Rui, Y. and Zhang, Z. 2018. Comparative effects of nano and bulk-Fe$_3$O$_4$ on the growth of cucumber (*Cucumis sativus*). Ecotoxicol. Environ. Saf. 165: 547–554. Doi: 10.1016/j.ecoenv.2018.09.053.

Kottegoda, N., Sandaruwan, C., Priyadarshana, G., Siriwardhana, A., Rathnayake, U.A., Berugoda Arachchige, D.M., Kumarasinghe, A.R., Dahanayake, D., Karunaratne, V. and Amaratunga, G.A.J. 2017. Urea-hydroxyapatite nanohybrids for slow release of nitrogen. ACS Nano. 11(2): 1214–1221. Doi: 10.1021/acsnano.6b07781.

Kumar, P., Lai, L., Battaglia, M.L., Kumar, S., Owens, V., Fike, J., Galbraith, J., Hong, C.O., Farris, R., Crawford, R., Crawford, J., Hansen, J., Mayton, H. and Viands, D. 2019. Impacts of nitrogen fertilization rate and landscape position on select soil properties in switchgrass field at four sites in the USA. Catena 180: 183–193.

Lahiani, M.H., Chen, J., Irin, F., Puretzky, A.A., Green, M.J. and Khodakovskaya, M. V. 2015. Interaction of carbon nanohorns with plants: Uptake and biological effects. Carbon 81: 607–619. Doi: 10.1016/j.carbon.2014.09.095.

Landa, P., Cyrusova, T., Jerabkova, J., Drabek, O., Vanek, T. and Podlipná, R. 2016. Effect of metal oxides on plant germination: Phytotoxicity of nanoparticles, bulk materials, and metal ions. Water Air Soil Pollut. 227: 448.

LaRue, C., Laurette, J., Herlin-Boime, N., Khodja, H., Fayard, B., Flank, A.M., Brisset, F. and Carriere, M. 2012. Accumulation, translocation and impact of TiO$_2$ nanoparticles in wheat (*Triticum aestivum* spp.): Influence of diameter and crystal phase. Sci. Total Environ. 431: 197–208.

Lee, D.R., Barrett, C.B. and McPeak, J.G. 2006. Policy, technology, and management strategies for achieving sustainable agricultural intensification. Agric. Econ. 34: 123–127. Doi: 10.1111/j.1574-0864.2006.00112.x.

León-Silva, S., Arrieta-Cortes, R., Fernández-Luqueño, F. and López-Valdez, F. 2018. Design and production of nanofertilizers. *In*: Agricultural Nanobiotechnology; Springer Science and Business Media LLC: Cham, Switzerland, 17–31.

Lesniak, A., Fenaroli, F., Monopoli, M.P., Åberg, C., Dawson, K.A. and Salvati, A. 2012. Effects of the presence or absence of a protein corona on silica nanoparticle uptake and impact on cells. ACS Nano. 6 (7): 5845–5857.

le Van, N., Ma, C., Shang, J., Rui, Y., Liu, S. and Xing, B. 2016. Effects of CuO nanoparticles on insecticidal activity and phytotoxicity in conventional and transgenic cotton. Chemosphere 144: 661–670.

Li, J.X., Wee, C.D. and Sohn, B.K. 2010. Growth response of hot pepper applied with ammonium (NH$_4^+$) and potassium (K$^+$)-loaded zeolite. Korean J. Soil Sci. Fert. 43(5): 741–747.

Li, X. and Lenhart, J.J. 2012. Aggregation and dissolution of silver nanoparticles in natural surface water. Environ. Sci. Technol. 46(10): 5378–5386.

Liscano, J.F., Wilson, C.E., Norman, R.J. and Jr. Slaton, N.A. 2000. Zinc availability to rice from seven granular fertilizers. Arkansas Agricultural Experiment Station: Fayetteville, CA, USA. 963.

Liu, R. and Lal, R. 2014. Synthetic apatite nanoparticles as a phosphorus fertilizer for soybean (*Glycine max*). Sci. Rep. 4: 5686. Doi: 10.1038/srep05686.

Liu, R. and Lal, R. 2015. Potentials of engineered nanoparticles as fertilizers for increasing agronomic productions. Sci. Total. Environ. 514: 131–139.

Liu, R., Zhang, H. and Lal, R. 2016. Effects of stabilized nanoparticles of copper, zinc, manganese, and iron oxides in low concentrations on lettuce (*Lactuca sativa*) seed germination: Nanotoxicants or Nanonutrients? Water, Air, and Soil Pollut. 227(1): 1–14.

Lough, T.J. and Lucas, W.J. 2006. Integrative plant biology: Role of phloem long-distance macromolecular trafficking. Annu. Rev. Plant Biol. 57: 203–232.

Lv, J., Christie, P. and Zhang, S. 2019. Uptake, translocation, and transformation of metal-based nanoparticles in plants: Recent advances and methodological challenges. Environ. Sci. Nano. 6(1): 41–59.

Lv, J., Zhang, S., Luo, L., Zhang, J., Yang, K. and Christied, P. 2015. Accumulation, speciation and uptake pathway of ZnO nanoparticles in maize. Environ. Sci. Nano. 2(1): 68–77.

Mahmoud, A.W.M. and Swaefy, H.M. 2020. Comparison between commercial and nano NPK in presence of nano zeolite on sage plant yield and its components under water stress. Agriculture 66(1): 24–39.

Mahmoud, M.A., Al-Sohaibani, S.A., Al-Othman, M.R., Abd El-Aziz, A.R.M. and Eifan, S.A. 2013. Synthesis of extracellular silver nanoparticles using *Fusarium semitectum* (ksu-4) isolated from Saudi Arabia. Dig. J. Nanomat. Biostruct. 8(2): 589–596.

Malusá, E., Sas-Paszt, L. and Ciesielska, J. 2012. Technologies for beneficial microorganisms inocula used as biofertilizers. Sci. World J. 2012: 1–12. Article ID 491206, Doi:10.1100/2012/491206 2012.

Martínez-Fernández, D., Barroso, D. and Komárek, M. 2016. Root water transport of *Helianthus annuus* L. under iron oxide nanoparticle exposure. Environ. Sci. Pollut. Res. 23(2): 1732–1741. Doi: 10.1007/s11356-015-5423-5.

Mastronardi, E., Tsae, P., Zhang, X., Monreal, C. and DeRosa, M.C. 2015. Strategic role of nanotechnology in fertilizers: Potential and limitations. *In*: Nanotechnologies in Food and Agriculture. Cham: Springer International Publishing, 25–67.

Mikhak, A., Sohrabi, A., Kassaee, M.Z. and Feizian, M. 2017. Synthetic nanozeolite/nanohydroxyapatite as a phosphorus fertilizer for German chamomile (*Matricaria chamomilla* L.). Ind. Crops Prod. 95: 444–452.

Millán, G., Agosto, F., Vázquez, M., Botto, L., Lombardi, L. and Juan, L. 2008. Use of clinoptilolite as a carrier for nitrogen fertilizers in soils of the Pampean regions of Argentina. Int. J. Agric. Nat. Resour. 35: 293–302.

Moghaddasi, S., Fotovat, A., Khoshgoftarmanesh, A.H., Karimzadeh, F., Khazaei, H.R. and Khorassani, R. 2017. Bioavailability of coated and uncoated ZnO nanoparticles to cucumber in soil with or without organic matter. Ecotoxicol. Environ. Saf. 144: 543–551.

Monreal, C.M., Derosa, M., Mallubhotla, S.C., Bindraban, P.S. and Dimkpa, C. 2016. Nanotechnologies for increasing the crop use efficiency of fertilizer-micronutrients. Biol. Fertil. Soils 52: 423–437.

Mozafari, A.A., Havas, F. and Ghaderi, N. 2018. Application of iron nanoparticles and salicylic acid in *in vitro* culture of strawberries (Fragaria x ananassa Duch.) to cope with drought stress. Plant Cell, Tissue Organ. Cult. (PCTOC) 132: 511–523.

Mukherjee, P., Ahmad, A., Mandal, D., Senapati, S., Sainkar, S.R., Khan, M.I., Parishcha, R., Ajaykumar, P.V., Alam, M., Kumar, R. and Sastry, M. 2001. Fungus-mediated synthesis of silver nanoparticles and their immobilization in the mycelial matrix: A novel biological approach to nanoparticle synthesis. Nano Lett. 1(10): 515–519. Doi: 10.1021/nl0155274.

Nagesh, A.K.B. 2019. Foliar application of nanofertilizers in agricultural crops—A review. J. Farm. Sci. 32: 239–249.

Nair, P.M.G. 2018. Toxicological impact of carbon nanomaterials on plants. *In*: Nanotechnology, Food Security and Water Treatment. Springer, Cham, 163–183.

Nair, P.M.G. and Chung, I.M. 2014. Impact of copper oxide nanoparticles exposure on *Arabidopsis thaliana* growth, root system development, root lignificaion, and molecular level changes. Environ. Sci. Pollut. Res. 21(22): 12709–12722. Doi: 10.1007/s11356-014-3210-3.

Naveen, H., Kumar, G., Karthik, L. and Roa, B. 2010. Extracellular biosynthesis of silver nanoparticles using the filamentous fungus *Penicillium* sp. Arch. Appl. Sci. Res. 2: 161–167.

Navya, P.N. and Daima, H.K. 2016. Rational engineering of physicochemical properties of nanomaterials for biomedical applications with nanotoxicological perspectives. Nano Converg. 3: 1–14. Doi. org/10.1186/s40580-016-0064-z.

Nawaz, H.R., Solangi, B.A. and Zehra, B. 2011. Preparation of nano zinc oxide and its application in leather as a retanning and antibacterial agent. Can. J. Sci. Ind. Res. 2: 164–170.

Naz, M.Y. and Sulaiman, S.A. 2016. Slow release coating remedy for nitrogen loss from conventional urea: a review. J. Control. Release 225: 109–120.

Ngo, Q.B., Dao, T.H., Nguyen, H.C., Tran, X.T., Van Nguyen, T., Khuu, T.D. and Huynh, T.H. 2014. Effects of nanocrystalline powders (Fe, Co and Cu) on the germination, growth, crop yield and product quality of soybean (Vietnamese species DT-51). Adv. Nat. Sci. Nanosci. Nanotechnol. 5(1): 015016. Doi: 10.1088/2043-6262/5/1/015016.

Noor, S., Shah, Z., Javed, A., Ali, A., Hussain, S.B., Zafar, S., Ali, H. and Muhammad, S.A. 2020. A fungal based synthesis method for copper nanoparticles with the determination of anticancer, antidiabetic and antibacterial activities. J. Microbiol. Methods 174: 105966. Doi: 10.1016/j.mimet.2020.105966.

Olkhovych, O., Volkogon, M., Taran, N., Batsmanova, L. and Kravchenko, I. 2016. The effect of copper and zinc nanoparticles on the growth parameters, contents of ascorbic acid, and qualitative composition of amino acids and acylcarnitines in *Pistia stratiotes* L. (Araceae). Nanoscale Res. Lett. 11: 218. Doi.org/10.1186/s11671-016-1422-9.

Palmqvist, N.G.M., Seisenbaeva, G.A., Svedlindh, P. and Kessler, V.G. 2017. Maghemite nanoparticles acts as nanozymes, improving growth and abiotic stress tolerance in *Brassica napus*. Nanoscale Res. Lett. 12: 1–9. Doi.org/10.1186/s11671-017-2404-2.

Patra, P., Choudhury, S.R., Mandal, S., Basu, A., Goswami, A., Gogoi, R., Srivastava, C., Kumar, R. and Gopal, M. 2013. Effect sulfur and ZnO nanoparticles on stress physiology and plant (*Vigna radiata*) nutrition. *In*: Advanced Nanomaterials and Nanotechnology. Springer: Guwahati, India, 301–309.

Pestovsky, Y.S. and Martínez-Antonio, A. 2017. The use of nanoparticles and nanoformulations in agriculture. J. Nanosci. Nanotechnol. 17: 8699–8730.

Pitambara, Archana and Shukla, Y.M. 2019. Nanofertilizers: A recent approach in crop production. pp. 25–28. *In*: Panpatte, D.G. and Jhala, Y.K. (eds.). Nanotechnology for Agriculture: Crop Production and Protection. Doi.org/10.1007/978-981-32-9374-8-2.

Pollard, M., Beisson, F., Li, Y. and Ohlrogge, J.B. 2008. Building lipid barriers: Biosynthesis of cutin and suberin. Trends Plant Sci. 13(5): 236–246.

Potbhare, A.K., Chouke, P.B., Mondal, A., Thakare, R.U., Mondal, S., Chaudhary, R.G. and Rai, A.R. 2020. *Rhizoctonia solani* assisted biosynthesis of silver nanoparticles for antibacterial assay. Materials Today: Proceedings 29: 939–945.

Pradhan, S., Patra, P., Das, S., Chandra, S., Mitra, S., Dey, K.K., Akbar, S., Palit, P. and Goswami, A. 2013. Photochemical modulation of biosafe manganese nanoparticles on *Vigna radiata*: A detailed molecular, biochemical, and biophysical study. Environ. Sci. Technol. 47(22): 13122–13131.

Prasad, B.B., Jauhari, D. and Tiwari, M.P. 2014. Doubly imprinted polymer nanofilm-modified electrochemical sensor for ultra-trace simultaneous analysis of glyphosate and glufosinate. Biosens. Bioelectron. 59: 81–88.

Prasad, R., Bhattacharyya, A. and Nguyen, Q.D. 2017. Nanotechnology in sustainable agriculture: Recent developments, challenges, and perspectives. Front. Microbiol. 8: 1014. Doi: 10.3389/fmicb.2017.01014.

Preetha, P.S. and Balakrishnan, N. 2017. A review of nano fertilizers and their use and functions in soil. Int. J. Curr. Microbiol. Appl. Sci. 6(12): 3117–3133.

Qureshi, A., Singh, D.K. and Dwivedi, S. 2018. Nano-fertilizers: A novel way for enhancing nutrient use efficiency and crop productivity. Int. J. Curr. Microbiol. App. Sci. 7(2): 3325–3335.

Rafiullah, Tariq, M., Khan, F., Shah, A.H., Fahad, S., Wahid, F., Ali, J., Adnan, M., Ahmad, M., Irfan, M., Zafar-ul-Hye, M., Battaglia, M.L., Zarei, T., Datta, R., Saleem, I.A., Hafeez-u-Rehman and Danish, S. 2020. Effect of micronutrients foliar supplementation on the production and eminence of plum (*Prunus domestica* L.). Qual. Assur. Saf. Crop Foods 12(SP1): 32–40.

Rai, M., Bonde, S., Golinska, P., Trzcińska-Wencel, J., Gade, A., Abd-Elsalam, K.A., Shende, S., Gaikwad, S. and Ingle, A.P. 2021. *Fusarium* as a novel fungus for the synthesis of nanoparticles: Mechanism and applications. J. Fungi. 7(2): 139. Doi.org/10.3390/jof7020139.

Rai, M., Gupta, I., Bonde, S., Ingle, P., Shende, S., Gaikwad, S., Razzaghi-Abyaneh, M. and Gade, A. 2022. Industrial Applications of Nanomaterials Produced from *Aspergillus* species [Online First], IntechOpen, Doi: 10.5772/intechopen.98780. Available from: https://www.intechopen.com/online-first/80232.

Rai, M., Ingle, A.P., Pandit, R., Paralikar, P., Shende, S., Gupta, I., Biswas, J.K. and da Silva, S.S. 2018. Copper and copper nanoparticles: Role in management of insect-pests and pathogenic microbes. Nanotechnol. Rev. 7(4): 303–315.

Rai, M., Yadav, A. and Gade, A. 2009a. Silver nanoparticles as a new generation of antimicrobials. Biotechnol. Adv. 27(1): 76–83. Doi: 10.1016/j.biotechadv.2008.09.002.

Rai, M., Yadav, A., Bridge, P. and Gade, A. 2009b. Myconanotechnology: A new and emerging science. pp. 258–267. *In*: Rai, M.K. and Bridge, P.D. (eds.). Applied Mycology; CABI: Wallingford, UK.

Rajput, S., Werezuk, R., Lange, R.M. and Mcdermott, M.T. 2016. Fungal isolate optimized for biogenesis of silver nanoparticles with enhanced colloidal stability. Langmuir 32(34): 8688–8697. Doi: 10.1021/acs.langmuir.6b01813.

Raju, D., Mehta, U.J. and Beedu, S.R. 2016. Biogenic green synthesis of monodispersed gum kondagogu (*Cochlospermum gossypium*) iron nanocomposite material and its application in germination and growth of mung bean (*Vigna radiata*) as a plant model. IET Nanobiotechnol. 10(3): 141–146.

Raliya, R. and Tarafdar, J.C. 2013. ZnO nanoparticle biosynthesis and its effect on phosphorous-mobilizing enzyme secretion and gum contents in clusterbean (*Cyamopsis tetragonoloba* L.). Agric. Res. 2: 48–57.

Raliya, R., Tarafdar, J.C. and Biswas, P. 2016. Enhancing the mobilization of native phosphorus in the mung bean rhizosphere using ZnO nanoparticles synthesized by soil fungi. J. Agric. Food Chem. 64: 3111–3118.

Raliya, R., Saharan, V., Dimkpa, C. and Biswas, P. 2017. Nanofertilizer for precision and sustainable agriculture: Current state and future perspectives. J. Agric. Food Chem. 66: 6487–503.

Ramyadevi, J., Jeyasubramanian, K., Marikani, A., Rajakumar, G. and Rahuman, A.A. 2012. Synthesis and antimicrobial activity of copper nanoparticles. Mater. Lett. 71: 114–116.

Rasool, S., Hameed, A., Azooz, M.M., Muneeb-U-Rehman, Siddiqi, T.O. and Ahmad, P. 2013. Salt stress: Causes, types and responses of plants. *In*: Ecophysiology and responses of plants under salt stress. Springer: New York, NY, USA, 1–24.

Rastogi, A., Tripathi, D.K., Yadav, S., Chauhan, D.K., Živčák, M., Ghorbanpour, M., El-Sheery, N.I. and Brestic, M. 2019. Application of silicon nanoparticles in agriculture. 3 Biotech 9(3): 1–11.

Rathnayaka, R., Iqbal, Y. and Rifnas, L. 2018. Influence of urea and nano-nitrogen fertilizers on the growth and yield of rice (*Oryza sativa* L.) Cultivar 'Bg 250'. Biol. Life Sci. 5: 7–17.

Remedios, C., Rosario, F. and Bastos, V. 2012. Environmental nanoparticles interactions with plants: Morphological, physiological and genotoxic aspects. J. Bot. 8: 1–8.

Reyes, L.R., Gómez, I. and Garza, M.T. 2009. Biosynthesis of cadmium sulfide nanoparticles by the Fungi Fusarium sp. Int. J. Green Nanotechnol. Biomed. 1(1): B90–B95.

Rico, C.M., Majumdar, S., Duarte-Gardea, M., Peralta-Videa, J.R. and Gardea-Torresdey, J.L. 2011. Interaction of nanoparticles with edible plants and their possible implications in the food chain. J. Agric. Food Chem. 59: 3485–3498.

Roppolo, D., de Rybel, B., Tendon, V.D., Pfister, A., Alassimone, J., Vermeer, J.E.M., Yamazaki, M., Stierhof, Y.D., Beeckman, T. and Geldner, N. 2011. A novel protein family mediates Casparian strip formation in the endodermis. Nature 473(7347): 381–384. Doi: 10.1038/nature10070.

Rui, M., Ma, C., Hao, Y., Guo, J., Rui, Y., Tang, X., Zhao, Q., Fan, X., Zhang, Z., Hou, T. and Zhu, S. 2016. Iron oxide nanoparticles as a potential iron fertilizer for peanut (*Arachis hypogaea*). Front. Plant Sci. 7: 815. Doi: 10.3389/fpls.2016.00815.

Sabet, H. and Mortazaeinezhad, F. 2018. Yield, growth and Fe uptake of cumin (*Cuminum cyminum* L.) affected by Fe-nano, Fe-chelated and Fe-siderophore fertilization in the calcareous soils. J. Trace Elements Med. Biol. 50: 154–160.

Sastry, M., Ahmad, A., Islam Khan, M. and Kumar, R. 2003. Biosynthesis of metal nanoparticles using fungi and actinomycete. Curr. Sci. 85(2): 162–170.

Savvas, D., Giotis, D., Chatzieustratiou, E., Bakea, M. and Patakioutas, G. 2009. Silicon supply in soilless cultivations of zucchini alleviates stress induced by salinity and powdery mildew infections. Environ. Exp. Bot. 65(1): 11–17. Doi: 10.1016/j.envexpbot.2008.07.004.

Sedghi, M., Hadi, M. and Toluie, S.G. 2013. Effect of nano zinc oxide on the germination parameters of soybean seeds under drought stress. Ann. West. Univ. Timisoara Ser. Biol. 16: 73.

Seleiman, M.F. and Abdelaal, M.S. 2018. Effect of organic, inorganic and bio-fertilization on growth, yield and quality traits of some chickpea (*Cicer arietinum* L.) varieties. Egypt. J. Agron. 40(1): 105–117.

Seleiman, M.F. and Kheir, A.M.S. 2018. Maize productivity, heavy metals uptake and their availability in contaminated clay and sandy alkaline soils as affected by inorganic and organic amendments. Chemosphere 204: 514–522. Doi: 10.1016/j.chemosphere.2018.04.073. Epub 2018 Apr 15. PMID: 29679872.

Seleiman, M.F., Santanen, A. and Mäkelä, P.S.A. 2020a. Recycling sludge on cropland as fertilizer—Advantages and risks. Resour. Conserv. Recycl. 155: 104647.

Seleiman, M.F., Ali, S., Refay, Y., Rizwan, M., Alhammad, B.A. and El-Hendawy, S.E. 2020b. Chromium resistant microbes and melatonin reduced Cr uptake and toxicity, improved physio-biochemical traits and yield of wheat in contaminated soil. Chemosphere 250: 126239. Doi: 10.1016/j. chemosphere.2020.126239.

Seleiman, M.F., Alotaibi, M.A., Alhammad, B.A., Alharbi, B.M., Refay, Y. and Badawy, S.A. 2020c. Effects of ZnO nanoparticles and biochar of rice straw and cow manure on characteristics of contaminated soil and sunflower productivity, oil quality, and heavy metals uptake. Agronomy 10(6): 790. Doi.org/10.3390/agronomy10060790.

Seleiman, M.F., Refay, Y., Al-Suhaibani, N., Al-Ashkar, I., El-Hendawy, S. and Hafez, E.M. 2019. Integrative effects of rice-straw biochar and silicon on oil and seed quality, yield and physiological traits of *Helianthus annuus* L. grown under water deficit stress. Agronomy 9(10): 637. Doi. org/10.3390/agronomy9100637.

Seleiman, M.F., Santanen, A., Jaakkola, S., Ekholm, P., Hartikainen, H., Stoddard, F.L. and Mäkelä, P.S.A. 2013. Biomass yield and quality of bioenergy crops grown with synthetic and organic fertilizers. Biomass-Bioenergy 59: 477–485.

Seleiman, M.F., Almutairi, K.F., Alotaibi, M., Shami, A., Alhammad, B.A. and Battaglia, M.L. 2021. Nano-fertilization as an emerging fertilization technique: Why can modern agriculture benefit from its use? Plants 10: 2. Doi.org/10.3390/plants10010002.

Seleiman, M.F., Santanen, A., Stoddard, F.L. and Mäkelä, P. 2012. Feedstock quality and growth of bioenergy crops fertilized with sewage sludge. Chemosphere 89(10): 1211–1217. Doi: 10.1016/j. chemosphere.2012.07.031.

Shalaby, T.A., Bayoumi, Y., Abdalla, N., Taha, H., Alshaal, T., Shehata, S., Amer, M., Domokos-Szabolcsy, É. and El-Ramady, H. 2016. Nanoparticles, soils, plants and sustainable agriculture. pp. 283–312. *In*: Ranjan, S., Dasgupta, N. and Lichtfouse, E. (eds.). Nanoscience in Food and Agriculture 1, Sustainable Agriculture Reviews 20© Springer International Publishing Switzerland. Doi: 10.1007/978-3-319-39303-2-10.

Shang, Y., Kamrul Hasan, M., Ahammed, G.J., Li, M., Yin, H. and Zhou, J. 2019. Applications of nanotechnology in plant growth and crop protection: A review. Molecules 24(14): 2558.

Sham, S. 2017. Effect of foliar application of nano zinc particles on growth, yield and qualities of sunflower (*Helianthus annuus* L.). Master's Thesis, University of Agricultural Sciences, Dharwad, Hubli, Karnataka, India.

Sharifi, R., Mohammadi, K. and Rokhzadi, A. 2016. Effect of seed priming and foliar application with micronutrients on quality of forage corn (*Zea mays*). Environ. Exp. Biology 14: 151–156.

Sharon, M. and Sharon, M. 2008. Carbon nanomaterials: Applications in physico-chemical systems and biosystems. Def. Sci. J. 58(4): 460–485.

Shenoy, V.V. and Kalagudi, G.M. 2005. Enhancing plant phosphorus use efficiency for sustainable cropping. Biotechnol. Adv. 23(7-8): 501–513.

Shukla, S.K., Kumar, R., Mishra, R.K., Pandey, A., Pathak, A., Zaidi, M., Srivastava, S.K. and Dikshit, A. 2015. Prediction and validation of gold nanoparticles (GNPs) on plant growth promoting rhizobacteria (PGPR): A step toward development of nano-biofertilizers. Nanotechnol. Rev. 4(5): 439–448. Doi.org/10.1515/ntrev-2015-0036.

Siddiqi, K.S. and Husen, A. 2016. Fabrication of metal nanoparticles from fungi and metal salts: Scope and application. Nanoscale Res. Lett. 11: 98. Doi.org/10.1186/s11671-016-1311-2.

Siddiqi, K.S. and Husen, A. 2017. Plant response to engineered metal oxide nanoparticles. Nanoscale Res. Lett. 12(92): 14–18. Doi: 10.1186/s11671-017-1861-y.

Simarmata, T., Hersanti Turmuktini, T., Fitriatin, B.N., Setiawati, M.R. and Purwanto. 2016. Application of bioameliorant and biofertilizers to increase the soil health and rice productivity. Hayati J. Biosci. 23(4): 181–184.

Singh, M.D. and Kumar, B.A. 2017. Bio efficacy of nano zinc sulphide (ZnS) on growth and yield of sunflower (*Helianthus annuus* L.) and nutrient status in the soil. Int. J. Agric. Sci. 9(6): 3795–3798.

Singh, S. and Husen, A. 2019. Role of nanomaterials in the mitigation of abiotic stress in plants. Nanomater. Plant. Potent. 441–471.

Sohair, E.E.D., Abdall, A.A., Amany, A.M., Hossain, M.F. and Houda, R.A. 2018. Effect of nitrogen, phosphorus and potassium nano fertilizers with different application times, methods and rates on some growth parameters of Egyptian cotton (*Gossypium barbadense* L.). Biosci. Res. 15: 549–564.

Soliman, A.S., Hassan, M., Abou-Elella, F., Hanafy Ahmed, A.H. and El-Feky, S.A. 2016. Effect of nano and molecular phosphorus fertilizers on growth and chemical composition of baobab (*Adansonia digitata* L.). J. Plant Sci. 11(4): 52–60.

Sotelo-Boyás, M.E., Bautista-Baños, S., Correa-Pacheco, Z.N., Jiménez-Aparicio, A. and Sivakumar, D. 2016. Biological activity of chitosan nanoparticles against pathogenic fungi and bacteria. *In*:

Chitosan in the Preservation of Agricultural Commodities. Elsevier Inc., 339–349. Doi.org/10.1016/B978-0-12-802735-6.00013-6.

Srivastava, G., Das, C.K., Das, A., Singh, S.K., Roy, M., Kim, H., Sethy, N., Kumar, A., Sharma, R.K., Singh, S.K., Philip, D. and Das, M. 2014. Seed treatment with iron pyrite (FeS$_2$) nanoparticles increases the production of spinach. RSC Adv. 4(102): 58495–58504.

Srivastava, S., Bhargava, A., Pathak, N. and Srivastava, P. 2019. Production, characterization and antibacterial activity of silver nanoparticles produced by *Fusarium oxysporum* and monitoring of protein-ligand interaction through in-silico approaches. Microb. Pathog. 129: 136–145.

Suriyaprabha, R., Karunakaran, G., Yuvakkumar, R., Rajendran, V. and Kannan, N. 2012. Silica nanoparticles for increased silica availability in maize (*Zea mays* L.) seeds under hydroponic conditions. Curr. Nanosci. 8: 902–908.

Taha, R.S., Seleiman, M.F., Alotaibi, M., Alhammad, B.A., Rady, M.M. and Mahdi, A.H.A. 2020. Exogenous potassium treatments elevate salt tolerance and performances of *glycine max* L. by boosting antioxidant defense system under actual saline field conditions. Agronomy 10(11): 1741. Doi.org/10.3390/agronomy10111741.

Taherian, M., Bostani, A. and Omidi, H. 2019. Boron and pigment content in alfalfa affected by nano fertilization under calcareous conditions. J. Trace Elements Med. Biol. 53: 136–143.

Tantawy, M., M Salama, Y.A., Abdel-Mawgoud, A. and Ghoname, A. 2014. Comparison of chelated calcium with nano calcium on alleviation of salinity negative effects on tomato plants. Middle East. J. Agric. Res. 3(4): 912–916.

Tantawy, A.S., Salama, Y.A.M., El-Nemr, M.A. and Abdel-Mawgoud, A.M.R. 2015. Nano silicon application improves salinity tolerance of sweet pepper plants. Int. J. Chem. Tech. Res. 8: 11–17.

Tarafdar, J.C., Raliya, R., Mahawar, H. and Rathore, I. 2014. Development of zinc nanofertilizer to enhance crop production in pearl millet (*Pennisetum americanum*). Agric. Res. 3: 257–262. Doi. org/10.1007/s40003-014-0113-y.

Taskın, M.B., Sahin, Ö., Taskin, H., Atakol, O., Inal, A. and Gunes, A. 2018. Effect of synthetic nano-hydroxyapatite as an alternative phosphorus source on growth and phosphorus nutrition of lettuce (*Lactuca sativa* L.) plant. J. Plant. Nutr. 41: 1148–1154.

Thunugunta, T., Reddy, A.C., Seetharamaiah, S.K., Hunashikatti, L.R., Chandrappa, S.G., Kalathil, N.C. and Reddy, L.R.D.C. 2018. Impact of zinc oxide nanoparticles on eggplant (*S. melongena*): Studies on growth and the accumulation of nanoparticles. IET Nanobiotechnol. 12: 706–713.

Torabian, S., Zahedi, M. and Khoshgoftar, A.H. 2015. Effects of foliar spray of two kinds of zinc oxide on the growth and ion concentration of sunflower cultivars under salt stress. J. Plant. Nutr. 39: 172–180.

Trujillo-Reyes, J., Majumdar, S., Botez, C., Peralta-Videa, J. and Gardea-Torresdey, J.L. 2014. Exposure studies of core–shell Fe/Fe$_3$O$_4$ and Cu/CuO NPs to lettuce (*Lactuca sativa*) plants: Are they a potential physiological and nutritional hazard? J. Hazard. Mater. 267: 255–263.

Usman, M., Farooq, M., Wakeel, A., Nawaz, A., Cheema, S.A., Rehman, H.U., Ashraf, I. and Sanaullah, M. 2020. Nanotechnology in agriculture: Current status, challenges and future opportunities. Sci. Total Environ. 721: 137778.

Vahabi, K., Mansoori, G.A. and Karimi, S. 2011. Biosynthesis of silver nanoparticles by fungus *Trichoderma reesei* (a route for large-scale production of AgNPs). Insciences J. 1: 65–79. Doi: 10.5640/insc.010165.

Vigneshwaran, N., Kathe, A.A., Varadarajan, P.V., Nachane, R.P. and Balasubramanya, R.H. 2006. Biomimetics of silver nanoparticles by white rot fungus, *Phaenerochaete chrysosporium*. Colloids Surf. B Biointerfaces 53(1): 55–59. Doi: 10.1016/j.colsurfb.2006.07.014.

Wahid, A., Gelani, S., Ashraf, M. and Foolad, M.R. 2007. Heat tolerance in plants: An overview. Environ. Exp. Bot. 61(3): 199–223.

Wang, W.N., Tarafdar, J.C. and Biswas, P. 2013. Nanoparticle synthesis and delivery by an aerosol route for watermelon plant foliar uptake. J. Nanoparticle Res. 15(1): 1417. Doi: 10.1007/s11051-013-1417-8.

Wang, Y., Lin, Y., Xu, Y., Yin, Y., Guo, H. and Du, W. 2019. Divergence in response of lettuce (var. ramosa Hort.) to copper oxide nanoparticles/microparticles as potential agricultural fertilizer. Environ. Pollut. Bioavailab. 31: 80–84.

Wiesner, M.R., Lowry, G.V., Jones, K.L., Hochella, J.M.F., Di Giulio, R.T. Casman, E. and Bernhardt, E.S. 2009. Decreasing uncertainties in assessing environmental exposure, risk, and ecological implications of nanomaterials. Environ. Sci. Technol. 43: 6458–6462.

Xia, T., Li, N. and Nel, A.E. 2009. Potential health impact of nanoparticles. Annu. Rev. Public Heal. 30: 137–150.

Xie, H., Huang, Y., Chen, Q., Zhang, Y. and Wu, Q. 2019. Prospects for agricultural sustainable intensification: A review of research. Land 8: 157. Doi: 10.3390/land8110157.

Yadav, A., Kon, K., Kratosova, G., Duran, N., Ingle, A.P. and Rai, M. 2015. Fungi as an efficient mycosystem for the synthesis of metal nanoparticles: Progress and key aspects of research. Biotechnol. Lett. 37: 2099–2120.

Yang, L. and Watts, D.J. 2005. Particle surface characteristics may play an important role in phytotoxicity of alumina nanoparticles. Toxicol. Lett. 158(2): 122–132.

Yao, K.S., Li, S.J., Tzeng, K.C., Cheng, T.C., Chang, C.Y., Chiu, C.Y., Liao, C.Y., Hsu, J.J. and Lin, Z.P. 2009. Fluorescence silica nanoprobe as a biomarker for rapid detection of plant pathogens. Adv. Mater. Res. 79-82: 513–516. Doi: 10.4028/www.scientific.net/AMR.79-82.513.

Yilma, A.N., Singh, S.R., Dixit, S. and Dennis, V.A. 2013. Anti-inflammatory effects of silver-polyvinyl pyrrolidone (Ag-PVP) nanoparticles in mouse macrophages infected with live Chlamydia trachomatis. Int. J. Nanomedicine 8: 2421–32. Doi: 10.2147/IJN.S44090. Epub 2013 Jul 8. PMID: 23882139; PMCID: PMC3709643.

Yin, N., Liu, Q., Liu, J., He, B., Cui, L., Li, Z., Yun, Z., Qu, G., Liu, S., Zhou, Q. and Jiang, G. 2013. Silver nanoparticle exposure attenuates the viability of rat cerebellum granule cells through apoptosis coupled to oxidative stress. Small 9(9-10): 1831–1841. Doi: 10.1002/smll.201202732.

Younes, N.A. and Nassef, D.M.T. 2016. Effect of silver nanoparticles on salt tolerancy of tomato trans-plants (*Solanum lycopersicom* L. Mill.). Assiut. J. Agric. Sci. 46: 76–85.

Zahra, Z., Arshad, M., Rafique, R., Mahmood, A., Habib, A., Qazi, I.A. and Khan, S.A. 2015. Metallic nanoparticle (TiO$_2$ and Fe$_3$O$_4$) application modifies rhizosphere phosphorus availability and uptake by *Lactuca sativa*. J. Agric. Food Chem. 63(31): 6876–6882. Doi: 10.1021/acs.jafc.5b01611.

Zareabyaneh, H. and Bayatvarkeshi, M. 2015. Effects of slow-release fertilizers on nitrate leaching, its distribution in soil profile, N-use efficiency, and yield in potato crop. Environ. Earth Sci. 74: 3385–3393. Doi: 10.1007/s12665-015-4374-y.

Zhao, L., Peralta-Videa, J.R., Rico, C.M., Hernandez-Viezcas, J.A., Sun, Y., Niu, G., Servin, A., Nunez, J.E., Duarte-Gardea, M. and Gardea-Torresdey, J.L. 2014. CeO$_2$ and ZnO nanoparticles change the nutritional qualities of cucumber (*Cucumis sativus*). J. Agri. Food Chem. 62: 2752–2759.

Zuverza-Mena, N., Medina-Velo, I.A., Barrios, A.C., Tan, W., Peralta-Videa, J.R. and Gardea-Torresdey, J.L. 2015. Copper nanoparticles/compounds impact agronomic and physiological parameters in cilantro (*Coriandrum sativum*). Environ. Sci. Process. Impacts 17: 1783–1793.

Zuverza-Mena, N., Martínez-Fernández, D., Du, W., Hernandez-Viezcas, J.A., Bonilla-Bird, N., López-Moreno, M.L., Komárek, M., Peralta-Videa, J.R. and Gardea-Torresdey, J.L. 2017. Exposure of engineered nanomaterials to plants: Insights into the physiological and biochemical responses—A review. Plant Physiol. Biochem. 110: 236–264.

10

Myconanotechnology in Food Preservation and Enhancement of Shelf-life of Agri-food and Fruits

Avinash P. Ingle,[1,*] *Pramod Ingle,*[2] *Magdalena Wypij,*[3]
Patrycja Golinska[3] and *Mahendra Rai*[2,3]

Introduction

The continuous increase in the global population has led to a significant rise in the demand for food. Therefore, providing fresh, healthy and nutritious food for every human being has become a great challenge (Calicioglu et al. 2019, Valoppi et al. 2021). Although due to recent advancements in the field of agriculture, enhanced food production becomes possible, which is sufficient to feed most of the world's population, but usually the available technology, food supply chains and infrastructure fail to provide the produced food to those in need. Hence, food wastage has become a key challenge in all food processing sectors. According to the report of Rethink Food Waste through Economics and Data (ReFED), different food products have different demands and their wastage percentages also vary depending on the food products. Fruits and vegetables (45%) are among the least expensive and fastest spoiling food items followed by fish and seafood (35%), cereals (30%), dairy products (20%), meat and poultry (20%), respectively. Figure 1 represents a schematic illustration of food wastage distribution for various types of food materials prepared from data obtained from ReFED (2016). Moreover, according to one of the analysis conducted

---

[1] Biotechnology Centre, Dr. Panjabrao Deshmukh Krishi Vidyapeeth, Akola, Maharashtra, India.
[2] Nanobiotechnology Lab., Department of Biotechnology, Sant Gadge Amravati University, Amravati, Maharashtra, India.
[3] Department of Microbiology, Nicolaus Copernicus University, 87-100 Torun, Poland.
* Corresponding author: ingleavinash14@gmail.com

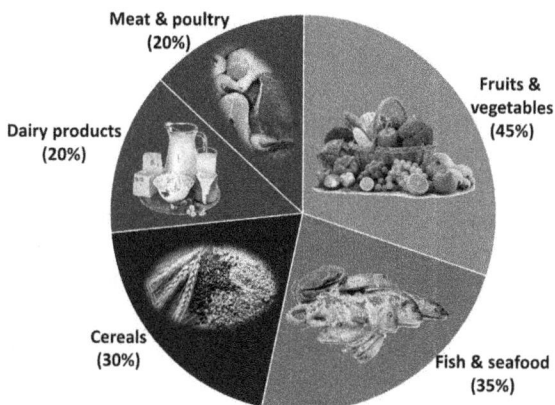

Figure 1. Food wastage distribution for various types of food materials.

in Finland in 2019 more than 50% of the food waste is from households (Filimonau and De Coteau 2019).

Moreover, it has also been observed that a significant amount of food gets wasted at various stages of food production and consumption chain. The amount of food wastage during these stages is also different. According to the report of Griffin et al. (2009), out of the total food waste generated, 20% food waste is associated with the production stage, 1% is generated during food processing, 19% waste is generated while its distribution, and 60% food gets wasted at the consumer level. However, the important reasons behind food wastage mainly include pre- and post-harvest losses in crops, shrinkage of food while cooking, manufacturing issues, supply chain barriers, high consumer standards, changing climatic conditions, soil runoffs, and policy constraints (Sridhar et al. 2021).

It is well-known that harvested food materials like harvested fruits and vegetables cannot be stored in natural conditions using conventional preservation methods for a longer period because of their perishable nature (Amit et al. 2017). To date, several conventional preservation methods such as thermal treatment, freezing, chilling, pasteurization, chemical preservation, etc., are being commonly used but these methods have certain limitations like higher cost, unsatisfactory shelf-life, and/or undesirable residue, etc. Therefore, there is an urgent need to develop most effective, economically viable, and environmentally friendly alternative approaches. In this context, it is believed that nanotechnology-based approaches have the potential to compensate for the shortcomings associated with the existing conventional preservation methods because of the unique and extraordinary properties of nanomaterials. Hence, nanotechnology-derived methods can be promisingly used in food preservation and enhancement of shelf-life of agri-food and fruits (Rai et al. 2019, Liu et al. 2020).

Nanotechnology is considered to be the rapidly emerging branch of science, usually dealing with the creation and manipulation of materials at the smallest scales, i.e., 1–100 nm (nanomaterials). Nanomaterials are the building blocks of nanotechnology and found to possess numerous applications in diverse sectors including food preservation so as to enhance the shelf-life of food materials like fruits, vegetables, etc. (Ashfaq et al. 2022, Couto and Almeida 2022). These nanomaterials or nanoparticles are generally synthesized using different physical, chemical, and biological methods. The physical and chemical methods are found to have some disadvantages like the involvement of radiation, hazardous chemicals, extremely high temperatures, and pressure, etc. (Ali et al. 2021, Chugh et al. 2021). However, the biological synthesis of nanoparticles is considered to be a greener and more promising approach. A variety of biological agents like plants, fungi, bacteria, actinomycetes, alga, etc., were successfully exploited for the synthesis of different metal nanoparticles (Pandit et al. 2022). Of these, the synthesis of nanoparticles using fungi (i.e., mycosynthesis) was reported to be the most effective and economically viable approach due to the easy scale of fungi and their tolerance toward metal (Adebayo et al. 2021). The term 'mycosynthesis' was coined by Dr. Avinash Ingle and his group for the first time in 2008 (Ingle et al. 2008). Currently, mycosynthesis of nanoparticles has attracted a great deal of attention from the scientific community because of their innovative applications in different sectors.

Considering these facts, the synthesis of nanoparticles using different methods and a food pipeline elaborating different stages involved in the food supply chain are discussed in the present chapter. Moreover, a special focus has been given on mycosynthesis of nanoparticles and their efficient application in food preservation and enhancement of shelf-life of agri-foods and fruits.

The Food Pipeline

Every food product after its production has to undergo various stages till it reaches the consumer and this process can be referred to as the food pipeline. The important stages involved in the food pipeline mainly include pre-processing, transport, storage, processing, and packaging, and the final stage is marketing so that it can reach the consumer in good quality (Suleiman and Rosentrater 2015). Moreover, different approaches are commonly used to achieve each of the above-mentioned stages.

Among these stages, pre-processing is usually performed through sun drying and using certain other improved equipment. It is essentially required to avoid post-harvest losses because post-harvest losses are responsible for both quantitative and qualitative damage. Inappropriate pre-harvest practices, inadequate farming equipment, poor grain handling, etc., resulted in post-harvest losses and ultimately in financial losses and food safety issues. The second important stage is transport. Prompt and better transport facility is required for the transportation of food products to a consumer within stipulated time to avoid losses. Another important stage is storage. Different approaches such as refrigeration, better preservatives, and improved drying technology will be promising in the management of food losses. As far as the processing and packaging stage is concerned, it is considered to be a crucial stage because appropriate processing and packaging of food materials will enhance their shelf-life period and also help to maintain original food quality (Suleiman and Rosentrater 2015). The final stage is marketing. It is also found to be very important, that finished food products should get prompt marketing so that they can reach to the consumer. Figure 2 represents the schematic illustration of the food pipeline.

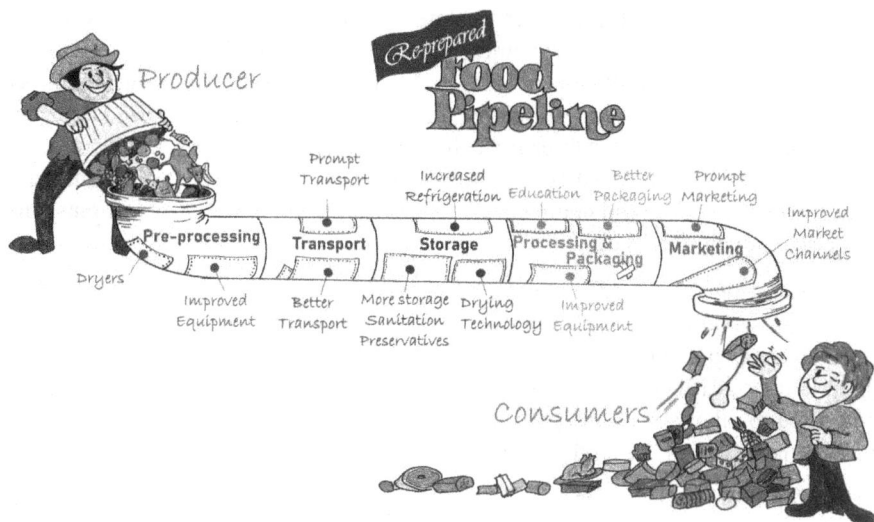

Figure 2. Schematic illustration of food pipeline (re-prepared and modified from the old Figure given in Bourne 1977).

Conventional Food Preservation Techniques

It is well-known that any harvested food material is prone to spoilage. Hence, its preservation using any suitable approach is necessary to counter the problem of food wastage. The concept of food preservation is being employed since ancient times when our ancestors were searching for ways to keep the food fresh and edible. In this context, different approaches like sun drying, salting, pasteurization, etc. were introduced depending on climatic and seasonal factors (Białkowska et al. 2020). In more recent times, rapid industrialization, globalization, and new research innovations introduced several other modified treatments for food preservation which mainly include heat or thermal treatment, use of radiations, freezing, one treatment, canning, etc. (Sridhar 2021). All these methods which are briefly discussed below can help to enhance the shelf life of food materials significantly by managing the pathogens responsible for food spoilage.

Sun-drying: It is most commonly used for the preservation of food items like cereals, fruits, and various other food items. This technique is the cheapest technique and hence widely used. In this technique food materials mentioned above are dried in sunlight and later stored. It promisingly helps to increase the shelf-life of these food items (Sridhar et al. 2021).

Thermal or Heat Treatment: Thermal or heat treatment is found to be one of the promising techniques commonly used for food preservation. Since, time immemorial, this treatment was reported to be most effective for various food materials like bakery products, dairy products, fruits, vegetables, etc. (Gharibi et al. 2020, Christiansen et al. 2020). The method is usually performed by heating foods at a comparatively higher temperature, i.e., between 75 and 90°C or even higher for a few seconds.

Cooling or Freezing: It is another treatment commonly used in the preservation of food materials like leafy vegetables, spices, meat, and milk products. As far as freezing is concerned, different freezing techniques like air blast, cryogenic, direct contact and immersion freezing, high-pressure freezing, ultrasound-assisted freezing, electromagnetic disturbance freezing, dehydration freezing, etc., are used depending on the type of food material (Barbosa de Lima et al. 2020). The food items with appropriate packaging and cooling will always inhibit the entry of microorganisms and also maintain food safety.

Use of radiations: Irradiation is also considered to be one of the important techniques of food preservation. In this technique, radiations like high intensity and frequency sound waves, gamma radiations, etc., are passed through the food materials (Dai and Mumper 2010). The efficient technology is found to be simple and low cost as compared to other advanced approaches. It is usually used to control the microbial pathogens present in the food can.

Ozone Treatment: The concept of ozone treatment technology is getting increased attention because of the growing demands of the consumer for food that is free from additives and preservatives. The ozone treatment is chosen over other approaches due to its diverse properties and quick disintegration. It will help to preserve food organically which ultimately provides healthy food and also help in maintaining a sustainable lifestyle (Sridhar et al. 2021).

Chemical Preservation: This technique involves the use of different kinds of chemical preservatives and it is usually used for the preservation of many food items like juices, sauces, jams, pickles, and many more. Chemicals such as benzoate, organic acids, parabens, etc., are commonly used in food industries (Sharif et al. 2017). Such food preservatives are commonly added to delay degradation in food products, by inhibiting the growth of bacteria, fungi, or antioxidants, and the oxidation of food constituents. However, some of the chemical preservatives showed adverse effects on human health.

Although all the above-mentioned techniques are effective in controlling the microbial spoilage in food and enhancing the shelf-life of food items, these techniques have certain disadvantages, e.g., heat treatment and freezing methods lead to food shrinkage, loss in texture and nutrient and organic properties leading to a huge overall loss in the food product (Sridhar et al. 2021). In the case of chemical preservation methods, used chemical preservatives can cause health hazards. Similarly, other methods also have some positive and some negative effects. Therefore, the development of nanotechnology-based methods having almost no limitation can be a boon for food industries.

Methods for Nanoparticle Synthesis

Ongoing advances in nanotechnology research have established a variety of methods to synthesize nanoparticles from a diverse range of materials, including metals, semiconductors, ceramics, metal oxides, polymers, etc. Depending on their origin and synthesis methods, nanoparticles possess unique physicochemical, structural, and morphological characteristics, which are important in a wide variety of applications concomitant to electronic, optoelectronic, optical, electrochemical, environmental, and biomedical fields (Dhand et al. 2015).

Physical Synthesis

Physical methods apply mechanical pressure, high energy radiations, thermal energy, or electrical energy to cause material abrasion, melting, evaporation, or condensation to generate nanoparticles. These methods mainly operate on a top-down strategy and are advantageous as they are free of solvent contamination and produce uniform monodisperse nanoparticles. At the same time, the abundant waste produced during the synthesis makes physical processes less economical. High energy ball milling, laser ablation, electro-spraying, inert gas condensation, physical vapor deposition, laser pyrolysis, flash spray pyrolysis, and melt mixing are some of the most commonly used physical methods to generate nanoparticles (Dhand et al. 2015). There are numerous physical methods reported to date for nanoparticle synthesis. Metal nanoparticles of Fe, Co, intermetallic FeCo, CeCO, $Nd_2Fe_{14}B$, other metal oxides, ferrites, carbonates, graphite, and carbon dots have been synthesized by high energy ball milling (Xing et al. 2013). Inert gas condensation employs inert gases (e.g., He or Ar) and liquid nitrogen cooled substrate holder for the preparation of NPs. IGC was found to be a highly efficient method for the synthesis of good quality silver and platinum nanoparticles (Ward et al. 2006, Perez-Tijerina et al. 2010, Maicu et al. 2014). Other physical methods include pulse vapor deposition, laser pyrolysis, flash spray pyrolysis, electrospraying, and melt mixing (Dhand et al. 2015).

Chemical Synthesis

Sol-gel method, microemulsion technique, hydrothermal synthesis, polyol synthesis, chemical vapor synthesis, and plasma-enhanced chemical vapor deposition technique are some of the most commonly used chemical methods for nanoparticles' synthesis (Dhand et al. 2015). In the sol-gel processing method, there are two types of components viz., 'sol' which is a colloidal suspension of solid particles in a liquid, and 'gel' which are polymers containing liquid. Thus, this process includes the creation of 'sols' in the liquid that lead to the formation of a network of discrete particles or network polymers by the connection of sol particles (Brinker and Scherer 1990, Aziz et al. 2013, Goncalves et al. 2014). The microemulsion method is also employed for the synthesis of many nanoparticles. Other chemical methods include hydrothermal synthesis (Abedini et al. 2013, Du et al. 2014, Ma et al. 2015), Polyol synthesis (Herricks et al. 2004, Park et al. 2007), chemical vapor synthesis and deposition (Lahde et al. 2011), etc. (Dhand et al. 2015). There are numerous methods available using various approaches including chemical, physical, and biological protocols for the synthesis of nanoparticles. The chemical method of synthesis is advantageous as it takes a short period for the synthesis of a large number of nanoparticles. However, in this method, capping agents are required for the size stabilization of the nanoparticles. Moreover, chemical reagents used generally for nanoparticles synthesis and stabilization are toxic and lead to byproducts that are not environmentally benign. The need for eco-friendly non-toxic methods for nanoparticles synthesis is developing interest in biological approaches which are free from the use of toxic chemicals as byproducts (Bhardwaj et al. 2020). The use of chemical and physical methods for the synthesis of nanoparticles has been regarded as not suitable for biochemical applications due to the generation of toxic chemicals (Abdeen et al. 2016). Carbon coated Fe, Co and Ni ferromagnetic nanoparticles of size 2–100 nm (Fe@C, Co@C, and Ni@C, respectively) have been produced by high-pressure chemical vapor deposition (El-Gendy et al. 2009).

Biological Synthesis

A variety of natural sources are there for metal nanoparticle synthesis including plants, fungi, yeast, actinomycetes, bacteria, etc. The unicellular and multicultural organisms can produce intracellular and extracellular inorganic nanoparticles (Sen 2011). Biomolecules derived from natural sources have been continuously applied for the synthesis and fabrication of eco-friendly nanoparticles (Osonga et al. 2020) which is the function of various physiological parameters (Jameel et al. 2020). Bio-assisted methods, biosynthesis, or green synthesis provide an environmentally benign, low-toxic, cost-effective, and efficient protocol to synthesize and fabricate nanoparticles. These methods employ biological systems like bacteria, fungi, viruses, yeast, actinomycetes, plant extracts, etc., for the synthesis of metal and metal oxide nanoparticles. Bio-assisted methods can be broadly divided into microorganisms assisted, biomolecules assisted and plant extracts assisted (Dhand et al. 2015).

Phytosynthesis of Nanoparticles

Plants offer a better option for the synthesis of nanoparticles as the protocols involving plant sources are free from toxic chemicals. Moreover, natural capping agents are readily supplied by the plants. Shankar et al. (2003, 2004) reported the synthesis of gold and silver nanoparticles using geranium extracts. Further, gold nano triangles and silver nanoparticles were synthesized using Aloe Vera plant extracts (Chandran et al. 2006). Most reports available on the synthesis of silver or gold nanoparticles use broths resulting from boiling fresh plant leaves. However, Huang et al. (2007) synthesized silver and gold nanoparticles using the sundried *Cinnamomum camphora* leaf extract. A simple green synthesis method for the production of well-defined silver nanowires was reported recently by Lin et al. (2010). The method involves the reduction of silver nitrate with the broth of sundried *Cassia fistula* leaf at room temperature without using any additive. Zarei et al. (2021) reported an environmentally friendly short method (EFSM) with high efficiency for the synthesis of amorphous silica oxide nanoparticles by using agricultural waste called rice husks. Green and cost-effective copper nanoparticles synthesized using extracts of non-edible and waste plant materials from *Vaccinium* species were reported for their antimicrobial activity (Benassai et al. 2021). Ingle et al. (2021) have shown the antimicrobial activity of *Moringa oleifera* mediated nanoparticles against pathogenic bacteria.

Bacterial Synthesis

Bacteria have been most extensively researched for the synthesis of nanoparticles because of their fast growth and relative ease of genetic manipulation. Slawson et al. (1992) recovered the silver-producing bacteria from the silver mines, *Pseudomonas stutzeri* AG259, in which the silver nanoparticles were accumulated in the periplasmic space but the particle's size ranged from 35 to 46 nm. The *Lactobacillus* strains present in the milk were exposed to a larger concentration of nanoparticles to produce silver, gold, and alloy crystals of defined morphology (Nair and Pradeep 2002). Bacteria have also been used to synthesize gold nanoparticles (Konishi et al. 2004). The pH was an important factor in controlling the morphology of bacteriogenic nanoparticles and the location of the deposition. These nanoparticles were used in many applications, e.g., direct electrochemistry of proteins (Liangwei et al. 2007). The most important application of bacterium would be in industrial silver recovery.

Yeast-mediated Synthesis

Yeasts are also known for their applicability for the bio-inspired synthesis of nanoparticles. Yeast is a eukaryotic microorganism belonging to the fungi kingdom. Carboxyl, hydroxyl and amide groups on the cell surface are anticipated to be responsible for silver nanoparticles synthesis. Yeast was also reported to be useful in the synthesis of Cd and PbS nanoparticles using *Candida glabrata* and *Rhodosporidium diobovatum* species, respectively (Dhand et al. 2015).

Algae in Nanoparticles Synthesis

There are few reports available that used algae as a "Biofactory" for nanoparticles synthesis. Marine algae is used for the biosynthesis of highly stable extracellular gold nanoparticles in a relatively short time compared to other biosynthesizing processes. Palladium and platinum nanoparticles starting with their corresponding metallic chloride-containing salts have been investigated. The extract of green alga *Botryococcus braunii* was used for the synthesis of copper and silver nanoparticles (Arya et al. 2018). Many algae have also been reported to synthesize nanoparticles using various precursors. Algae like *Portieria hornemannii* (Arya et al. 2019), *Botryococcus braunii* (Manikandakrishnan et al. 2019), *Caulerpa racemosa* (Salem et al. 2019), *Colpomenia sinuosa* and *Pterocladia capillacea* (Colin et al. 2018), *Egregia* sp. (González-Ballesteros et al. 2017), *Cystoseira baccata* (Mourdikoudis et al. 2018) can reduce the precursor ions from salts into nanomaterial form.

Actinomycetes in Nanoparticles Synthesis

The monodisperse silver and gold nanoparticles produced either intracellularly or extracellularly are of interest to scientists. But it was not very high and was far inferior to that obtained by conventional methods. Actinomycetes are microorganisms that can share important characteristics with fungi and bacteria (Okami et al. 1988).

Many members of actinomycetes are known to synthesise nanoparticles (Golinska et al. 2014, Anasane et al. 2016, Wypij et al. 2016, 2019). The thermophilic *Thermomonospora* sp. was exposed to gold and silver irons and the metals got reduced extracellularly (Sastry et al. 2003). These micro-organisms have developed numerous special adaptations to survive in such extreme habitats which include a new mechanism of enzyme transduction, regulating metabolism, and maintaining the structure and function of the membrane. The actinomycetes can produce secondary metabolites that can reduce metals into nanoparticles. The nanoparticles synthesised by actinomycetes were also found to be non-toxic to the cells which continued to multiply even after the formation of the nanoparticles.

Viruses in Nanoparticle Synthesis

The biological synthesis of nanoparticles has been extended to biological particles like viruses, proteins, peptides, and enzymes. Cowpea chlorotic mottle virus and cowpea mosaic virus have been used for the mineralization of inorganic materials. Tobacco mosaic virus has been shown to successfully direct the mineralization of sulphide and crystalline nanowires. One step further, peptides capable of nucleating Nanocrystal growth have been identified from combinatorial screens and displayed on the surface of the M13 bacteriophage (Douglas and Young 1993).

Mycosynthesis of Nanoparticles

The fungi are taking the centre stage in studies on the biological generation of nanoparticles because of their tolerance and bioaccumulation. There are various advantages of using fungi in their scale-up process (e.g., using a thin solid substrate fermentation method). Fungi are efficient secretors of extracellular enzymes which can be produced at large-scale. Further advantages of using a fungal-mediated green approach for the synthesis of metallic nanoparticles include economic viability and ease in handling biomass. The main drawback of biosynthesis of nanoparticles synthesis in eukaryotic organisms lies in the problem of genetic manipulation of the organisms as a means to over-express the enzymes compared to prokaryotes. Mukherjee et al. (2008) demonstrated "green synthesis" of highly stabilized nanocrystalline silver particles by a non-pathogenic and agriculturally important fungus, *Trichoderma asperellum*.

The synthesis of nanoparticles using fungi offers certain advantages in comparison to bacteria. Thus, research in this area has seen an increasing interest over the last decade. Some of the advantages of mycelium include easy scale-up and downstream process, economically feasible and ecologically friendly so they can cover a large surface (Syed and Ahmad 2012). Fungi contain enzymes in the cytoplasm and cell wall, which transform metal ions into nanoparticles (Chatterjee et al. 2020). The metal ions have a positive charge, which is attracted by the fungi and triggers biosynthesis. Certain proteins are also induced by metal ions, and these hydrolyze the ions.

Fungi are able to secrete high amounts of protein, thus expediting the process of nanoparticle synthesis. The mechanism behind the process of reduction of silver ions by the fungus has been partially demonstrated. Duran et al. (2005) have worked on *Fusarium oxysporum* for the extracellular synthesis of silver nanoparticles. The primary confirmation of silver nanoparticle synthesis was performed using a UV-Visible spectrophotometer. It showed an absorption peak at 265 nm which is mainly due to electronic excitations in tryptophan and tyrosine residues present in the proteins, confirming the role of proteins in the reduction of the metal ions to nanoparticles. The presence of hydrogenase in fungus *F. oxysporum* was demonstrated with washed cell suspension that had been grown aerobically and anaerobically in a medium containing glucose and nitrate. This enzyme was thought to be involved in the reduction process. Besides this enzyme, several naphthoquinones and anthraquinones which have excellent redox properties were believed to be involved in the electron shuttle. Fluorescence emission spectra of fungal filtrate show an emission band centered around 340 nm. The nature of emission shows that the proteins are in their native form and also are related to reductase and/or quinone generation. It appears that the reductase is responsible for the reduction of Ag+ ions and the subsequent formation of nanoparticles (Duran et al. 2005, Acharya et al. 2009).

The mycosynthesis of gold nanoparticles was first reported by Mukherjee et al. (2001). Gold nanoparticles are of interest, mainly due to their stability under atmospheric conditions, resistance to oxidation, and biocompatibility. These unique properties can potentially be exploited in a diverse range of industrial applications using their optical and electronic properties. Many studies reported that the controlled synthesis of gold nanoparticles of well-defined size and shape is a major challenge and numerous chemical methods were developed to control the physical properties of the particles (Burda et al. 2005, Molnár et al. 2018, Mikhailova 2021, Santhosh et al. 2022).

The fungal biomass when exposed to aqueous cadmium sulphate (CdS) solution leads to the formation of extremely stable CdS nanoparticles extracellularly. The change in color of cadmium sulphate solution to bright-yellow color signifies the development of CdS nanoparticles which were further verified spectrophotometrically. The formation of extracellular CdS nanoparticles is due to the enzymatic reduction of sulphate ions by the fungus and the long-term stability of nanoparticles is due to the presence of the proteins in a solution that binds to the surface of the nanoparticle and prevents aggregation (Ahmad et al. 2002). Transition metal platinum nanoparticles have attracted attention for industrial applications due to their greater catalytic properties. Due to the antioxidant properties of the platinum nanoparticles, they are the subject of substantial research with applications in a wide variety of areas, including nanotechnology, medicine, and the synthesis of novel materials with unique properties (Kim et al. 2008, Onizawa et al. 2009). The first report on the biosynthesis of platinum nanoparticles was shown by Lengke et al. (2006) using cyanobacterium *Plectonema boryanum* UTEX485. Mycosynthesis of platinum nanoparticles was first reported by Riddin et al. (2006) using a tomato wilt pathogenic fungus *F. oxysporum* sp. *lycopersici*.

Zirconia (ZrO_2) is gaining interest as a ceramic material with a useful optical, electrical, thermal hardness, and other characteristics. Zirconia is also emerging as an important class of catalyst. Recent application of zirconium nanocrystals includes nano-crystalline zirconium alloys for space applications as alternatives for titanium in components of liquid rocket engines (e.g., lines and turbo pumps), since they are lighter and less susceptible to embrittlement by hydrogen, and in the nanowire, nanofiber and in certain other alloy and catalyst applications (Acharya et al. 2009). For the first time, Bansal et al. (2004) have reported the mycosynthesis of zirconium nanoparticles. Mycosynthesis of silica was reported by Bansal and his co-workers in 2005 (Bansal et al. 2005). Table 1 showed the list of nanoparticles synthesized from different fungi.

Table 1. Synthesis of different nanoparticles using fungi.

Sr. No.	Fungus used for Synthesis	Type of Synthesis	Type of Nanoparticle	Size (nm)	Reference
1.	*Fusarium oxysporum*	Intracellular	Ag	5–15	Ahmad et al. (2003)
2.	*Verticillium* sp.	Intracellular	Ag	25 ± 12	Mukherjee et al. (2001)
3.	*Aspergillus fumigatus*	Intracellular	Ag	5–25	Bhainsa and D'Souza (2006)
4.	*Fusarium oxysporum* and *Verticillium* sp.	Intracellular	Magnetite	20–50	Bharde et al. (2006)
5.	*Candida glabrata* and *Schizosaccharomyces pombe*	Intracellular	CdS	**	Dameron et al. (1989)
6.	*Phanerochaete chrysosporium*	Extracellular	Ag	50–200	Vigneshwaran et al. (2006)
7.	*Fusarium solani*	Extracellular	Ag	5–35	Ingle et al. (2009)
8.	*Fusarium acuminatum*	Extracellular	Ag	5–40	Ingle et al. (2008)
9.	*Aspergillus niger*	Extracellular	Ag	8–10	Gade et al. (2008)
10.	*Coriolus versicolor*	Extracellular & Intracellular	Ag	25–75	Sanghi and Verma (2009)
11.	*Trichoderma asperellum*	Extracellular	Ag	13–18	Mukherjee et al. (2008)
12.	*Penicillium* sp.	Extracellular	Ag	~ 6.97	Sadowski et al. (2008)
13.	*Penicillium* sp.	Extracellular	Ag	15–150	Maliszewska et al. (2009)
14.	*Fusarium semitectum*	Extracellular	Ag	10–60	Basavaraja et al. (2008)
15.	*Pleurotus sajor-caju*	Extracellular	Ag	30.5 ± 4.0	Vigneshwaran et al. (2009)
16.	*Fusarium oxysporum*	Extracellular	CdSe quantum dots	11	Kumar et al. (2007)
17.	*Fusarium oxysporum* sp. *lycopersici*	Intra and extracellularly	Pt	10–50	Syed and Ahmad (2012)
18.	*Fusarium oxysporum*	Extracellular	Pt	5–30	Bansal et al. (2004)

Table 1 contd. ...

...Table 1 contd.

Sr. No.	Fungus used for Synthesis	Type of Synthesis	Type of Nanoparticle	Size (nm)	Riddin et al. (2006)
19.	*Fusarium oxysporum*	Extracellular	Zirconia	3–11	Bansal et al. (2005)
20.	*Fusarium oxysporum*	Extracellular	Silica	5–15	Bansal et al. 2005)
21.	*Fusarium oxysporum*	Extracellular	TiO_2	6–13	Adeleye et al. (2020)
22.	*Rhizopus stolonifer*	Extracellular	Fe	**	Abdeen et al. (2016)
23.	*Aspergillus niger* YESM1	Extracellular	Magnetic Fe and Fe_3O_4	18 & 50	Chatterjee et al. (2020)
24.	*Aspergillus niger* BSC-1	Extracellular	FeO	20–40	Manjunath and Joshi (2019)
25.	*Cladosporium cladosporioides*	Extracellular	Ag	30–60	Zhang et al. (2019a)
26.	*Aspergillus niger*	Extracellular	Cu	500	Zhang et al. (2019b)
27.	*Mariannaea* sp. HJ	Intra and extracellularly	Se	45.19 & 212.65	Vijayanandan and Balakrishnan (2018)
28.	*Aspergillus nidulans*	Extracellular	CoO	20.29	Bhargava et al. (2016)
29.	*Cladosporium oxysporum*	Extracellular	Au	72.32 ± 21.80	Barnett (2003), Das et al. (2012)
30.	*Rhizopus oryzae*	Extracellular	Au	**	Kowshik et al. (2003)
31.	Yeast strain MKY3	Extracellular	Ag	2–5	Kowshik et al. (2003)

Note: ** = information not provided/available

Emerging Role of Nanotechnology in Food Industry

Recent years have seen rapid growth in the world population which has created a state of urgency to meet the growing demand in the food sector. Therefore, it is essential to increase the production of food in a safe, efficient, and sustainable manner (Ameta et al. 2020). The quality of the food can be compromised at any stage of the food chain due to chemical, physical and biological contaminants. The presence of these contaminants may pose a threat to the health of people (Thompson and Darwish 2019). The main role of packing materials includes, protection of food product from external damaging sources by acting as a barrier, preservative, retention of food quality, and showcasing its nutritional content to consumers (Farhoodi 2016, Alamri et al. 2021).

Food packaging is the very first application of nanotechnology in the food industry, employing various materials like nano-fibers, nanoparticles, nanocomposites, and nanoplates (Schmidt et al. 2002, Thostenson et al 2005, Sharma et al. 2017, Babu 2021). Both organic and inorganic nanomaterials, in combination with various types of polymers have been employed to design novel packing materials having various geometries (Thakur and Gupta 2016). Nanoplates can be formed by the combination of clay and silica lead which increases the permeability of infiltrate molecules. Nanoplates have been reported to promote mechanical stability and inhibit the exchange of gases, and water vapor (Mirzadeh and Kokabe 2007, Adame and Beall 2009).

Nanoparticle-based consumer products include new electronics and computers, health trackers, and other miscellaneous home products. It is predicted that nanotechnology will eventually be the next breakthrough across several sectors, including food manufacturing and packaging. Resonant energy transfer (RET) systems incorporating noble metal nanoparticles are growing in the fields of optics and material science so that nanoparticles can be used in future commercial products. The antimicrobial agents are essential in water disinfection, medicinal applications, textile, and food packaging. Currently, nanoparticles are functionalized by different groups with the purpose of overcoming the varied bacterial populations. Materials such as ZnO, TiO_2, Cu and Ni-based nanoparticles, are being used for specific functions because of their ideal antibacterial activities (Mughal et al. 2021). In addition to food packaging, several nanomaterials have been used in other sectors of food industries like food processing, food safety, delivery of inhibitory chemicals, development of edible coatings, etc. However, in the present chapter, a special focus has been given to the application of mycosynthesized nanoparticles in food preservations and enhancement of the shelf-life of agri-food and fruits.

Myconanotechnology in Food Preservation and Enhancement of its Shelf-life

Food preservation and enhancement of food products' shelf-life can be achieved by employing mycosynthesized nanomaterials including nanoparticles in food industries for various purposes like food packaging, food processing, food safety, shelf-life improvement, and management of food pathogens, etc. Figure 3 showed the role of myconanotechnology in different food sectors.

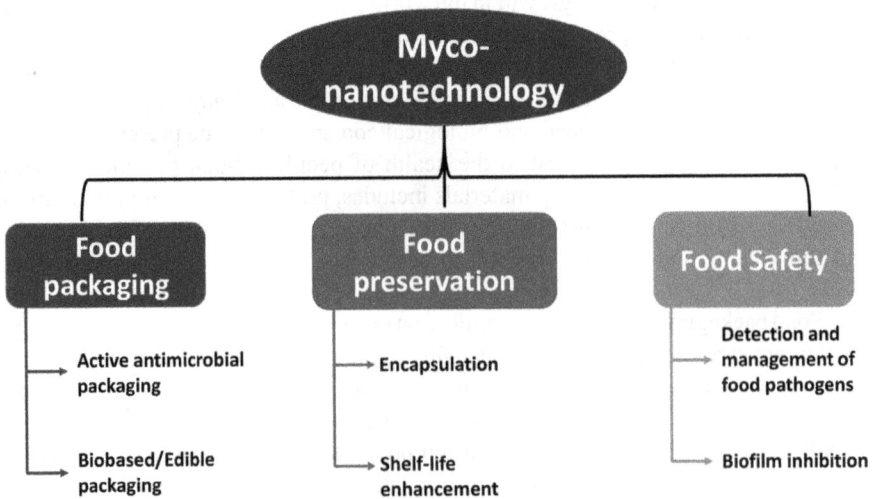

Figure 3. Role of myconanotechnology in the different food sectors.

Myconanotechnology in Food Packaging

It is well-known that agricultural food materials usually go through various chains of processing and contamination with spoilage and pathogenic microorganisms. Therefore, there is always a risk of widespread foodborne disease. Hence, detection and management of such foodborne pathogens is the only option to achieve food preservation. The use of nanotechnology in the development of innovative packaging materials has had remarkable growth in the last few years and is expected to have an important impact on the food market in the near future. This growth is a consequence of the increasing knowledge about nanotechnology applications in food packaging, which brought to the academia and industry new tools for the development of new nanotechnology-based products with improved technological functionalities and properties, as well as the corresponding advances in materials science, processing technology, and analytical techniques (Cerqueira et al. 2018).

Nanotechnological advancements in food packaging have the potential to improve various properties which mainly include, the mechanical strength, structural properties, heat resistance, and barrier against ultraviolet radiation, oxygen, carbon dioxide, moisture, etc., of existing materials, in terms of maintaining the quality of food and preserving its freshness. It also helps in monitoring biochemical or microbial changes within packaging and aids in the enhancement of food shelf-life (Babu 2021). Numerous nanoparticles such as titanium dioxide nanoparticles, zinc oxide nanoparticles, titanium nitride nanoparticles, silver nanoparticles, and nanoclays have been utilized in food packaging materials as an additive (Tager 2014). Moreover, such nanomaterials can also extend the shelf-life of the product by the release of molecules like enzymes, antioxidants, nutraceuticals, fragrances, and flavors (Cha and Chinnan 2004, LaCoste et al. 2005). It has been reported that most of the above-mentioned nanoparticles, e.g., titanium dioxide nanoparticles, zinc oxide nanoparticles, titanium nitride nanoparticles, silver nanoparticles, etc., have been successfully synthesized using a variety of fungi (Ingle et al. 2008, Rajakumar et al. 2012, Sharma et al. 2021). Therefore, mycosynthesized nanoparticles thus developed can be promisingly applied in food packaging applications and enhancement of food shelf-life by using different coatings. Some of the important studies are discussed below-

Myconano-based Coatings in Shelf-life Improvement

In recent years, a number of edible nano-coatings have been developed exhibiting better moisture, lipid, and gas barrier properties that can be potentially used for coating a variety of food products such as fruits, vegetables, chocolate, bakery items, meats, etc. (Zambrano-Zaragoza et al. 2018). Therefore, the application of edible coatings on food materials, particularly fruits is gaining considerable interest due to their safe nature. In this context, nano chitosan-based edible coatings can be the potential alternative for extending the postharvest life of fruit. Melo et al. (2018) studied the effects of fungal chitosan nanoparticles as eco-friendly edible coatings on the quality of post-harvest table grapes. The microdilution method was used to determine the minimum inhibitory and bactericidal concentration of chitosan nanoparticles

(128.3 nm) against food-borne pathogenic bacteria. The recorded findings showed that these nanoparticles demonstrated an inhibitory effect against pathogenic food-borne bacteria. Moreover, the application of edible chitosan nanoparticle coatings on grapes showed a significant delay in the ripening process. The delay in ripening of the grapes resulted in many positive outputs such as decreased weight loss of grapes, soluble solids, and reduced sugar contents as well as increased moisture retention and preservation of the treatable acidity values and sensory characteristics.

Knowing the importance of edible coatings in food preservations, researchers are looking forward to using natural and biodegradable polymers. In this line, apart from chitosan, pullulan is one of the most important biopolymers that have attracted much interest over the period of last few years due to its peculiar characteristics. This polymer is usually obtained from the fermentation medium of the fungus *Aureobasidium pullulans* under suitable conditions. Nowadays, pullulan (a myco-product) is being used in the synthesis of different nanoparticles which is a kind of mycosynthesis. In this context, recently, Khan et al. (2022) fabricated pullulan-based active packaging films incorporated with curcumin and pullulan mediated silver nanoparticles to maintain the quality and shelf life of broiler meat. Actually, the environment of broiler meat is very suitable for the growth of microbes The microbial intensification normally leads to the oxidation of lipid and protein and the degradation of its physicochemical properties. It is proved that the utilization of the antioxidants can prevent meat degradation and hence, pullulan-based active packaging incorporated with green silver nanoparticles, as anti-oxidants, was developed for the packaging of broiler meat, and its efficacy was determined by 14 days of refrigerated storage at $4 \pm 1°C$. The results revealed that the pullulan-based active packaging incorporated with curcumin and pullulan mediated silver nanoparticles can maintain the textural and physicochemical properties of meat in these 14 days of storage, confirming the alternative approach for the preservation of broiler meat for prolonged shelf life.

In another report, pullulan/chitosan-based multifunctional edible composite films were prepared by using mushroom-mediated zinc oxide nanoparticles (ZnONPs). Thus, used nanoparticles were mycosynthesized using extract of enoki mushroom. The characterization of these ZnONPs confirmed that the nanoparticles are irregular in shape having an average size of 26.7 ± 8.9 nm. Further, edible composite films were fabricated using ZnONPs and propolis and pullulan/chitosan as biodegradable polymers. It has been observed that the reinforcement with ZnONPs and propolis significantly improved the mechanical strength of the pullulan/chitosan-based film (by ~ 25%). In addition, other properties like the water vapor barrier and hydrophobicity of the film was also found to be slightly increased. Moreover, these biocomposite films are reported to possess potent antioxidant activity due to the propolis and promising antibacterial activity against foodborne pathogens due to the ZnONPs. Finally, edible pullulan/chitosan-based film thus developed, was used for pork belly packaging (Roy et al. 2021). Moreover, Fayaz et al. (2009) demonstrated the synthesis of silver nanoparticles from *Trichoderma viride* and the development of silver nanoparticles incorporated in sodium alginate films. Further, the antibacterial efficacy of thus developed nano-films was evaluated against *Escherichia coli*

(ATCC 8739) and *Staphylococcus aureus* (ATCC 6538). The results obtained showed that the silver nanoparticle incorporated sodium alginate thin films exhibit promising antibacterial activity against both the tested bacterial strain. Moreover, the application of this film on the surface of carrot and pear was reported to enhance their shelf life as compared to control (without application of film) with respect to weight loss and soluble protein content. Hence, it is believed that nanoparticles can be effectively used in the preservation of fruits and vegetables.

The application of edible coatings has an advantage over the direct use of antioxidant/antibacterial ingredients in foods because these kinds of coatings used in packaging allow the controlled release of active ingredients from the food packaging surface. Therefore, it is proposed that the use of edible coatings containing various natural polymers like chitosan, pullulans, etc., can be a promising strategy to improve the shelf-life and quality of food products.

Control of Food-Spoilage Pathogens

As far as food safety is concerned, management of food-spoilage pathogens is extremely necessary to avoid the harmful health effects on human beings (Babu 2021). To date, different nanoparticles are being used for this purpose. However, there is limited literature available on the application of mycosynthesized nanoparticles in food safety.

The nanoparticles synthesised by fungi can be used as nanobiosensor for the detection of food-borne pathogens. The silver nanoparticles synthesised by *Trichoderma viride, T. reesei, Aspergillus flavus* NJP08 and *Rhizopus stolonifer* can be used for the development of biosensor and bioimaging (Boroumandmoghaddam et al. 2015, Adebayo et al. 2021).

In this context, Vidya and Subramani (2017) studied the antibacterial potential of mycosynthesized silver nanoparticles against foodborne pathogens. Here, the authors synthesized silver nanoparticles using *Aspergillus flavus* and evaluated its activity against *Escherichia coli* (MTCC 118), *Bacillus subtilis* (MTCC 441), *S. aureus* (MTCC 737), and *Enterobacter aerogenes* (MTCC 111). The observations revealed that the mycosynthesized silver nanoparticles exhibit significant antibacterial activity against all the tested foodborne pathogens. The maximum activity was reported against *B. subtilis* followed by *S. aureus* and the minimum activity was found against *E. aerogenes*. The management of foodborne pathogens ultimately helps to increase the shelf-life of food. Similarly, Suryavanshi et al. (2017) reported the synthesis of sulfur and aluminium oxide nanoparticles from *Colletotrichum* sp. and studied the activity of these nanoparticles singly and in combination with essential oils extracted from *Citrus medica* and *Eucalyptus globulus* (nanofuctionalized oils) against foodborne pathogens, viz., *Listeria monocytogenes, Salmonella typhi, Chromobacterium violaceum, F. oxysporum,* and *A. flavus*. It was observed that both the tested nanoparticles showed promising activity against all the foodborne pathogens. However, nano-functionalized oils demonstrated superior antimicrobial activity against these pathogens as compared to nanoparticles. The nano-functionalized oil containing sulfur nanoparticles was found to be most effective against *S. typhi,* followed by *F. oxysporum, C. violaceum, A. flavus* whereas, the least activity was recorded against *L. monocytogenes*. On the other hand, nano-functionalized oil containing aluminium oxide nanoparticles showed maximum activity against *F. oxysporum* and minimum activity against *L. monocytogenes*.

Moreover, Mohanta et al. (2018) developed an economically viable and environmental friendly method for the synthesis of silver nanoparticles using wild mushroom *Ganoderma sessiliforme* and further tested its activity against common food-borne bacteria, namely, *E. coli*, *B. subtilis*, *Streptococcus faecalis*, *Listeria innocua* and *Micrococcus luteus*. Among the tested all five bacterial species, four gram-positive bacterial strains (i.e., *B. subtilis*, *S. faecalis*, *L. innocua* and *M. luteus*) showed more than 90% growth inhibition whereas gram-negative *E. coli* exhibited growth inhibition below 90%. Apart from the films and coatings developed for food packaging, currently, novel materials with innovative and appealing performance have been conceived. Pullulan (fungal biopolymer) was used in combination with silver to synthesize nanoparticles with potent antimicrobial efficacy, where the pullulan acts as both a stabilizing and reducing agent. Further, the antibacterial and antibiofilm activities of these myco-based silver nanoparticles was evaluated against food-borne and multidrug-resistant pathogens, viz., *E. coli*, *L. monocyogenes*, *B. cereus*, *K. pneumoniae*, *P. aeruginosa*, *Aspergillus* spp., and *Penicillium* spp. The results obtained revealed that all bacterial and fungal pathogens were inhibited in a dose-dependent manner (Kanmani and Lim 2013).

In addition, a study performed by Al-Owasi et al. (2019) demonstrated that pullulan-mediated silver nanoparticles can be promisingly used for the control of different fungal pathogens including food spoilage fungi like *Penicillium* and *Rhizopus*. In this study, the authors used fungal-derived pullulan for the mycosynthesis of silver nanoparticles and further evaluated its efficacy against food spoilage fungi. The results obtained revealed that these mycosynthesized nanoparticles showed promising antifungal activity against both the above-mentioned food spoilage fungi. Moreover, nowadays researchers are focusing on the development of nano-based emulsions (nanoemulsions) using different plant-based essential oils due to their extraordinary applications in food industries for food storage. In this context, a silver-based nanoemulsion was prepared. Here, silver nanoparticles were synthesized from the extract of endophytic fungus (i.e., *Colletotrichum siamense*), and then it was blended with tea tree oil to develop a silver-based nanoemulsion. Further, the antibacterial efficacy of this silver-based nanoemulsion was evaluated against foodborne pathogens, viz., *Enterobacter kobei* SFV 1 and *Klebsiella oxytoca* SFV 3 using antibacterial assays. From the findings recorded, it was observed that both nanoparticles and nanoemulsions significantly controlled the biofilm formation in both these bacterial strains (Ranjani et al. 2022).

All the studies discussed in the above sections clearly indicated that mycosynthesized nanoparticles can potentially be used in the management of a wide range of food pathogens. Although limited studies have been performed so far on the application of mycosynthesized nanoparticles in the food industry for food preservation and enhancement of food products' shelf life, there is huge scope for research in this field. Considering the broad range of bioactivities including the antimicrobial potential of mycosynthesized nanoparticles, we believed that further extensive studies are extremely necessary for novel nanotechnology-based solutions for food preservation and management of food wastage.

Conclusions

Preservation of food products such as fruits and vegetables and increase in their shelf-life is a need of the hour because million tons of food get wasted every year all over the world. Several conventional approaches and technologies have been developed and employed for food preservation in food industries. But available conventional approaches have certain limitations. Hence, there is still a necessity to take major strides to develop economically viable and sustainable technology for effective food preservation and to avoid the losses that happened at different stages of the food pipeline. Management of food wastage and proper utilization of available food will definitely help to solve the crisis associated with food security. Hence, constant efforts need to be made in this area to find safe preservatives. In this context, nanotechnology is attracting a great deal of attention from researchers working in the area. Different nanomaterials including nanoparticles have been found to have a remarkable role in food preservation and its shelf life enhancement. Considering the novel antimicrobial activities of mycogenically synthesized nanoparticles, such nanoparticles can be used as a safe alternative for chemical preservatives. Although limited reports are available on the application of mycosynthesized nanoparticles in food industries, there is a necessity to perform more and more studies on these aspects so as to develop sustainable technology for food preservation.

Acknowledgment

API is highly thankful to the Science and Engineering Research Board (SERB), Department of Science and Technology, Government of India, New Delhi for providing financial assistance in the form of the Ramanujan Fellowship. M.R. would like to thank the Polish National Agency for Academic Exchange (NAWA) for financial support under the grant PPN/ULM/2019/1/00117/DEC/1 2019-10-02.

References

Abdeen, M., Sabry, S., Ghozlan, H., El-Gendy, A.A. and Carpenter, E.E. 2016. Microbial-physical synthesis of Fe and Fe_3O_4 magnetic nanoparticles using *Aspergillus niger* YESM1 and supercritical condition of ethanol. J. Nanomater. 2016: 9174891. https://doi.org/10.1155/2016/9174891.

Abdelrahman, M.S., Nassar, S.H., Mashaly, H., Mahmoud, S., Maamoun, D., El-Sakhawy, M., Khattab, T.A. and Kamel, S. 2020. Studies of polylactic acid and metal oxide nanoparticles-based composites for multifunctional textile prints. Coatings 10: 58. https://doi.org/10.3390/coatings10010058.

Abedini, A., Daud, A.R., Hamid, M.A.A., Othman, N.K. and Saion, E. 2013. A review on radiation-induced nucleation and growth of colloidal metallic nanoparticles. Nanoscale Res. Lett. 8: 474. https://doi.org/10.1186/1556-276X-8-474.

Acharya, K., Sarkar, J. and Deo, S.S. 2009. Mycosynthesis of nanoparticles. pp. 204–215. *In*: Bhowmik Basu, S.K. and Goyal, A. (eds.). Advances in Biotechnology. Bentham Publisher, USA.

Adame, D. and Beall, G.W. 2009. Direct measurement of the constrained polymer region in polyamide/clay nanocomposites and the implications for gas diffusion. Appl. Clay Sci. 42: 545–552.

Adebayo, E.A., Azeez, M.A., Alao, M.B., Oke, A.M. and Ain, D.A. 2021. Fungi as veritable tool in current advances in nanobiotechnology. Heliyon 7(11): e08480. https://doi.org/10.1016/j.heliyon.2021.e08480.

Adeleye, T.M., Kareem, S.O. and Kekere-Ekun, A.A. 2020. Optimization studies on biosynthesis of iron nanoparticles using *Rhizopus stolonifer*. Nanotechnology applications in Africa: Opportunities and constraints. IOP Conf. Series: Materials Science and Engineering 805: 012037. Doi: 10.1088/1757-899X/805/1/012037.

Ahmad, A., Mukherjee, P., Mandal, D., Senapati, S., Khan, M.I., Kumar, R. and Sastry, M. 2002. Enzyme mediated extracellular synthesis of CdS nanoparticles by the fungus, *Fusarium oxysporum*. J. Am. Chem. Soc. 124: 12108–12109.

Ahmad, A., Mukherjee, P., Senapati, S., Mandal, D., Khan, M.I., Kumar, R. and Sastry, M. 2003. Extracellular biosynthesis of silver nanoparticles using the fungus *Fusarium oxysporum*. Colloids Surf. B 28: 313–318.

Alamri, M.S., Qasem, A.A.A., Mohamed, A.A., Hussain, S., Ibraheem, M.A., Shamlan, G., Alqah, H.A. and Qasha, A.S. 2021. Food packaging's materials: A food safety perspective. Saudi Journal of Biological Sciences 28: 4490–4499.

Ali, A., Shah, T., Ullah, R., Zhou, P., Guo, M., Ovais, M., Tan, Z. and Rui, Y. 2021. Review on recent progress in magnetic nanoparticles: synthesis, characterization, and diverse applications. Front. Chem. 9: 629054. Doi: 10.3389/fchem.2021.629054.

Al-Owasi, Y., Elshafie, A., Sivakumar, N. and Al-Bahry, S.N. 2019. Synthesis of pullulan-mediated silver nanoparticles (AgNPs) and their antimicrobial activities. SQU Journal for Science 24(2): 88–94.

Ameta, S.K., Rai, A.K., Hiran, D., Ameta, R. and Ameta, S.C. 2020. Use of nanomaterials in food science. pp. 457–488. *In*: Ghorbanpour, M., Bhargava, P., Varma, A. and Choudhary, D.K. (eds.). Biogenic NanoParticles and their Use in Agro-ecosystems. Springer, Singapore.

Amit, S.K., Uddin, M.M., Rahman, R, Islam S.M.R. and Khan, M.S. 2017. A review on mechanisms and commercial aspects of food preservation and processing. Agriculture & Food Security 6: 51. https://doi.org/10.1186/s40066-017-0130-8.

Anasane, N., Golińska, P., Wypij, M., Rathod, D., Dahm, H. and Rai, M. 2016. Acidophilic actinobacteria synthesised silver nanoparticles showed remarkable activity against fungi-causing superficial mycoses in humans. Mycoses 59(3): 157–166.

Arya, A., Gupta, K., Chundawat, T.S. and Vaya, D. 2018. Biogenic synthesis of copper and silver nanoparticles using green alga *Botryococcus braunii* and its antimicrobial activity. Bioinorg. Chem. Appl. 1–9. Doi: 10.1155/2018/7879403.

Arya, A., Mishra, V. and Chundawat, T.S. 2019. Green synthesis of silver nanoparticles from green algae (*Botryococcus braunii*) and its catalytic behaviour for the synthesis of benzimidazoles. Chem. Data Collect. 20: 1–7.

Ashfaq, A., Khursheed, N., Fatima, S., Anjum, Z. and Younis, K. 2022. Application of nanotechnology in food packaging: Pros and Cons. J. Agri. Food Res. 7: 100270. https://doi.org/10.1016/j.jafr.2022.100270.

Aziz, M., Abbas, S.S. and Baharom, W.R.W. 2013. Size-controlled synthesis of SnO_2 nanoparticles by sol-gel method. Mater. Lett. 91: 31–34.

Babu, P.J. 2021. Nanotechnology mediated intelligent and improved food packaging. Inter. Nano Lett. https://doi.org/10.1007/s40089-021-00348-8.

Bansal, V., Rautaray, D., Ahmad, A. and Sastry, M. 2004. Biosynthesis of zirconia nanoparticles using the fungus *Fusarium oxysporum*. J. Mater. Chem. 14: 3303–3305.

Bansal, V., Rautaray, D., Bharde, A., Ahire, K., Sanyal, A., Ahmad, A. and Sastry, M. 2005 Fungus-mediated biosynthesis of silica and titania particles. J. Mater. Chem. 15: 2583–2589.

Barbosa de Lima, W.C.P., Nascimento, L.P.C., Lima Dantas, R., Lima-Tresena, N., Silva Júnior, J.B., Santos de Lima, G. and Barbosa de Lima, A.G. 2020. Heat transfer in the cooling, freezing and post-freezing of liquid food: Modeling and simulation. Diffus Found. 25: 37–53.

Barnett, J.A. 2003. Beginnings of microbiology and biochemistry: The contribution of yeast research. Microbiology 149(3): 557–567.

Basavaraja, S., Balaji, S.D., Lagashetty, A., Rajasab, A.H. and Venkataraman, A. 2008 Extracellular biosynthesis of silver nanoparticles using the fungus *Fusarium semitectum*. Mat. Res. Bull. 43: 1164–1170.

Behnajady, M.A., Eskandarloo, H., Modirshahla, N. and Shokri, M. 2011. Investigation of the effect of sol–gel synthesis variables on structural and photocatalytic properties of TiO_2 nanoparticles. Desalination 278: 10–17.

Benassai, E., Del Bubba, M., Ancillotti, C., Colzi, I., Gonnelli, C., Calisi, N., Salvatici, M.C., Casalone, E. and Ristori, S. 2021. Green and cost-effective synthesis of copper nanoparticles by extracts of non-edible and waste plant materials from *Vaccinium* species: Characterization and antimicrobial activity. Mater. Sci. Eng. C 119: 111453.

Bhainsa, K.C. and D'Souza, S.F. 2006. Extracellular biosynthesis of silver nanoparticle using the fungus *Aspergillus fumigates*. Colloids Surf. B Biointerf. 47: 160–4.

Bharde, A., Rautaray, D., Bansal, V., Ahmad, A., Sarkar, I. and Yusuf, S.M. 2006 Extracellular biosynthesis of magnetite using fungi. Small 2(1): 135–141.

Bhargava, A., Jain, N., Khan, M.A., Pareek, V., Dilip, R.V. and Panwar, J. 2016. Utilizing metal tolerance potential of soil fungus for efficient synthesis of gold nanoparticles with superior catalytic activity for degradation of rhodamine B. J. Environ. Manag. 183: 22–32.

Białkowska, A., Majewska, E., Olczak, A. and Twarda-clapa, A. 2020. Ice binding proteins: Diverse biological roles and applications in different types of industry. Biomolecules 10: 274. https://doi.org/10.3390/biom10020274.

Boroumandmoghaddam, A., Namvar, F., Moniri, M., Md Tahir, P., Azizi, S. and Mohamad, R. 2015. Nanoparticles biosynthesized by fungi and yeast: A review of their preparation, properties, and medical applications. Molecules (Basel, Switzerland) 20(9): 16540–16565.

Bourne, M.C. 1977. Post-harvest Food Losses: The Neglected Dimension in Increasing the World Food Supply. Cornell University International Agriculture, Mimeo, New York.

Brinker, C.J. and Scherer, G.W. 1990. Sol-Gel Science: The Physics and Chemistry of Sol-Gel Processing, Academic Press, USA.

Brust, M. and Kiely, C.J. 2002. Some recent advances in nanostructure preparation from gold and silver particles: A short topical review. Coll. Surf. A. Physicochem. Eng. Asp. 202: 175–186.

Burda, C., Chen, X., Narayanan, R. and El-Sayed, M.A. 2005. Chemistry and properties of nanocrystals of different shapes. Chem. Rev. 105: 1025–1102.

Calicioglu, O., Flammini, A., Bracco, S., Bellù, L. and Sims, R. 2019. The future challenges of food and agriculture: An integrated analysis of trends and solutions. Sustainability 11: 222. Doi: 10.3390/su11010222.

Cerqueira, M.A., Vicente, A.A. and Pastrana, L.M. 2018. Nanotechnology in food packaging: Opportunities and challenges nanomaterials for food packaging. pp. 577–611. *In*: Malik, A., Erginkaya, Z. and Erten, H. (eds.). Health and Safety Aspects of Food Processing Technologies. Springer, Switzerland.

Cha, D.S. and Chinnan, M.S. 2004. Biopolymer-based antimicrobial packaging: A review. Crit. Rev. Food Sci. Nutr. 44: 223–237.

Chandran, S.P., Chaudhary, M., Pasricha, R., Ahmad, A. and Sastry, M. 2006. Synthesis of gold nano triangles and silver nanoparticles using Aloe Vera plant extract. Biotechnol. Prog. 22: 577–583.

Chatterjee, S., Mahanty, S., Das, P., Chaudhuri, P. and Das, S. 2020. Biofabrication of iron oxide nanoparticles using manglicolous fungus *Aspergillus niger* BSC-1 and removal of Cr(VI) from aqueous solution. Chem. Eng. J. 385: 123790.

Christiansen, M.V., Pedersen, T.B. and Brønd, J.N. 2020. Physical properties and storage stability of reverse osmosis skim milk concentrates: Effects of skim milk pasteurisation, solid content and thermal treatment. J. Food Eng. 278: 109922. https://doi.org/10.1016/j.jfoodeng.2020.109922.

Chugh, D., Viswamalya, V.S. and Das, B. 2021. Green synthesis of silver nanoparticles with algae and the importance of capping agents in the process. J. Genet. Eng. Biotechnol. 19: 126. https://doi.org/10.1186/s43141-021-00228-w.

Colin, J.A., Pech-Pech, I.E., Oviedo, M., Águila, S.A., Romo-Herrera, J.M. and Contreras, O.E. 2018. Gold nanoparticles synthesis assisted by marine algae extract: Biomolecules shells from a green chemistry approach. Chem. Phys. Lett. 708: 210–215.

Couto, C. and Almeida, A. 2022. Metallic nanoparticles in the food sector: A mini-review. Foods 11: 402. https://doi.org/10.3390/foods11030402.

Dai, J. and Mumper, R.J. 2010. Plant phenolics: extraction, analysis and their antioxidant and anticancer properties. Molecules 15: 7313–7352.

Dameron, C.T., Reese, R.N., Mehra, R.K., Kortan, A.R., Carroll, P.J., Steigerwald, M.L. and Winge, D.R. 1989. Biosynthesis of cadmium sulphide quantum semiconductor crystallites. Nature 338(6216): 596–597.

Das, S.K., Dickinson, C., Lafir, F., Brougham, D.F. and Marsili, E. 2012. Synthesis, characterization and catalytic activity of gold nanoparticles biosynthesized with *Rhizopus oryzae* protein extract. Green Chem. 14(5): 1322–1334.

Dhand, C., Dwivedi, N., Loh, X.J., Ying. A.N.J., Verma, N.K., Beuerman, R.W., Lakshminarayanan, R and Ramakrishna, S. 2015. Methods and strategies for the synthesis of diverse nanoparticles and their applications: A comprehensive overview. RSC Adv. 5: 105003–105037.

Douglas, T. and Young, M. 1993. Host-guest encapsulation of materials by assembled virus protein cages. Nature 393: 152–155.

Du, J.J., Chen, C., Gan, Y.L., Zhang, R.H., Yang C.Y. and Zhou, X.W. 2014. Facile one-pot hydrothermal synthesis of Pt nanoparticles and their electrocatalytic performance. Int. J. Hydrogen Energy 39: 17634–17637.

Duran, N., Marcato, P.D., Alves, O.L., D'Souza, G. and Esposito, E. 2005. Mechanistic aspects of biosynthesis of silver nanoparticles by several *Fusarium oxysporum* strains. J. Nanobiotechnol. 3: 08–14.

El-Gendy, A.A., Ibrahim, E.M., Khavrus, V.O., Krupskaya, Y., Hampel, S., Leonhardt, A., Büchner, B. and Klingeler, R. 2009 The synthesis of carbon coated Fe, Co and Ni nanoparticles and an examination of their magnetic properties. Carbon 47(12): 2821–2828.

Farhoodi, M. 2016. Nanocomposite materials for food packaging applications: Characterization and safety evaluation. Food Eng. Rev. 8: 35–51.

Fayaz, M.A., Balaji, K., Girilal, M., Kalaichelvan, P.T. and Venkatesan, R. 2009. Mycobased synthesis of silver nanoparticles and their incorporation into sodium alginate films for vegetable and fruit preservation. J. Agri. Food Chem. 57(14): 6246–6252.

Filimonau, V. and De, Coteau, D.A. 2019. Food waste management in hospitality operations: A critical review. Tour Manag. 71: 234–245.

Gade, A.K., Bonde, P.P., Ingle, A.P., Marcato, P., Duran, N. and Rai, M.K. 2008. Exploitation of *Aspergillus niger* for synthesis of silver nanoparticles. J. Biobased Mater. Bioeng. 2: 243–7.

Gharibi, V., Khanjani, N. and Heidari, H. 2020. The effect of heat stress on hematological parameters and oxidative stress among bakery workers. Toxicol. Ind. Health 36: 1–10.

Golinska, P., Wypij, M., Ingle, A.P., Gupta, I., Dahm, H. and Rai, M. 2014. Biogenic synthesis of metal nanoparticles from Actinomycetes: Biomedical applications and cytotoxicity. Applied Microbiology and Biotechnology 98: 8083–8097.

Goncalves, L.F.F.F., Kanodarwala, F.K., Stride, J.A., Silva, C.J.R., Pereira, M.R. and Gomes, M.J.M. 2014. One-pot synthesis of CdSe nanoparticles exhibiting quantum size effect within a sol-gel derived ureasilicate matrix. J. Photochem. Photobiol. A 285: 21–29.

González-Ballesteros, N., Prado-López, S., Rodríguez-González, J.B., Lastra, M. and Rodríguez-Argüelles, M.C. 2017. Green synthesis of gold nanoparticles using brown algae *Cystoseira baccata*: Its activity in colon cancer cells. Colloids Surf. B Biointerf. 153: 190–198.

Griffin, M., Sobal, J. and Lyson, T. A. 2009. An analysis of a community food waste stream. Agri. Human Values 26: 67–81.

Herricks, T., Chen, J. and Xia, Y. 2004. Polyol synthesis of platinum nanoparticles: Control of morphology with sodium nitrate. Nano Lett. 4(12): 2367–2371.

Huang, J., Li, Q., Sun, D., Lu, Y., Su, Y., Yang, X., Wang, H., Wang, Y., Shao, W., He, N. Hong, J. and Chen, C. 2007. Biosynthesis of silver and gold nanoparticles by novel sundried *Cinnamomum camphora* leaf. Nanotechnology 18: 105104–105114.

Ingle, A., Gade, A., Pierrat, S., Sonnichsen, C. and Rai, M. 2008. Mycosynthesis of silver nanoparticles using the fungus *Fusarium acuminatum* and its activity against some human pathogenic bacteria. Curr. Nanosci. 4: 141–144.

Ingle, A., Rai, M., Gade, A. and Bawaskar, M. 2009. *Fusarium solani*: A novel biological agent for the extracellular synthesis of silver nanoparticles. J. Nanopart. Res. 11(8): 2079–2085.

Ingle, P.U., Biswas, J.K., Mondal, M., Rai, M.K, Kumar, P.S. and Gade, A.K. 2021. Assessment of *in vitro* antimicrobial efficacy of biologically synthesized metal nanoparticles against pathogenic bacteria. Chemosphere 132676. Doi: 10.1016/j.chemosphere.2021.132676.

Jameel, M.S., Aziz, A.A. and Dheyab, M.A. 2020. Green synthesis: Proposed mechanism and factors influencing the synthesis of platinum nanoparticles. Green Process. Synth. 9: 386–398.

Kanmani, P. and Lim, S.T. 2013. Synthesis and characterization of pullulan-mediated silver nanoparticles and its antimicrobial activities. Carbohydr. Polym. 97(2): 421–428.

Khan, M.J., Ramiah, S.K., Selamat, J., Shameli, K., Sazili, A.Q. and Mookiah, S. 2022. Utilisation of pullulan active packaging incorporated with curcumin and pullulan mediated silver nanoparticles to maintain the quality and shelf life of broiler meat. Italian J. Animal Sci. 21(1): 244–262.

Kim, J., Takahashi, M., Shimizu, T., Shirasawa, T., Kaita, M., Kanayama, A. and Miyamoto, Y 2008. Effects of a potent antioxidant, platinum nanoparticle, on the lifespan of *Caenorhabditis elegans*. Mech. Age Dev. 129: 322–331.

Konishi, Y., Ohno, K., Saitoh, N., Nomura, T. and Nagamine, S. 2004. Microbial synthesis of gold nanoparticles by metal reducing bacterium. Trans. Mater. Res. Soc. Japan 29: 2341–2343.

Kowshik, M., Ashtaputre, S., Kharrazi, S., Vogel, W., Urban, J., Kulkarni, S. and Paknikar, K. 2003. Extracellular synthesis of silver nanoparticles by a silver tolerant yeast strain MKY3. Nanotechnology 14: 95–100.

Krumov, N., Oder, S., Perner-Nochta, I., Angelov, A. and Posten, C. 2007. Accumulation of CdS nanoparticles by yeasts in a fed-batch bioprocess. J. Biotechnol. 132(4): 481–486.

Kumar, S.A., Ansary, A.A., Ahmad, A. and Khan, M.I. 2007. Extracellular biosynthesis of CdSe quantum dots by the fungus, *Fusarium oxysporum*. J. Biomed. Nanotechnol. 3: 190–194.

Kumari, R.M., Kumar, V., Kumar, M., Agrawal, A., Pareek, N. and Nimesh, S. 2020. Extracellular biosynthesis of silver nanoparticles using *Aspergillus terreus*: Evaluation of its antibacterial and anticancer potential. Mater. Today Proc. https://doi.org/10.1016/j.matpr.2020.04.494.

LaCoste, A., Schaich, K.M., Zumbrunnen, D. and Yam, K.L. 2005. Advancing controlled release packaging through smart blending. Pack. Technol. Sci. Int. J. 18: 77–87.

Lahde, A., Kokkonen, N., Karttunen, A.J., Jaaskelainen, S., Tapper, U., Pakkanen, T.A. and Jokiniemi, J. 2011. Preparation of copper-silicon dioxide nanoparticles with chemical vapor synthesis. J. Nanopart. Res. 13: 3591–3598.

Lengke, M.F., Fleet, M.E. and Southam, G. 2006. Synthesis of platinum nanoparticles by reaction of filamentous cyanobacteria with platinum (IV)-chloride complex. Langmuir 22: 7318–23.

Liangwei, D., Hong, J., Xiaohua, L. and Erkang, W. 2007. Biosynthesis of gold nanoparticles assisted by Escherichia coli DH5α and its application on direct electrochemistry of hemoglobin. Electrochem. Commun. 9: 1165–1170.

Lin, L., Wang, W., Huang, J., Li, Q., Sun, D., Yang, X., Wang, H., He, N. and Wang. Y. 2010. Nature factory of silver nanowires: Plant-mediated synthesis using broth of *Cassia fistula* leaf. Chem. Eng. J. 162: 852–858.

Liu, W., Zhang, M. and Bhandari, B. 2020. Nanotechnology—A shelf life extension strategy for fruits and vegetables. Crit. Rev. Food Sci. Nutr. 60(10): 1706–1721.

Ma, Y., Chen, M. and Li, M. 2015. Hydrothermal synthesis of hydrophilic NaYF4: Yb, Er nanoparticles with bright up conversion luminescence as biological label. Mater. Lett. 139: 22–25.

Maicu, M., Schmittgens, R., Hecker, D., Glob, D., Frach, P. and Gerlach G., 2014. Synthesis and deposition of metal nanoparticles by gas condensation process. J. Vac. Sci. Technol. A 32: 02B113. https://doi.org/10.1116/1.4859260.

Maliszewska, I., Szewczyk, K. and Waszak, K. 2009. Biological synthesis of silver nanoparticles. J. Phys. Confer. Ser. 146: 1–6.

Manikandakrishnan, M., Palanisamy, S., Vinosha, M., Kalanjiaraja, B., Mohandoss, S., Manikandan, R., Tabarsa, M., You, S.G. and Prabhu, N.M. 2019. Facile green route synthesis of gold nanoparticles using *Caulerpa racemosa* for biomedical applications. J. Drug Deliv. Sci. Technol. 54: 101345.

Manjunath, H. and Joshi, C.G. 2019. Characterization, antioxidant and antimicrobial activity of silver nanoparticles synthesized using marine endophytic fungus—*Cladosporium cladosporioides*. Process Biochem. 82: 199–204.

Melo, N.F.C.B., MendonçaSoares, B.L.D., Diniz, K.M., Leal, C.F., Canto, D., Flores, M.A.P., Tavares-Filho, J.H.D.C., Galembeck, A., Stamford, T.L.M., Stamford-Arnaud, T.M. and Stamford, T.C.M. 2018. Effects of fungal chitosan nanoparticles as eco-friendly edible coatings on the quality of postharvest table grapes. Postharvest Biol. Technol. 139: 56–66.

Mikhailova, E.O. 2021. Gold nanoparticles: Biosynthesis and potential of biomedical application. Journal of Functional Biomaterials 12: 70. https:// doi.org/10.3390/jfb12040070.

Mirzadeh, A. and Kokabi, M. 2007. The effect of composition and draw-down ratio on morphology and oxygen permeability of polypropylene nanocomposite blown films. Eur. Polymer J. 43: 3757–3765.

Mohanta, Y.K., Nayak, D., Biswas, K., Singdevsachan, S.K., Abd-Allah, E.F., Hashem, A., Alqarawi, A.A., Yadav, D. and Mohanta, T.K. 2018. Silver nanoparticles synthesized using wild mushroom show potential antimicrobial activities against food borne pathogens. Molecules 23: 655. Doi: 10.3390/molecules23030655.

Molnár, Z., Bódai, V., Szakacs, G., Erdélyi, B., Fogarassy, Z., Sáfrán, G., Varga, T., Kónya, Z., Tóth-Szeles, E., Szűcs, R. and Lagzi, I. 2018. Green synthesis of gold nanoparticles by thermophilic filamentous fungi. Scientific Reports 8: 3943. https://doi.org/10.1038/s41598-018-22112-3.

Mourdikoudis, S., Pallares, R.M. and Thanh, N.T.K. 2018. Characterization techniques for nanoparticles: Comparison and complementarity upon studying nanoparticle properties. Nanoscale 10: 12871–12934.

Mughal, B., Zaidi, S.Z.J., Zhang, X. and Hassan, S.U. 2021. Biogenic nanoparticles: Synthesis, characterisation and applications. Appl. Sci. 11: 2598. https:// doi.org/10.3390/app11062598.

Mukherjee, P., Ahmad, A., Mandal, D., Senapati, S., Sainkar, S.R. and Khan, M.I. 2001. Fungus mediated synthesis of silver nanoparticles and their immobilization in the mycelial matrix: A novel biological approach to nanoparticle synthesis. Nano Lett. 1: 515–519.

Mukherjee, P., Roy, M., Dey, G.K., Mukherjee, P.K, Ghatak, J., Tyagi, A.K. and Kale, S.P. 2008. Green synthesis of highly stabilized nanocrystalline silver particles by a non-pathogenic and agriculturally important fungus *Trichoderma. asperellum*. Nanotechnology 19(7): 075103. Doi: 10.1088/0957-4484/19/7/075103.

Nair, B. and Pradeep, T. 2002. Coalescence of nanoclusters and formation of submicron crystallites assisted by *Lactobacillus* strains. Cryst. Growth Des. 2: 293–298.

Okami, Y., Beppu, T. and Ogawara, H. 1988. Biology of actinomycetes, Japan. Sci. Soc. Press, Tokyo 88: 508.

Onizawa, S., Aoshiba, K., Kajita, M., Miyamoto, Y. and Nagai, A. 2009. Platinum nanoparticle antioxidants inhibit pulmonary inflammation in mice exposed to cigarette smoke. Pulm. Pharmacol. Ther. 22: 340–349.

Osonga, F.J., Kalra, S., Miller, R.M., Isika, D. and Sadik, O.A. 2020. Synthesis, characterization and antifungal activities of eco-friendly palladium nanoparticles. RSC Adv. 10: 5894–5904.

Pandit, C., Roy, A., Ghotekar, S., Khusro, A., Islam, M.N., Emran, T.B., Lam, S.E., Khandaker, M.U. and Bradley, D.A. 2022. Biological agents for synthesis of nanoparticles and their applications. Journal of King Saud University: Science 34: 101869. https://doi.org/10.1016/j.jksus.2022.101869.

Park, B.K., Jeong, S., Kim, D., Moon, J., Lim, S. and Kim, J.S. 2007. Synthesis and size control of monodisperse copper nanoparticles by polyol method. J. Colloid Interface Sci. 311: 417–424.

Perez-Tijerina, E., Mejía-Rosales, S., Inada, H. and Yacaman, M. 2010. Effect of temperature on AuPd nanoparticles produced by inert gas condensation. J. of Phys. Chem. C 114: 6999–7003.

Rai, M., Ingle, A.P., Gupta, I., Pandit, R., Paralikar, P., Gade, A., Chaud, M.C. and Santos, C.A.D. 2019. Smart nanopackaging for the enhancement of food shelf life. Environ. Chem. Lett. 17: 277–290.

Rajakumar, G., Rahuman, A.A., Roopan, S.M., Khanna, V.G., Elango, G., Kamaraj, C., Zahir, A.A. and Velayutham, K. 2012. Fungus-mediated biosynthesis and characterization of TiO2 nanoparticles and their activity against pathogenic bacteria. Spectrochim. Acta A: Mol. Biomol. Spectros. 91: 23–29.

Ranjani, S., Janani, P.G., Karunya, J.R. and Hemalatha, S. 2022. Differential actions of nanoparticles and nanoemulsion synthesized from *Colletotrichum siamense* on foodborne pathogen. LWT: Food Sci. Technol. 155: 112995. https://doi.org/10.1016/j.lwt.2021.112995.

ReFED. 2016. A Roadmap to Reduce U.S. Food Waste by 20 Percent (available at: https://staging.refed. org/downloads/ReFED_Report_2016.pdf, accessed on 5 April, 2022).

Riddin, T.L., Gericke, M. and Whiteley, C.G. 2006. Analysis of the intra- and extracellular formation of platinum nanoparticles by *Fusarium oxysporum* f. sp. *lycopersici* using response surface methodology. Nanotechnology 17: 3482–3489.

Roy, S., Priyadarshi, R. and Rhim, J.W. 2021. Development of multifunctional pullulan/chitosan-based composite films reinforced with ZnO nanoparticles and propolis for meat packaging applications. Foods 10: 2789. https://doi.org/10.3390/ foods10112789.

Sadowski, Z., Maliszewska, I.H., Grochowalska, B., Polowczyk, I. and Kozlecki, T. 2008. Synthesis of silver nanoparticles using microorganisms. Mat. Sci. Pol. 26: 419–424.

Salem, D.M.S.A., Ismail, M.M. and Aly-Eldeen, M.A. 2019. Biogenic synthesis and antimicrobial potency of iron oxide (Fe_3O_4) nanoparticles using algae harvested from the Mediterranean Sea, Egypt. Egypt. J. Aquat. Res. 45: 197–204.

Sanghi, R. and Verma, P. 2009. Biomimetic synthesis and characterization of protein capped silver nanoparticles. Bioresource Technol. 100: 501–504.

Santhosh, P.B., Genova, J. and Chamati, H. 2022. Green synthesis of gold nanoparticles: An eco-friendly approach. Chemistry 4: 345–369.

Saravanakumar, K., Sathiyaseelan, A., Mariadoss, A.V.A., Xiaowen, H. and Wang, M.H. 2020. Physical and bioactivities of biopolymeric films incorporated with cellulose, sodium alginate and copper oxide nanoparticles for food packaging application. Int. J. Biol. Macromol. 153: 207–214.

Sastry, M., Ahmad, A., Khan, I. and Kumar, R. 2003. Biosynthesis of metal nanoparticles using fungi and actinomycete. Curr. Sci. 85(2): 162–170.

Schmidt, D., Shah, D. and Giannelis, E.P. 2002. New advances in polymer/ layered silicate nanocomposites. Curr. Opin. Solid State Mater. Sci. 6: 205–212.

Shankar, S., Ahmad, A. and Sastry, M. 2003. Geranium leaf assisted biosynthesis of silver nanoparticles Biotechnol. Prog. 19: 1627–1631.

Shankar, S., Rai, A., Ahmad, A. and Sastry, M. 2004. Rapid synthesis of Au, Ag, and bimetallic Au core–Ag shell nanoparticles using neem (*Azadirachta indica*) leaf broth. J. Colloid Interface Sci. 275: 496–502.

Sharif, Z., Mustapha, F., Jai, J., Mohd-Yusof, N. and Zaki, N. 2017. Review on methods for preservation and natural preservatives for extending the food longevity. Chem. Eng. Res. Bull. 19: 145–153.

Sharma, C., Dhiman, R., Rokana, N. and Panwar, H. 2017 Nanotechnology: An untapped resource for food packaging. Front. Microbiol. 8: 1735.

Sharma, J.L., Dhayal, V. and Sharma, R.K. 2021. White-rot fungus mediated green synthesis of zinc oxide nanoparticles and their impregnation on cellulose to develop environmental friendly antimicrobial fibers. 3 Biotech 269. Doi: 10.1007/s13205-021-02840-6.

Slawson, R.M., Van Dyke, M.I., Lee, H. and Trevor, J.T. 1992. Germanium and silver resistance, accumulation and toxicity in microorganisms. Plasmid 27: 73–79.

Sridhar, A., Ponnuchamy, M., Kumar, P.S. and Kapoor, A. 2021. Food preservation techniques and nanotechnology for increased shelf life of fruits, vegetables, beverages and spices: A review. Environ. Chem. Lett. 19(2): 1715–1735.

Suleiman, R. and Rosentrater, K. 2015. Current maize production, postharvest losses and the risk of mycotoxins contamination in Tanzania. An ASABE Meeting Presentation, Paper Number: 152189434.

Suryavanshi, P., Pandit, R., Gade, A., Derita, M., Zachino, S. and Rai, M. 2017. *Colletotrichum* sp. mediated synthesis of sulfur and aluminium oxide nanoparticles and its *in vitro* activity against selected food-borne pathogens. LWT-Food Sci. Technol. 81: 188–194.

Syed, A. and Ahmad, A. 2012. Extracellular biosynthesis of platinum nanoparticles using the fungus Fusarium oxysporum. Colloids Surf. B Biointerfaces 97: 27–31.

Tager, J. 2014. Nanomaterials in food packaging: FSANZ fails consumers again. Chain React. 16: 17.

Thakur, V.K. and Gupta, R.K. 2016. Recent progress on ferroelectric polymer-based nanocomposites for high energy density capacitors: synthesis, dielectric properties, and future aspects. Chem. Rev. 116: 4260–4317.

Thompson, L.A. and Darwish, W.S. 2019. Environmental chemical contaminants in food: Review of a global problem. J. Toxicol. 2019: 2345283. https://doi.org/10.1155/2019/2345283.

Thostenson, E.T., Li, C. and Chou, T.W. 2005. Nanocomposites in context. Composites Sci. Technol. 65(3-4): 491–516.

Valoppi, F., Agustin, M., Abik, F., de Carvalho, D.M., Sithole, J., Bhattarai, M., Varis, J.J., Arzami, A.N.A.B., Pulkkinen, E. and Mikkonen, K.S. 2021. Insight on current advances in food science and technology for feeding the world population. Front. Sustain. Food Syst. https://doi.org/10.3389/fsufs.2021.626227.

Vidya, P. and Subramani, G. 2017. Fungus mediated synthesis of silver nanoparticles using *Aspergillus flavus* and its antibacterial activity against selective food borne pathogens. Indo. Am. J. P. Sci. 4(12): 4627–4634.

Vigneshwaran, N., Kathe, A.A., Varadarajan, P.V., Nachane, R.P. and Balasubramanya, R.H. 2006. Biomimetics of silver nanoparticles by white rot fungus, *Phaenerochaete chrysosporium*. Coll. Surf. B Biointerf. 53: 55–59.

Vigneshwaran, N., Kathe, A.A., Varadarajan, P.V., Nachane, R.P. and Balasubramanya, R.H. 2009. Silver-protein (core-shell) nanoparticles production using spent mushroom substrate. Langmuir 23: 7113–7117.

Vijayanandan, A.S. and Balakrishnan, R.M. 2018. Biosynthesis of cobalt oxide nanoparticles using endophytic fungus *Aspergillus nidulans*. J. Environ. Manag. 218: 442–450.

Ward, M.B., Brydson, R. and Cochrane, R.F. 2006. Mn nanoparticles produced by inert gas condensation. J. Phys.: Conf. Ser. 26: 296–299.

Wypij, M., Golinska, P., Dahm, H. and Rai, M. 2016. Actinobacterial-mediated synthesis of silver nanoparticles and their activity against pathogenic bacteria. IET Nanobiotechnology 11(3): 336–342.

Wypij, M., Świecimska, M., Dahm, H., Rai, M. and Golinska, P. 2019. Controllable biosynthesis of silver nanoparticles using actinobacterial strains. Green Processing and Synthesis 8(1): 207–214.

Xing, T., Sunarso, J., Yang, W., Yin, Y., Glushenkov, A.M., Li, L.H., Howlett, P.C. and Chen, Y. 2013. Ball milling: A green mechanochemical approach for synthesis of nitrogen doped carbon nanoparticles. Nanoscale 5: 7970–7976.

Zambrano-Zaragoza, M.L., González-Reza, R., Mendoza-Muñoz, N., Miranda-Linares, V., Bernal-Couoh, T.F., Mendoza-Elvira, S. and Quintanar-Guerrero, D. 2018. Nanosystems in edible coatings: A novel strategy for food preservation. Int. J. Mol. Sci. 19(3): 705. Doi: 10.3390/ijms19030705.

Zarei, V., Mirzaasadi, M., Davarpanah, A., Nasiri, A., Valizadeh, M. and Hosseini, M.J.S. 2021. Environmental method for synthesizing amorphous silica oxide nanoparticles from a natural material. Processes 9: 334.

Zhang, C., Liu, J., Li, H., Qin, L., Cao, F. and Zhang, W. 2019a. A fungal based synthesis method for copper nanoparticles with the determination of anticancer, antidiabetic and antibacterial activities Sadaf. Appl. Catal. B Environ. 261: 118224.

Zhang, H., Zhou, H., Bai, J., Li, Y., Yang, J., Ma, Q. and Qu, Y. 2019b. Biosynthesis of selenium nanoparticles mediated by fungus *Mariannaea* sp. HJ and their characterization. Colloids Surf. A Physicochem. Eng. Asp. 571: 9–16.

11

Fungal Nanopesticides as Next Generation Insecticides for Insect Control

Murugan Arunthirumeni,[#] Kandhasamy Lalitha[#] and Muthugounder Subramanian Shivakumar*

Introduction

Myconanotechnology has emerged as the combination of mycology and nanotechnology for the production of eco-friendly nanomaterials. Though myconanotechnology is in its infancy, potential applications can lead to transformation in agriculture and can provide incremental solutions through green chemistry approaches for food security (Rai et al. 2009). The current interest in metallic nanoparticles is due to the ability of fungal mycelia and secondary metabolites in reducing metallic ions from the salt. Nanotechnology has revolutionized our approaches in material science, chemical, biomedical, and biology. Nanosized particles have been used by humans in the form of finely ground powders for several hundred years not understanding its functional importance. Today, nanomaterials find their use in every walk of life ranging from cosmetics, food products, electronics, biomedical equipment, drug delivery systems to nanoemulsions used in paints and for synthesizing efficient chemical pesticides (Togola et al. 2018, Abd-Elsalam 2021). Several types of materials are used to make nanoparticles which include metals, metal oxides, carbon quantum dots, carbon, ceramics, silicates, lipids, polymers, proteins, dendrimers, nitrogen-doped nanoparticles, and emulsions (Puoci et al. 2008). Nanoparticles have multifunctional characters and have unique properties, such as physicochemical, optical, mechanical, magnetic, etc., that are helpful to

Molecular Entomology Lab., Department of Biotechnology, Periyar University, Salem-636011, Tamilnadu, India.
* Corresponding author: skentomology@gmail.com, sk24@periyaruniversity.ac.in
[#] These authors have contributed equally in this Chapter.

make an appropriate product for their use in various applications (Yadav et al. 2015). Nanoparticles have great advantage of ultra-small size and large surface area (Prasad et al. 2017).

The synthesis of nanoparticles is done using two fundamental approaches namely top-down and bottom-up methods. The top-down approach usually involves the breaking of van der Waals forces between the bulk components for the attainment of thin layer crystals. In the bottom-up method, ionic or covalent bonding are involved (Song et al. 2019). Top-down approaches include laser ablation, ion sputtering, mechanical milling, etc. (Rajput 2015). Besides there are three ways of synthesizing nanoparticles, which include physical, chemical, and biological or green synthesis approaches. While physical and chemical approaches produce relatively uniform-sized particles, their production has several constraints as it requires high-end equipment which is costly and leaves large environmental footprint. However, these methods are currently used for large-scale production of nanomaterials for their application in material science, pesticide, formulations, controlled release, and in the biomedical industry (Prasad et al. 2017, Zielinska et al. 2020, Chaud et al. 2021). Green synthesis or biological synthesis of nanoparticles is an upcoming field and in the past decade, several techniques for the synthesis have been under development. The green synthesis has several advantages as compared to their counterparts, as they are environmental friendly, easy to produce, and cheap. At the same time, there are certain shortcomings as the particles produced have variations in their size, though this variation can be reduced by optimization of the methods depending on the type of biological molecule/organism and type of nanomaterial used. Green synthesized nanoparticles find their applications in antimicrobial products, biopesticides, gene, and drug delivery systems. Yet their potential is not fully realized. Among the various types of materials used in biological synthesis, novel metals and metal oxides have been studied in detail. In addition, carbohydrate polymers are now being increasingly researched for their use in biological synthesis.

In agriculture, nanotechnology has played a significant role in the production of controlled release of agrochemicals as well as producing formulations which have resulted in increasing their efficiency. Nanofertilizers are now increasingly being used for maximizing the bioavailability of minerals to aid in faster plant growth and development. Nanopesticides are used to suppress the development of phytopathogens (El-Ashry et al. 2021) and in the control of insect pests of agriculture and medically important insect pests like mosquitoes.

Insects are an important group of arthropods that are economically important to humans. Insect competes with humans for common resources like food production, damage stored food produce, and also are vectors of several diseases in humans and domesticated animals (Dhaliwal et al. 2010, Rai et al. 2014). For several centuries there has been a constant struggle by humans for controlling the damage caused by insect pests. Humans have used several methods to protect, prevent and kill insect pests of agriculture. With the development of organic chemistry during the 20th century, various organochemicals like DDT, BHC, organophosphates, and synthetic pyrethroids (Muthusamy and Shivakumar 2015) were produced and used for the control of insect pests. These insecticides though effective initially, have several drawbacks like they were highly toxic to humans, domesticated animals, and non-target organisms (Sanchez-Bayo 2012). In addition, regular use of organochemicals was harmful, as these were non-biodegradable and led to contamination of soil and water bodies causing environmental biomagnification leading to their accumulation in various levels of foodweb and resulting in environmental pollution. After a decade of use of DDT, the effectiveness of DDT in control of mosquitoes and houseflies diminished (Abbas et al. 2014). Similar observations were made for organophosphates and synthetic pyrethroid insecticides whose effectiveness in controlling insects diminished over time (Ramkumar and Shivakumar 2015). This phenomenon was linked to the ability of insect pests to overcome the pesticides, which was later termed as Insecticide Resistance (Opiyo et al. 2021). Today, insecticide resistance is recognized as a major hindrance in insect management programmes for both agricultural and medically important insects (Zhu et al. 2016). This resulted in a search for novel insecticides, which are effective, environmentally safe, and are specific to target insect pests.

Entomopathogenic microbes like bacteria, fungi, viruses, nematodes, and plant secondary metabolites are now identified as potential insecticides for the management of insecticide resistance in insects and at the same time, their environmental footprints are low. A few of these microbial pesticides have been commercialized for, e.g., *Bacillus thrungiensis*, *Pseudomonas fluorescence*, *Nucleo polyhedral virus*, *Beauveria bassiana*, *Trichoderma viridae*, and a few entomopathogenic nematodes are being mass-produced (Senthil-Nathan 2015). However, there are a few constraints in their large-scale production and storage especially with fungal and nematode biopesticides, as their effectiveness decreases with storage at low humidity and higher temperatures. This limits their use in subtropical and tropical agroecosystems. Plant and fungi produce a variety of secondary metabolites having bioactive properties. They are also increasingly being researched for their application in insect control and a few of them like azadirachtin, nimbin, pyrethrin, nicotine are mass-produced (Enyiukwu et al. 2014).

The constraints in the use of entomopathogenic microbe and to increase the effectiveness of secondary metabolites have been solved by their combination with nanotechnology-based approaches. Entomopathogenic microbes and secondary metabolites have several function groups which can interact with metals and their oxides. This property has led to a tremendous explosion in its use in the biological synthesis of nanoparticles. In this chapter we will look into various types of nanomaterials used in biological synthesis and methods of their characterization. Among the microbial pesticides and secondary metabolites, fungal-based insecticides are an important class that holds a lot of potential in insect control, so the special focus is given to fungal-based nanoinsecticides, their use in agroecosystems, and for mosquito control. In addition, recent developments in the formulation of nanoinsecticides and their regulatory approval are also discussed.

Types of Nanomaterials used in Biological Synthesis

Metal and Metal Oxide Nanoparticles

Metal-based nanoparticles are synthesized using various types of noble metal salts such as silver, gold, platinum, copper, zinc, titanium, and magnesium. These metals have gained considerable attention for biomedical applications due to their multifunctional abilities (Nivetha et al. 2021, Salem and Fouda 2021). Gold nanoparticles have been used since olden times for curing various diseases. Silver nanoparticles (AgNPs) has been widely used in nanoparticle synthesis due to their unique antimicrobial property and its ability to interact with biomolecules. Although silver is toxic at higher concentrations, several studies have established that a lower concentration of $AgNO_3$ has higher chemical stability, catalytic activity, biocompatibility, and intrinsic therapeutic potential activity (Fahimirad et al. 2019). Zinc oxide nanoparticles (ZnONPs) are widely used in the fields of electrochemistry, medical devices, cosmetics, and the textile industry due to their high specific surface area, biocompatibility, ultraviolet light absorption, scattering, and antibacterial properties (Xu et al. 2021). Copper nanoparticles (CuNPs) are widely used for biomedicine, as it is an important micronutrient and is an essential component in proteins and enzymes. Besides these metals, other metals and their oxides such as titanium, selenium, iron, and platinum are also widely used in the biological synthesis for drug delivery, antimicrobial creams, and insecticidal properties (Arunthirumeni et al. 2021, Kumaravel et al. 2021).

Carbon-based Nanoparticles

The carbon-based nanomaterials (CNMs) are classified as zero-dimensional nanomaterials of fullerenes, graphene quantum dots (GQDs), and carbon dots (CDs). One-dimensional nanomaterials are carbon nanotubes (CNTs) and carbon fibers (CFs). Graphene and graphitic carbon nitride (g-CN) are classified as two-dimensional nanomaterials that show strong broad-spectrum biological activity. Nano-diamonds (NDs) are classified as three-dimensional nanomaterials (Liang et al. 2021, Rilley and Narayan 2021). The size of the carbon-based nanomaterial is < 10 nm. The QDs have unique properties intermediate between single atoms and bulk materials, resulting in the very small size of QDs in the common range of 2–6 nm.

Green Synthesis of Nanoparticles and their Characterization

Broadly two different types of nanoparticles are used, namely inorganic and organic nanoparticles. The inorganic nanoparticles include metal and metal oxides including silver (Ag), iron oxide (Fe_3O_4), titanium oxide (TiO_2), copper oxide (CuO), and zinc oxide (ZnO), etc. Green synthesis of nanoparticles has been adopted to accommodate various biological sources like bacteria, fungi, algae and plant extracts (Singh et al. 2018). This biological material can be synthesized by modifying parameters such as solvent, temperature, pressure, and pH conditions for nanoparticle synthesis. These nanoparticles have a wide range of applications in agriculture, food industry, and medicine, because of the presence of biological materials which contain rich sources of active compounds like metabolites, protein, flavonoids, phenolic compounds, etc. These metabolites have hydroxyl, carbonyl, and amine functional groups that react with metal ions and reduce their size into the nano range. Specifically, flavonoids play an important role in the reduction of metal ions in nanoparticle synthesis and are also involved in the capping of nanoparticles (Naseer et al. 2020). Green biogenic metallic nanoparticle synthesis can be divided into two categories. One is bioreduction, in this process, the metal ions are chemically reduced and take a more stable form. Another one is biosorption, which involves the binding of metal ions (Pantidos and Horsfall 2014).

Nanoparticles Synthesized by using Plant Extract

Plants are a rich source of active biomolecules, metabolic compounds like proteins, vitamins, coenzymes-based intermediates, phenols, flavonoids, and carbohydrates. These molecules have the ability to reduce metal ions and hence are used in the synthesis of nanoparticles. Plants such as *Azadirachta indica, Ocimum tenuiflorum* (Banerjee et al. 2014), *Ficus benghalensis* (Saware and Venkataraman 2014) *Polyalthia longifolia* (Kaviya et al. 2011) have been used for the synthesis of AgNPs. Plant-mediated nanoparticle synthesis mainly uses metals such as silver, gold, platinum, selenium, iron, etc. The nanoparticles synthesized in this are widely used in antimicrobial creams and for insect control.

Nanoparticles Synthesized by using Bacteria

Bacterial mediated nanoparticle synthesis is a very promising field in which the bacteria have the ability to reduce heavy metal ions. Biological synthesis of metal nanoparticles using bacteria or bacterial supernatant is easy (Wijnhoven et al. 2009). Bacterial nanoparticle biosynthesis can take place either intracellularly or extracellularly. Extracellular synthesis is preferred, as prior removal of the bacterial cells is not required which simplifies the recovery of nanoparticles. Different metals and metal oxides such as gold, silver, platinum, palladium, titanium, titanium dioxide, and magnetite are used in nanoparticles synthesis (Kalaimurugan et al. 2019). Zinc oxide nanoparticles (ZnNPs) are widely popular for their antibacterial activity (Shankar and Rhim 2017). *Bacillus* sp., *E. coli, Lactobacillus* sp., *Klebsiella* sp., and *Pseudomonas* sp. are few bacteria that can reduce ZnNPs (Pereira et al. 2015). Silver and gold nanoparticles were synthesized from *Pseudomonas stutzeri, Bacillus megaterium* (Saravanan et al. 2011). *Pseudomonas aeruginosa, Cyanobacteria, Brevibacterium casei, Rhodopseudomonas capsulata, Escherichia coli,* and *Bacillus subtilis* (Rajeshkumar et al. 2016) have been used for nanoparticle synthesis.

Nanoparticles Synthesized by using Algae

Green algae contain a rich source of bioactive compounds like proteins, lipids, carbohydrates, carotenoids, vitamins, and other secondary metabolites and also have biological activity (Mahajan et al. 2019). Various algae species namely *Bifurcaria bifurcata, Caulerpa peltata, Hypnea valencia, Sargassum myriocystum, Sargassum muticum, Acanthophora spicifera, Chlorella pyrenoidusa, Kappaphycus alvarezii, Sargassum wightii* and *Laminaria japonica* have been used with metal and metal oxides like as Ag, Au, Zn, and CuO for nanoparticles synthesis. The advantages of algae-based NPs synthesis are algae are fast-growing, easy to handle, cost-effective and non-toxic (AINandhari et al. 2021).

Nanoparticles Synthesized using Symbiotic Bacteria from Nematodes

Metabolites from symbiotic bacteria from entomopathogenic nematodes contain hydrolytic exoenzymes, bacteriocins, and antibiotic, intra, and extracellular proteins. These are the compounds having insecticidal and antimicrobial properties (Shapiro-Ilan and Gaugler 2002). These proteins hold the potential to be synthesized as nanoparticles with the help of metals. Production of biosyntheised nanoparticles from symbiotic bacteria extracted from entomopathogenic nemamtodes have few study reports, viz., synthesis of gold and silver nanoparticles from symbiotic bacteria *Photorhabdus luminescens* (Aiswarya et al. 2019) and also Copper nanoparticles synthesized using *Morganella morgonii* have good mosquitocidal activity (Lalitha et al. 2020). With developments in green synthesis, it may be possible to load these metabolites for an effective delivery system and thus present with a lot of potential in mosquito control.

Nanoparticles Synthesized using Fungi

The entomopathogenic fungus possesses the ability to infect and cause death in all insect species. *Beauveria bassiana, Metarizam anisopliae,* and *Isaria fumosorosea* are mainly used for biocontrol programs. Nowadays, many fungi such as *Fusarium oxysporum, Beauveria bassiana* (Vivekanandhan et al. 2018) *M. anisopliae* (Kumaravel et al. 2021) *Aspergillus niger,* and *Aspergillus flavus* have been used for synthesized nanoparticles (Balumahendhiran et al. 2019, Karthi et al. 2018). These fungi produce distinct extracellular serine proteases, such as subtilisin-like proteases, trypsin-like proteases, metalloproteases, and eco-acting proteases that are believed to be important for host cuticle degradation (Yosri et al. 2018). *Aspergillus terreus* and *M. anisopliae* have been reported as promising entomopathogenic fungi for the control of many vectors larvae. The nanoparticles synthesized from fungal metabolites and extracellular enzymes are an alternative biocontrol agent and obtain nanoparticles with insecticidal activity. The fungus organism mentioned in (Figure 1) is routinely used to synthesize nanoparticles. (Santos et al. 2021, Bawin et al. 2016).

The advantage of synthesizing fungal nanoparticles is due to the ease of handling compared with other microorganisms, which makes the biosynthesis of nanoparticles using fungi potentially exciting. Previously, chemical and physical methods were achieved for the synthesis of metal nanoparticles (Shankar et al. 2004), but these methods have certain disadvantages due to the involvement of toxic chemicals and radiation. Biological synthesis using entomopathogenic fungi has been preferably used for the synthesis of a variety of nanoparticles. Fungi produce and secrete a high amount of active metabolic compounds, enzymes, and mycotoxins, which have the ability to reduce metal salts and form nanoparticles, and can be handled easily. Entomopathogenic fungi are increasingly studied in a biological control context regarding their ability to infect and cause the death of their host with more or less selectivity (Bawin et al. 2016). The entomopathogenic fungi produce distinct extracellular serine proteases, such as subtilisin-like proteases, trypsin-like

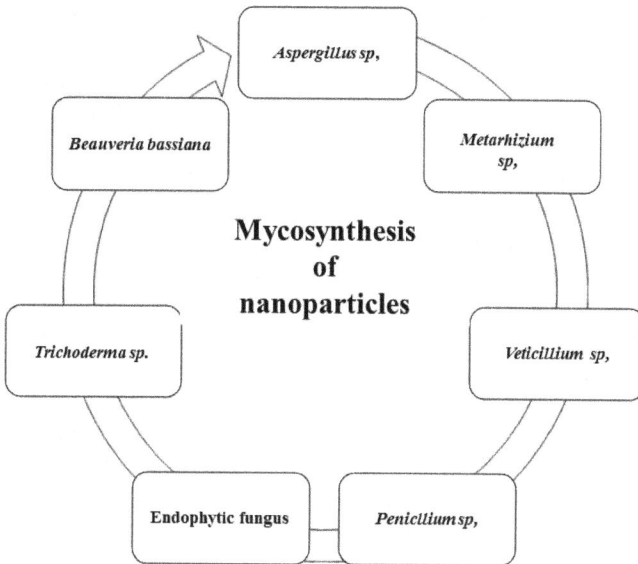

Figure 1. Fungus used for mycosynthesis of nanoparticles for insect control.

proteases, metalloproteases, and several families of exo-acting peptidases that are believed to be important for host cuticle degradation (Leger and Wang 2010). These organic compounds are necessary both for maintaining the vital functions of fungi and for effectively infecting pathogens by damaging the nervous system or reducing insect resistance.

Fungal Biopesticides and their Types

According to the United States Environmental Protection Agency (USEPA), biopesticides are pesticides prepared from natural sources such as animals, plants, microbes, and minerals, also including live organisms that obliterate agricultural pests. Biopesticides target specific pests and are considered to be an effective solution for the management of insect pests. Fungal biopesticides are potentially the most adaptable biological control means having a broad host range. These fungi comprise various groups of over 90 genera with approximately 750 species. Entomopathogenic fungi are among the most important group which are considered as potential biological control agents. Entomopathogenic fungi like *Beauveria bassiana*, *Metarhizium* sp., *Nomuraea rileyi*, *Verticillium* sp., and *Trichoderma* sp. infect a wide range of insect pests and have a cosmopolitan distribution. Fungal spores and fungal metabolites are considered as potential biological control agents. Their use includes biofungicides (*Trichoderma*), bioherbicides (*Phytophthora*), and bioinsecticides (*Beauveria* and *Metarhizium*).

Beauveria bassiana, a white muscardine fungus that grows naturally in soil has a worldwide distribution and is pathogenic on various insect species (Sandhu et al. 2004). Several insect pests of agriculture such as the Colorado potato beetle, the codling moth, termites, *Helicoverpa armigera* are controlled using *B. bassiana*

(Thakur et al. 2020). Oil formulated *B. bassiana* conidia is considered for use in the management of forest-defoliating caterpillars such as the eastern spruce budworm (Hicks 2016). *Metarhizium* causes a disease known as "green muscardine" in insect hosts, named after the the green color of its conidial cells. Several countries use *M. anisopliae* and *M. acridum* as an active ingredient in the development of mycoinsecticide and mycoacaricides, to target pests. Several commercial products such as BIO1020, a mycoinsecticide developed by Bayer, Germany, and Green Guard, a mycoacaricide developed by Becker Underwood Inc. (Ames, IA, United States) contain *M. acridum* as an active ingredient to target pests (Aw and Hue 2017). *Trichoderma* species are frequent in soil and root ecosystems of several plants and they are easily isolated from soil, decaying wood, and other organic material (Zeilinger and Omann 2007). *Trichoderma* species have been very popular as mycofungicides as they are fast-growing and have high reproductive capability. *Trichoderma*, have strong aggressiveness against phytopathogenic fungi and also promote plant growth (Vinale et al. 2006). Several studies have shown that *Trichoderma* spores and its metabolites have the ability to control insect pests (Rodriguez-Gonzalez et al. 2020, Vijayakumar and Alagar 2017, Chinnaperumal et al. 2018). *Verticillium lecanii* is a widely distributed fungus, which can cause huge epizootic in tropical and subtropical regions (Nunez et al. 2008). *V. lecanii* is primarily marketed for control of greenhouse aphids, whiteflies, and thrips (Wraight et al. 2001) *V. lecanii* is commercialized under the name of Mycotal only for use against whitefly larvae. *Paecilomyces fumosoroseus* is one of the most important natural enemies of whiteflies worldwide and causes yellow muscardine disease (Nunez et al. 2008). It is high epizootic against *Trialeurodes* and *Bemisia* spp. in both greenhouse and open field environments (Kim et al. 2002). *P. fumosoroseus* is currently used for the biological control of the whiteflies *Bemisia tabaci* and *B. argentifolii. Nomuraea rileyi*, is a dimorphic hyphomycete that can cause epizootic on various plant infesting insects. Insects belonging to Lepidoptera and coleopteran are susceptible (Srisukchayakul et al. 2005). *Isaria fumosorosea* and some other *Isaria* spp., are important entomopathogenic fungi with a worldwide distribution. Their host range includes aphids, leafhoppers, whiteflies, citrus psyllid, subterranean termites, rice weevils, and leaf beetles (Majeed et al. 2017).

Endophytic Fungi and their Metabolites

Endophytic fungi are a group of fungi which inhabit the tissue of living plants deprived of causing any apparent disease symptoms (Strobel and Daisy 2003). Endophytic fungi have been comprehensively studied and discovered to produce a significant variety of natural products (Maxwell et al. 2018). Tropical, semi-arid areas and rainforests have the ultimate diversity and richness of endophytes due to their enormous plant diversity (Deng et al. 2013). The association between host plant and endophytes vary based on symbiosis and pathogenesis. Many bioactive metabolites are synthesized by endophytes which include phenolic acids, steroids, flavonoids, quinones, terpenoids, glycosides, and alkaloids (Kusari et al. 2014). These metabolites act as herbicidal, immunomodulatory, insecticidal, antidiabetic agents and bio-active compounds have been isolated from them (Kaul et al. 2012).

Many entomopathogenic fungal (EPF) species, such as *Beauveria bassiana*, *Isaria fumosorosea*, and *Metarhizium anisopliae*, naturally occurring as endophytes, have been recorded to provide effective protection against herbivores.

Mode of Action of Fungal Biopesticides

Entomopathogenic fungi have the same basic mode of action when infecting insect hosts and are unique in that they can infect insect hosts through the cuticle (Bidochka and Small 2005). Fungal conidia come into direct contact with the host and adhere to the cuticle via nonspecific hydrophobic mechanisms (Inglis et al. 2001). Under specific environmental conditions the conidia germinate and a germ tube, or appressorium (penetration-peg structure), is produced (Inglis et al. 2001). The cuticle is penetrated by mechanical pressure from the appressorium and the action of cuticle-degrading enzymes, such as trypsin, metalloproteases, and aminopeptidases. The fungus grows by vegetative growth in the host hemocoel and external conidia are produced upon the death of the host when fungal hyphae exit through the less sclerotised areas of the cuticle (Inglis et al. 2001). These fungal hyphae grow into the insect, feed on its body tissue, produce toxins, and reproduce. It takes up to seven days for the insect to die (Roberts and St. Leger 2004). During favorable (moist) conditions (92% humidity or greater), *B. bassiana* release more spores into the environment to repeat the cycle on other pest insects. Entomopathogenic fungi produce secondary metabolites, including bassianin, bassiacridin, beauvericin, bassianolide, beauverolides, tenellin and oosporein (Quesada-Moraga and Alain 2004). It also produces insecticidal enzymes proteases, chitinases, and lipases, which can degrade the insect cuticle. EPF species, such as *Metarhizium* and *Beauveria*, toxic metabolite inhibit the action of chitin synthesis, degrades the insect midgut tissue and cuticle (Inglis et al. 2001, Kim et al. 2002).

Fungi-based Nano-insecticides

The synthesis of metal nanoparticles by using a fungus means that the fungal mycelium is exposed to the metal salt solution, which stimulates the fungus to produce enzymes and metabolites for its own survival. In this process, the toxic metal ions are reduced to the non-toxic metallic solid nanoparticles through the catalytic effect of the extracellular enzyme and metabolites of the fungus (Vahabi et al. 2011). Chemical and physical methods for nanomaterial synthesis need energy, employ toxic chemicals and require higher temperatures. Therefore, nontoxic and environment friendly process for the synthesis of nanoparticles is in great demand. Biological systems are characterized by the processes that occur at ambient temperature and pressure. In microbial synthesis, bacteria were mostly studied and considered as a potential source for nanoparticle synthesis. The use of fungi, as compared to bacteria is more advantageous in many ways. Fungi usually grow on simple medium and their biomass is easier to handle in the laboratory and can be easily scaled up. Fungal filtration is fairly easy and cheap as it involves simpler set-ups like filter-press to clear the fungal broths. Likewise, fungi secrete metabolic compounds, enzymes such as oxidative and hydrolytic enzymes and is capable of reducing metal ions from aqueous solution and proteins. Extracellular synthesis of nanoparticles could be highly advantageous from the point of view of synthesis in large quantities and easy downstream processing (Riddin et al. 2006). Hence, the use of fungus as a novel material for the nanomaterials synthesis significantly increases the productivity of high stability along with preventing agglomeration of the particles (Noor et al. 2020).

Different kinds of materials are used to synthesize nanoparticles in different forms and chemical compositions (Puoci et al. 2008, Athanassiou et al. 2018). The nanoparticles can be synthesized both intracellularly and extracellularly in nanoscale dimensions with exquisite morphology (Gholami-Shabani et al. 2013). In intracellular synthesis, nanoparticles are synthesized inside the fungal cell. In this method, the fungal biomass is treated with a metal salt solution and incubated for 24 hours in the dark (Vala 2014). In extracellular synthesis, the fungal filtrate is treated with a metal salt solution and observed for the synthesis of nanoparticles (Duran et al. 2011, Nanda and Majeed 2014). The synthesis of metal nanoparticles using the extracellular method is much faster as compared to the intracellular method (Narayanan and Sakthivel 2010). Recently nanotechnology has been included in the realm of pesticides which has the capacity to transform modern-day agriculture. Different types of nanoparticles like aluminum oxide nanoparticles (ANP), silver nanoparticles (SNP), titanium dioxide, and zinc oxide were experimented for the control of grasserie disease in silkworm and rice weevil (Rai and Ingale 2012). Gold and silver nanoparticles of *Chrysosporium tropicum,* was highly effective in control of mosquito larvae (Soni and Prakash 2011). Gold nanoparticles synthesized using *Aspergillus niger* were found to produce 100% mortality after 48 hours of exposure (Soni and Prakash 2012). Silver nanoparticles synthesized using fungi *Chrysoporium tropicum, Chrysosporium keratinophilum, Verticillium lecanii,* and *Fusarium oxysporum* have been shown to produce mortality in larvae and adults *Culex quinquefasciatus.* The NPs synthesized from *C. keratinophilum* and *V. lecanii* shows higher efficacy against *Culex quinquefasciatus* (Soni and Prakash 2012a). SeNPs synthesized by *Trichoderma* sp. were found to be the highest larval mortality of *S. litura* larvae (Arunthirumeni et al. 2021). Table 1 provide mycosynthesized nanoparticles against insect pest and Mosquitoes.

Types of Nanomaterial used for the Synthesis of Fungal Nanoparticles

Fusarium oxysporum is a fungus that has been often used for the biosynthesis of nanoparticles. Extracellular gold and silver nanoparticles synthesized using *F. oxysporum* are spherical or triangular in shape and have high dispersibility and do not aggregate even after a month (Vivekanandhan et al. 2018). A wide variety of bio-nanopesticides are developed using metals such as Silver, Copper, Titanium, Zinc, Selenium and Palladium (Manimaran et al. 2020). Green synthesis microbial nanoparticles of gold, silver, and other metals can be synthesized with metabolites produced from fungi that show effectiveness on mosquito larvae (Soni and Prakash 2013). Recent studies have shown that nano-formulation of biopesticides greatly enhances their efficacy (Togola et al. 2018). Bimetallic nanoparticles (BMNPs), in

Table 1. Mycosynthesized nanoparticles used in the control of Agricultural insect pest and mosquitoes.

Fungal Culture	Nanomaterials	Target Pest	References
Fusarium oxysporum	Silver nanoparticles	*Anopheles stephensi* *Aedes aegypti* *Culex quinquefasciatus*	Vivekanandhan et al. 2018
Fusarium proliferatum	Copper nanoparticles	*Ae. aegypti* *An. stephensi* *Cx. quinquefasciatus*	Kalaimurugan et al. 2019
Metarhizium anisopliae	Titanium nanoparticles	*Galleria mellonella*	Yosri et al. 2018
Metarhizium sp.	Titanium Oxide and Zinc Oxide Bimetallic nanoparticle	*Spodoptera furgiperda*	Kumaravel et al. 2021
Cochliobolus lunatus	Silver nanoparticles	*Ae. aegypti* *An. stephensi*	Salunkhe et al. 2011
Isaria fumosorosea	Silver nanoparticles	*Ae. aegypti* *Cx. quinquefasciatus*	Banu and Balasubramanian 2014
Penecillium verucosum	Silver nanoparticles	*Cx. quinquefasciatus*	Kamalakannan et al. 2014
Chrysosporium keratinophilum, Verticillium lecanii	Silver and Gold nanoparticles	*An. stephensi* *Ae. aegypti* *Cx. quinquefasciatus*	Soni and Prakash 2014
Aspergillus niger	Cerium oxide nanoparticles	*Ae. aegypti*	Gopinath et al. 2015
Beauveria bassiana	Silver nanoparticles	*Ae. aegypti*	Banu and Balasubramanian 2014
Beauveria bassiana	Silver nanoparticles	*Ae. aegypti* *An. stephensi* *Cx. quinquefasciatus*	Prabakaran et al. 2016
Nomuraea rileyi	Chitosan nanocomposite	*Spodoptera litura*	Namasivayam et al. 2018
Pleurotus djamor	Zinc Oxide, Titanium dioxide nanoparticles	*Ae. aegypti* *Cx. quinquefasciatus*	Manimaran et al. 2020
Trichoderma harzianum	Silver nanoparticles	*Ae. aegypti*	Sundaravadivelan and Padmanabhan 2014
Trichoderma viride	Titanium dioxide nanoparticle	*Helicoverpa armigera*	Chinnaperumal et al. 2018
Trichoderma sp.	Selenium nanoparticle	*Spodoptera litura*	Arunthirumeni et al. 2021

particular, have shown to be more effective as compared to monometallic particles due to greater stability, selectivity and catalytic activity (Kumaravel et al. 2021). Due to high enzymatic activity and the ability to produce secondary metabolites, nanoparticle synthesis using metals is easy. As metal oxides can be easily reduced by the functional groups present in these metabolites. The metal and metal oxide mentioned in Figure 2 is routinely used for mycosynthesized nanoparticles.

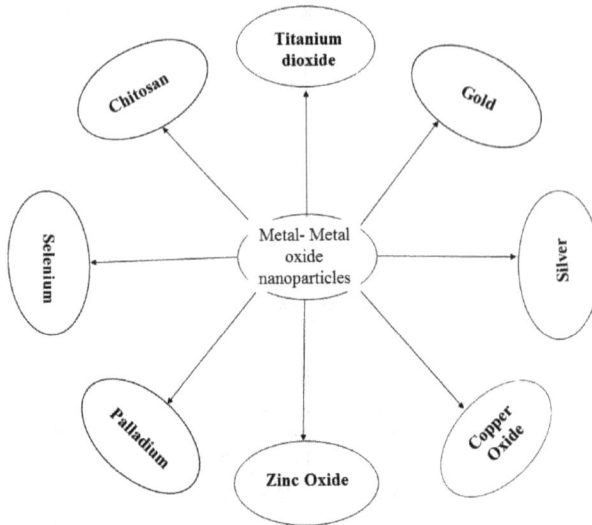

Figure 2. Types of nanomaterial used for synthesis of mycosynthesized nanoparticles.

Mode of Action of Fungal Nanoparticles in Insects

The ability of nanoparticles to pass through the cell membrane due to their small size has made it an ideal choice for delivery of drugs to precise location. The fungal nanoparticles containing secondary metabolites can enter the insect as a result of ingestion. Following ingestion, the metabolites can bind to the gut cells and cause a variety of symptoms in the insect. The insect may stop feeding (antifeedant) and have abnormal behavioral effects (Kumaravel et al. 2021). In addition, the ingested nanoparticle can conjugate with the larval detoxification enzymes leading to their increased susceptibility to other pathogens. Considerable damage to insect hemocytesis also observed following treatment with fungal conjugated nanoparticles (Chinnaperumal et al. 2018, Xu et al. 2020, Tuncsoy and Mese 2021). For example, Destruxins a fungal secondary metabolite causes immunosuppressivity leading to insect mortality. Nanosilica based insecticides are another interesting class for controlling insect pests. When applied to the surface of the leaf and stem surface of the plant, nano-silica gets adsorbed to cuticular lipids thereby functioning as a water barrier resulting in insect mortality (Rai et al. 2014).

The insect exoskeleton and the peritrophic membrane in the midgut are two barriers that protect the insects from environmental hazards and ingested toxins. Entomopathogenic fungi have overcome these kinds of barriers by producing multiple extracellular enzymes, like chitinases and proteolytic enzymes that help to penetrate the cuticle and facilitate infection (Jaidev and Narasimha 2010, Sandhu et al. 2012). Fungal nanoparticles are formed by bioreduction by the trapping of the metal ions on the cell wall surfaces which occurs due to electrostatic interaction between the metal ions and positively charged groups of enzymes in the cell wall. This leads to their aggregation and formation of nanoparticles. Once these nanoparticles settle on the insect cell wall, the fungi start excreting enzymes such as proteases, chitinases, quitobiases, lipases, and lipoxygenases (Bansal et al. 2006). These enzymes help in penetrating the insect cuticle and produce specific infection hyphae originating at appressoria. After the successful penetration, the fungus is then distributed into the haemolymph by formation of blastospores (Figure 3) (Sandhu et al. 2012, Jaidev and Narasimha 2010).

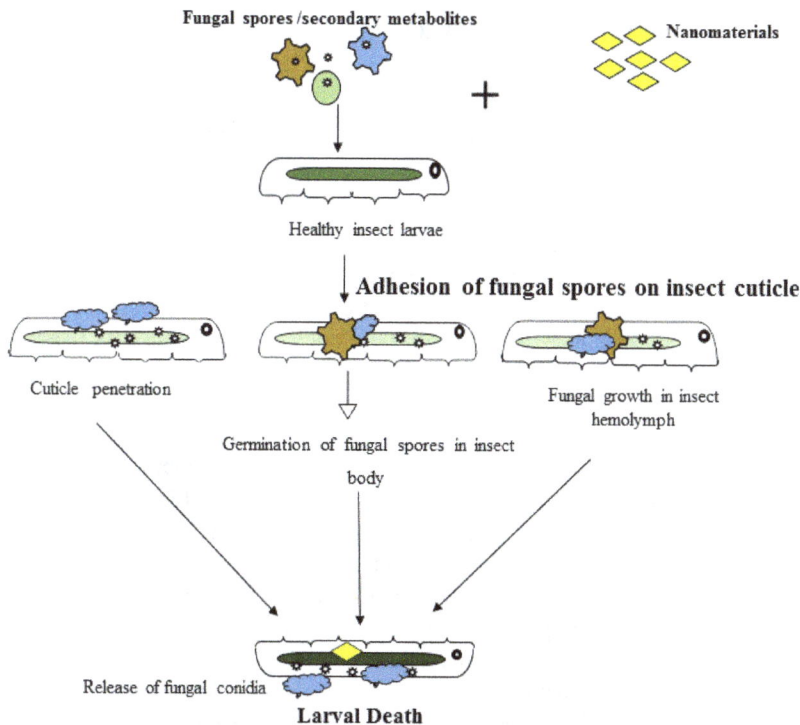

Figure 3. Mode of action mycosynthesized nanoparticles.

Role of Myco-Nanoinsecticides in Agricultural Insect Pest

Two major groups of insects such as Coleoptera and Lepidoptera comprise the most economically important insect pests. Several species of Coleoptera and Lepidoptera attack crops both in the field and stored grains. A chitin derivate (N-(2-chloro-6-flurobanzyl-chitosan), chitosan was tested as an insecticide in laboratory conditions showing strong larvicidal activity (Rabea et al. 2005). Chitosan nanoparticle-coated fungal metabolites, uncoated fungal metabolites, fungal spores of entomopathogenic fungi *Nomuraea rileyi* evaluated against *S. litura* resulted in the chitosan-coated fungal metabolites showing higher mortality. Chitosan (CS)-g-poly (acrylic acid) PAA nanoparticles reduced egg-laying of *Aphis gossypii* in the laboratory and under semi-field conditions. *Spodoptera litura* and *Achaea janata* larvae were fed with PVP coated-AgNPs treated castor leaf at different concentrations resulting in decreased larval and pupal body weights. A low amount of silver was accumulated in the larval guts, but a major portion of it was eliminated through the feces. *Trichoderma*, selenium nanoparticles are very effective as larvicidal and as antifeedants on *Spodoptera litura* larvae (Arunthirumeni et al. 2021). Mycosynthesized iron nanoparticles of *Beauveria brongniartii* caused significant reductions in feeding and growth parameters in *Spodoptera litura* larvae (Xu et al. 2021). Mycosynthesized Titanium nanoparticles from the extracellular filtrate of *T. viride* produced high mortality in *Helicoverpa armigera larvae* (Chinnaperumal et al. 2018). Entomopathogenic fungi show great potential since they secrete large amounts of enzymes and are easy to use in the laboratory, which leads to higher yields of nanoparticles (Amersan et al. 2016).

Studies on Mosquito Control using Fungal Nanoparticles

Mosquitoes (Diptera: Culicidae) act as vectors for important pathogens and parasites and hence are a huge threat to public health and socio-economic development in the developing world. Mosquito control using conventional chemical insecticides is a constant challenge due to the rapid development of insecticide resistance (Ranson and Lissenden 2016). The number of insecticides approved for use in mosquito control is limited as human contact with these chemicals is immediate (Rani et al. 2021). Further, there is a limitation to their use for mosquito larval control as it may contaminate water bodies (Mulla and Mian 1981, Hare 1992). Personal protection creams and mosquito repellants are another domain in mosquito control where the use of chemical insecticides are heavily regulated apart from a few like DEET used in personal protection creams (Mitra et al. 2020) and a few narrow spectrum synthetic pyrethroids which are used as mosquito repellants and in coils (Khater et al. 2019). In this area, biological molecules hold a huge untapped potential that could be combined with nanotechnology for producing safe mosquito repellants. In mosquito control, fungal-based nanoparticles hold huge potential as they can be produced in large amounts and are cost-effective. Many species of fungi have been used for nanoparticle production including *Cochliobolus lunatus*, *Trichoderma harzianum*, *Neurospora intermedia*, *Aspergillus niger*, *Cryptosporiopsis ericae*, *Pleurotus ostreatus*, *Fusarium oxysporum*, *Candida albicans* and *Epicoccum nigrum*. AgNPs synthesized using culture filtrate *Trichoderma harzianum* produced concentration-dependent mortality in larvae and pupae of *Aedes aegypti* (Sundaravadivelan and Padmanabhan 2014). AgNPs synthesized using *Isaria fumosorosea*, *Beauveria bassiana* produced high larval mortality in *Aedes aegypti*, *Beauveria bassiana* (Govindarajan et al. 2016). *Fusarium oxysporum*, *Aspergillus fumigatus*, *Verticillium lecanii*, *Beauveria* sp., *Trichoderma* sp., and *Chrysosporium tropicum* have been conjugated with nanoparticles and have shown great potential to be used as mosquito larvicides (Arunthirumeni et al. 2021). *M. robertsii* extracellular metabolites synthesized CuNPs is highly toxic to mosquito larvae and shows lower toxicity on non-target organisms (Vivekanandhan et al. 2021). Mycosynthesized silver nanoparticles can be proposed as effective tools for eco-friendly control of the rural malaria vector *A. culicifacies* (Amerasan et al. 2016).

Formulation and Release of Fungal NPs

Pesticides, when encapsulated, are likely to have a more gradual release over time, which requires their application less often as compared with very highly concentrated and perhaps toxic initial applications followed by repeated applications. Several different concepts for loading of active pesticide molecules on NPs may include adsorption, covalent attachment mediated by different ligands, encapsulation, and entrapment inside NP. Controlled and slow release of the active molecules can be achieved based on degradation properties of the nanocarrier (e.g., polymer), bonding of the ingredients to the material, and the environmental conditions. The most attractive NPs that are considered as carriers for delivery of pesticides are based on polymers (soft NPs), synthetic silica, titanium, alumina, Ag, Cu, and natural minerals with nanoscale dimensions (inorganic or solid NPs) (Athanassiou et al. 2018).

Fungal bioactive secondary metabolites, though effective, may have some limitations as they may have low environmental persistence, low solubility, high volatility, instability at high temperatures, or low dispersivity in solution. These limitations can be overcome by the use of nanotechnology, which can greatly improve their physicochemical characteristics (Monterio et al. 2021). The development of nanocarriers for addressing this solution is an upcoming field of research. Insect gut enzymes and their pH can be utilized for delivering the biopesticides encapsulated in the nanoparticles (Camara et al. 2019). In their digestive system, insects have enzymes with proteolytic activities, such as trypsin, chymotrypsin, carboxypeptidase, and elastase, which can be used as targets for sustained/targeted release of these pesticides (Wang et al. 2020). Nanoencapsulated chemical pesticides like avermectin utilize mesoporous organosilica which is considered efficient in stimulating the insect α-amylase for release of the pesticide in the insect gut (Smith and Gilbertson 2018). Similarly, carboymethylcellulose loaded avermectin uses insect gut esterases for its release (Zhu et al. 2020). Polydopamine microspheres have also been shown to be used as carriers for the release of insecticides based on changes in the temperature and near-infrared light (Xu et al. 2017). The pH of the insect gut can also be utilized for releasing the pesticide contents into the gut. One example is poly (glycidyl methacrylate-co-acrylic acid) grafted on hollow mesoporous silica to release abamectin in the insect gut. In near future, we expect that protein bases will be used for the preparation of nanocarriers for the delivery of insecticides into insect systems.

Recent Developments and Innovation in Myconanotechnology

Fungi represent an important group of entomopathogenic organisms, which have a huge potential application in the field of insect control. There are several species of fungi, where studies have been conducted to know their potential in the control of agricultural pests and mosquitoes (Aw et al. 2017). Among all fungi groups, *Beauveria bassiana* and *Metarhizium anisopliae* are widely studied fungi. They have a broad host range which includes insects from Lepidoptera, Coleoptera, Hymenoptera, Diptera, and Hemiptera (Kirkland et al. 2005). The field/commercial application of fungi include their culturing, formulation, compatibility with other pest control practices. In order to commercialize fungi for insect control, several aspects of fungal biology and their insect host, the mode of infection, multiplication in the host, host response to fungal infection mainly the defense mechanisms which includes composition of exoskeleton, peritrophic membrane of gut, enzymes, and immune system needs to be understood in detail on a species specific basis. Several studies have shown that fungi are ideal candidates which are very effective in insect control in laboratory conditions (Ortiz-Urquiza et al. 2015). However, when applied in field conditions, their virulence and pathogenesis are affected by a variety of environmental factors. These include temperature, humidity, and UV radiation which affect the spore viability and virulence of fungi (Qin et al. 2014).Various methods have been used to address these issues like the use of UV protectants, humidity stabilizers, etc., but the cost and their effect on spore germination and viability is a major constraint (Behle et al. 2009). Similarly, growth media has been tweaked with salicylic acid and hydrocarbons has been instrumental in providing thermotolerance and increased virulence of spores (Crespo et al. 2002). Though formulations have addressed the constraints of fungal application in fields, a significant breakthrough in the commercial application of these fungi is hindered by decreased fitness of these fungal strains and reduced conidiation (Ortz et al. 2014). Besides the interaction of fungi with the host enzymes and immune system is another aspect that limits the faster progression of toxicoses. Nanotechnology has the potential to address these issues. Studies have shown that the insect physiology and the cuticular structure can be used to our advantage by selecting a suitable nanomaterial depending on the type of insect host targeted. Metals and metal oxides can be used when the insect cuticle is targeted as they can interact with the cuticle and provide a strong attachment following which the spores can grow (Litwin et al. 2020). Carbohydrate polymer-based nanoparticles can be used for the delivery of spores and secondary metabolites into the insect gut directly (Yu et al. 2021). This is advantageous for example, when cellulose nanopolymers are relatively stable and can be digested by the cellulases produced by the microbial symbionts of the insect gut which degrade cellulose and help in releasing the content in the nanoencapsulation. Similar is true for protein-based nanoparticles which can utilize host proteases for the release of the contents (Yang et al. 2014). This is advantageous in targeting a specific type of insect pest and at the same time effects on the non-target organisms can be reduced. A small quantity may be sufficient for application which can also reduce the cost. Research needs to be undertaken in this area, to produce efficient fungal nanoparticles which are cost-effective, target-specific and environmentally safe.

Conclusion

Fungi are a diverse group of organisms that have the potential to be used in the biological control of insect pests of agriculture and for mosquitoes. The fungal conidia and their secondary metabolites have been shown to be effective in the control of insects in laboratory. Studies have also shown that large scale production of these fungi is easy and cost effective. Fungi also produce a huge quantity of extracellular metabolites and their extraction is easy. However, their application in field is low due to decreased virulence and viability in field condition. Formulation technology has tried to address these issues, however still very few formulations have been prepared which are compatible with fungi. Further research is needed in this area. These can be solved by the use of appropriate nanomaterial for the development of fungal nanopesticides. Currently, the majority of the studies have used metals and metal oxides for the synthesis of fungal nanopesticides. The use of biopolymers like cellulose, chitin and proteins are still in their infancy and hold a key to effectively addressing the limitations in the field application of fungi. In the future, with the development in formulation technology and use of biopolymers for the synthesis of fungal nanoparticles, we expect to see several commercial fungal nanopesticides.

References

Abbas, A., Abbas, R.Z., Khan, J.A., Iqbal, Z., Hayat Bhatti, M.M., Sindhu, Z.U.D. and Zia, M.A. 2014. Integrated strategies for the control and prevention of dengue vectors with particular reference to *Aedes aegypti*. Pak. Vet. J. 34(1).

Abd-Elsalam, K.A. 2021. Fungal nanotechnology. J. Fungi. 7(8): 583.

Aiswarya, D., Raja, R.K., Kamaraj, C., Balasubramani, G., Deepak, P., Arul, D. and Perumal, P. 2019. Biosynthesis of gold and silver nanoparticles from the symbiotic bacterium, *Photorhabdus luminescens* of entomopathogenic nematode: Larvicidal properties against three mosquitoes and *Galleria mellonella* Larvae. J. Clust. Sci. 30(4): 1051–1063.

AlNadhari, S., Al-Enazi, N.M., Alshehrei, F. and Ameen, F. 2021. A review on biogenic synthesis of metal nanoparticles using marine algae and its applications. Environ. Res. 194: 110672.

Amerasan, D., Nataraj, T., Murugan, K., Panneerselvam, C., Madhiyazhagan, P., Nicoletti, M. and Benelli, G. 2016. Myco-synthesis of silver nanoparticles using *Metarhizium anisopliae* against the rural malaria vector *Anopheles culicifacies* Giles (Diptera: Culicidae). J. Pest. Sci. 89(1): 249–256.

Arunthirumeni, M., Veerammal, V. and Shivakumar, M.S. 2021. Biocontrol efficacy of mycosynthesized selenium nanoparticle using *Trichoderma* sp. on insect pest *Spodoptera litura*. J. Clust. Sci. 1–9.

Athanassiou, C.G., Kavallieratos, N.G., Benelli, G., Losic, D., Rani, P.U. and Desneux, N. 2018. Nanoparticles for pest control: current status and future perspectives. J. Pest. Sci. 91(1): 1–15.

Aw, K.M.S. and Hue, S.M. 2017. Mode of infection of *Metarhizium* spp. fungus and their potential as biological control agents. J. Fungus 3(2): 30.

Balumahendhiran, K., Vivekanandhan, P. and Shivakumar, M.S. 2019. Mosquito control potential of secondary metabolites isolated from *Aspergillus flavus* and *Aspergillus fumigatus*. Biocatal. Agric. Biotechnol. 21: 101334.

Banerjee, P., Satapathy, M., Mukhopahayay, A. and Das, P. 2014. Leaf extract mediated green synthesis of silver nanoparticles from widely available Indian plants: Synthesis, characterization, antimicrobial property and toxicity analysis. Bioresour. Bioprocess. 1(1): 1–10.

Banu, A.N. and Balasubramanian, C. 2014. Myco-synthesis of silver nanoparticles using *Beauveria bassiana* against dengue vector, *Aedes aegypti* (Diptera: Culicidae). Parasitol. Res. 113(8): 2869–2877.

Banu, A.N. and Balasubramanian, C. 2014. Optimization and synthesis of silver nanoparticles using *Isaria fumosorosea* against human vector mosquitoes. Parasitol. Res. 113(10): 3843–3851.

Bansal, A., Yang, H., Li, C., Benicewicz, B.C., Kumar, S.K. and Schadler, L.S. 2006. Controlling the thermomechanical properties of polymer nanocomposites by tailoring the polymer–particle interface. J. Polym. Sci. B Polym. Phys. 44(20): 2944–2950.

Bawin, T., Seye, F., Boukraa, S., Zimmer, J.Y., Raharimalala, F.N., Zune, Q. and Francis, F. 2016. Production of two entomopathogenic *Aspergillus* species and insecticidal activity against the mosquito *Culex quinquefasciatus* compared to *Metarhizium anisopliae*. Biocontrol Sci. Technol. 26(5): 617–629.

Behle, R.W., Compton, D.L., Laszlo, J.A. and Shapiro-Ilan, D.I. 2009. Evaluation of soyscreen in an oil-based formulation for UV protection of *Beauveria bassiana* conidia. J. Econ. Entomol. 102(5): 1759–1766.

Bidochka, M.J. and Small, C.L. 2005. Phylogeography of *Metarhizium*, an insect pathogenic fungus. Insect-fungal Associations, 28–49.

Camara, M.C., Campos, E.V.R., Monteiro, R.A., Santo Pereira, A.D.E., de Freitas Proença, P.L. and Fraceto, L.F. 2019. Development of stimuli-responsive nano-based pesticides: Emerging opportunities for agriculture. J. Nanobiotechnology 17(1): 1–19.

Chaud, M., Souto, E.B., Zielinska, A., Severino, P., Batain, F., Oliveira-Junior, J. and Alves, T. 2021. Nanopesticides in agriculture: Benefits and challenge in agricultural productivity, toxicological risks to human health and environment. Toxics 9: 131.

Chinnaperumal, K., Govindasamy, B., Paramasivam, D., Dilipkumar, A., Dhayalan, A., Vadivel, A. and Pachiappan, P. 2018. Bio-pesticidal effects of *Trichoderma viride* formulated titanium dioxide nanoparticle and their physiological and biochemical changes on *Helicoverpa armigera* (Hub.). Pestic. Biochem. Physiol. 149: 26–36.

Crespo, R., Júrez, M.P., Dal Bello, G.M., Padin, S., Fernández, G.C. and Pedrini, N. 2002. Increased mortality of *Acanthoscelides obtectus* by alkane-grown *Beauveria bassiana*. BioControl 47(6): 685–696.

Deng, Z., Zhang, R., Shi, Y., Tan, H. and Cao, L. 2013. Enhancement of phytoremediation of Cd- and Pb-contaminated soils by self-fusion of protoplasts from endophytic fungus Mucor sp. CBRF59. Chemosphere 91(1): 41–47

Dhaliwal, G.S., Jindal, V. and Dhawan, A.K. 2010. Insect pest problems and crop losses: Changing trends. Indian J. Ecol. 37(1): 1–7.

Durán, N., Marcato, P.D., Durán, M., Yadav, A., Gade, A. and Rai, M. 2011. Mechanistic aspects in the biogenic synthesis of extracellular metal nanoparticles by peptides, bacteria, fungi, and plants. Appl. Microbiol. Biotechnol. 90(5): 1609–1624.

El-Ashry, R.M., El-Saadony, M.T., El-Sobki, A.E., El-Tahan, A.M., Al-Otaibi, S., El-Shehawi, A.M. and Elshaer, N. 2021. Biological silicon nanoparticles maximize the efficiency of nematicides against biotic stress induced by *Meloidogyne incognita* in eggplant. Saudi J. Biol. Sci. 29(2): 920–932.

Enyiukwu, D.N., Awurum, A.N. and Nwaneri, J.A. 2014. Efficacy of plant-derived pesticides in the control of myco-induced postharvest rots of tubers and agricultural products: A review. Net. J. Agric. Sci. 2(1): 30–46.

Fahimirad, S., Ajalloueian, F. and Ghorbanpour, M. 2019. Synthesis and therapeutic potential of silver nanomaterials derived from plant extracts. Ecotoxicol. Environ. Saf. 168: 260–278.

Gholami-Shabani, M.H., Akbarzadeh, A., Mortazavi, M. and Emadzadeh, M.K. 2013. Evaluation of the antibacterial properties of silver nanoparticles synthesized with *Fusarium oxysporum* and *Escherichia coli*. Int. J. Life Sci. Bt. Pharm. Res. 2: 342–348.

Gopinath, K., Karthika, V., Sundaravadivelan, C., Gowri, S. and Arumugam, A. 2015. Mycogenesis of cerium oxide nanoparticles using *Aspergillus niger* culture filtrate and their applications for antibacterial and larvicidal activities. J. Nanostructure. Chem. 5(3): 295–303.

Govindarajan, M., Rajeswary, M., Veerakumar, K., Muthukumaran, U., Hoti, S.L. and Benelli, G. 2016. Green synthesis and characterization of silver nanoparticles fabricated using *Anisomeles indica*: Mosquitocidal potential against malaria, dengue and Japanese encephalitis vectors. Exp. Parasitol. 161: 40–47.

Hare, L. 1992. Aquatic insects and trace metals: Bioavailability, bioaccumulation, and toxicity. Crit. Rev. Toxicol. 22(5-6): 327–369.

Hicks, B.J. 2016. Optimization of *Beauveria bassiana* in a spray formulation against *Choristoneura fumiferana*. Can. J. For. Res. 46(4): 543–547.

Hsiao, W.F., Bidochka, M.J. and Khachatourians, G.G. 1992. Effect of temperature and relative humidity on the virulence of the entomopathogenic fungus, *Verticillium lecanii*, toward the oat-bird berry aphid, *Rhopalosiphum padi* (Hom., Aphididae). J. Appl. Entomol. 114(1-5): 484–490.

Inglis, I.M. and Gray, A.J. 2001. An evaluation of semiautomatic approaches to contour segmentation applied to fungal hyphae. Biometrics 57(1): 232–239.

Jaidev, L.R. and Narasimha, G. 2010. Fungal mediated biosynthesis of silver nanoparticles, characterization and antimicrobial activity. Colloids Surf. B 81(2): 430–433.

Kalaimurugan, D., Vivekanandhan, P., Sivasankar, P., Durairaj, K., Senthilkumar, P., Shivakumar, M. S. and Venkatesan, S. 2019. Larvicidal activity of silver nanoparticles synthesized by *Pseudomonas fluorescens*YPS3 isolated from the Eastern Ghats of India. J. Clust. Sci. 30(1): 225–233.

Karthi, S., Vaideki, K., Shivakumar, M.S., Ponsankar, A., Thanigaivel, A., Chellappandian, M. and Senthil-Nathan, S. 2018. Effect of *Aspergillus flavus* on the mortality and activity of antioxidant enzymes of *Spodoptera litura* Fab.(Lepidoptera: Noctuidae) larvae. Pestic. Biochem. Physiol. 149: 54–60.

Kaul, S., Gupta, S., Ahmed, M. and Dhar, M.K. 2012. Endophytic fungi from medicinal plants: A treasure hunt for bioactive metabolites. Phytochem. Rev. 11(4): 487–505.

Kaviya, S., Santhanalakshmi, J. and Viswanathan, B. 2011. Green synthesis of silver nanoparticles using *Polyalthia longifolia* leaf extract along with D-sorbitol: Study of antibacterial activity. J. Nanotechnol. 2011.

Khater, H.F., Selim, A.M., Abouelella, G.A., Abouelella, N.A., Murugan, K., Vaz, N.P. and Govindarajan, M. 2019. Commercial mosquito repellents and their safety concerns. In Malaria. IntechOpen.

Kim, Y.K., Wang, Y., Liu, Z.M. and Kolattukudy, P.E. 2002. Identification of a hard surface contact-induced gene in *Colletotrichum gloeosporioides* conidia as a sterol glycosyltransferase, a novel fungal virulence factor. Plant J. 30(2): 177–187.

Kirkland, B.H., Eisa, A. and Keyhani, N.O. 2005. Oxalic acid as a fungal acaricidal virulence factor. J. Med. Entomol. 42(3): 346–351.

Kumaravel, J., Lalitha, K., Arunthirumeni, M. and Shivakumar, M.S. 2021. Mycosynthesis of bimetallic zinc oxide and titanium dioxide nanoparticles for control of *Spodoptera frugiperda*. Pestic. Biochem. Physiol. 178: 104910.

Kamalakannan, S., Gobinath, C. and Ananth, S. 2014. Synthesis and characterization of fungus mediated silver nanoparticle for toxicity on filarial vector, *Culex quinquefasciatus*. Int. J. Pharm. Sci. Rev. Res. 24(2): 124–132.

Kusari, S., Singh, S. and Jayabaskaran, C. 2014. Biotechnological potential of plant-associated endophytic fungi: hope versus hype. Trends Biotechnol. 32(6): 297–303.

Lalitha, K., Kalaimurgan, D., Nithya, K., Venkatesan, S. and Shivakumar, M.S. 2020. Antibacterial, antifungal and mosquitocidal efficacy of copper nanoparticles synthesized from entomopathogenic nematode: Insect–host relationship of bacteria in secondary metabolites of *Morganella morganii* sp. (PMA1). Arab. J. Sci. Eng. 45(6): 4489–4501.

Leger, R.J.S. and Wang, C. 2010. Genetic engineering of fungal biocontrol agents to achieve greater efficacy against insect pests. Appl. Microbiol. Biotechnol. 85(4): 901–907.

Liang, X., Li, N., Zhang, R., Yin, P., Zhang, C., Yang, N. and Kong, B. 2021. Carbon-based SERS biosensor: From substrate design to sensing and bioapplication. NPG Asia Mater. 13(1): 1–36.

Litwin, A., Nowak, M. and Rozalska, S. 2020. Entomopathogenic fungi: Unconventional applications. Rev. Environ. Sci. Biotechnol. 19(1): 23–42.

Mahajan, P., Kaushal, J., Upmanyu, A. and Bhatti, J. 2019. Assessment of phytoremediation potential of Chara vulgaris to treat toxic pollutants of textile effluent. J. Toxicol. 2019. Doi: 10.1155/2019/8351272.

Majeed, A., Muhammad, Z., Ullah, Z., Ullah, R. and Ahmad, H. 2017. Late blight of potato (*Phytophthorainfestans*) I: Fungicides application and associated challenges. Turk. J. Sci. Technol. 5(3): 261–266.

Manimaran, K., Murugesan, S., Ragavendran, C., Balasubramani, G., Natarajan, D., Ganesan, A. and Seedevi, P. 2020. Biosynthesis of Tio2 nanoparticles using edible mushroom (pleurotusdjamor) extract: Mosquito larvicidal, histopathological, antibacterial and anticancer effect. J. Clust. Sci. 1–12.

Maxwell, C.S., Sepulveda, V.E., Turissini, D.A., Goldman, W.E. and Matute, D.R. 2018. Recent admixture between species of the fungal pathogen Histoplasma. Evol. Lett. 2(3): 210–220.

Mitra, S., Rodriguez, S.D., Vulcan, J., Cordova, J., Chung, H.N., Moore, E. and Hansen, I.A. 2020. Efficacy of active ingredients from the EPA 25 (B) list in reducing attraction of *Aedes aegypti* (Diptera: Culicidae) to humans. J. Med. Entomol. 57(2): 477–484.

Monteiro, G.P., Tavares, I.M.D.C., de Carvalho, M.C.F., Carvalho, M.S., Pimentel, A.B., Santos, P.H., Vilas Boas, E.V.D.B., de Oliveira, J.R., Capelossi, V.R., Bilal, M. and Franco, M. 2022. Evaluation of fungal biomass developed from cocoa by-product as a substrate with corrosion inhibitor for carbon steel. Chemical Engineering Communications, pp. 1–16.

Mulla, M.S. and Mian, L.S. 1981. Biological and environmental impacts of the insecticides malathion and parathion on nontarget biota in aquatic ecosystems. Residue Rev. 101–135.

Muthusamy, R. and Shivakumar, M.S. 2015. Involvement of metabolic resistance and F1534C kdr mutation in the pyrethroid resistance mechanisms of *Aedes aegypti* in India. Acta Trop. 148: 137–141.

Namasivayam, S.K.R., Bharani, R.A. and Karunamoorthy, K. 2018. Insecticidal fungal metabolites fabricated chitosan nanocomposite (IM-CNC) preparation for the enhanced larvicidal activity-An effective strategy for green pesticide against economic important insect pests. Int. J. Biol. Macromol. 120: 921–944.

Nanda, A. and Majeed, S. 2014. Enhanced antibacterial efficacy of biosynthesized AgNPs from *Penicillium glabrum* (MTCC1985) pooled with different drugs. Int. J. Pharm. Tech. Res. 6: 217–223.

Narayanan, K.B. and Sakthivel, N. 2010. Biological synthesis of metal nanoparticles by microbes. Adv. Colloid Interface Sci. 156(1-2): 1–13.

Naseer, M., Aslam, U., Khalid, B. and Chen, B. 2020. Green route to synthesize zinc oxide nanoparticles using leaf extracts of *Cassia fistula* and *Melia azadarach* and their antibacterial potential. Sci. Rep. 10(1): 1–10.

Nivetha, N., Asha, A.D., Thakur, J.K., Dukare, A.S., Paul, B. and Paul, S. 2021. Microbial-based nanoparticles as potential approach of insect pest management. In Microbes for Sustainable Insect Pest Management. Springer, Cham, pp. 135–157.

Noor, S., Shah, Z., Javed, A., Ali, A., Hussain, S.B., Zafar, S. and Muhammad, S.A. 2020. A fungal based synthesis method for copper nanoparticles with the determination of anticancer, antidiabetic and antibacterial activities. J. Microbiol. Methods 174: 105966.

Nunez del Prado, E., Iannacone, J. and Gómez, H. 2008. Efecto de Dos Hongos Entomopatógenosen el Control de *Aleurodicus cocois* (Curtis, 1846) (Hemiptera: *Aleyrodidae*). Chil. J. Agric. Res. 68(1): 21–30.

Opiyo, M.A., Ngowo, H.S., Mapua, S.A., Mpingwa, M., Nchimbi, N., Matowo, N.S. and Okumu, F.O. 2021. Sub-lethal aquatic doses of pyriproxyfen may increase pyrethroid resistance in malaria mosquitoes. PLos One 16(3): e0248538.

Ortiz-Urquiza, A., Luo, Z. and Keyhani, N.O. 2015. Improving mycoinsecticides for insect biological control. Appl. Microbiol. Biotechnol. 99(3): 1057–1068.

Pantidos, N. and Horsfall, L.E. 2014. Biological synthesis of metallic nanoparticles by bacteria, fungi and plants. J. Nanomed. Nanotechnol. 5(5): 1.

Pereira, L., Mehboob, F., Stams, A.J., Mota, M.M., Rijnaarts, H.H. and Alves, M.M. 2015. Metallic nanoparticles: Microbial synthesis and unique properties for biotechnological applications, bioavailability and biotransformation. Crit. Rev. Biotechnol. 35(1): 114–128.

Prabakaran, K., Ragavendran, C. and Natarajan, D. 2016. Mycosynthesis of silver nanoparticles from *Beauveria bassiana* and its larvicidal, antibacterial, and cytotoxic effect on human cervical cancer (HeLa) cells. RSC Adv. 6(51): 44972–44986.

Prasad, R., Bhattacharyya, A. and Nguyen, Q.D. 2017. Nanotechnology in sustainable agriculture: Recent developments, challenges, and perspectives. Front. Microbial. 8: 1014.

Perinotto, W.M.S., Angelo, I.C., Golo, P.S., Camargo, M.G., Quinelato, S., Santi, L. and Bittencourt, V.R.E.P. 2014. *Metarhizium anisopliae* (Deuteromycetes: Moniliaceae) Pr1 activity: Biochemical marker of fungal virulence in *Rhipicephalusmicroplus* (Acari: Ixodidae). Biocontrol Sci. Technol. 24(2): 123–132.

Puoci, F., Iemma, F., Spizzirri, U.G., Cirillo, G., Curcio, M. and Picci, N. 2008. Polymer in agriculture: A review. Am. J. Agric. Biol. Sci. 3(1): 299–314

Qin, Y., Ortiz-Urquiza, A. and Keyhani, N.O. 2014. A putative methyltransferase, mtrA, contributes to development, spore viability, protein secretion and virulence in the entomopathogenic fungus *Beauveria bassiana*. Microbiology 160(11): 2526–2537.

Quesada-Moraga, E. and Alain, V.E.Y. 2004. Bassiacridin, a protein toxic for locusts secreted by the entomopathogenic fungus *Beauveria bassiana*. Mycol. Res. 108(4): 441–452.

Rabea, E.I., Badawy, M.E., Rogge, T.M., Stevens, C.V., Höfte, M., Steurbaut, W. and Smagghe, G. 2005. Insecticidal and fungicidal activity of new synthesized chitosan derivatives. Pest. Manag. Sci. 61(10): 951–960.

Rai, M. and Ingle, A. 2012. Role of nanotechnology in agriculture with special reference to management of insect pests. Appl. Microbiol. Biotechnol. 94(2): 287–293.

Rai, M., Kon, K., Ingle, A., Duran, N., Galdiero, S. and Galdiero, M. 2014. Broad-spectrum bioactivities of silver nanoparticles: The emerging trends and future prospects. Appl. Microbiol. Biotechnol. 98(5): 1951–1961.

Rai, M., Yadav, A., Bridge, P. and Gade A.K. 2009. Myconanotechnology: A new and emerging science. pp. 258–267. *In*: Rai, M. and Bridge, P. (eds.). Appl. Mycol.

Rajeshkumar, S., Malarkodi, C., Vanaja, M. and Annadurai, G. 2016. Anticancer and enhanced antimicrobial activity of biosynthesizd silver nanoparticles against clinical pathogens. J. Mol. Struct. 1116: 165–173.

Rajput, N. 2015. Methods of preparation of nanoparticles—A review. Int. J. Adv. Eng. 7(6): 1806.

Ramkumar, G. and Shivakumar, M.S. 2015. Laboratory development of permethrin resistance and cross-resistance pattern of *Culex quinquefasciatus* to other insecticides. Parasitol. Res. 114(7): 2553–2560.

Rani, L., Thapa, K., Kanojia, N., Sharma, N., Singh, S., Grewal, A.S. and Kaushal, J. 2021. An extensive review on the consequences of chemical pesticides on human health and environment. J. Clean. Prod. 283: 124657.

Ranson, H. and Lissenden, N. 2016. Insecticide resistance in African Anopheles mosquitoes: A worsening situation that needs urgent action to maintain malaria control. Trends Parasitol. 32(3): 187–196.

Riddin, T.L., Gericke, M. and Whiteley, C.G. 2006. Analysis of the inter-and extracellular formation of platinum nanoparticles by *Fusarium oxysporum* f. sp. *lycopersici* using response surface methodology. Nanotechnology 17(14): 3482.

Riley, P.R. and Narayan, R.J. 2021. Recent advances in carbon nanomaterials for biomedical applications: A review. Curr. Opin. Biomed. Eng. 17: 100262.

Roberts, D.W. and St. Leger, R.J. 2004. *Metarhizium* spp., cosmopolitan insect-pathogenic fungi: Mycological aspects. Adv. Appl. Microbiol. 54(1): 1–70.

Rodríguez-González, Á., Campelo, M.P., Lorenzana, A., Mayo-Prieto, S., González-López, Ó., Álvarez-García, S. and Casquero, P.A. 2020. Spores of *Trichoderma* strains sprayed over *Acanthoscelides obtectus* and *Phaseolus vulgaris* L. beans: Effects in the biology of the bean weevil. J. Stored Prod. Res. 88: 101666.

Rodríguez-González, V., Obregón, S., Patrón-Soberano, O.A., Terashima, C. and Fujishima, A. 2020. An approach to the photocatalytic mechanism in the TiO_2-nanomaterials microorganism interface for the control of infectious processes. Appl. Catal. B 270: 118853.

Salem, S.S. and Fouda, A. 2021. Green synthesis of metallic nanoparticles and their prospective biotechnological applications: An overview. Biol. Trace Elem. Res. 199(1): 344–370.

Salunkhe, R.B., Patil, S.V., Patil, C.D. and Salunke, B.K. 2011. Larvicidal potential of silver nanoparticles synthesized using fungus *Cochliobolus lunatus* against *Aedes aegypti* (Linnaeus, 1762) and *Anopheles stephensi* Liston (Diptera; Culicidae). Parasitol. Res. 109(3): 823–831.

Sánchez-Bayo, F. 2012. Insecticides mode of action in relation to their toxicity to non-target organisms. J. Environ. Anal. Toxicol. 4: S4–002.

Sandhu, S.S., Sharma, A.K., Beniwal, V., Goel, G., Batra, P., Kumar, A. and Malhotra, S. 2012. Myco-biocontrol of insect pests: Factors involved mechanism, and regulation. J. Pathog. 2012.

Sandhu, P., Xu, X., Bondiskey, P.J., Balani, S.K., Morris, M.L., Tang, Y.S. and Pearson, P.G. 2004. Disposition of caspofungin, a novel antifungal agent, in mice, rats, rabbits, and monkeys. Antimicrob. Agents Chemother. 48(4): 1272–1280.

Santos, T.S., Silva, T.M., Cardoso, J.C., de Albuquerque-Júnior, R.L., Zielinska, A., Souto, E.B. and Mendonça, M.D.C. 2021. Biosynthesis of silver nanoparticles mediated by entomopathogenic fungi: Antimicrobial resistance, nanopesticides, and toxicity. Antibiotics 10(7): 852.

Saravanan, M., Vemu, A.K. and Barik, S.K. 2011. Rapid biosynthesis of silver nanoparticles from *Bacillus megaterium* (NCIM 2326) and their antibacterial activity on multi drug resistant clinical pathogens. Colloids Surf. B 88(1): 325–331.

Saware, K. and Venkataraman, A. 2014. Biosynthesis and characterization of stable silver nanoparticles using *Ficus religiosa* leaf extract: A mechanism perspective. J. Clust. Sci. 25(4): 1157–1171.

Senthil-Nathan, S. 2015. A review of biopesticides and their mode of action against insect pests. J. Environ. Sustain. 49–63.

Shankar, S.S., Rai, A., Ahmad, A. and Sastry, M. 2004. Rapid synthesis of Au, Ag, and bimetallic Au core–Ag shell nanoparticles using Neem (*Azadirachta indica*) leaf broth. J. Colloid Interface Sci. 275(2): 496–502.

Shankar, S. and Rhim, J.W. 2017. Preparation and characterization of agar/lignin/silver nanoparticles composite films with ultraviolet light barrier and antibacterial properties. Food Hydrocoll. 71: 76–84.

Shapiro-Ilan, D.I. and Gaugler, R.A.N.D.Y. 2002. Production technology for entomopathogenic nematodes and their bacterial symbionts. J. Ind. Microbiol. Biotechnol. 28(3): 137–146.

Singh, J., Dutta, T., Kim, K.H., Rawat, M., Samddar, P. and Kumar, P. 2018. 'Green' synthesis of metals and their oxide nanoparticles: Applications for environmental remediation. J. Nanobiotechnology 16(1): 1–24.

Smith, A.M. and Gilbertson, L.M. 2018. Rational ligand design to improve agrochemical delivery efficiency and advance agriculture sustainability. ACS Sustain. Chem. Eng. 13599–13610.

Song, C., Sun, W., Xiao, Y. and Shi, X. 2019. Ultrasmall iron oxide nanoparticles: Synthesis, surface modification, assembly, and biomedical applications. Drug. Discov. 24(3): 835–844.

Soni, N. and Prakash, S. 2011. *Aspergillus niger* metabolites efficacies against the mosquito larval (*Culex quinquefasciatus, Anopheles stephensi* and *Aedes aegypti*) population after column chromatography. Am. J. Microbiol. 2(1): 15–20.

Soni, N. and Prakash, S. 2012. Efficacy of fungus mediated silver and gold nanoparticles against *Aedes aegypti* larvae. Parasitol. Res. 110(1): 175–184.

Soni, N. and Prakash, S. 2012a. Larvicidal effect of *Verticillium lecanii* metabolites on *Culex quinquefasciatus* and *Aedes aegypti* larvae. Asian Pac. J. Trop. Dis. 2(3): 220–224.

Soni, N. and Prakash, S. 2013. Possible mosquito control by silver nanoparticles synthesized by soil fungus (*Aspergillus niger* 2587).

Soni, N. and Prakash, S. 2014. Microbial synthesis of spherical nanosilver and nanogold for mosquito control. Ann. Microbiol. 64(3): 1099–1111.

Srisukchayakul, P., Wiwat, C. and Pantuwatana, S. 2005. Studies on the pathogenesis of the local isolates of *Nomuraea rileyi* against *Spodoptera litura*. Sci. 31: 273–276.

Strobel, G. and Daisy, B. 2003. Bioprospecting for microbial endophytes and their natural products. Microbiol. Mol. Biol. Rev. 67(4): 491–502.

Sundaravadivelan, C. and Padmanabhan, M.N. 2014. Effect of mycosynthesized silver nanoparticles from filtrate of *Trichoderma harzianum* against larvae and pupa of dengue vector *Aedes aegypti* L. Environ. Sci. Pollut. Res. 21(6): 4624–4633.

Thakur, N., Kaur, S., Tomar, P., Thakur, S. and Yadav, A.N. 2020. Microbial biopesticides: Current status and advancement for sustainable agriculture and environment. In New and Future Developments in Microbial Biotechnology and Bioengineering (pp. 243–282). Elsevier.

Togola, A., Meseka, S., Menkir, A., Badu-Apraku, B., Boukar, O., Tamò, M. and Djouaka, R. 2018. Measurement of pesticide residues from chemical control of the invasive *Spodoptera frugiperda* (Lepidoptera: Noctuidae) in a maize experimental field in Mokwa, Nigeria. Int. J. Environ. Res. Public Health 15(5): 849.

Tuncsoy, B. and Mese, Y. 2021. Influence of titanium dioxide nanoparticles on bioaccumulation, antioxidant defense and immune system of *Galleria mellonella* L. Environ. Sci. Pollut. Res. 1–9.

Vahabi, K., Mansoori, G.A. and Karimi, S. 2011. Biosynthesis of silver nanoparticles by fungus *Trichoderma reesei* (a route for large-scale production of AgNPs). Insciences J. 1(1): 65–79.

Vala, A.K. 2014. Intra-and extracellular biosynthesis of gold nanoparticles by a marine-derived fungus *Rhizopus oryzae*. Synth. React. Inorg Met. Org. Chem. 44(9): 1243–1246.

Vijayakumar, N. and Alagar, S. 2017. Consequence of Chitinase from *Trichoderma viride* integrated feed on digestive enzymes in *Corcyra cephalonica* (Stainton) and antimicrobial potential. Biosci. Biotechnol. Res. Asia 14(2): 513–519.

Vinale, F., Marra, R., Scala, F., Ghisalberti, E.L., Lorito, M. and Sivasithamparam, K. 2006. Major secondary metabolites produced by two commercial *Trichoderma* strains active against different phytopathogens. Lett. Appl. Microbiol. 43(2): 143–148.

Vivekanandhan, P., Kavitha, T., Karthi, S., Senthil-Nathan, S. and Shivakumar, M.S. 2018. Toxicity of *Beauveria bassiana*-28 mycelial extracts on larvae of *Culex quinquefasciatus* mosquito (Diptera: Culicidae). Int. J. Environ. Res. Public Health 15(3): 440.

Vivekanandhan, P., Swathy, K., Thomas, A., Kweka, E.J., Rahman, A., Pittarate, S. and Krutmuang, P. 2021. Insecticidal efficacy of microbial-mediated synthesized copper nano-pesticide against insect pests and non-target organisms. Int. J. Environ. Res. 18(19): 10536.

Wang, A., Mahai, G., Wan, Y., Yang, Z., He, Z., Xu, S. and Xia, W. 2020. Assessment of imidacloprid related exposure using imidacloprid-olefin and desnitro-imidacloprid: Neonicotinoid insecticides in human urine in Wuhan, China. Environ. Int. 141: 105785.

Wijnhoven, S.W., Peijnenburg, W.J., Herberts, C.A., Hagens, W.I., Oomen, A.G., Heugens, E.H. and Geertsma, R.E. 2009. Nano-silver—A review of available data and knowledge gaps in human and environmental risk assessment. Nanotoxicology 3(2): 109–138.

Wraight, S.P., Jackson, M.A. and De Kock, S.L. 2001. Production, stabilization and formulation of fungal biocontrol agents. Fungi as Biocontrol Agents: Progress, Problems and Potential, 253–287.

Xu, J., Chen, H.B. and Li, S.L. 2017. Understanding the molecular mechanisms of the interplay between herbal medicines and gut microbiota. Med. Res. Rev. 37(5): 1140–1185.

Xu, J., Huang, Y., Zhu, S., Abbes, N., Jing, X. and Zhang, L. 2021. A review of the green synthesis of ZnO nanoparticles using plant extracts and their prospects for application in antibacterial textiles. J. Eng. Fibers Fabr. 16: 15589250211046242.

Xu, J., Zhang, K., Cuthbertson, A.G., Du, C. and Ali, S. 2020. Toxicity and biological effects of *Beauveria brongniartii* Fe0 nanoparticles against *Spodoptera litura* (Fabricius). Insects 11(12): 895.

Yadav, A., Kon, K., Kratosova, G., Duran, N., Ingle, A.P. and Rai, M. 2015. Fungi as an efficient mycosystem for the synthesis of metal nanoparticles: Progress and key aspects of research. Biotechnol. Lett. 37(11): 2099–2120.

Yang, C., Neshatian, M., van Prooijen, M. and Chithrani, D.B. 2014. Cancer nanotechnology: Enhanced therapeutic response using peptide-modified gold nanoparticles. J. Nanosci. Nanotechnol. 14(7): 4813–4819.

Yosri, M., Abdel-Aziz, M.M. and Sayed, R.M. 2018. Larvicidal potential of irradiated myco-insecticide from *Metarhizium anisopliae* and larvicidal synergistic effect with its mycosynthesized titanium nanoparticles (TiNPs). J. Radiat. Res. Appl. Sci. 11(4): 328–334.

Yu, J., Wang, D., Geetha, N., Khawar, K.M., Jogaiah, S. and Mujtaba, M. 2021. Current trends and challenges in the synthesis and applications of chitosan-based nanocomposites for plants: A review. Carbohydr. Polym. 117904.

Zeilinger, S. and Omann, M. 2007. Trichoderma biocontrol: Signal transduction pathways involved in host sensing and mycoparasitism. Gene. Regul. Syst. Biol. 1: GRSB–S397.

Zielińska, A., Carreiró, F., Oliveira, A.M., Neves, A., Pires, B., Venkatesh, D.N., Durazzo, A., Lucarini, M., Eder, P., Silva, A.M. and Santini, A., 2020. Polymeric nanoparticles: Production, characterization, toxicology and ecotoxicology. Molecules 25(16): 3731.

Zhu, F., Lavine, L., O'Neal, S., Lavine, M., Foss, C. and Walsh, D. 2016. Insecticide resistance and management strategies in urban ecosystems. Insects 7(1): 2.

Zhu, L., Qi, S., Xue, X., Niu, X. and Wu, L. 2020. Nitenpyram disturbs gut microbiota and influences metabolic homeostasis and immunity in honey bee (*Apismellifera* L.). Environ. Pollut. 258: 113671.

12

Application of Mycosynthesized Nanoparticles in Veterinary Sciences

Ranjit Ingole,[1,]* Pramod Ingle[2] and Aniket Gade[3]

Introduction

Nanotechnology is described as a technology and science that acts on a nanoscale and adheres to scientific principles as well as new properties that can be discovered and mastered when working in this range. Nanobiotechnology has sparked a lot of interest in recent decades and is often regarded as the next industrial revolution. It is an important field of modern study that comprises investigations on particle structure creation, design, and manipulation between 1–100 nm (Aboul-Nasr 2018). Most typical nanoparticle (NPs) synthesis procedures like chemical methods involve concerns, such as the usage of harmful chemicals and waste emergence, which could result in environmental pollution (Iravani et al. 2014, Ahmed et al. 2016). As a result, there has been a surge in interest in green nanoparticle production in recent years. Biosynthesis of nanoparticles using microorganisms is replacing chemical and physical methods. It is an important technique for producing clean, non-toxic, and environment friendly procedures for fabricating metal nanoparticles with an inherent ability to reduce metals by specific metabolites from various biochemical pathways. Extracts from various plants, microorganisms, yeast, fungi, or any living organism can be used to produce biogenic nanoparticles (Durán et al. 2011). As different biomolecules react differently with metal ions, the basic mechanism of synthesis

[1] Department of Veterinary Pathology, Post Graduate Institute of Veterinary and Animal Sciences, Akola, Maharashtra, India - 444104.

[2] Nanobiotechnology Lab, Department of Biotechnology, Sant Gadge Baba Amravati University, Amravati, Maharashtra, India – 444602.

[3] Department of Biological Science and Biotechnology, Institute of Chemical Technology (Formerly UDCT), Nathalal Parekh Marg, Mumbai-400019, India.

* Corresponding author: ingoleranjit@rediffmail.com

of nanoparticles by using biological agents is not fully understood. The capping of nanoparticles with biomolecules produced from the living organism or their extracts and byproducts during green nanoparticle production can improve stability and preserve the biological interest for what is being synthesized (Ballotin et al. 2016).

Myco-synthesis is the process of making metal nanoparticles using fungi/ mushroom secretions rather than other bio or chemical materials. Microorganisms manufacture inorganic materials either inside the cell (intracellular) or outside the cell (extracellular), according to the literature. However, the mechanism of intracellular and extracellular nanoparticle formation differs depending on the biological agent. Mycosynthesis of nanoparticles has several advantages over nanoparticle synthesis using other microorganisms. This is because fungi have high tolerance to high metal ion concentrations in the medium, good particle dispersion, ease of management in large-scale production, much higher protein expression, and environmental friendliness. Fungi may also be grown on a large scale (called "nanofactories") and create nanoparticles with controlled length and morphology (Gade et al. 2008, Khan et al. 2017). Moreover fabrication of nanoparticles by fungi is known to be an ecofriendly or green approach (Gade and Rai 2021). Mycosynthesis process coats the nanoparticles with biomolecules present in fungus and their extracts, which improve stability and may confer bio catalytic activity. Recently, the fungal system has been proved as "Bionanofactories" synthesizing nanoparticles from various metals like gold, silver, platinum and CdS. Actually the fungi are rich source of reducing enzymes and play a pivotal role in nanoparticle synthesis. Recently, an elaborated account on potential role of *Phoma* spp. for the synthesis of silver nanoparticles was studied by Gade et al. (2022).

Despite the fact that many studies have been published regarding the biogenic synthesis of silver nanoparticles using fungi, the precise mechanism has not yet been elucidated. Earlier studies revealed that extracellular synthesis of nanoparticles is consistent with reactions in which the enzymes present within the fungal filtrate act to catalytically reduce silver and gold ions, producing elemental silver (Ag^0) and gold at a nanometric scale (Guilger-Casagrande and Lima 2019). Synthesis of silver nanoparticles with the usage of *Fusarium oxysporum* suggested reduction of silver ions due to the action of the nitrate reductase enzyme and anthraquinones (Durán et al. 2005). Purified nitrate reductase and phytochelatins from the same fungus were used in one investigation and demonstrated that extracellular NADPH-dependent nitrate reductase enzymes and quinones are primarily responsible for nanoparticle production (Kumar et al. 2007).

The metal precursors are delivered to the mycelial culture and internalized in the biomass in case of intracellular synthesis. As a result, after synthesis, the nanoparticles must be extracted using chemical treatment, centrifugation, and filtration by denaturing the fungal biomass to liberate the nanoparticles (Rajput et al. 2016, Molnár et al. 2018). The metal precursor is added to an aqueous filtrate containing solely fungal biomolecules in extracellular synthesis, leading in the creation of free nanoparticles in the dispersion. This method is the most popular since it does not require any treatments to release the nanoparticles from the cells

(Azmath et al. 2016, Costa Silva et al. 2017, Gudikandula et al. 2017). However, in order to remove fungal residues and contaminants, the nanoparticle dispersion must be purified, which can be accomplished using methods such as simple filtering, membrane filtration, gel filtration, dialysis, and ultracentrifugation (Ashrafi et al. 2013, Qidwai et al. 2018, Yahyaei and Pourali 2019). The cell wall plays a key role in heavy metal absorption in intracellular production, allowing metal ions to be trapped and reduced by enzymes within the cell wall and cytoplasm, resulting in metal ion aggregation and nanoparticle formation. However, studies have shown that the reductase enzyme is responsible for metal ion reduction and consequent nanoparticle creation in extracellular myconanosynthesis.

The method of manufacturing silver nanoparticles with fungi begins with cultivating the fungus on agar and then transferring it to a liquid media. Later, the biomass is suspended into water for the release of the components involved in the synthesis. The biomass is discarded after filtration, and silver nitrate is added to the filtrate which are immediately reduced to silver nanoparticles by secondary metabolites secreted in the extract (Guilger et al. 2017, Mekkawy et al. 2017, Ottoni et al. 2017).

Advantages of Mycosynthesized NPs

Fungi have a great deal of potential for producing a wide range of chemicals that can be employed in a variety of applications. Microscopic filamentous fungi (ascomycetes and imperfect fungi) and other fungal species are known to produce around 6400 bioactive compounds (Bérdy 2005). Nanoparticles made with fungi have a wide range of potential applications in health, agriculture, and pest management. There are no findings on whether biogenic nanoparticles generated from other sources, such as bacteria, or plants, have better or poorer activity. However, fungi-based synthesis has the following advantages.

(1) Because of their enormous surface to volume proportion, higher reactivity, bioavailability, control particle size, control drug release, site specific targeting, and controlled arrival of medications, large quantities of metabolites are produced and have inventive physicochemical characteristics superior to bulk counterparts (Youssef et al. 2019).

(2) Fungi can produce antibiotics, which can be contained in the capping and work in tandem with the nanoparticle core.

(3) Previous research has demonstrated that fungus based biogenic synthesis can be used to control pathogenic fungi and bacteria, fight cancer cells and viruses, and provide larvicidal and insecticidal properties.

(4) Nanoparticles have the ability to affect *in vivo* and *in vitro* biological research and applications.

(5) Because nanoparticles may enter cells, tissues, and organs more easily than macroparticulate substances, they offer the potential to overcome current drugs limited bio accessibility and high toxicity (Mohanraj and Chen 2006, Mohsen and Zahra 2008).

(6) Nanotechnology has played a significant role in improving medicine delivery (Scott 2007).

(7) Nanotheranostics is a therapeutic technique that incorporates both drugs and diagnostics.

Various NPs used in Veterinary Science

There are various types of nanoparticles that are applied in veterinary sciences based upon the desired size, shape, composition, use and resulting activity *in vivo* and/ or *in vitro*. The following Figure 1, represents the range of nanoparticles used in veterinary sciences.

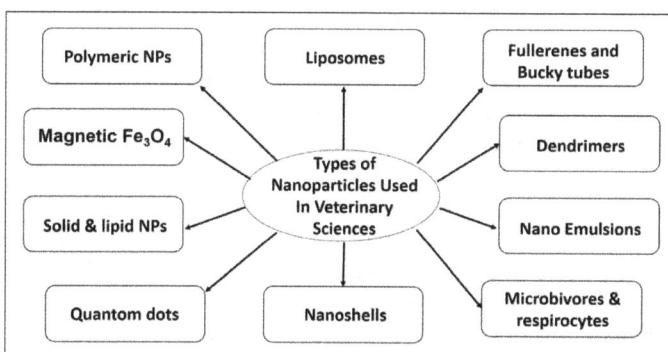

Figure 1. Types of nanoparticles used in veterinary sciences.

Polymeric NPs

The particle size of polymeric nanoparticles ranges from 1 to 1000 nm. Various biodegradable polymers, such as poly-(lactide-co-glycolide) (PLGA) copolymers, poly (lactide) (PLA), poly (-caprolactone) (PCL), and poly-(amino acids), as well as natural polymers such as chitosan, gelatin, alginate, and albumin, are often utilized to produce polymeric nanoparticles. Easy to produce, biocompatible, biodegradable, affordable, non-toxic, non-immunogenic, and water soluble are just a few of the qualities of polymeric nanoparticles. Active chemicals can be introduced into polymeric nanoparticles and entrapped within or surface adsorbed onto the polymeric core. Polymeric nanoparticles show peculiar features such as adsorption, dissolution, entrapment, encapsulation, or chemical interaction with drug molecules on their surface, and are employed in the formulation of various therapeutic molecules. Polymeric nanoparticles improve drug bioavailability or specific delivery at the site of action. There are currently two types of polymers: manufactured polymers, like polyethylene glycol (PEG), and trademark polymers, such as inulin and chitosan, in which nanoparticles are structured on polysaccharides. Polymer nanoparticles are useful for meeting the criteria of specific drug delivery systems due to their adaptability. Biodegradable nanoparticles provide excellent drug delivery in a regulated release profile, are stable in the circulatory system, and are non-toxic and non-immunogenic to the reticuloendothelial system, which is part of the immune system. In veterinary medicine, polymeric nanoparticles are used for targeted medication administration via encapsulation to treat diseases like Newcastle disease in poultry (Yang 2000, Mohantya et al. 2014).

Liposomes

In the 1960s, the term "liposome" was coined after it was found that closed lipid bilayer vesicles formed spontaneously in water. The phrase "lipid nanoparticles" was first used in the early 1990s, coinciding with the dawn of the nanoscience and nanotechnology era.

Liposomes are commercially available nanoparticles that have been utilised to deliver cytotoxic medicines, antifungal medications, and vaccinations in clinical trials. Liposomes and lipid nanoparticles are designed similarly, although their composition and function differ slightly. They are circular PEGylated particles that are biodegradable and non-destructive. They could also be utilised to administer medications for cancer and other disorders with lower systemic toxicity (Torres-Sangiao et al. 2016). Multidrug-resistant tumors can now be treated more effectively, with less cardiotoxicity. Liposomes, a precursor to lipid nanoparticles, are a very adaptable nano-carrier platform that can transport both hydrophobic and hydrophilic molecules such as small molecules, proteins, and nucleic acids. The exterior surface of liposomes is coated with an external poly ethylene glycol (PEG) layer, which protects the particles from being attacked by the body's immune system. Chelated antibodies are fixed on their exterior surface using liposomes. Liposome-encapsulated drugs target specific cells or tissues by covalently attaching monoclonal antibodies or other proteins to the liposomes' outer surface. Liposomes can be used to deliver both hydrophobic and hydrophilic medications. Because of their biodegradable structure, they are regarded as being prosperous. Liposomes have biocompatibility and biodegradability, and their nanosize allows them to be used in nanomedicine, cosmetics, and the food sector. Liposomes are used as adjuvants in vaccinations, drug delivery vehicles in medicine, signal enhancers or carriers in medical diagnostics and analytical biochemistry, solubilizing agents for various ingredients, support matrices for various ingredients, and penetration enhancers, among other industrial applications. Liposomes are the first nanomedicine carrier and delivery system for clinical application transition, as well as a well-established technical platform with widespread clinical acceptability. Immuno-liposomes are a type of liposome that can be linked to antibodies that attack harmful bodily cells, and they can also be attached to external antigens for vaccination purposes (Bakker-Woudenberg et al. 2005). Liposomes are used as therapeutic agents in cattle, horses, dogs, birds, cats and sheep (Underwood and van Eps 2012).

Fullerenes and Bucky Tubes

Cylindrical fullerenes, also known as carbon nanotubes or buckytubes, have the shape of a little ball and can communicate with pathogens and cells effectively. Buckytubes are carbon allotropes with a hollow structure and exceptional electrical, thermal, and mechanical capabilities. Buckyballs refer to spherical fullerenes, while nanotubes refer to cylindrical fullerenes. Richard Smalley discovered fullerene compounds in 1985. The buckyball molecules are large and form C-C bond and produce a structure like trigonal establishing spheroids. Due to its symmetrical structure and reactivity to several types of reagents, fullerene is an interesting molecule. There are 12 pentagons and 20 hexagons in a fullerene. It tends to take on many shapes depending on its size. They can be used as biosensors for identifying various components, such as immunoglobulins. Nanotubes can be administered into cells for disease treatment due to their needle-like shape (Madani et al. 2011, Jurj et al. 2017, Meena et al. 2018). In medicine, fullerenes are employed as light-activated antibacterial agents. It is also used in the development of high-performance MRI contrast agents, photodynamic treatment, X-ray imaging contrast agents, and medication and gene delivery, among other biological uses. Fullerene has been found to work as a possible antioxidant, reducing oxidative stress in proteins. The addition of an OH group to fullerene is expected to boost fullerenol's antioxidant activity in comparison to the parent molecule. Fullerenes have the ability to intercalate DNA molecules, interfering with pathogen replication and inhibiting infections. In case of animals, these fullerenes interact with intracellular signal processing and catalytic activities, without hampering the metabolic processes. Thus, fullerenes are used in antitumor and anticancer treatments with minimal alterations in metabolic activities (Bourassa et al. 2019).

Microbivores and Respirocytes

Microbivores, also known as nanorobotics phagocytes, are nanorobots that act as artificial white blood cells (Feritas Jr. 2005). Microbiovores are nanorobotic devices that could safely deliver rapid elimination of blood pathogens with relatively modest dosages of devices, adding to the physician's therapeutic armament. Microbivores grab pathogens in the bloodstream and break them down into smaller molecules. The primary role of the microbiovore is to absorb and digest pathogens in the bloodstream *via* phagocytosis. Microbivores could be useful in the treatment of infections of meninges and cerebrospinal fluid by quickening the kill rate as opposed to typical antibiotic treatments (Shabnashmi et al. 2016). Microbiovores could be employed to treat bacterial infections in various fluids and tissues, as well as to regulate systemic inflammatory cytokine levels (Costerton et al. 1995). Toxemia, systemic dissemination of harmful products of bacteria growing in a focal region, and other biochemical sequelae of sepsis are all combated by the addition of microbivores. In the microbivores, all the bacteria components are internalized and digested into harmless non antigenic molecules followed by discharge from the device. Nano robots represent a complete antimicrobial therapy without the risk of sepsis or septic shock. Microbivores could also be used to rid the blood of viral pathogens, which are typically present during viremia at titers similar to those found in bacteremia. Because viruses are smaller than bacteria, the processing time per virion may be lowered to as little as 5–10 seconds. Microbivores clear fungemias with particle loads ranging from 1–1000 CFU/ml. Fungal particles may be up to ~ 400 micron^3 in volume, requiring ~ 100 min for complete digestion using a microbivores' protocol that employs careful piecewise digestion involving ~ 800 "bites" (Shabnashmi et al. 2016). Nanorobotics have been discovered to have a lot of potential for diagnosing and treating diseases like cancer, cardiac blockages, diabetes, arteriosclerosis, kidney stones, and so on. The nano robot can provide tailored treatment, resulting in excellent efficacy against a variety of ailments. Respirocytes are artificial Nano medicinal erythrocytes (Feritas Jr. 2002) that are designed to duplicate all of the main functions of red blood cells, retain live tissue, and give treatment for cardiac blockage, anaemia, lung illness, hypoxia, and other respiratory issues (Shabnashmi et al. 2016). Respirocytes have properties that are similar to both red and white blood cells. Microbivores catch diseases while efficiently supplying tissues with oxygen and avoiding the accumulation of CO_2 by deploying excellent management sensors in a well-controllable manner. By an enzymatic impact, the cleaned microbes were transformed into building components that included amino acids, nucleotides, and unsaturated lipids (Feritas Jr. 2005). Respirocytes contain an internal pressure of 1000 atmospheres of trodden oxygen and carbon dioxide. The intense pressure would be safely contained in two separate high pressure vessels likely made of pure diamond. At this intense pressure, a respirocyte could hold 236 times more oxygen and carbon dioxide than our natural red blood cells (Feritas Jr. 2009). Respirocytes in oxygenated form help to treat all types of anemia, including acute anemia caused by a sudden loss of blood after injury or surgical intervention (Youngson et al. 1993). Respirocyte production has been used to measure tissue oxygen tension. Respirocyte technology could help regulate fatty acids, serum glycerides, lipoproteins, diabetic ketosis, gestational diabetes, and other dietary disorders by providing a precisely metered ingestible or injectable drug delivery system (Shabnashmi et al. 2016).

Nanoshells

Nanoshells are made up of a spherical or rod-shaped compound core surrounded by a thin layer of another material that is 1–20 nm thick. Semiconductors, metals, and insulators can all be used to make nanoshell materials. These are used to irradiate malignant tumors using an infrared laser for analysis. The nanoshell can be triggered by infrared light, in which case the nanoshell absorbs the energy of the light and generates heat, and nanoshells have the potential to destroy bacteria and tumor cells through this technique. Nanoshells are nanoscale beads and rods with varying thicknesses of gold that can absorb specific wavelengths of light and convert them to heat. This approach is very useful in radiation devices that are used for imaging and selecting cell death. Nanoshells reduce the force of an X-beam and can thus be used as a radiotherapy adjuvant. Furthermore, gold nanoparticles are biocompatible and non-toxic to the body (Freitas 2005, Manuja et al. 2012). Nanoshells can be linked to antibodies for targeted delivery into cells with specific features, and they've been demonstrated in cell line models to cause tumour death while leaving healthy cells unharmed. Nanocrystals, nanospheres, and nanofibers are utilised alone or in combination with stem cells for a variety of applications, including improving scaffold bioactivity to regulate infection and medication, gene, and cell delivery. Nanoshells have a variety of uses, including treating severe gingival and periodontal infections, removing oral and dental cancer cells, and possibly welding soft tissue in inaccessible areas. Nanoshells have been employed for biochemical imaging and cancer treatment due to their optical and chemical qualities. The gold shell's

PEG coating reduced non-specific macrophage absorption and enhanced *in vivo* bioavailability. Nanoshells can also contain anticancer medications like doxorubicin and combretastatin, allowing the two treatments to work together at the tumour location. Antitumor medications (e.g., doxorubicin, paclitaxel, small interfering RNA, and single-stranded DNA) are delivered into cancer cells using AuNSs (gold nanoshells), which improves therapeutic efficacy. Active targeting ligands, such as aptamers, antibodies, and peptides, can be added to AuNSs to boost the particles' specific binding to the targeted targets (Mudshinge et al. 2011). Colorimetry, photothermal therapy, biosensing, gene screening, chemical libraries, barcoding and biological imaging, and colloidal stability are just a few of the clinical and therapeutic applications for nanoshells. Cancer therapy employs magnetic nanoprobes. An alternating magnetic field applied externally to iron nanoparticles coated with monoclonal antibodies targeting at tumour cells can generate significant levels of heat and then accumulate in specific areas. The enzyme-linked immunosorbant assay uses the coupling of AuNSs with isopropyl acrylamide and acrylamide with a mix of enzymes and antibodies to diagnose immunological disorders (ELISA) (Zhang et al. 2016). The above mentioned attributes makes nanoshells an excellent material for the non-invasive detection of tumors and fastidious diagnosis (Mohanty et al. 2014).

Quantum Dots

Quantum dots (QDs) are nanometer size luminescent semiconductor crystals having unique chemical and physical properties due to their size and their highly compact structure. QDs are extremely tiny, measuring between 1.5 and 10.0 nm in size. Because of their small size, QDs produce physically confined electrons, also known as "quantum confinements." Due to its unique qualities, such as size-dependent optical properties, exceptional photostability, high extinction coefficient and brightness, and significant Stokes shift, quantum dots have been successfully demonstrated. Cadmium, zinc, and selenium make up the framework of semiconductor parts (Torres-Sangiao 2016). Organic dyes have been shown to be limited in their ability to exhibit all of these qualities, making them unsuitable for many imaging and biosensing applications. Due to their enormous surface area *in vitro/in vivo*, optical trackability, QDs' ultra-small size may have aided in the development of multimodal or multifunction probes. QDs being photostable have become an ideal candidate for multicolored imaging and scrutinizing various processes in the living cells. Cancer cells and blood vessels were also targeted using quantum dots by conjugating surfaces with specific petptides. QDs are used in a wide range of devices, including solar cells, transistors, LEDs, medical imaging, and quantum computing, due to these features. Many studies have proved the lab-scale synthesis of QDs, However, the scale-up process for QD manufacture is problematic. By adjusting the size of QDs, the size dependence indicated attributes may be easily regulated. QDs will become more affordable once new methods for reliably synthesising varying sizes of monodisperse nanocrystals are established. These unique qualities have given QDs potential for a lot of scientific and industrial applications. QDs are generally employed in diagnostics and immunodiagnostics (Mohanty et al. 2014). These materials have a wide range of uses, such as novel types of fluorescence probes and active components in nanostructure complexes. QDs are made up of elements from the periodic table's groups III–V, II–VI, or IV–VI, such as CdSe, CdS, CdTe, CdSe@ZnS, CdS@ZnS, and CdSeTe@ZnS. These QDs have excellent fluorescence properties and are widely used in nanosensing and biosensing applications. Biocompatible QDs as CFQDs have been used due to their nontoxic property. These are SiQDs, C-dots, and GQDs. With the exception of silicone QDs, which have superior biocompatibility in cells, most QDs have been discovered to pose considerable dangers to cells. Toxic elements, such as cadmium and lead, are commonly found in the most widely used QDs, posing a serious threat to normal cellular metabolism. To address this issue, core-shell structures have been designed to prevent hazardous ion leakage. QDs are used in biological imaging systems such as *in vitro* and *in vivo* animal research because of their exceptional optical characteristics. Self-assembling nanostructures, genetic editing, protein binding, and biomarkers are just a few of the biotechnological applications of DNA. QDs are classified according to their chemical makeup. The binary QDs made up of elements from groups II–VI, including such CdSe, CdS, ZnS, ZnTe, and HgS, are the most common. However, stability, and surface defects are the two major problems relating to QDs' crystal-structure faults, which may result in QD deactivation and changes in the emission profile. In animal sciences, QDs are used as fluorescent contrast agents and good medication carriers in *in vitro* diagnostics (Kim et al. 2010).

Solid Lipid NPs

Colloidal drug conveyance systems called solid liquid nanoparticles (SLNs) were invented in the early 1980s. It's a hybrid of liposomes and niosomes that include phospholipids and surfactants. Solid lipid nanoparticles are colloidal drug carriers with diameters ranging from 50 nm to 1 μm and are made up mostly of fatty acids or mono-, di-, or triglycerides. It is possible to encapsulate lipophilic and hydrophilic medicines in solid lipid nanoparicles. It's a hybrid of liposomes and niosomes that include phospholipids and surfactants. For the delivery of several drugs, SLNs demonstrated to be more favourable than liposomes and niosomes. SLNs offer increased stability and provide superior protection against pharmaceutical deterioration. The safety profile of solid lipid nanoparticles is based on biocompatible lipids that are well tolerated by the body. It can be mass-produced in vast quantities. It is possible to sterilize it. Drugs, vitamins, chemicals, and practically all xenobiotics can be encapsulated in solid lipid nanoparticles. Solid lipid nanoparticles could also be used to transport genes and anticancer medicines. SLNs have proven to be the most effective skin delivery vehicle. Furthermore, SLNs have various advantages over other nano-sized carriers. Different hydrophilic medications or antibodies can have their outer hydrophilic shell conjugated to SLNs. The outer shell also increases the bio profitability of the medicine. Furthermore, cationic solid lipid nanoparticles can legally tie nucleic acid components by electrostatic linkage, enabling their employment for quality treatment. This type of nanoparticles can be administered in a variety of ways, including topical, oral, and subcutaneous injection. They can efficiently deliver drugs inside the central nervous system by crossing the blood-brain barrier. Apart from strong lipid nanoparticles, the use of fluid lipid nanoparticles is now being investigated (Elgqvist 2017). Biotoxicity is not evident in solid lipid nanoparticles since the lipids utilised are biocompatible and biodegradable. Without the use of organic solvents, solid lipid nanoparticles can be made. Physically, it is extremely stable. Solid lipid nanoparticles can be used to accomplish regulated drug release and drug targeting. The addition of active compounds in solid lipid nanoparticles can improve their stability. Solid lipid nanoparticles are a new type of colloidal delivery method that combine the benefits of polymeric nanoparticles and liposomes while avoiding acute and long term toxicity. Solid lipid nanoparticles have several advantages, including excellent bioavailability, CNS targeting, and biodegradability without hazardous degradation products. Surface functionalization of SLNs with ligands allows for effective drug administration across the blood-brain barrier. As a result, they can be used in a variety of methods for delivering medications to animal brain cells (Muller and Keck 2004).

Magnetic Iron Oxide NPs

Magnetic iron oxide nanoparticles are appealing nanoparticles with outstanding magnetic properties that can be used in biomedical applications. Metal oxides such as TiO_2, SnO_2, ZnO, SiO_2, ZrO_2, and others have been utilized to interact with a variety of biopolymers to create metal oxide based biopolymer composites. Co, Cu, and Mn oxides, along with the Pt group and Ag, are the most active single metal oxides for full oxidation. This group is distinguished by an external magnetic field that allows them to be directed to their target cells *via* the circulation. They're better for things like imaging, thermal therapy, and drug delivery (Manuja et al. 2012). *In vivo* applications for supramagnetic iron oxide nanoparticles with proper surface chemistry include MRI contrast enhancement, immune assay, tissue repair, biological fluid detoxification, drug delivery, hyperthermia, and cell separation. High magnetization values, a size lower than 100 nm, and a tight particle size distribution are required for all of these applications. Magnetic nanoparticles can bind to drug molecules, antibodies, proteins, enzymes, or nucleotides and be directed to an organ, tissue, or tumor with the help of an external magnetic field.

The structure of magnetic NPs consists of an iron core surrounded by a silica outer fluorescent layer where drug attachment takes place. The outer shell, which is made of polymer, aids in particle adjustment. They are used in a few beautiful reverberation medical applications for disease diagnosis and treatment as multi helpful theranostic structures because of their suitable attractive ties. To prevent particle aggregation and protect them from the immunological response, polyethylene glycol is utilized to coat the particles. By enhancing light retention, the presence of silica coatings enhances malignancy imaging. Furthermore, by limiting the systemic distribution of medications and lowering the dosages of harmful substances, magnetic IONP-based drug targeting is a viable cancer treatment strategy for avoiding the adverse effects of conventional chemotherapy (Cao et al. 2008). The magnetic properties of IONPs can be used in a variety of *in vivo* applications, which can be divided into three categories: (i) magnetic vectors, which can be directed by a magnetic field gradient towards a specific location, such as in targeted drug delivery; (ii) magnetic contrast agents in MRI; and (iii) hyperthermia or thermoablation agents, which heat magnetic particles selectively by applying high-frequency magnetic field (Arruebo et al. 2007). Thus, the magnetic iron oxide nanoparticles can be used as MRI contrast agents and drug delivery agents, for example in cats (Kim et al. 2010, Underwood and van Eps 2012).

Dendrimer

Dendrimers are nano sized, radially symmetric molecules with well-defined homogenous and monodisperse structure consisting of tree like arms or branches (Srinavasa and Yarema 2007). Dendrimers are hyper branching nanomaterials that are excellently dissolved in aqueous solution and are made up of polymers that are much smaller than human cells (Rodríguez-Burneo et al. 2017, Awate et al. 2013, Chakravarthi and Balaji 2010). Dendrimers are artificial macromolecules with a highly defined molecular structure and a large number of functional groups (Tomalia and Frechet 2002). When compared to other nanoscale synthetic structures such as conventional polymers, buck balls, or carbon nanotubes, dendrimers are either very non-defined or have limited structural variation.

Biomedical applications benefit from dendritic polymers. These dendritic polymers are functionalized similarly to enzymes, proteins, and viruses. Dendrimers and other molecules can be connected to the outside of their voids or encased within them (Patel and Patel 2013). Dendrimers administer medications in a controlled and precise manner while also improving the pharmacokinetic features of cancer treatments. The drug dendrimer conjugates have a high solubility, low systemic toxicity, and selective solid tumor accumulation. Encapsulation, complexation, and conjugation have all been proposed as ways to contain medicinal molecules, genetic materials, targeting agents, and colors within the dendrimer structure. The chemistry of a host guest is based on the reaction of a substrate molecule interacting with a receptor molecule. Dendrimers have also been used in the development of transdermal medicine delivery devices. Dendrimers are a suitable alternative in the field of effective delivery system for bioactive pharmaceuticals with hydrophobic moieties in their structure and low water solubility (Cheng et al. 2007). Although dendrimers are single molecules, they can have a lot of functional groups on their surfaces. This qualifies them for applications requiring covalent bonding or close contact of a large number of species. The unimolecular micelle characteristic of dendrimers is due to their hydrophilic exteriors and interiors. Oral bioavailability of problematic medications can be improved using dendrimer-based carriers. Since the first dendrimers were made, there has been a tremendous increase in the importance of dendrimer chemistry (Abbasi et al. 2014). They are mostly used in the treatment of cancer. Dendrimers are distinguished by their massive, branching, and complex structure. Dendrimers can be added to medications that are used for imaging. *Pseudomonas aeruginosa*, *E. coli*, and *Staphylococcus aureus* are all resistant to dendrimer-loaded nanocomposites, according to numerous investigations. The landing of their store of medications or radioactive substances within the tumor allows for migration. Finally, if the treatment is successful, signals are transmitted together with the execution of dangerous cells. Thus these dendrimers are successfully applied for vaccine and microbiocide delivery in pigs (Underwood and van Eps 2012).

Nanoemulsion

Nanoemulsions are colloidal particle systems with submicron sizes (10 to 1000 nm) that operate as drug carriers and are thermodynamically stable. Microfluidization, high-pressure homogenization, and the phase-inversion temperature approach are the three most used procedures for preparing nanoemulsions. Nanoemulsions have a solid sphere shape, and their surface is amorphous, lipophilic, and negatively charged. Nanoemulsions are being researched to improve intranasal drug delivery. Oils with anticancer effect against breast cancer, such as spearmint oil, can be conveniently given using nanoemulsions. Nanoemulsions have excellent stability, degradation resistance, fast digestion, controlled release, and a high ability to increase medication bioavailability. They are stable at ambient temperature, but when exposed to ultrasonic waves, they clump together to produce micro-bubbles. In some applications, nanoemulsions have a higher loading capacity for lipophilic active substances than microemulsions, which can be advantageous. Nanoemulsions can be used to create both hydrophobic and hydrophilic medicines. As a result, nanoemulsions possess properties that make them acceptable for use as delivery methods in clear liquid formulations. Nanoemulsions are nontoxic and nonirritating systems that can be used for skin or mucous membranes, parenteral and non-parenteral delivery, and have been used in the beauty industry. Caffeine, plasmid DNA, aspirin, methyl salicylate, gamma tocopherol, insulin, and nimesulide are some of the medications that use nanoemulsions for transdermal drug administration. Increased drug loading and bioavailability are two advantages of nanoemulsions. It produces a repeatable plasma drug profile. It can also be utilized to distribute drugs consistently and precisely. The oily phase of the nanoemulsion was reported to influence its wound healing potential, since it affects the release of the ingredients from the oily core, and the inclusion of actives within nano/microemulsion formulations was shown to exhibit enhanced wound healing potential. These novel properties makes them a promising drug delivery and therapeutic agent in dogs and cats (Underwood and van Eps 2012).

Metallic Nanoparticles

Metallic nanoparticles are solid colloidal metal particles with a therapeutic molecule that is either disseminated in the matrix of the polymer carrier, entrapped within a polymer shell, or attached or adsorbed to the surface of the particles through covalent bonding. Chemical reduction of metal salts or complexes with chemical reductants such as hydrazine, sodium borohydride, sodium citrate or ascorbate, sugars, or even natural extracts produces metallic nanoparticles. Metallic nanoparticles can be used for diagnostic and therapeutic purposes. As drug carriers, metallic nanoparticles such as quantum dots, gold nanoparticles, and magnetic nanoparticles have benefits. Metals of various sorts are employed in the nanosystem for various functions. Silver, gold, manganese, and platinum nanoparticles are examples of metallic nanoparticles that are commonly utilized in veterinary medicine for diagnostic and therapeutic purposes. In addition, metallic nanoparticles are loaded with antibodies and chelated radionuclides. Metallic nanoparticles' physicochemical features, high stability, high reactivity, photo dermal, and plasmonic capabilities make them effective medicinal carriers. Metallic nanoparticles have attracted the attention of most researchers worldwide due to their flexible nanostructure and capacity to adjust shape, composition, size, structure, assembly, and optical properties. Microorganisms such as bacteria, algae, yeast, and fungus, as well as plant extracts, are employed in the biosynthesis of green metal nanoparticles to manufacture nontoxic, safe, and biodegradable nanoparticles using cobalt, iron oxide, silver, gold, platinum, zinc oxide, quantum dots, and sulphides. Natural products and their subparts have been used as reducing and capping agents in the green synthesis of nanoparticles. Metal nanoparticle synthesis is a straightforward, natural, and environmentally benign technique. However, it has some drawbacks, including toxicity, particle instability, carcinogenicity, contaminants, and synthesis complexity. Metal nanoparticles are employed in veterinary medicine for biomedicine, tumor therapy, medical treatment, drug delivery, and anti-infective agents, among other things. Bio probes made of silica-coated gold NPs are utilized in calorimetric DNA and protein detection. Metallic NPs can be made from a variety of materials, but the most frequent are gold, silver, and platinum, which have been utilized in effective nanomedicine. Gold nanoparticles (AuNPs) are the most prevalent and are used for diagnostic, imaging, and targeted therapy (Jain et al. 2007, Mishra et al. 2009).

Advantages of Mycosynthesized NPs

Nanoparticles production mainly takes place either by chemical or physical methods which mainly involves toxic chemicals and radiations. Production of nanoparticles using microorganism is increasing day by day. The initial approach for production of nanoparticles using pathogenic fungus of *Taxus* plant, *Verteicillium* sp. with silver nitrate was carried out by Mukherjee et al. (2001) and produced silver nanoparticles of about 25 nm diameters intracellularly within the biomass. Xiaowen et al. (2019) synthesized Tp-AgNPs using mycelial extract of endophytic fungus *Talaromyces purpureogenus* (MEEF), and analyzed for antibacterial, anti-proliferation and cell wound healing activities. They reported biosynthesized AgNPs with strong antibacterial, anticancer and cell wound healing properties using endophytic fungus *T. purpureogenus*. Tp-AgNPs significantly inhibited the growth of Gram-positive or Gram-negative pathogens at the minimal inhibitory concentration of 16.12 μg/mL and 13.98 μg/mL, respectively. Moreover, TpAgNPs showed higher zone of inhibition against *S. aureus* (9 mm), *P. aeruginosa* (13 mm) and *E. coli* (11 mm) and it is similar to positive antibiotic control vancomycin.

Applications in Veterinary Sciences

This section elaborates some of the important applications of nanoparticles in veterinary sciences. Figure 2, represents the various aspects of veterinary sciences in which the nanoparticles are applied.

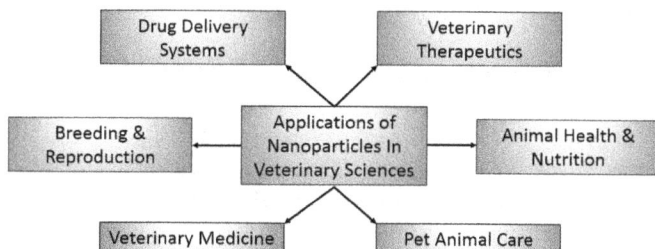

Figure 2. Applications of nanoparticles in veterinary sciences.

Veterinary Medicine

Manipulation of materials to their nanoscale, or nanoparticles, has had an evolutionary impact on a variety of medical fields. Veterinary practices have recently begun to include diverse nanotechnology applications. Nanoparticles are employed in veterinary medicine for diagnostic as well as therapeutic applications, tissue engineering for vaccine manufacture and contemporary disinfectant production, animal breeding and reproduction, and animal nutrition (El-Sayed and Kamel 2020). Due to their surface to volume proportion, bioavailability, bioactivity, more reactivity, more stability, controlled particle size, controlled drug release, site specific targeting, and controlled arrival of medicine, nanoparticles have a number of physicochemical characteristics superior to many more old drug materials (Mohanraj and Chen 2006, Moudgil and Ying 2007, Buzea et al. 2007, Num and Useh 2013). Nanotechnology facilitates medication delivery directly into target cells, allowing for the use of very low doses, reducing drug residue and withdrawal period in animals (Troncarelli et al. 2013). Most ailments, such as tuberculosis, brucellosis, food and mouth disease, and other infections, have found new remedies thanks to nanotechnology. Many other trials against various animal diseases are also in the works or under investigation. Nanoparticles combined with antibodies or nucleic acids can be used to create diagnostic assays that are quick, sensitive, specific, and portable. Nano and biochips have aided in the diagnosis of pathogens as well as the knowledge of genetic predisposition factors. Nanoparticles are also useful in the development of nanovaccines and nanoadjuvants. Their application in cancer treatment and gene therapy ushered forth a new era in medicine (El-Sayed and Kamel 2020). Other potential applications include delivery of drugs, vitamins, probiotics and nutritional supplements and for removal of causes of infections. Nanoparticles containing tumor-specific antibodies enable the destruction of metastatic malignant growth cells far from the site of the initial injury (Lee et al. 2013). Indiscriminate use of antibiotics in veterinary practices can leave residues of drugs in meat and milk. Considering the residual effect of antibiotics and drug, Governments make policies to restrict or ban the use of antibiotics in veterinary practices. Because of their nano size, using nanotechnology in veterinary procedures minimizes the number of antibiotics administered (Youssef et al. 2019). Nanoparticles also help to reduce the problem of drug resistance in humans and animals, as well as drug resistance in milk and meat. Nanotechnology is also employed to make pet care goods and sanitary items (El-Sayed and Kamel 2020).

Drugs and Drug Delivery Systems

Because of their improved stability, high bioavailability, and low toxicity, nanomaterials and nano-drug delivery systems are becoming increasingly popular in human and veterinary medicine. Low bioavailability and higher drug accumulation than dangerous levels are significant difficulties in veterinary operations, hence nanoparticles' drug delivery systems aim to deliver medication molecules to the application site (Oguzhan and Arslan 2021). In mastitis and pulmonary infections in cattle, liposomes containing antibacterial drugs such as tobramycin, gentamycin, and other aminoglycosides have been proven to be more effective than single antibiotics. A review of the literature shows that liposomes are more effective than their traditional counterparts in the treatment of cancer. As compared to any other drug delivery systems, nanomaterial based systems allows appropriate intake of small doses of antimicrobials. A sub-atomic coded 'address mark' in the bundle could enable the bundle to be conveyed to the particular right site in the body. Nano and microscale mechanical frameworks would fill in as the 'transporters' in such a framework. Smart drug delivery systems can likewise have the capacity to screen the impacts of the conveyance of nutraceuticals, pharmaceuticals, supplements, bioactive mixes, sustenance supplements, probiotics, synthetic concoctions and antibodies (Tomànek and Enbody 2000).

Native Antimicrobials

A biochip (or microarray) is a gadget regularly made of hundreds or thousands of short strands of artificial DNA saved clearly on a silicon circuit. Biochips are used to trace the original source of food and feeds to sense the presence of animal products from diverse species as a means to regulate the source of pathogens such as avian flu and mad cow disease. Besides, there are different varieties identifying minute amounts of proteins and synthetic materials in an example, making them convincing for identifying bio warfare specialists or malady. Biochips based on blood cells, tissues and semen can be used to discriminate or recognize microbial infections in domesticated animals. Nanoshells are another sort of optically tunable nanoparticles made out of a dielectric (for instance, silica) center covered with an ultra-thin metallic (for instance, gold) layer (Hirsch et al. 2003). Others have been exploring diverse paths regarding 'keen' super paramagnetic nanoparticles, which when pervaded in the circulation system target tumor receptor cells. These nanoparticles are produced using iron oxides that upgrade the capacity of the nanoparticles to find tumor cells when they are exposed to an attractive field. At the site of the tumor, the nanoparticles release the loaded medication to decimate the malignancy cells. Other type of nanomaterials are QDs which are nanometer-scale gems that were initially shaped for optoelectronic applications. QDs might be infused into the circulation system of animals and they may distinguish cells that are breaking down. Nucleic corrosive building based tests and techniques propose better approaches to convey helpful or protection treatment for specific ailments. These diverse techniques for nanotechnology can be an eccentric remedial device in thrashing the health problems of animals (Luo 2003).

Medical Diagnostics and Therapeutics

Livestock is a key economic component in many countries around the world, as well as a source of foreign exchange, social amenities, and overall development. Livestock provides for the growing population and the ever-increasing demand for protein-rich diet. Diseases which represent a major setback in animal production, are the principal limiting factor for the surplus food supply. As a result, the demand for better livestock disease diagnoses, methods, and therapies is shifting to molecular and, more recently, nanotechnology. Nanotechnology should be coupled with molecular diagnostics and treatments to improve the efficiency in the diagnosis and treatment of animal diseases for greater protein supply and food security, according to recent advances in nanotechnology (Num and Useh 2013).

Nanotechnology will have a big impact on veterinary medicine around the world (Feneque 2003, Scott 2005). For quick and specific detection of respiratory syncytial virus *in vivo* and *in vitro*, functionalized nanoparticles (NPs) such as quantum dots (QD) for *in vivo* imaging in small animal models (Bentolila et al. 2009) and nanoparticle conjugated monoclonal antibodies have been used. The nanoparticles thus bridge the gap between current ambiguous virus detection assays and the timely need for more sensitive and rapid detection of viral agents (Tripp et al. 2007, Schlachter et al. 2011, Huang et al. 2012). NPs are currently employed in treatment of African animal trypanosomosis (Kroubi et al. 2010) leading to a new drug delivery system instead of diaminazene (DMZ), a trypanocidal veterinary drug. Protein cage nanoparticles (PCN) for pre-treatment of mice was effective to protect subject mice against a mouse-adapted SARS-coronavirus, or mouse pneumovirus. The effectiveness was indicated by increased survival and virus clearance (Wiley et al. 2009). Biocompatible and nontoxic gelatin nanoparticle bound CpG-ODN 2216 were used for treatment of equine recurrent airway obstruction resulting in triggering of anti-inflammatory and anti-allergic cytokine IL-10 expression (Klier et al. 2012). With NPs smaller than 70 nm, when administered intra-tumoral under the external energy field, and paclitaxel (Tx)-loaded biodegradable nanoparticles (Sahoo et al. 2004) the tumors can be destroyed specifically through localized physical and chemical effect (Chauhan et al. 2010). Antigen covalently coupled to nano-beads induced substantial cell mediated responses along with moderate humoral responses in sheep model without any adverse effect (Scheerlinck et al. 2006). Inert nano-beads used to target antigen to dendritic cells (DCs) to induce immune responses against FMDV (foot and mouth disease virus) specific synthetic peptides in sheep (Greenwood et al. 2008).

Biocidal Nanoparticles

Nanoparticles offer unique qualities such as controlled size, ultra-small size, large surface area, and high reactivity, making antimicrobial pharmaceutical administration, therapeutic viability, and dissolvability easier. NPs conjointly enhance therapeutic effectiveness and overcome the limited undesirable aspects of pharmaceutical compounds. They have numerous advantages, including the ability to extend the systemic circulation lifetime of drugs, improve bioavailability, reduce sedate digestion, and enable a more controllable delivery of helpful mixes, as well as the delivery of at least two medications at the same time for mistreatment (Youssef et al. 2019). They're also utilized to refresh old pharmaceutical bases that are still in use. This is done to minimize the chances of multidrug resistance in animal pathogens. NPs have the ability to reduce the toxicity and side effects of conventional pharmaceutical chemicals, as well as eliminate the dangers of drug residues in animal products. One of the most significant advantages of employing nanoparticles is the reduction of required doses, dosage frequency, and medication concentration during treatment (Peer et al. 2007, Zhang et al. 2008, Gao et al. 2014).

Sperm Nanopurification

To ensure optimal conception rates, sperm variability within or between ejaculates and between individuals necessitates a variety of post-collection management of semen. This section focuses on future nanotechnology-based sperm selection, tagging, tracking, and imaging approaches, with a focus on nanotechnology's recent influence. Because of their diverse features such as charge, polarity, and ligand attachment, inorganic nanoparticles such as iron oxide, semiconductor quantum dot nanocrystals, silver, and gold nanoparticles, and mesoporous silica are suitable candidates for sperm labelling. These characteristics can be adjusted for a variety of uses (Alivisatos et al. 2005, Albanese et al. 2012, Tan and Cheong 2013). Viable mammalian spermatozoa and embryos have limitation over accumulation of biocompatible metal and quantum dot nanoparticles. Although sperm functionality can be affected by high exposures depending upon the doses (Feugang et al. 2012, Barchanski et al. 2015, Taylor et al. 2015, Yoisungnern et al. 2015).

Magnetic iron oxide nanoparticles (Fe_3O_4), attributes facilitation of sperm purification: biocompatibility, physical nature, and magnetism (Pankhurst et al. 2003). The Fe_3O_4 based nanopurification is comparable to commercial magnetic-assisted cell sorting technology (MACS), which is applied in combination with cryopreservation in aquaculture (Valcarce et al. 2016) and clinics for infertile patients (Gil et al. 2013). The effectiveness of semen nanopurification is still partially investigated. Its implementation in herds with low population with high vigor can help reconstitute the quality traits *via* cross breeding.

Breeding and Reproduction

Management of animal rearing is a tedious issue for dairy and swine ranchers. The nanotube is utilized as an apparatus for ensuing estrus in animals on the grounds due to its capability to tie and differentiate the estradiol counter acting agent at the season of estrus by close infrared fluorescence. Microfluidics is used today in animal science to fundamentally streamline customary *in vitro* treatment techniques applied as a part of animal reproduction. It is being exploited as a part of farm animals reproducing to physically sort sperm and eggs. Micro and nano fluidic are the frameworks to examine by governing the stream of fluids or gases through a movement of little channels and valves. With the mapping of the human genome behind them, geneticists are currently quickly sequencing the genomes of poultry, pig, steers, sheep, and other animals deficient to recognize quality successions that identify with monetarily lucrative characteristics (O'Connell et al. 2002, Hanafy 2018).

Pet Animal Care

Nanotechnology has also been linked to the development of novel goods for pet animal care. Because of their physicochemical qualities, they are utilized to improve surface refreshing and disinfectants. Shampoos, for example, include silver nanoparticles for topical usage (Feneque 2012).

Current Limitation and Safety of NPs

Most nanoparticles are benign in general, but some can have dangerous side effects, such as prolonged lung exposure to carbon nanotubes, which can cause reproductive difficulties in pharmaceutical employees. Furthermore, the creation of enticing nanoparticles made of iron oxide within the edge, or through damages hastened by an inconsistency in the interaction between the medicine and the particles, may further dispatch the medication in solid tissues rather than objective tissues. The partial arrival of the direction away from its objective tissue or organ will now result in healthy tissue toxicity as well as sub therapeutic dosage delivery to the objective component. They have the capacity to move various organic limitations, such as the blood–brain barrier, within the casing (Manuja et al. 2012, Mohanty et al. 2014, Johansson et al. 2017, Wu et al. 2018).

Conclusion

Nanomaterials due to their increased surface area to volume ratio exhibit novel or unique physical, chemical, biological properties. As a result, they enable nanotechnology to provide innumerable applications in different sectors of industry, agriculture and medicine. Although nanotechnology is making inroads in almost each and every sector because of the progressive research and promising results, still increased and exhaustive studies are needed to prove the reliability of Nanotechnology without having any adverse effect on plant, animal and environment in terms of its toxicity. Considering the issues and demand of nanomaterials in the consumer sectors, more emphasis is given on the green synthesis of nanomaterials. Biological system in general and fungal system in particular are preferred systems for the synthesis of nanomaterials in eco-friendly manner. One of the neglected sectors for the nanotechnology application is veterinary science, but slowly and steadily nanotechnology is making its presence because of its tremendous potential in animal disease diagnosis and therapeutics. Therefore, extensive research is required in the area of veterinary science demonstrating several applications of Nanotechnology in diagnosis and therapeutics.

References

Abbasi, E., Aval, S., Akbarzadeh, A., Milani, M., Nasrabadi, H., Joo, S.W., Hanifehpour, Y., Nejati-Koshki, K. and Pashaei-Asl, R. 2014. Dendrimers: Synthesis, applications, and properties. Nanoscale. Res. Lett. 9: 247. https://doi.org/10.1186/1556-276X-9-247.

Aboul-Nasr, M.B. 2018. Mycogenesis of silver nanoparticles using nonpathogenic opportunistic and pathogenic fungi. J. Ecosys. Ecograp. 8.

Ahluwalia, V., Kumar, J., Sisodia, R., Shakil, N.A. and Walia, S. 2014. Green synthesis of silver nanoparticles by Trichoderma harzianum and their bioefficacy evaluation against *Staphylococcus aureus* and *Klebsiella pneumonia*. Ind. Crops Prod. 55: 202–206. Doi: 10.1016/j.indcrop.2014.01.026.

Ahmed, S., Ahmad, M., Swami, B.L. and Ikram, S. 2016. A review on plants extract mediated synthesis of silver nanoparticles for antimicrobial applications: A green expertise. J. Adv. Res. 7: 17–28. Doi: 10.1016/j.jare.2015.02.002.

Albanese, A., Tang, P.S. and Chan, W.C. 2012. The effect of nanoparticle size, shape, and surface chemistry on biological systems. Ann. Rev. Biomed. Engg. 14: 1–16.

Alivisatos, A.P., Gu, W. and Larabell, C. 2005. Quantum dots as cellular probes. Ann. Rev. Biomed. Engg. 7: 55–76.

Arruebo, M., Fernández-Pacheco, R., Ibarra, M.R. and Santamaría, J. 2007. Magnetic nanoparticles for drug delivery. Nano Today 2: 22. 10.1016/S1748-0132(07)70084-1.

Ashrafi, S.J., Rastegar, M.F., Ashrafi, M., Yazdian, F., Pourrahim, R. and Suresh, A.K. 2013. Influence of external factors on the production and morphology of biogenic silver nanocrystallites. J. Nanosci. Nanotechnol. 13: 2295–2301. Doi: 10.1166/jnn.2013.6791.

Awate, S., Babiuk, L.A.B. and Mutwiri, G. 2013. Mechanisms of action of adjuvants. Front. Immunol. 4: 114.

Azmath, P., Baker, S., Rakshith, D. and Satish, S. 2016. Mycosynthesis of silver nanoparticles bearing antibacterial activity. Saudi Pharm. J. 24: 140–146. Doi: 10.1016/j.jsps.2015.01.008.

Ballotin, D., Fulaz, S., Souza, M.L., Corio, P., Rodrigues, A.G., Souza, A.O., Gaspari, P.M., Gomes, A.F., Gozzo, F. and Tasic, L. 2016. Elucidating protein involvement in the stabilization of the biogenic silver nanoparticles. Nanoscale Res. Lett. 11: 313. Doi: 10.1186/s11671-016-1538-y.

Barchanski, A., Taylor, U., Sajti, C.L., Gamrad, L., Kues, W.A., Rath, D. and Barcikowski, S. 2015. Bioconjugated gold nanoparticles penetrate into spermatozoa depending on plasma membrane status. J. Biomed. Nanotech. 11(9): 1597–1607.

Bentolila, L.A., Ebenstein, Y. and Weiss, S. 2009. Quantum dots for *in vivo* small-animal imaging. Journal of Nuclear Medicine 50(4): 493–496.

Bérdy, J. 2005. Bioactive microbial metabolites. J. Antibiot. 58: 1–26. Doi: 10.1038/ja.2005.1.

Bourassa, D.J., Kerna, N.A. and Desantis, M. 2019. A retrospective investigation of novel catenated multi-shelled fullerene-like material for biocompatibility after prolonged inhalation, dermal, and ingestion exposure in human subjects. Part 3 in a series: Will Nanocarbon Onion-Like Fullerenes (NOLFs) Play a Decisive Role in the Future of Molecular Medicine? EC Pharmacol. Toxicol. 7: 577–584. https://www.ecronicon.com/ecpt/pdf/ECPT-07-00313.pdf.

Buzea, C., Pacheco, B.I. and Robbie, K. 2007. Nanomaterials and nanoparticles: Sources and toxicity. Biointerphases 2: 1–103.

Cao, S.W., Zhu, Y.J., Ma, M.Y., Li, L. and Zhang, L. 2008. Hierarchically nanostructured magnetic hollow spheres of Fe_3O_4 and gamma-Fe_2O_3: Preparation and potential application in drug delivery. J. Phys. Chem. C 112: 1851. 10.1021/jp077468.

Chakravarthi, V.P. and Balaji, S.N. 2010. Applications of nanotechnology in veterinary medicine. Vet. World 3: 477–480.

Chauhan, R.S., Sharma, G. and Rana, J.M.S. 2010. Nanotechnology in health and disease. Bytes and Bytes, Bareilly, UP, India, 1–11.

Cheng, Y., Man, N., Xu, T., Fu, R., Wang, X., Wang, X. and Wen, L. 2007. Transdermal delivery of nonsteroidal anti-inflammatory drugs mediated by polyamidoamine (PAMAM) dendrimers. J. Pharm. Sci. 96: 595–602.

Costa Silva, L.P., Oliveira, J.P., Keijok, W.J., Silva, A.R., Aguiar, A.R., Guimarães, M.C.C., Ferraz, C.M., Araújo, J.V., Tobias, F.L. and Braga, F.R. 2017. Extracellular biosynthesis of silver nanoparticles using the cell-free filtrate of nematophagus fungus *Duddingtonia flagans*. Int. J. Nanomed. 12: 6373–6381. Doi: 10.2147/IJN.S137703.

Costerton, J.W. Lewandowski, Z., Caldwell, D.E., Korber, D.R. and Lappin-Scott, H.M. 1995. Microbial Biofilms. Annu. Rev. Microbiol. 49: 711–745.

Durán, N., Marcato, P.D., Alves, O.L., De Souza, G.I.H. and Esposito, E. 2005. Mechanistic aspects of biosynthesis of silver nanoparticles by several *Fusarium oxysporum* strains. J. Nanobiotechnol. 3: 8. Doi: 10.1186/1477-3155-3-8.

Durán, N., Marcato, P.D., Durán, M., Yadav, A., Gade, A. and Rai, M. 2011. Mechanistic aspects in the biogenic synthesis of extracelular metal nanoparticles by peptides, bacteria, fungi and plants. Appl. Microbiol. Biotechnol. 90: 1609–1624. Doi: 10.1007/s00253-011-3249-8.

Elgqvist, J. 2017. Nanoparticles as theranostic vehicles in experimental and clinical applications—Focus on prostate and breast cancer. Int. J. Mol. Sci. 18: 110.

El-Sayed, A. and Kamel, M. 2020. Advanced applications of nanotechnology in veterinary medicine. Env. Sci. Poll. Res. 27(16).

Feneque, J. 2003. Brief introduction to the veterinary applications of nanotechnology. Nanotechnology Now www.nanotech-now.com/Jose-Feneque/Veterinary-Applications-Nanotechnology.htm, retrieved 2012-12-15.

Feneque, J. 2012. Brief introduction to the veterinary applications of nanotechnology. Nanotechnol. No wwwwnanotech-nowcom/Jose-Feneque/Veterinary - applications. Nanotechnologyhtm 7: 12–15.

Feugang, J.M., Youngblood, R.C., Greene, J.M., Fahad, A.S., Monroe, W.A., Willard, S.T. and Ryan, P.L. 2012. Application of quantum dot nanoparticles for potential non-invasive bio-imaging of mammalian spermatozoa. J. Nanobiotech. 10: 45.

Freitas, R.A. Jr. 2001. Nanomedicine: Microbivores, Artificial Mechanical Phagocytes—Institute for Molecular Manufacturing, Report No. 25.

Freitas, R.A. Jr. 2002. The future of nanofabrication and molecular scale devices in nanomedicine. pp. 45–59. *In*: Renata G. Bushko (ed.). Future of Health Technology. IOS Press, Amsterdam, The Netherlands.

Freitas, R.A. Jr. 2005. Microbivores: Artificial mechanical phagocytes using digest and discharge protocol. J. Evol. Tech. 14(1): 55–106. ISSN 1541-0099.

Freitas, R.A. Jr. 2009. Medical nanorobotics: The long term goal for nanomedicine. pp. 367–392. *In*: Mark, J., Schulz Vesselin and Shanov,N. (eds.). Nanomedicine Design of Particles, Sensors, Motors, Implants, Robots and Devices. Artech. House, Norwood Ma.

Gade, A. and Rai, M. 2021. Mycofabrication of metal nanoparticles: A green approach. *In*: Satyanarayana, T., Deshmukh, S.K. and Deshpande, M.V. (eds.). Progress in Mycology. Springer, Singapore. https://doi.org/10.1007/978-981-16-3307-2_9.

Gade, A., Shende, S. and Rai, M. 2022. Potential role of *Phoma* spp. for mycogenic synthesis of silver nanoparticles. *In*: Rai, M., Zimowska, B. and Kövics, G.J. (eds.). Phoma: Diversity, Taxonomy, Bioactivities, and Nanotechnology. Springer, Cham. https://doi.org/10.1007/978-3-030-81218-8_17.

Gade, A.K., Bonde, P., Ingle, A.P., Marcato, P.D., Durán, N. and Rai, M.K. 2008. Exploitation of Aspergillus niger for synthesis of silver nanoparticles. J. Biobased Mater. Bioenergy 2: 243–247. Doi: 10.1166/jbmb.2008.401.

Gao, W., Thamphiwatana, S., Angsantikul, P. and Zhang, L. 2014. Nanoparticle approaches against bacterial infections. Wiley Interdiscip. Rev. Nanomed. Nanobiotechnol. 6: 532–547.

Gil, M., Sar-Shalom, V., Sivira, Y.M., Carreras, R. and Checa, M.A. 2013. Sperm selection using magnetic activated cell sorting (MACS) in assisted reproduction: A systematic review and meta-analysis. J. Assist. Reprod. Genet. 30(4): 479–485.

Greenwood, D.L.V., Dynonc, K., Kalkanidis, M., Xiang, S., Plebanskid, M. and Scheerlinck, J.Y. 2008. Vaccination against foot-and-mouth disease virus using peptides conjugated to nano-beads. Vaccine 26(22): 2706–2713.

Gudikandula, K., Vadapally, P. and Charya, M.A.S. 2017. Biogenic synthesis of silver nanoparticles from white rot fungi: Their characterization and antibacterial studies. Open Nano. 2: 64–78. Doi: 10.1016/j.onano.2017.07.002.

Guilger, M., Pasquoto-Stigliani, T., Bilesky-Jose, N., Grillo, R., Abhilash, P.C., Fraceto, L.F. and de Lima, R. 2017. Biogenic silver nanoparticles based on Trichoderma harzianum: Synthesis, characterization, toxicity evaluation and biological activity. Sci. Rep. 7: 44421. Doi: 10.1038/srep44421.

Guilger-Casagrande, M. and Lima, R. 2019. Synthesis of silver nanoparticles mediated by fungi: A review. Front. Bioeng. Biotechnol. 7: 287. Doi: 10.3389/fbioe.2019.00287.

Hanafy, M.H. 2018. Myconanotechnology in veterinary sector: Status quo and future perspectives. Int. J. Vet. Sci. Med. 6(2): 270–273. Doi: 10.1016/j.ijvsm.2018.11.003.

Hirsch, L.R., Stafford, R.J., Bankson, J.A., Sershan, S.R., Rvera, B., Price, R.E., Hazel, J.D., Halas, N.J. and West, J.L. 2003. Nanoshell mediated near infrared thermal therapy of tumors under magnetic resonance guidance. PNAS 100: 13549–54.

Huang, J., Zhong, X., Wang, L., Yang, L. and Mao, H. 2012. Improving the magnetic resonance imaging contrast and detection methods with engineered magnetic nanoparticles. Theranostics 2(1): 86–102.

Iravani, S., Korbekandi, H., Mirmohammadi, S.V. and Zolfaghari, B. 2014. Synthesis of silver nanoparticles: Chemical, physical and biological methods. Res. Pharm. Sci. 9: 385–406.

Jain, P.K., El-Sayad, I.H. and El-Sayed, M.H. 2007. Au nanoparticles target cancer. Nano Today 2: 18–29.

Johansson, H.K.L., Hansen, J.S., Elfving, B., Lund, S.P., Kyjovska, Z.O., Loft, S., Barford, K.K., Jackson, P., Vogel, U. and Hougaard, K.S. 2017. Airway exposure to multi-walled carbon nanotubes disrupts the female reproductive cycle without affecting pregnancy outcomes in mice. Part. Fibre. Toxicol. 30(14): 17.

Jurj, A., Braicu, C., Pop, L.A., Tomulesa, C., Gherman, C.D. and Neagoe, I.D. 2017. The new era of nanotechnology, an alternative to change cancer treatment. Drug. Dis. Dev. Ther. 11: 2871–2890.

Khan, N.T., Khan, M.J., Jameel, J., Jameel, N. and Rheman, S.U.A. 2017. An overview: Biological organisms that serves as nanofactories for metallic nanoparticles synthesis and fungi being the most appropriate. Bioceram. Dev. Appl. 7: 101. Doi: 10.4172/2090-5025.1000101.

Kim, B.Y.S., Rutka, J.T. and Chan, W.C.W. 2010. Nanomedicine. New Eng. J. Med. 363: 2434–443.

Klier, J., Fuchs, S., May, A., Schillinger, U., Plank, C., Winter, G., Gehlen, H. and Coester, C. 2012. A nebulized gelatin nanoparticle-based CpG formulation is effective in immunotherapy of allergic horses. Pharmaceutical Research 29(6): 1650–1657.

Kroubi, M., Daulouede, S., Karembe, H., Jallouli, Y., Howsam, M., Mossalayi, D., Vincendeau, P. and Betbeder, D. 2010. Development of a nanoparticulate formulation of diminazene to treat African trypanosomiasis. Nanotechnology 21(50): 1–8.

Kumar, S.A., Abyaneh, M.K., Gosavi, S.W., Kulkarni, S.K., Pasricha, N., Ahmad, A. and Khan, M.I. 2007. Nitrate reductase-mediated synthesis of silver nanoparticles from $AgNO_3$. Biotechnol. Lett. 29: 439–445. Doi: 10.1007/s10529-006-9256-7.

Lee, S.J., Yhee, J.Y., Kim, S.H., Kwon, I.C. and Kim, K. 2013. Biocompatible gelatin nanoparticles for tumor-targeted delivery of polymerized siRNA in tumor-bearing mice. J. Control. Release 172: 358–366.

Luo, D. 2003. Yearbook of Science and Technology. New York: McGraw-Hill, 93–5.

Madani, S.Y., Naderi, N., Dissanayake, O., Tan, A. and Seifalian, A.M. 2011. A new era of cancer treatment: Carbon nanotubes as drug delivery tools. Int. J. Nanomedicine 6: 2963–79. Doi: 10.2147/IJN.S16923. Epub 2011 Nov 22. PMID: 22162655; PMCID: PMC3230565.

Manuja, A., Kumar, B. and Singh, R.K. 2012. Nanotechnology developments: Opportunities for animal health and production. Nanotechnol. Dev. 2: 4.

Meena, N., Sahni, Y., Thakur, D. and Singh, R.P. 2018. Applications of nanotechnology in veterinary therapeutics. J. Entomol. Zool. 6: 167–175.

Mekkawy, A.I., El-Mokhtar, M.A., Nafady, N.A., Yousef, N., Hamad, M.A., El-Shanawany, S.M., Ibrahim, E.H. and Elsabahy, M. 2017. *In vitro* and *in vivo* evaluation of biologically synthesized silver nanoparticles for topical applications: Effect of surface coating and loading into hydrogels. Int. J. Nanomed. 12: 759–777. Doi: 10.2147/IJN.S124294.

Mishra, B., Patel, B.B. and Tiwari, S. 2009. Colloidal nanocarriers: A review on formulation technology, types and applications toward targeted drug delivery. Nanomed. 6: 9–24.

Mohanraj, V.J. and Chen, Y. 2006. Nanoparticles—A review. Trop. J. Pharma. 5: 561–573.

Mohanty, N., Palai, T.K., Prusty, B.R. and Mohapatra, J.K. 2014. An overview of nanomedicine in veterinary science. Vet. Res. Int. 2(4): 90–95.

Mohsen, J. and Zahra, B. 2008. Protein nanoparticle a unique system as drug delivery vehicles. Afric. J. Biotech. 7: 4926–4934.

Molnár, Z., Bódai, V., Szakacs, G., Erdélyi, B., Fogarassy, Z., Sáfrán, G., Varga, T., Kónya, Z., Tóth-Szeles, E., Szűcs, R. and Lagzi, I. 2018. Green synthesis of gold nanoparticles by thermophilic filamentous fungi. Sci. Rep. 8: 3943. Doi: 10.1038/s41598-018-22112-3.

Moudgil, B.S. and Ying, J.Y. 2007. Calcium doped organosilicate nanoparticles nanoparticles as gene delivery vehicles for bone cells. Sci. Technol. Adv. Mat. 19: 3130–3135.

Mudshinge, S.R., Deore, A.B., Patil, S. and Bhalgat, C.M. 2011. Nanoparticles: Emerging carriers for drug delivery. Saudi Pharm. J. 19: 129–41.

Mukherjee, P., Ahmad, A., Mandal, D., Senapati, S., Sainkar, S.R., Khan, M.I., Ramani, R., Parischa, R., Kumar, P.A.V., Alam, M., Sastry, M. and Kumar, R. 2001. Bioreduction of AuCl4-ions by the fungus, Verticillium sp. and surface trapping of the gold nanoparticles formed. Angew. Chem. Int. Ed. 40: 3585–8.

Muller, R.H. and Keck, C.M. 2004. Challenges and solutions for the delivery of biotech drugs—A review of drug nanocrystal technology and lipid nanoparticles. J. Biotech. 113: 151–170.

Num, S.M. and Useh, N.M. 2013. Nanotechnology applications in veterinary diagnostics and therapeutics. Sokoto. J. Vet. Sci. 11: 10–14.

O'Connell, M.J., Bachilo, S.M., Huffman, C.B., Moore, V.C., Strano, M.S., Haroz, E.H., Rialon, K.L., Boul, P.J., Noon, W.H. and Kittrel, C. 2002. Band gap fluorescence from individual single walled carbon nanotubes. Science 297: 593–6.

Oguzhan, Y. and Arslan, H.H. 2021. Nano drug delivery systems and their applications in veterinary medicine. Clin. Res. Anim. Sci. 1(4): 1–3.

Ottoni, C.A., Simões, M.F., Fernandes, S., Santos, J.G., Silva, E.S., Souza, R.F.B. and Maiorano, A.E. 2017. Screening of filamentous fungi for antimicrobial silver nanoparticles synthesis. AMB Express 7: 31. Doi: 10.1186/s13568-017-0332-2.

Pankhurst, Q.A., Connolly, J., Jones, S.K. and Dobson, J. 2003. Applications of magnetic nanoparticles in biomedicine. J. Phy. D: App. Phy. 36(13): R167.

Patel, H.N. and Patel, D.R.P.M. 2013. Dendrimer applications—A review. Int. J. Pharm. Bio. Sci. 4(2): 454–463.

Peer, D., Karp, J.M., Hong, S., Farokhzad, O.C., Margalit, R. and Langer, R. 2007. Nanocarriers as an emerging platform for cancer therapy. Nat. Nanotechnol. 2: 751–760.

Qidwai, A., Pandey, A., Kumar, R., Shukla, S.K. and Dikshit, A. 2018. Advances in biogenic nanoparticles and the mechanisms of antimicrobial effects. Indian J. Pharm. Sci. 80: 592–603. Doi: 10.4172/pharmaceutical-sciences.1000398.

Rajput, S., Werezuk, R., Lange, R.M. and Mcdermott, M.T. 2016. Fungal isolate optimized for biogenesis of silver nanoparticles with enhanced colloidal stability. Langmuir 32: 8688–8697. Doi: 10.1021/acs.langmuir.6b01813.

Rodríguez-Burneo, N., Busquets, M.A. and Estelrich, J. 2017. Magnetic nanoemulsions: Comparison between nanoemulsions formed by ultrasonication and by spontaneous emulsification. Nanomaterials 7: 190.

Scheerlinck, J.P.Y., Gloster, S., Gamvrellis, A., Mottram, P.L. and Plebanski, M. 2006. Systemic immune responses in sheep, induced by a novel nano-bead adjuvant. Vaccine 24(8): 1124–1131.

Schlachter, E.K., Widmer, H.R., Bregy, A., Lonnfors-Weitzel, T., Vajtai, I., Corazza, N., Bernau, V.J., Weitzel, T., Mordasini, P., Slotboom, J., Herrmann, G., Bogni, S., Hofmann, H., Frenz, M. and Reinert, M. 2011. Metabolic pathway and distribution of superparamagnetic iron oxide nanoparticles: *In vivo* study. International Journal of Nanomedicine 6: 1793–1800.

Scott, N.R. 2005. Nanotechnology and animal health. OIE Scientific and Technical Review 24(1): 425–432.

Scott, N.R. 2007. Nanoscience in veterinary medicine. Vet. Res. Comm. 31: 139–144.

Shabnashmi, P.S., Naga, K.S., VIthya, V., Vijaya Lakshmi, B. and Jasmine, R. 2016. Therapeutic applications of nanrobots-respirocytes and microbivores. J. Chem. Pharma. Res. 8(5): 605–609.

Srinivasa-Gopalan, S. and Yarema, K.J. 2007. Nanotechnologies for the Life Sciences: Dendrimers in Cancer Treatment and Diagnosis. 7. New York: Wiley.

Tan, K.S. and Cheong, K.Y. 2013. Advances of Ag, Cu, and Ag-Cu alloy nanoparticles synthesized via chemical reduction route. J. Nanopart. Res. 15(4): 1–29.

Taylor, U., Tiedemann, D., Rehbock, C., Kues, W.A., Barcikowski, S. and Rath, D. 2015. Influence of gold, silver and gold–silver alloy nanoparticles on germ cell function and embryo development. Beilst. J. Nanotech. 6(1): 651–664.

Tomalia, D.A. and Frechet, J.M.J. 2002. Discovery of dendrimers and dendritic polymers: A brief historical perspective. J. Polym. Sci. Part A 40: 2719.

Tomànek, D. and Enbody, R.J. 2000. Science and Application of Nanotubes. 1st ed. Springer, 24–43.

Torres-Sangiao, E., Holban, A.M. and Gestal, M.C. 2016. Advanced nanobiomaterials: Vaccines, diagnosis and treatment of infectious diseases. Molecules 21: 867.

Tripp, R.A., Alvarez, R., Anderson, B., Jones, L., Weeks, C. and Chen, W. 2007. Bioconjugated nanoparticle detection of respiratory syncytial virus infection. International Journal of Nanomedicine 2(1): 117–124.

Troncarelli, M.Z., Brandão, H.M., Gern, J.C., Guimarães, A.S. and Langoni, H. 2013. Nanotechnology and antimicrobials in veterinary medicineBadajoz, Spain. FORMATEX available at http://www. formatex.info/microbiology4/vol1/543-556.pdf.

Underwood, C. and van Eps, A.W. 2012. Nanomedicine and veterinary science: The reality and the practicality. Vet. J. 193: 12–23.

Valcarce, D.G., Herráez, M.P., Chereguini, O., Rodríguez, C. and Robles, V. 2016. Selection of nonapoptotic sperm by magnetic-activated cell sorting in Senegalese sole (*Solea senegalensis*). Theriogenology 86(5): 1195–1202.

Wiley, J.A., Richert, L.E., Swain, S.D., Harmsen, A., Barnard, D.L., Randall, T.D., Jutila, M., Douglas, T., Broomell, C., Young, M. and Harmsen, A. 2009. Inducible bronchus-associated lymphoid tissue elicited by a protein cage nanoparticle enhances protection in mice against diverse respiratory viruses. PLoS One 4(9): 1–10 e7142.

Woudenberg, I.A., Schiffelers, R.M., Storm, G., Becker, M.J. and Guo, L. 2005. Long circulating sterically stabilized liposomes in the treatment of infections. Methods Enzymol. 391: 228–260.

Wu, T.J., Chiu, H.Y. and Yu, J. 2018. Nanotechnologies for early diagnosis, *in situ* disease monitoring, and prevention. pp. 1–92. *In*: Uskoković, V. and Uskoković, D.P. (eds.). Nanotechnologies in Preventive and Regenerative Medicine. Chapter 1. Amsterdam: Elsevier.

Xiaowen, H., Saravanakumar, K., Jin, T. and Wang, M.H. 2019. Mycosynthesis, characterization, anticancer and antibacterial activity of silver nanoparticles from endophytic fungus *Talaromyces purpureogenus*. Int. J. Nanomed. 14: 3427–3438.

Yahyaei, B. and Pourali, P. 2019. One step conjugation of some chemotherapeutic drugs to the biologically produced gold nanoparticles and assessment of their anticancer effects. Sci. Rep. 9: 10242. Doi: 10.1038/s41598-019-46602-0.

Yang, L.A.P. 2000. Physiochemical aspects of drug delivery and release from polymer-based colloids. Curr. Opin. Coll. Interf. Sci. 51: 132–143.

Yoisungnern, T., Choi, Y.J., Han, J.W., Kang, M.H., Das, J., Gurunathan, S., Kwon, D.N., Cho, S.G., Park, C., Chang, W.K., Chang, B.S., Parnpai, R. and Kim, J.H. 2015. Internalization of silver nanoparticles into mouse spermatozoa results in poor fertilization and compromised embryo development. Scient. Rep. 5: 11170.

Youngson, C., Nurse, C., Yeger, H. and Cutz, E. 1993. Oxygen sensing in airway chemoreceptors. Nature, 363–155.

Youssef, F.S., El-Banna, H.A., Elzorba, H.Y. and Galal, A.M. 2019. Application of some nanoparticles in the field of veterinary medicine. Int. J. Vet. Sci. Med. 7(1): 78–93. Doi: 10.1080/23144599.2019.1691379.

Zhang, L., Gu, F.X., Chan, J.M., Wang, A.Z., Langer R.S. and Farokhzad, O.C. 2008. Nanoparticles in medicine: Therapeutic applications and developments. Clin. Pharmacol. Ther. 83: 761–769.

Zhang, Y., Huang, Y., Kang, Y., Miao, J. and Lai, K. 2021. Selective recognition and determination of malachite green in fish muscles via surface-enhanced Raman scattering coupled with molecularly imprinted polymers. Food Control 130: 108367. Doi: 10.1016/j.foodcont.2021.108367.

Section IV
Miscellaneous

13

Fungal-mediated Zinc Oxide Nanoparticles and their Applications

Chandrakant V. Pardeshi,[1,*] *Swapnil N. Jain*[2] *and Nitin R. Shirsath*[2]

Introduction

Nanotechnology has become one of the important areas in the field of modern material sciences. NPs are considered potential scientific tools which are widely used in various biotechnological and pharmacological industries. Nanoscience is the study of substances at the molecular and atomic levels. The basic component in the manufacture of nanostructures is the matter at the nanoscale of the order of 10^{-9} m = 1 nm, which is one-billionth of a meter. This technology is being employed in many areas and not limited to textile industries, agriculture, food processing, sophisticated diagnostic and medicinal techniques, and drug delivery, etc. (Boroumand Moghaddam et al. 2015, Pandurangan and Kim 2015). NPs are usually utilized for the exploration, synthesis, and characterization of materials in the range of 1–100 nm (Li et al. 2011).

NPs make a part of a large group of nanomaterials, with 'nano' indicating a tiny physical unit of dimension. Hence, NPs have properties that are quantitatively or qualitatively distinct from other physical forms of the same material (Saravanan et al. 2021). The large surface area to mass ratio, resulting in an increased ratio of surface to core atoms and an increased set of corner and edge atoms is sufficient to explain the size-related modifications in particle properties. This may not only result in greater reactivity but also enhanced related physical properties, allowing new applications to be developed (Medhi et al. 2020).

[1] Department of Pharmaceutics and Pharmaceutical Technology, R. C. Patel Institute of Pharmaceutical Education and Research, Shirpur, Maharashtra, India.
[2] Department of Pharmaceutics, H. R. Patel Institute of Pharmaceutical Education and Research, Shirpur, Maharashtra, India.
* Corresponding author: chandrakantpardeshi11@gmail.com

Zinc (Zn) is one of the important essential trace elements in the human body and it is required for many functions like the expression of genes, a few enzymatic reactions, learning and memory, synthesis of proteins, and immune functions (Kambe et al. 2015). The recommended intake of Zn is about 11 and 8 mg/day for an adult male and woman, respectively (Roohani et al. 2013). Meat, shellfish, nuts, legumes, mushrooms, seeds, and vegetables like broccoli and spinach are the natural sources of Zn and are also commercially available as a diet supplement in the form of Zn salts like (gluconate, acetate, and sulfate, etc.). The imbalance of Zn in the body can lead to the alteration of various normal physiological activities (Keerthana and Kumar 2020).

Many studies are currently focusing on the ZnO NPs due to their physical properties like chemical and optical behavior, which can be easily manipulated by altering their shape and size, as well as a wide range of applications in optics, electronics, and medicinal systems. ZnO has a wide variety of properties, including piezoelectricity, semi conductivity, and pyroelectricity, in addition to their antibacterial, antioxidant, antifungal, anticancer, and antidiabetic properties (Mohamed et al. 2021). Fungi are being used in the field of nanotechnology for the last few years to prepare NPs. The fungal-based synthesis has demonstrated that this environment-friendly and renewable source could be used effectively as a reducing agent for the synthesis of ZnO NPs. Also known as myconanotechnology, the fungal-based biosynthetic approach investigates the possibilities of NPs' synthesis using fungi. Because fungi are easy to culture, their potential utility has achieved a lot of attention (El-Saadony et al. 2021).

Out of the two methods, the extracellular release of enzymes to generate NPs has significantly improved downstream biomass processing. This method of synthesis involves the adsorption of metal ions on the cell surface and reducing them in the presence of enzymes, with the NPs being easily recovered. On the contrary, intracellular synthesis involves ions being transported into the microbial cell and thus forming NPs in the presence of enzymes, but these NPs are difficult to recover in pure form. The reduction of metal ions is thought to be aided by the fungal cell wall and cell wall carbohydrates. Fungi are one of the most effective biological agents for the production of metal nanoparticles. Thus, the extracellular synthesis of ZnO NPs from the fungus is highly efficient due to convenient downstream processing, large-scale production, and economic sustainability (Sumanth et al. 2020, Pomastowski et al. 2020).

Fungal strains are preferred over bacteria because of their excellent tolerance and metal bio-accumulation property. ZnO NPs were synthesized by Pavani and coworkers using the mycelia of *Aspergillus fumigatus*. Here, the fungal strain was isolated from the soil sample taken from the metal plating industry (Pavani et al. 2012). In another study, Ahmad and others reported a discovery in which they used *Fusarium oxysporum* fungus in the process of extracellular synthesis of technologically important semiconductor Cadmium sulfide NPs purely by enzymatic process (Ahmad et al. 2002).

Myconanotechnology deals with the synthesis of NPs using fungi, and is the interface between mycology and nanotechnology with a considerable potential, due to the wide range and diversity of the fungal strains. In this chapter, the biosynthesis of ZnO NPs using green technology has been discussed along with other physical and chemical methods. The mechanism of the formation of nanostructures is also briefly discussed along with the significance of various physical properties during the development stages. These nanomaterials have a wide range of applications and are discussed with recent findings and advancements. The formulation and development of nanostructures into a suitable delivery vehicle needs to undergo many challenges and has to follow strict regulatory guidelines making it worth a ratio of high risk to reward.

Methods and Mechanism of Synthesis

NPs can be prepared by a large number of methods such as physical, chemical, biological, and hybrid methods (Liu et al. 2011, Kato 2011). The biological synthesis of NPs has achieved extensive attention in recent years due to the use of various biological resources like plant extracts as reducing and microorganisms as stabilizing agents (Ghomi et al. 2019).

a. Physical methods

In physical methods like the colloidal dispersion method, the physical forces are involved in the formation of stable and well-defined nanostructures (Vidya et al. 2013). It requires one or more basic techniques such as vapor condensation, physical fragmentation, crystallization, and others. The physical method also involves the use of high vacuum in processes like pulsed laser deposition, molecular beam epitaxy, and thermal evaporation, etc. (Aladpoosh and Montazer 2015). The costly equipment, high processing temperature, and pressure, and large area requirement for setting up of machines are a few drawbacks of this method. The application of microwave irradiation from a technical point of view is of great interest to improve the morphological control of ZnO nanostructures along with the decrease in synthesis times. The microwave irradiation method can be used to synthesize both pure and tin-doped ZnO nanostructures. (Cho et al. 2008, Prakash et al. 2014, Saloga and Thünemann 2019).

b. Chemical methods

Common chemical methods of NP synthesis are micro-emulsion, hydrothermal, chemical reduction, sol-to-gel, and precipitation, etc. The sol-gel synthesis is the most commonly used method and it uses a few chemical reagents along with the Zn precursor salt (Brintha and Ajitha 2015). A small amount of stabilizers like polyvinylpyrrolidone or citrate is usually added during the synthesis to obtain the optimum morphological properties and to avoid the agglomeration of NPs. The solution needs to be heated at high temperatures. During the synthesis, the amount of Zn precursor and other reagents used have been reported to significantly affect the shape and size of ZnO NPs produced (Sahai and Goswami 2014).

Chemical methods are among the widely employed methods, but the major concern in the synthesis of NPs is the use of toxic chemicals (Li et al. 2011). And it considerably limits their biomedical applications in clinical fields. It is therefore a challenge for the researchers to develop non-toxic, reliable, and environment-friendly methods for the synthesis of NPs (Suresh et al. 2011). It has been observed that the use of microorganisms has the potential to achieve this goal (Quester et al. 2013).

c. Green technology

In green technology, expensive and toxic chemicals are not required and an environment-friendly, safe, cost-effective, and biocompatible approach is employed. This technology is more feasible because the majority of the microorganisms inhabit ambient conditions of pH, temperature, and pressure (Aladpoosh and Montazer 2015). Further, the particles generated by these processes have measurable catalytic reactivity, a more specific surface area, and an enhanced contact between the metal salt and an enzyme in the reaction mixture (Bhattacharya and Mukherjee 2008).

Fungal species like *Fusarium oxysporum* can readily synthesize metal NPs extracellularly by using secreted proteins and/or enzymes that not only provide stability but also give enhanced yield over an intracellular one. When it comes to downstream processing, an extracellular synthesis has demonstrable advantages over an intracellular process, since there would be much less handling of the fungal biomass. In contrast, sophisticated instruments would be necessary to isolate the NPs from the biomass and into the cell-free filtrate, with an intracellular synthesis (Riddin et al. 2006, Gade et al. 2010).

More recently, Preeti and others demonstrated mycosynthesis of ZnO NPs by using the aqueous extract of *Agarius bisporus* as a reducing agent. The outcomes in this study illustrated that the mycosynthesized ZnO NPs formed a protective film that reduced the corrosion rate on copper used in cooling water towers (Preethi et al. 2019). Baskar and his group used *Aspergillus terreus* to prepare metalloproteins of L-asparaginase conjugated with ZnO NPs for the treatment of acute lymphoblastic leukemia. Here, the authors also suggest that ZnO has better electrostatic property, which assists them to have different charges on the surface in acidic and basic pH. And this property can be utilized for the conjugation of therapeutic agents and also internalize NPs to tumor cells as their surface is loaded with negatively charged phospholipids (Baskar et al. 2015). A general stepwise procedure of fungal-mediated ZnO NPs synthesis is shown in Figure 1.

Some researchers propose a mechanism based on the chemical constituents of plant extract used for the preparation of ZnO NPs. It was stated that the Zn ions get reduced to metallic Zn by plant extract instead of coordinated complex formation (Gupta et al. 2018). The complete reduction of the Zn precursor is followed by a reaction between metallic Zn and dissolved oxygen in the solution that leads to the formation of Zn oxide nuclei (Tade et al. 2020). Here, the phytoconstituents work as stabilizing agents to prevent the agglomeration of particles (Sutradhar and Saha 2016, Singh et al. 2018).

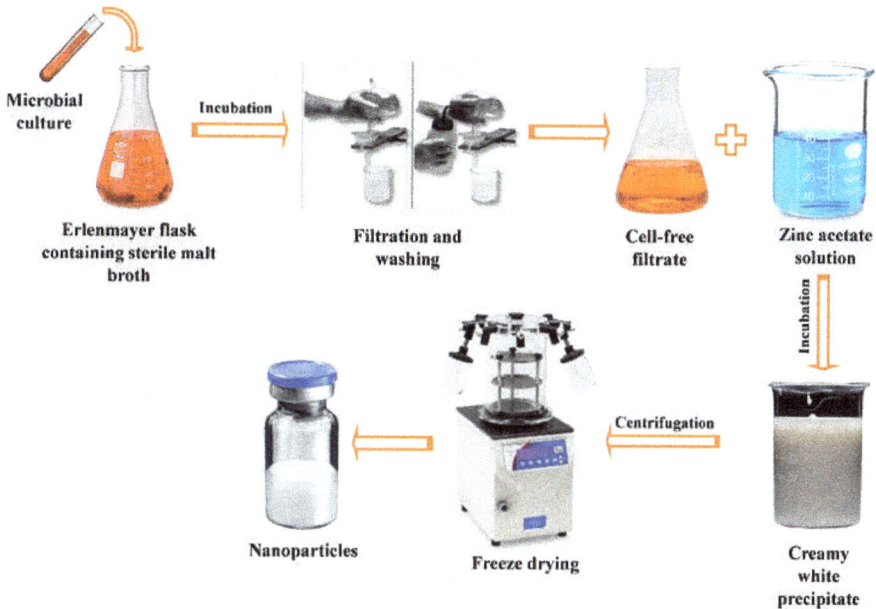

Figure 1. Fungal-mediated synthesis of ZnO NPs.

Importance of Physical Properties

ZnO has many characteristics which can considerably increase its use in various fields like biomedical applications, drug delivery, and others. The main attributes are UV absorption, semi conductivity, antimicrobial activity, etc. (Ginjupalli et al. 2018). The increased research work related to ZnO NPs is due to the non-specific activity of inorganic antimicrobial agents that may be useful to optimize the fight against microbial resistance. This may be because of small particle size and high surface area that ultimately can enhance antimicrobial activity, causing an improvement in surface reactivity (de Aragao et al. 2019). The shape of NPs is another important aspect. The most commonly prepared NPs exhibit spherical or near-spherical shapes but the recent advances in nanotechnology have inspired exhaustive studies to determine whether the modulation in the shape can improve the therapeutic efficacy of NPs (Kiio and Park 2021). ZnO NPs also have photodynamic properties, where illumination causes the generation of a large number of reactive oxygen species that result in cell apoptosis (Ryter et al. 2007).

In another study, it was found that the presence of ZnO NPs significantly improved the swelling ability of the nanohybrid hydrogels as compared to pure chitosan–cellulose hydrogel. This may be due to the capability of ZnO NPs in the hydrogel matrices that aids in the hydration process by holding more water in its lattice structure (George et al. 2020). The manipulation in the shape and size of NPs can be done by changing the pH of the reaction mixture. The change in the pH value affects the electrical charges of biomolecules that might alter the reducing and stabilizing ability of the medium and also the frequent growth of NPs. It may further assist in obtaining the desired shape and size of the NPs (Khandel and Shahi 2018).

Applications of Fungal-mediated ZnO NPs

Researchers have already begun a broad-spectrum investigation to find out the possibility of new applications of green manufactured NPs in various fields. Different optical and structural features of ZnO NPs are generating a lot of interest in nanoparticle-based drug delivery, bio-imaging, and therapeutic uses for mankind (Bhunia 2017). Some important applications (Figure 2) of ZnO NPs are discussed in this section.

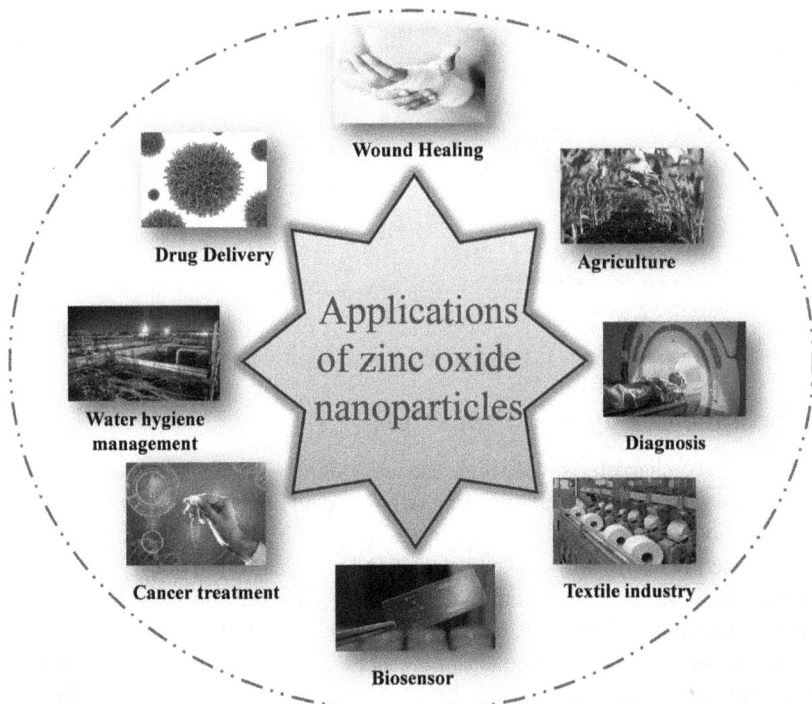

Figure 2. Applications of ZnO NPs in various fields.

a. Drug delivery

Many novel chemical entities are now being synthesized, but they are facing problems related to solubility and absorption (Karagianni et al. 2018). Various methods have been presented to overcome this issue, but each one has their own set of difficulties. As a result, a carrier-based nanoparticulate drug delivery system is being developed, which will employ a range of techniques and nanocarriers to provide improved efficacy, drug loading capacity, and a modified release profile (Shirsath and Goswami 2019). Only a few of the advantages of the new drug delivery approach over conventional/traditional dosage forms include excellent biocompatibility, sustained-controlled drug release, and greater bioavailability (Sharma et al. 2019).

ZnO NPs are widely used as carriers for the delivery of different therapeutic molecules like drugs and nucleic acids, etc., to the target sites. Owing to the presence of a significant number of hydroxyl groups on the ZnO NPs surface, it is easier to functionalize them with various ligands, probes, and therapeutic molecules, which make them ideal candidates as delivery vehicles (Simonelli and Arancibia 2015, Chiang and Roberts 2011, AbouAitah et al. 2021). Other goals for fabricating ZnO NPs based delivery systems are to increase the therapeutic efficacy of the biomolecules, enhance systemic circulation, and decrease the dose of the therapeutic agents, thereby reducing their adverse side effects. Umrani and Paknikar prepared the NPs at the Centre for Nanobiosciences, Pune, and evaluated them for antidiabetic effect. They investigated antidiabetic activity in streptozotocin-induced type 1 and 2 diabetic rats and conducted a single-dose pharmacokinetic investigation, as well as cytotoxicity, hemolysis, acute and sub-acute toxicity testing. The outcomes indicated that the oral treatment of NPs had significant anti-diabetic effects, including increased glucose tolerance, greater serum insulin, lower blood glucose, lower non-esterified fatty acids, and lower triglycerides (Umrani and Paknikar 2014).

Arvanag and colleagues reported the use of a green microwave-assisted technique using *Silybum marianum* L. seed extract for the production of ZnO NPs for the treatment of diabetes. The levels of blood glucose, insulin, total cholesterol, total triglyceride, and high-density lipoprotein were measured and compared before and after therapy with the investigated therapeutic agents. Furthermore, the antibacterial properties of both ZnO samples were tested against *E. coli*. According to his findings, ZnO NPs showed remarkable performance in resolving diabetic problems as well as good antibacterial activity against the bacteria examined (Arvanag et al. 2019).

Researchers have made several attempts that are aimed towards finding out the potential of ZnO NPs for anti-inflammatory activity. Agarwal and Shanmugam worked on the green synthesis of ZnO NPs. In their study, they discussed the interaction of ZnO NPs with cells as well as their pharmacokinetic behavior inside the cells. This study underlines ZnO NPs' potential as an anti-inflammatory therapeutic molecule as well as a drug delivery vector (Agarwal and Shanmugam, 2020). The following Table 1 gives an insight into the fungal-mediated synthesis of NPs, their physical properties, and their applications.

Zamani and others prepared titanium dioxide (TiO_2)@ZnO NPs core-shell nanostructured and TiO_2@ mesoporous NPs–graphene oxide hybrid nanocomposites by a facile sonochemical method. A novel mesoporous and core-shell structure was prepared as a drug nanocarrier for the loading and pH-dependant characteristics of the curcumin. The microscopic images showed an average grain size of about 190 nm with uniform hexagonal mesoporous morphology. As expected, the nanocarriers showed pH-dependent drug release behavior and MTT (3-(4, 5-dimethyl thiazolyl-2)-2, 5-diphenyl tetrazolium bromide) assay results revealed that the curcumin-loaded nanocarriers showed significant toxicity that lead to a reduction in cell viability, thereby suggesting its anticancer effects (Zamani et al. 2018).

b. Diagnosis

Biosensors (also known as photometric, calorimetric, electrochemical, and piezoelectric sensors) are widely accepted in healthcare, chemical and biological analysis, environmental monitoring, and the food industries (Chatterjee et al. 2014).

Table 1. Physical properties and applications of ZnO NPs prepared by fungal species.

Fungus	Size	Morphology	Application	References
Aspergillus terreus	28–63 nm	Spherical	Anti-cancer activity	(Baskar et al. 2015)
Alternaria alternata	45–150 nm	Spherical, triangular, and hexagonal	Biosafety evaluation	(Sarkar et al. 2014)
Fusarium species	5–27 nm	Spherical	Bioremediation	(Velmurugan et al. 2010)
Aspergillus fumigatus	1.2–6.8 nm	Spherical and hexagonal	Industrial, medical and agricultural sectors	(Raliya and Tarafdar 2013)
Aspergillus aeneus	100–140 nm	Spherical	–	(Jain et al. 2013)
Aspergillus species	50–120 nm	Spherical	–	(Pavani et al. 2011)
Candida albicans	15–25 nm	Quasi-spherical	Catalyst in the synthesis of steroidal pyrazoline	(Mashrai et al. 2017)
Rhodococcus pyridinivorans	100–120 nm	Quasi-spherical and hexagonal	Colon carcinoma	(Kundu et al. 2014)
Xylaria acuta	40–50 nm	Hexagonal	Anticancer	(Sumanth, et al. 2020)

Nanomaterials, alone or in conjunction with biologically active compounds, are gaining a lot of attention because of their unique features, they can provide a good platform for the creation of high-performance biosensors (Khan et al. 2008, Ren et al. 2009). Nanomaterials' high surface area can be used to immobilize biomolecules like enzymes, antibodies, and other proteins (Verma et al. 2017). They can also enable direct electron transport between biomolecule active sites and the electrodes. Apart from semiconducting properties (Bai et al. 2006), ZnO nanomaterials have several desirable biosensing characteristics, including high catalytic efficiency, strong adsorption capability, and a high isoelectric point, which is suitable for electrostatic adsorption of certain proteins (e.g., enzymes and antibodies with low isoelectric points).

Venkatesan performed a study on the utilization of *Boswellia ovalifoliolata* bark extract for the preparation of ZnO NPs and evaluated the compatibility with medicinal and biological applications such as picric acid detection. The fluorescence emission of ZnO NPs is highly selective for picric acid and can be employed as a fluorescent probe for the detection via fluorescence quenching. The electron transfer process between ZnO NPs and picric acid may be responsible for the fluorescence quenching mechanism of picric acid, as demonstrated by cyclic voltammetry. According to this research, the fluorescent feature of ZnO NPs can be used for cell imaging (Venkatesan et al. 2019).

The ZnO NPs are also attractive nanomaterials for biosensors because of their good biocompatibility and stability, low toxicity, and high electron transfer capability (Schrauben et al. 2012). The bulk of ZnO-based biosensors that have been published are for the detection of small molecules and analytes such as glucose, phenol, hydrogen peroxide, cholesterol, urea, and so on (Kaur et al. 2016, Agarwal and Shanmugam 2020, Shanmugam et al. 2017). The combination of (3-aminopropyl) triethoxysilane and ZnO nanorods were prepared by Mohammed and his team using microwave-assisted chemical bath deposition on thermally oxidized SiO_2 thin films that resulted in the preparation of an electrochemical DNA biosensor (Mohammed et al. 2017). In another research, the ZnO NPs embedded nitrogen-doped carbon sheets were developed by Muthuchamy and others as a highly selective, sensitive, and stable enzymatic glucose sensors. It was built on a glassy carbon electrode and had a high and repeatable sensitivity. These nanocomposites could also be valuable in other domains such as solar cells and optoelectronic devices. These encouraging results suggest a simple and effective method to obtain electrode material for the enzymatic glucose sensor (Muthuchamy et al. 2018).

c. Biomedical research

Diseases, healthcare concerns, and medical costs have increased as a result of the deteriorating environment and increasing aging population, particularly in developing nations, resulting in a significant need for better and lower-cost biomedical devices with innovative bio-functionalities (Wang et al. 2019). For the diagnosis of many diseases and disorders, bioimaging techniques such as Magnetic resonance imaging (MRI), computed tomography (CT), positron emission tomography (PET), and ultrasounds are frequently used (Ehman et al. 2017). These methods are non-invasive and can yield high-resolution images of interior organs in some cases (Decazes et al. 2021).

Contrast chemicals are commonly employed in bioimaging techniques to identify the organ or tissue of interest, as well as to distinguish healthy tissue from a sick one (Penen et al. 2016). The fundamental problems with the contrast chemicals for MRI and CT imaging are their toxicity, as well as their short retention period and imaging time (Grabherr et al. 2015). Core-shell NPs have been studied as a possible contrasting agent to improve imaging time and boost the biocompatibility of contrast agents, as they can offer increased biocompatibility and imaging time (Tartaj et al. 2003).

Many publications on the use of ZnO NPs for cellular imaging exist in the literature due to their efficient excitonic blue and near-UV emission, which can also feature green luminescence attributable to oxygen vacancies (Liu et al. 2020b, Pan et al. 2018). The study by Zhu and others suggests that the penetration of ZnO NPs in human skin is possible and can be tested *in vitro* and *in vivo* using their intrinsic fluorescence. As per the study, the majority of ZnO NPs remained in the stratum corneum, posing a low risk of causing safety problems.

The ZnO-based nanostructures seem to have a wide range of structures and characteristics, giving them a wide range of capabilities (Zhu et al. 2016). The significant potential of ZnO-based nanostructures for biomedical applications has been recognized in the previous decade due to their unique electrical, optical, catalytical, and antibacterial capabilities, as well as their exceptional biocompatibility (Zhang et al. 2011, Gordon et al. 2011, Rashmi et al. 2015). In another research,

Butler and others worked on the bioimaging properties of ZnO. In this case, ZnO NPs were encapsulated in an n-isopropyl acrylamide hydrogel polymer network using the arc discharge process and homogeneously disseminated in water. The resultant size of the ZnO NPs found to be dramatically enlarged after binding with the protein, according to dynamic light scattering studies.

Several types of cross-link functionalized polymers considerably changed the photoemission from ZnO but the amine-functionalized polymer hydrogel was most suited for surface modification of ZnO and bioimaging applications, according to optical characterization. Continuous-wave photoluminescence experiments demonstrated that this sample emits more ultraviolet (UV)-blue light and are more stable. Increased scattering owing to nanoparticle dispersion, which leads to increased absorption cross-section in ZnO, is required for photoluminescence enhancement in the hydrogel network, according to time-resolved photoluminescence measurements (Butler et al. 2017, Chatterjee et al. 2014). ZnO embedded in a plant cell and a plant leaf has also been imaged using fluorescence microscopy (Lv et al. 2015).

The use of therapeutic genes in gene delivery is primarily used to treat various genetic problems. The negative charge of the plasma membrane, on the other hand, restricts nucleic acid transfection. An attempt was made to synthesize the three-dimensional functionalized silica-coated amino-modified tetrapod-like ZnO nanostructures as novel carriers for mammalian cell transfections. The tetrapods adhere to cell membranes and can stand on the cells using three needle-shaped legs for DNA delivery when mixed with cells. The results reveal that the functionalized nanostructures' charge concentration, tetrapod-like form, and biocompatibility provide a foundation for mammalian cell transfection (Nie et al. 2006).

d. Cancer treatment

Cancer is a condition of uncontrolled malignant cell proliferation that has been typically treated by chemotherapy, radiotherapy, and surgery in the past several decades. Although all these therapies appear to be very effective for killing cancer cells from a theoretical aspect, these non-selective therapy methods also cause a lot of serious side effects. In recent years, nanomaterial-based medicines, with high biocompatibility, easy surface functionality, cancer-targeting, and drug delivery capacity, have demonstrated a new path to overcome these side effects (Jiang et al. 2018, Hameed et al. 2019).

In vitro experiments have shown that ZnO NPs are particularly toxic to cancer cells, bacteria, and leukemic T cells. As a result, ZnO NPs are being studied as cancer therapy as well as drug/gene delivery vehicles. Yuan and others developed a new approach of combining anti-cancer drug therapy with quantum dots technology. They carried out the preliminary research on the preparation of blue-light emitting ZnO quantum dots combined with biodegradable chitosan. Their findings suggested that the proposed new generation of quantum dots loaded with anti-cancer drugs and encased in biocompatible polymer could be used to achieve tumor-targeted drug delivery (Yuan and Hein 2010). Some studies have indicated that ZnO NPs affected functions of different cells or tissues, biocompatibility, and neural tissue engineering (Osmond and Mccall 2010).

Abouaitah and colleagues prepared hybrid nanostructures with inorganic NPs such as ZnO and natural antibacterials. ZnO NPs have antibacterial properties, and they were utilized in this study to investigate their action in combination with protocatechuic acid which is a natural antibacterial agent. The nanostructure's *in vitro* release kinetics and antibacterial properties against *Staphylococcus aureus* were studied (Abouaitah et al. 2021). This study suggests that, in comparison to the separate components, the hybrid nanoformulation displays effective antibacterial and bactericidal activities that are dependent on surface modification.

L. He and coworkers developed a ZnO quantum dots-based nano-drug delivery system by altering the surface of the hydrophilic copolymer poly(methacrylate-co-N-isopropyl acrylamide-co-polyethylene glycol methyl acrylate) (ZnOPMNE), which was made using a one-step copolymerization technique. The photoluminescence performance of ZnOPMNE NPs in water is reported to be good, temperature-sensitive and biodegradable in an acidic environment. In an acidic tumor microenvironment, ZnOPMNE NPs loaded with doxorubicin break down to Zn ions and doxorubicin. Due to the demonstration of cytotoxicity by ZnO quantum dots after dissolution, compared to normal cells, the cytotoxicity of cancer cells was greatly amplified, resulting in a synergistic anticancer effect that improved the therapeutic index (He et al. 2019).

Another chemotherapeutic agent-loaded ZnO NPs for the intracellular delivery of doxorubicin was attempted by Sharma and others. Because ZnO NPs have inherent anticancer qualities, it was hypothesized that loading an anticancer medication into them will result in increased anticancer efficacy (Sharma et al. 2016).

Recently, Sadhukhan and his team synthesized phenylboronic acid conjugated ZnO NPs, loaded with quercetin. The presence of phenylboronic acid moieties over the nanoparticle surface leads to the targeted delivery of quercetin to the sialic acid over-expressed cancer cells. Also, quercetin-loaded phenylboronic acid conjugated ZnO NPs provided pH-responsive drug release behavior. The outcomes suggested that the prepared NPs induced apoptotic cell death in human breast cancer cells through increased oxidative stress and mitochondrial damage. Overall, this data emphasizes the chemotherapeutic potential of the novel nanohybrid that can be useful for clinical cancer treatment (Sadhukhan et al. 2019).

e. Wound healing

With its wound healing and anti-inflammatory characteristics, Zn is a classic metal that exists in trace amounts in our bodies and has numerous biological functions (Lin et al. 2018). Because of their strong antibacterial capabilities and Zn's epithelialization-stimulating impact, ZnO NPs have also been successfully used in wound dressings (Jamnongkan et al. 2015). Wang and others fabricated the ZnO NPs using the leaf extract of Coleus amboinicus (a capping and reducing agent). The ZnO NPs were found to have remarkable antibacterial activity against a variety of gram-positive and gram-negative microbial infections. Furthermore, wound healing investigations in rats demonstrated that the wound closure rate was higher in the ZnO NPs treated rats at all doses when compared to the untreated group, indicating that manufactured ZnO NPs had an effective wound healing potential (Wang et al. 2020).

Generally, gels are mostly used as primary dressings whereas hydrogels are used as primary or secondary dressings (Figure 3). Hydrogel dressings retain significant amounts of water and they cannot absorb too much exudate, so they are used for moderately exuding wounds. In a study, Batool reported that the biogenically fabricated ZnO NPs using Aloe barbadensis leaves successfully lead to the development of ZnO NPs silica gel dressing that was having the non-irritant and healing ability. The ZnO NPs silica gel dressing exhibited a wound contraction of mices' skin with enhanced wound skin elasticity, blood clotting, insignificant inflammation, and rapid skin repairing (Batool et al. 2021).

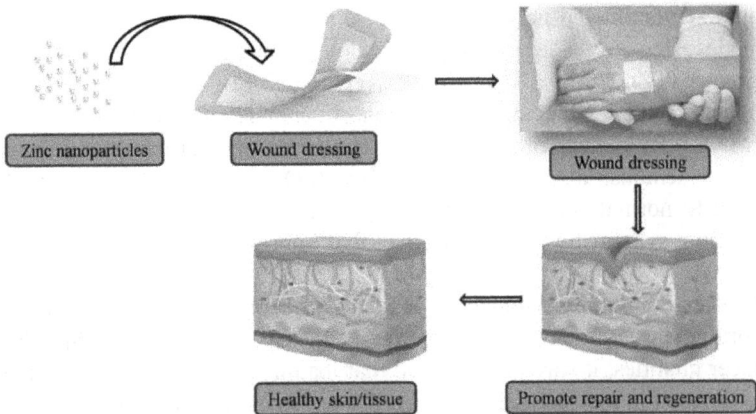

Figure 3. Zinc nanoparticles in wound healing.

In recent years, collagen-based wound dressings have been designed and a few of them contain NPs. A research study reports the synthesis of novel wound dressings based on collagen and essential oil functionalized ZnO NPs intended to aid the treatment of burns and to reduce the risk for developing wound sepsis in patients with burns or chronic wounds. These dressings also reported exhibiting a plethora of distinguishable properties like good biocompatibility with dermal fibroblasts, biodegradability, and enhanced regenerative capacity due to collagen component along with strong bactericidal and antioxidant activity imparted by ZnO NPs (Balaure et al. 2019). Another possibility of using electrospun nonwoven fiber mats of polyvinyl acetate-ZnO hybrids as antibacterial wound dressings has been explained by Jamnongka and colleagues (Jamnongkan et al. 2015).

Khorasani and others used the freeze-thaw method to prepare polyvinyl alcohol/ chitosan/nano ZnO nanocomposite hydrogels and by maintaining essential process parameters like the number of freeze-thaw cycles, thawing time, and thawing temperature. The initial investigation was about the antibacterial properties of ZnO NPs suspensions by estimating the fraction of bacterial cells that survived the exposure to ZnO. Later, it was about polyvinyl acetate/ZnO fiber mats to be used as wound dressings and directly investigated the antibacterial properties of the nonwoven fiber mats using the disk diffusion method. The biocompatibility, antibacterial properties, and in vitro wound healing tests suggested that the preparation showed no toxicity and sufficient ability to treat the wounds (Khorasani et al. 2019).

f. Agriculture

Nanotechnology has a dominant position in transforming agriculture and food production with a great potential to modify conventional agricultural practices (Sabir et al. 2014, Pestovsky and Martínez-Antonio 2017). Most of the agrochemicals applied to the crops are lost and do not reach the target site due to several factors including leaching, drifting, hydrolysis, photolysis, and microbial degradation (Singh et al. 2020). NPs and nanocapsules provide an efficient means to distribute pesticides and fertilizers in a controlled fashion with high site-specificity thus reducing collateral damage. Farm application of nanotechnology is gaining attention by efficient control and precise release of pesticides, herbicides, and fertilizers. Nanosensors development can help in determining the required amount of farm inputs such as fertilizers and pesticides and also to detect the amount of moisture and nutrients in the soil (Figure 4). High sensitivity, low detection limits, extreme selectivity, fast responses, and compact diameters are all features of nanosensors for pesticide residue detection (Chhipa 2017).

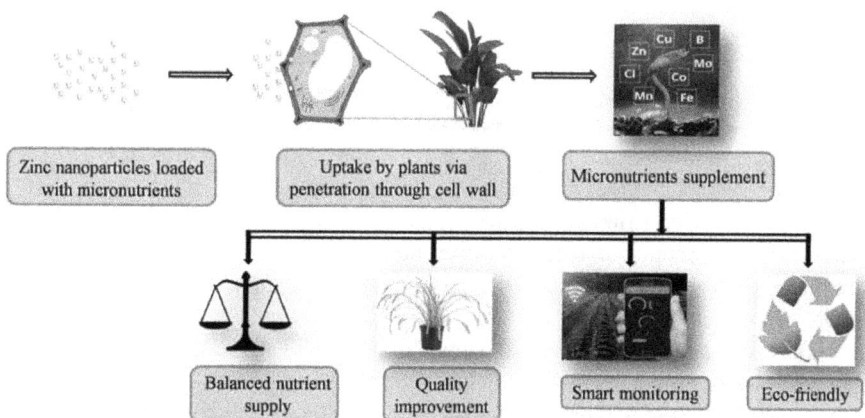

Figure 4. Role of zinc nanoparticles in agriculture.

Yusefi and others investigated the effects of ZnO NPs on seed yield in soil-grown soybean. The study was focused on particle size, shape, and concentration-dependent responses of several antioxidant defense indicators. ZnO NPs evoked strong oxidative stress responses in soybean and it suggests that ZnO NPs could be employed as a nano fertilizer for crops cultivated in Zn-deficient soils to boost agricultural output, enhance food quality, and combat global malnutrition (Yusefi-Tanha et al. 2020).

Nanofertilizers are quickly absorbed by plants. Slow-release nano encapsulated fertilizers can reduce fertilizer use while also reducing pollution. ZnO NPs have the potential to increase food crop productivity and growth (Duhan et al. 2017). Seed germination, seedling vigor, and plant growth were all improved with a ZnO nanoscale treatment (25 nm mean particle size) at 1000 ppm concentration, and these ZnO NPs were also beneficial in enhancing stem and root growth in peanuts. ZnO NPs colloidal solution is used as fertilizer. In agriculture, this form of nanofertilizer is quite useful. Nano-fertilizer is a plant nutrient that is more than a fertilizer in that it not only provides nutrients to the plant but also restores the soil to an organic condition without the adverse effects of chemical fertilizers. Nano-fertilizers have the advantage of being able to be applied in small doses. Nanopowders can be utilized as fertilizers and pesticides with success. (Singh et al. 2013, Milani et al. 2012).

Nanomaterials can be very useful in effective disease management methods to reduce crop spoiling alongside increasing shelf life. In a study, ZnO NPs were found to be safe against Rhizopus soft rot of sweet potatoes. According to the findings by Nafady and his team, tubers treated with ZnO NPs had fewer fungal populations than tubers that were not treated. The tubers treated with ZnO-NPs and infected with *Rhizopus stolonifer* showed no obvious degradation for two weeks, confirming that ZnO NPs coating layer on the tuber surface. These findings show that this technique is an efficient way to protect sweet potatoes against infection, maintain their quality, and extend their shelf life for up to several weeks in storage (Nafady et al. 2019).

g. Miscellaneous

Apart from the healthcare, diagnostic, and agricultural-related applications of ZnO NPs, it is also being tested in areas like water hygiene management, textile industries, and vegetable and food preservation.

I. Water hygiene management

Due to the rapidly-growing demand for clean water, wastewater management systems for the supply of safe water are becoming complex. Thus, it is vital to comprehend water treatment technologies, which are primarily geared at resolving water contamination issues (Salgot and Folch 2018). Adsorption/absorption, photocatalysis, and microbiological disinfection are all used in the water treatment approach. Adsorption is the most widely utilized technology out of these, because of its particular advantages, which include energy saving, cost-effectiveness, simplicity, and a wide working range of parameters such as pH, concentration, and temperature. ZnO and TiO_2 reported as having good sorption properties for organic and inorganic contaminants (Crini and Lichtfouse 2019). Baruah and colleagues discussed heterogeneous photocatalytic systems via metal oxide semiconductors like ZnO and TiO_2 that operate effectively for the treatment of wastewater. Multifunctional photocatalytic membranes using ZnO nanostructures are undoubtedly advantageous compared to the freely suspended NPs due to the ease of their removal from the purified water (Baruah et al. 2012).

The organic dyes are carcinogenic and need to be separated from wastewater. In this regard, the work by Jain explains the increased dye degradation by the heterogeneous Fenton process by using ZnO-graphene oxide nanohybrid to generate hydroxyl radical from hydrogen peroxide (Jain et al. 2020). In another study, Dehaghi and others synthesized chitosan–ZnO NPs composite beads by a polymer-based method. Adsorption properties for the removal of pesticide pollutants were studied and the optimum conditions were investigated, including adsorbent dose, initial concentration of pesticide, agitating time, and pH after the adsorption of pesticide by chitosan loaded with ZnO NPs beads (Dehaghi et al. 2014). These nanocomposites provide more active surface sites and minimize NPs aggregation, but they are prone to leaching. Filtration may become important for the removal and prevention of fouling in wastewater to overcome the issue. This can be accomplished by employing nanocomposites to fabricate nano-based filters for the removal of wastewater (Chouchene et al. 2017).

II. Textile industry

The application of NPs in textile materials has been the focus of several studies aiming at the production of finished fabrics with altered characteristics. The use of nanotechnology has increased rapidly. This may be because traditional methods used to impart the properties to fabrics did not lead to permanent effects, and started losing functions after laundering or wearing (Yadav et al. 2006). Becheri and others reported the synthesis of nanosized ZnO particles and their use on cotton and wool fabrics for UV shielding. After the application of ZnO NPs to cotton and wool samples to impart sunscreen activity to the treated textiles, the effectiveness was assessed through UV-visible spectrophotometry. This study indicated a significant increment of the UV absorbing activity in the fabrics treated with ZnO (Becheri et al. 2008).

Another research by Pulit-Prociak and colleagues presented a method for functionalization of textile materials by using fabric dyes modified with ZnO NPs. It is worth noting that *in situ* NPs were obtained in the reaction of preparing indigo dye. The modified dyes were further used for the dyeing of cotton fibers. Therefore, it can be understood that the NPs here not only had the potential to produce fabrics with antimicrobial properties but were also effective in blocking harmful UV radiation (Pulit-Prociak et al. 2016).

III. Vegetables and fruits preservation

Fruits and vegetables are high on customer purchase lists because of their vitamins, minerals, antioxidants, and fibers, which play a key part in proper nutrition. The main issue with preserving fruits and vegetables is their short shelf life due to their high moisture content, which causes quick degradation and decomposition, as well as an unappealing look. The implementation of appropriate packaging and preservation procedures could be used to extend the shelf life of fruits and vegetables (Liu et al. 2020a, Anugrah et al. 2020). The nanocomposite antimicrobial packaging technologies offer increased mechanical, barrier, thermal, and antibacterial qualities. Although the type of commodity determines the proper packaging or coating, biocomposites containing nano TiO_2 or ZnO are good choices for improving the shelf life of fruits and vegetables. The antimicrobial activity of ZnO NPs from the nanocomposite has also proven to be effective on coated fruits. ZnO NPs have been integrated into food coating materials due to their biocompatibility and public health safety. The addition of ZnO NPs improved the coating mixtures' mechanical and structural qualities (Xing et al. 2019).

ZnO NPs synthesized by *Citrus sinensis* peel extract was compared with commercial ZnO NPs, and if applied as nanocoatings on fresh strawberries, they showed a significant preservation effect. Therefore, ZnO NPs prepared by the green method also have great potential in food packaging application (Gao et al. 2020). Similarly, Rajamanickam suggested that developing coated paper with antibacterial properties of ZnO NPs may be an alternative to other food preservation methods. Nanocoated glass bottles could be used for the preservation of the quality of fruits and vegetable products by protecting them from light and thereby improving their shelf life (Rajamanickam et al. 2012).

Challenges for the Development of ZnO NPs

Apart from their ample merits, NPs are reported to be toxic to the living organisms due to the ease of penetration across biomembranes and interference with basal metabolic reactions within the cell. Also, these particles are easily translocated throughout the body not only via the circulatory system but also the nerve cells (Buzea et al. 2007). The concentration of NPs could not always be reduced by macrophage clearance and detoxification processes in the liver. The spleen cannot often effectively bring down the NPs concentration and hence it may further lead to incurable diseases such as Alzheimer's and Parkinson's (Ghosh Chaudhuri and Paria 2012). The unexampled development and advancements in nanotechnologies have led to the production of large quantities of NPs in laboratories and industries. But it is also necessary to address the huge challenges for us in terms of the bio-safety and biocompatibility issues (Chen and Liang 2020).

It is necessary to improve the dissolution of NPs because it directly affects the bioavailability of the active ingredient. Zhang and others reported the premature release from the antimicrobial drug-loaded NPs. They solved it by developing an infection microenvironment-sensitive drug release NP to minimize the drug loss before it reaches the infectious site. It is reported that a very less amount of drug got released from NPs in the bloodstream whereas rapid drug release occurred after the NPs reach the infectious cells or tissues (Zhang et al. 2010).

More efforts are required to deliver the optimum quantity of NPs into the cells and tissues *in vivo* and development of animal models to evaluate the efficacy of drug-loaded NPs (Zhou et al. 2018). Recently, many researchers are also working on NPs' toxicity and therefore the design of degradation studies and stability protocols has become even more necessary (Ramos-Soriano et al. 2019).

Regulatory Aspects and Clinical Translational Updates

Despite many attempts that have already been made, the clearly defined global regulatory trend is yet not available, and therefore a closer collaboration between regulatory agencies is needed. Currently, the strategies employed for designing conventional medicinal products have been frequently accepted to evaluate the safety and toxicity profile along with the compatibility testing of the nanomedicines (Dorbeck-Jung and Chowdhury 2011, Ehmann et al. 2013). Most of the regulators consider that the active pharmaceutical ingredient of the NPs guides the specifications to be analyzed within the regulatory framework. In the case of biological moieties such as proteins, peptides, enzymes, or antibodies, the innovative product is supposed to follow the guidelines defined for biological and medicinal products.

The mechanism of synthesis of NPs by fungi is unclear and not deeply understood yet. Therefore, more research and experimental trials are required to illustrate the exact pathway to identify the responsible biomolecules and other mediators involved in the reduction and stabilization of NPs. Adapting the manufacturing processes is one of the important obstacles in the clinical translation and further development of nanomedicines. Identification and control of the critical points during each manufacturing process is of utmost importance. Major questions related to the manufacturing of nano-formulations include recent innovations at production facilities and challenges to the scale-up potential. A concept of quality-by-design such as process analytical technology will help to ensure an on-line/at-line quality assessment approach. Further, understanding and anticipating the critical points of production will facilitate the implementation of troubleshooting procedures to problems, as they occur (Sainz et al. 2015).

In the last few years, we have witnessed the translation of several applications of nanomedicine in clinical practice that range from medical devices to nanopharmaceuticals. However, there is still a long journey to achieve the complete regulation of such formulations, from deriving the harmonized definitions to the development of characterization protocols, and process control parameters (Soares et al. 2018).

Patents on Zn Nanoparticle-based Formulations

Many publications have demonstrated the novelty of ZnO NPs in different areas over the past few decades, each of them untangling a multitude of possibilities of improving the existing ones. A method for preparing a composition of colloidal NPs of various metals including Zn, silver, copper, gold, mercury, platinum, copper, palladium, or bismuth with bacteria is described in the US Patent US20090239280 (Abu-Saied et al. 2014).

Another invention related to the Zn NPs was described in patent WO2014206969A1 in which the lactic acid bacterium and a bacterium of the genus Bifidobacterium were selected by the inventors. This study presents a simple and efficient method to prepare artificial magnetic bacteria by the deposition of magnetic NPs on a bacterial species such as the surface of gram-positive bifidobacteria. The positive charge of the NPs and the negative one of the external bacteria surface allow interaction. The methodology employs mild conditions and may be further scaled up, which will allow the production of low-cost, environmentally friendly components for devices (Vera et al. 2019).

The invention in a patent US20120097068A1 relates to a method for producing modified ZnO NPs. This reported method can be used to prepare modified ZnO NPs that comprise of Si-O-alkyl groups and can be dissolved in organic solvents. The invention further explains a method for providing stability to organic materials against the effects of light, radicals, or heat, wherein modified ZnO NPs may additionally contain UV absorbers and stabilizers. This invention further provides useful insights, especially in automobile construction and agriculture as wrapping material (Riggs et al. 2012).

The ZnO NPs have been used in a dermal drug delivery platform that comprises a primary wound dressing with three-dimensional polymer protuberances that extends in an upward direction from the dressing surface to cover the wound (VanDelden 2018). Another invention in patent US9718860B2 relates to a complex of ZnO NPs and a protein comprising ZnO-binding peptides. It states the use of NPs as a drug delivery vehicle for manufacturing medicines, as a vaccine component and a contrast agent in the form of a composite. The protein comprising ZnO-binding peptides significantly enhances the *in vivo* availability of ZnO-binding peptides. And therefore the preparation can be used not only as a drug delivery vehicle but also for *in vivo* imaging or cell imaging. The complex can also be used for producing separating agents for effective separation of biological materials, a therapeutic agent for hyperthermia, a contrast agent for magnetic resonance imaging, and beads used in the biosensors (Cho et al. 2017).

Conclusion and Future Perspectives

Microorganisms have an excellent ability for the synthesis of zinc NPs and in this concern, fungal-mediated synthesis of NPs has obtained a great interest in the last few years. This is one of the green technologies which is found to be efficient and suitable for the synthesis of NPs with huge potential and is relatively inexpensive and safe.

The design and development of smart delivery systems for targeted drug release will be the future of green technology and the metal NPs for disease diagnosis as well as treatment. The development of smart biosensors and detection systems to protect crops against insects and pathogens is already in pipeline. In brief, it can be said that myconanotechnology is still a work in progress. This field also needs harmonized guidelines and related process parameters by seeking and encouraging collaboration between various regulatory agencies.

The applications of NPs will continue to grow but still, we need to assess their toxicity profile, accumulation in the environment, and their hazardous effects on the health of humans and animals. The available literature suggests that considerable work has been carried out in *in-vitro* applications of NPs but not much data available on *in-vivo* applications. Also, the exhaustive study is necessary to elaborate the knowledge, applications, and functions of nanomaterials to reach several milestones in the areas of medicines, agriculture, electronics, cosmetics, environment, etc. Although the fungal-mediated synthesis of ZnO NPs has many advantages, still there are several restrictions and challenges to look out before it can be used clinically. There is hope that NPs will be used in the treatment of various diseases and disorders and open a new avenue in the biomedical field in the coming years.

References

Abouaitah, K., Piotrowska, U., Wojnarowicz, J., Swiderska-Sroda, A., El-Desoky, A. and Lojkowski, W. 2021. Enhanced activity and sustained release of protocatechuic acid, a natural antibacterial agent, from hybrid nanoformulations with zinc oxide nanoparticles. Inter. J. of Molecular Sci. 22: 5287.

Abu-Saied, M.A., Hafez, E.E., Taha, T.H. and Khalil, K.A. 2014. Bionanosilver-poly (methyl 2-methylpropenoate) electrospun nanofibre as a potent antibacterial against multidrug resistant bacteria. Int. J. Nanoparticles 7: 190–202.

Agarwal, H. and Shanmugam, V. 2020. A review on anti-inflammatory activity of green synthesized zinc oxide nanoparticle: Mechanism-based approach. Bioorganic Chem. 94: 103423.

Ahmad, A., Mukherjee, P., Mandal, D., Senapati, S., Khan, M.I., Kumar, R. and Sastry, M. 2002. Enzyme mediated extracellular synthesis of CdS nanoparticles by the fungus, *Fusarium oxysporum*. J. American Chem. Soc. 124: 12108–12109.

Aladpoosh, R. and Montazer, M. 2015. The role of cellulosic chains of cotton in biosynthesis of ZnO nanorods producing multifunctional properties: Mechanism, characterizations and features. Carbohydrate Polymers 126: 122–129.

Anugrah, D.S., Alexander, H., Pramitasari, R., Hudiyanti, D. and Sagita, C.P. 2020. A review of polysaccharide-zinc oxide nanocomposites as safe coating for fruits preservation. Coatings 10: 988.

Arvanag, F.M., Bayrami, A., Habibi-Yangjeh, A. and Pouran, S.R. 2019. A comprehensive study on antidiabetic and antibacterial activities of ZnO nanoparticles biosynthesized using *Silybum marianum* L. seed extract. Mater. Sci. and Engi.: C 97: 397–405.

Bai, H.J., Zhang, Z.M. and Gong, J. 2006. Biological synthesis of semiconductor zinc sulfide nanoparticles by immobilized *Rhodobacter sphaeroides*. Biotech Letters 28: 1135–1139.

Balaure, P.C., Holban, A.M., Grumezescu, A.M., Mogoşanu, G.D., Bălşeanu, T.A., Stan, M.S., Dinischiotu, A., Volceanov, A. and Mogoantă, L. 2019. *In vitro* and *in vivo* studies of novel fabricated bioactive dressings based on collagen and zinc oxide 3D scaffolds. I. J. Pharmaceutics 557: 199–207.

Baruah, S.K., Pal, S. and Dutta, J. 2012. Nanostructured zinc oxide for water treatment. Nanosci. & Nanotech-Asia 2: 90–102.

Baskar, G., Chandhuru, J., Fahad, K.S., Praveen, A.S., Chamundeeswari, M. and Muthukumar, T. 2015. Anticancer activity of fungal L-asparaginase conjugated with zinc oxide nanoparticles. Journal of Materials Science: Materials in Med. 26: 43.

Batool, M., Khurshid, S., Qureshi, Z. and Daoush, W.M. 2021. Adsorption, antimicrobial and wound healing activities of biosynthesised zinc oxide nanoparticles. Chemical Paper 75: 893–907.

Becheri, A., Dürr, M., Nostro, P.L. and Baglioni, P. 2008. Synthesis and characterization of zinc oxide nanoparticles: application to textiles as UV-absorbers. J. of Nanoparticle Research 10: 679–689.

Bhattacharya, R. and Mukherjee, P. 2008. Biological properties of "naked" metal nanoparticles. Adv. Drug Del. Reviews 60: 1289–1306.

Bhunia, A.K. 2017. ZnO nanoparticles: Recent biomedical applications and interaction with proteins. Curr. Trends Biomed. Eng. Biosci. 6.

Boroumand Moghaddam, A., Namvar, F., Moniri, M., Azizi, S. and Mohamad, R. 2015. Nanoparticles biosynthesized by fungi and yeast: A review of their preparation, properties, and medical applications. Molecules 20: 16540–16565.

Brintha, S.R. and Ajitha, M. 2015. Synthesis and characterization of ZnO nanoparticles via aqueous solution, sol-gel and hydrothermal methods. IOSR J. Appl. Chem. 8: 66–72.

Butler, S., Neogi, P., Urban, B., Fujita, Y., Hu, Z. and Neogi, A. 2017. ZnO nanoparticles in hydrogel polymer network for bio-imaging. Glob. J. Nanomed. 1: 555–572.

Buzea, C., Pacheco, I.I. and Robbie, K. 2007. Nanomaterials and nanoparticles: Sources and toxicity. Biointerphases 2: MR17–MR71.

Chatterjee, K., Sarkar, S., Rao, K.J. and Paria, S. 2014. Core/shell nanoparticles in biomedical applications. Adv. in Colloid and Interface Sci. 209: 8–39.

Chen, L. and Liang, J. 2020. An overview of functional nanoparticles as novel emerging antiviral therapeutic agents. Materials Sci. and Engi.: C 112: 110924.

Chhippa, H. 2017. Nanofertilizers and nanopesticides for agriculture. Environmental Chem Letters 15: 15–22.

Chiang, C.T. and Roberts, J.T. 2011. Surface functionalization of zinc oxide nanoparticles: An investigation in the aerosol state. Chem. of Materials 23: 5237–5242.

Cho, N.H., Cheong, T.C., Seong, S.Y., Min, J.H., Wu, J.H. and Kim, Y.K. 2017. Complex of a protein comprising zinc oxide-binding peptides and zinc oxide nanoparticles, and use thereof. Google Patents.

Cho, S., Jung, S.-H. and Lee, K.-H. 2008. Morphology-controlled growth of ZnO nanostructures using microwave irradiation: from basic to complex structures. The J. of Physical Chem. C 112: 12769–12776.

Chouchene, B., Chaabane, T.B., Mozet, K., Girot, E., Corbel, S., Balan, L., Medjahdi, G. and Schneider, R. 2017. Porous Al-doped ZnO rods with selective adsorption properties. Applied Surface Sci. 409: 102–110.

Crini, G. and Lichtfouse, E. 2019. Advantages and disadvantages of techniques used for wastewater treatment. Environmental Chem. Letters 17: 145–155.

de Aragao, A.P., de Oliveira, T.M., Quelemes, P.V., Perfeito, M.L.G., Araujo, M.C., Santiago, J.D.A.S. and da Silva, D.A. 2019. Green synthesis of silver nanoparticles using the seaweed Gracilaria birdiae and their antibacterial activity. Arab. J. Chem. 12: 4182–4188.

de Oliveira, C.T., Junior, J.P. and Tavares, M.I. 2019. Relationship between structure and antimicrobial activity of zinc oxide nanoparticles: An overview. Inter. J. of Nanomed. 14: 9395.

Decazes, P., Hinault, P., Veresezan, O., Thureau, S., Gouel, P. and Vera, P. 2021. Trimodality PET/CT/MRI and radiotherapy: a mini-review. Frontiers in Oncology 10: 3392.

Dehaghi, S.M., Rahmanifar, B., Moradi, A.M. and Azar, P.A. 2014. Removal of permethrin pesticide from water by chitosan–zinc oxide nanoparticles composite as an adsorbent. J. Saudi Chemical Soc. 18: 348–355.

Dorbeck-Jung, B.R. and Chawdury, N. 2011. Is the European medical products authorisation regulation equipped to cope with the challenges of nanomedicines? Law & Policy 33: 276–303.

Duhan, J.S., Kumar, R., Kumar, N., Kaur, P., Nehra, K. and Duhan, S. 2017. Nanotechnology: The new perspective in precision agriculture. Biotech Reports 15: 11–23.

Ehman, E.C., Johnson, G.B., Villanueva-Meyer, J.E., Cha, S., Leynes, A.P., Larson, P.E. and Hope, T.A. 2017. PET/MRI: where might it replace PET/CT? J. Magnetic Resonance Imaging 46: 1247–1262.

Ehmann, F., Sakai-Kato, K., Duncan, R., Pérez de la Ossa, D.H., Pita, R., Vidal, J.M., Kohli, A., Tothfalusi, L., Sanh, A., Tinton, S. and Robert, J.L. 2013. Next-generation nanomedicines and nanosimilars: EU regulators' initiatives relating to the development and evaluation of nanomedicines. Nanomed. 8: 849–856.

El-Saadony, M.T., Alkhatib, F.M., Alzahrani, S.O., Shafi, M.E., Abdel-Hamid, S.E., Taha, T.F., Aboelenin, S.M., Soliman, M.M. and Ahmed, N.H. 2021. Impact of mycogenic zinc nanoparticles on performance, behavior, immune response, and microbial load in Oreochromis niloticus. Saudi J. Biological Sci. 28: 4592–4604.

Gade, A., Ingle, A., Whiteley, C. and Rai, M. 2010. Mycogenic metal nanoparticles: Progress and applications. Biotech Letters 32: 593–600.

Gao, Y., Xu, D., Ren, D., Zeng, K. and Wu, X. 2020. Green synthesis of zinc oxide nanoparticles using Citrus sinensis peel extract and application to strawberry preservation: A comparison study. LWT 126: 109297.

George, D., Maheswari, P.U. and Begum, K.M. 2020. Chitosan-cellulose hydrogel conjugated with L-histidine and zinc oxide nanoparticles for sustained drug delivery: Kinetics and *in-vitro* biological studies. Carbohydrate Polym. 236: 116101.

Ghomi, A.R., Mohammadi-Khanaposhti, M., Vahidi, H., Kobarfard, F., Reza, M.A. and Barabadi, H. 2019. Fungus-mediated extracellular biosynthesis and characterization of zirconium nanoparticles using standard penicillium species and their preliminary bactericidal potential: A novel biological approach to nanoparticle synthesis. Iranian J. of Pharmaceutical Res.: IJPR 18: 2101.

Ghosh Chaudhuri, R. and Paria, S. 2012. Core/shell nanoparticles: classes, properties, synthesis mechanisms, characterization, and applications. Chem. Reviews 112: 2373–2433.

Ghosh, S., Bhagwat, T., Chopade, B.A. and Webster, T.J. 2013. Method for producing metal nanoparticles. Google Patents.

Ginjupalli, K., Alla, R., Shaw, T., Tellapragada, C., Gupta, L.K. and Upadhya, P.N. 2018. Comparative evaluation of efficacy of zinc oxide and copper oxide nanoparticles as antimicrobial additives in alginate impression materials. Mater, Today: Proceedings 5: 16258–16266.

Gordon, T., Perlstein, B., Houbara, O., Felner, I., Banin, E. and Margel, S. 2011. Synthesis and characterization of zinc/iron oxide composite nanoparticles and their antibacterial properties. Colloids and Surfaces A: Physicochem. and Engineering Aspect. 374: 1–8.

Grabherr, S., Grimm, J., Baumann, P. and Mangin, P. 2015. Application of contrast media in post-mortem imaging (CT and MRI). La Radiologia Medica 120: 824–834.

Gupta, M., Tomar, R.S., Kaushik, S., Mishra, R.K. and Sharma, D. 2018. Effective antimicrobial activity of green ZnO nano particles of Catharanthus roseus. Frontiers in Microbio. 9: 2030.

Hameed, S., Iqbal, J., Ali, M., Khalil, A.T., Abbasi, B.A., Numan, M. and Shinwari, Z.K. 2019. Green synthesis of zinc nanoparticles through plant extracts: Establishing a novel era in cancer theranostics. Materials Research Express 6: 102005.

He, L., Sun, X., Nan, X., Wang, T. and Bai, P. 2019. Thermal and pH responsive ZnO-based nanoparticles for efficient drug delivery. AIP Advances 9: 125026.

Jain, B., Hashmi, A., Sanwaria, S., Singh, A.K., Susan, M.A. and Singh, A. 2020. Zinc oxide nanoparticle incorporated on graphene oxide: An efficient and stable photocatalyst for water treatment through the Fenton process. Advanced Composites and Hybrid Materials 3: 231–242.

Jain, N., Bhargava, A., Tarafdar, J.C., Singh, S.K. and Panwar, J.A. 2013. A biomimetic approach towards synthesis of zinc oxide nanoparticles. Applied Microbio. and Biotech. 97: 859–869.

Jamnongkan, T., Sukumaran, S.K., Sugimoto, M., Hara, T., Takatsuka, Y. and Koyama, K. 2015. Towards novel wound dressings: antibacterial properties of zinc oxide nanoparticles and electrospun fiber mats of zinc oxide nanoparticle/poly (vinyl alcohol) hybrids. J. Polymer Engineering 35: 575–586.

Jiang, J., PI, J. and Cai, J. 2018. The advancing of zinc oxide nanoparticles for biomedical applications. Bioinorganic Chem. and Applications 2018: 1062562 .

Kambe, T., Tsuji, T., Hashimoto, A. and Itsumura, N. 2015. The physiological, biochemical, and molecular roles of zinc transporters in zinc homeostasis and metabolism. Physiological Rev. 95: 749–784.

Karagianni, A., Kachrimanis, K. and Nikolakakis, I. 2018. Co-amorphous solid dispersions for solubility and absorption improvement of drugs: Composition, preparation, characterization and formulations for oral delivery. Pharmaceutics 10: 98.

Kato, H. 2011. Tracking nanoparticles inside cells. Nature Nanotechnology 6: 139–140.

Kaur, N., Kaur, S., Singh, J. and Rawat, M. 2016. A review on zinc sulphide nanoparticles: From synthesis, properties to applications. J. Bioelectron. Nanotechnol. 1: 1–5.

Keerthana, S. and Kumar, A. 2020. Potential risks and benefits of zinc oxide nanoparticles: A systematic review. Critical Rev. in Toxicology 50: 47–71.

Khan, R., Kaushik, A., Solanki, P.R., Ansari, A.A., Pandey, M.K. and Malhotra, B.D. 2008. Zinc oxide nanoparticles-chitosan composite film for cholesterol biosensor. Analytica Chimica Acta 616: 207–213.

Khandel, P. and Shahi, S.K. 2018. Mycogenic nanoparticles and their bio-prospective applications: Current status and future challenges. J. of Nanostructure in Chemistry 8: 369–391.

Khorasani, M.T., Joorabloo, A., Adeli, H., Mansoori-Moghadam, Z. and Moghaddam, A. 2019. Design and optimization of process parameters of polyvinyl (alcohol)/chitosan/nano zinc oxide hydrogels as wound healing materials. Carbohydr. Polymers 207: 542–554.

Kiio, T.M. and Park, S. 2021. Physical properties of nanoparticles do matter. J. of Pharmaceutical Invest. 51: 35–51.

Kundu, D., Hazra, C., Chatterjee, A., Chaudhari, A. and Mishra, S. 2014. Extracellular biosynthesis of zinc oxide nanoparticles using *Rhodococcus pyridinivorans* NT2: Multifunctional textile finishing, biosafety evaluation and in vitro drug delivery in colon carcinoma. Journal of Photochemistry and Photobiology B: Biology 140: 194–204.

Li, X., Xu, H., Chen, Z.-S. and Chen, G. 2011. Biosynthesis of nanoparticles by microorganisms and their applications. J. of Nanomaterials 2011: 1–16.

Lin, P.H., Sermersheim, M., Li, H., Lee, P.H., Steinberg, S.M. and Ma, J. 2018. Zinc in wound healing modulation. Nutrients 10: 16.

Liu, J., Qiao, S.Z., Hu, Q.H. and Lu, G.Q. 2011. Magnetic nanocomposites with mesoporous structures: Synthesis and applications. Small 7: 425–443.

Liu, W., Zhang, M. and Bhandari, B. 2020a. Nanotechnology—A shelf life extension strategy for fruits and vegetables. Critical Rev. in Food Sci. and Nutrition 60: 1706–1721.

Liu, Z.Y., Shen, C.L., Lou, Q., Zhao, W.B., Wei, J.Y., Liu, K.K., Zang, J.H., Dong, L. and Shan, C.X. 2020b. Efficient chemiluminescent ZnO nanoparticles for cellular imaging. J. of Luminescence 221: 117111.

Lv, J., Zhang, S., Luo, L., Zhang, J., Yang, K. and Christie, P. 2015. Accumulation, speciation and uptake pathway of ZnO nanoparticles in maize. Environmental Sci.: Nano. 2: 68–77.

Mashrai, A., Khanam, H. and Aljawfi, R.N. 2017. Biological synthesis of ZnO nanoparticles using C. albicans and studying their catalytic performance in the synthesis of steroidal pyrazolines. Arabian Journal of Chemistry 10: S1530–S1536.

Medhi, R., Marquez, M.D. and Lee, T.R. 2020. Visible-light-active doped metal oxide nanoparticles: Review of their synthesis, properties, and applications. ACS Appl. Nano Materials 3: 6156–6185.

Milani, N., McLaughlin, M.J., Stacey, S.P., Kirby, J.K., Hettiarachchi, G.M., Beak, D.G. and Cornelis, G. 2012. Dissolution kinetics of macronutrient fertilizers coated with manufactured zinc oxide nanoparticles. J. of Agricultural and Food Chemistry 60: 3991–3998.

Mohamed, A.A., Abu-Elghait, M., Ahmed, N.E. and Salem, S.S. 2021. Eco-friendly mycogenic synthesis of ZnO and CuO nanoparticles for in vitro antibacterial, antibiofilm, and antifungal applications. Biological Trace Element Res. 199: 2788–2799.

Mohammed, A.M., Ibraheem, I.J., Obaid, A.S. and Bououdina, M. 2017. Nanostructured ZnO-based biosensor: DNA immobilization and hybridization. Sensing and Bio-sensing Research 15: 46–52.

Muthuchamy, N., Atchudan, R., Edison, T.N., Perumal, S. and Lee, Y.R. 2018. High-performance glucose biosensor based on green synthesized zinc oxide nanoparticle embedded nitrogen-doped carbon sheet. J. of Electroanalytical Chem. 816: 195–204.

Nafady, N.A., Alamri, S.A., Hassan, E.A., Hashem, M., Mostafa, Y.S. and Abo-Elyousr, K.A. 2019. Application of ZnO-nanoparticles to manage Rhizopus soft rot of sweet potato and prolong shelf-life. Folia Horticulturae 31: 319–329.

Nie, L., Gao, L., Feng, P., Zhang, J., Fu, X., Liu, Y., Yan, X. and Wang, T. 2006. Three-dimensional functionalized tetrapod-like ZnO nanostructures for plasmid DNA delivery. Small 2: 621–625.

Osmond, M.J. and Mccall, M.J. 2010. Zinc oxide nanoparticles in modern sunscreens: An analysis of potential exposure and hazard. Nanotoxicol. 4: 15–41.

Pan, U.N., Sanpui, P., Paul, A. and Chattopadhyay, A. 2018. Surface-complexed zinc ferrite magnetofluorescent nanoparticles for killing cancer cells and single-particle-level cellular imaging. ACS Applied Nano Materials 1: 2496–2502.

Pandurangan, M. and Kim, D.H. 2015. In vitro toxicity of zinc oxide nanoparticles: A review. Journal of Nanoparticle Research 17: 1–8.

Pavani, K.V., Balakrishna, K. and Cheemarla, N.K. 2011. Biosynthesis of zinc nanoparticles by Aspergillus species. Int. J. Nanotechnol. Appl. 5: 27–36.

Pavani, K.V., Kumar, N.S. and Sangameswaran, B.B. 2012. Synthesis of lead nanoparticles by Aspergillus species. Polish. J. Microbio. 61: 61–63.

Penen, F., Malherbe, J., Isaure, M.P., Dobritzsch, D., Bertalan, I., Gontier, E., Le Coustumer, P. and Schaumlöffel, D. 2016. Chemical bioimaging for the subcellular localization of trace elements by high contrast TEM, TEM/X-EDS, and NanoSIMS. J. of Trace Elements in Med. and Bio. 37: 62–68.

Pestovsky, Y.S. and Martínez-Antonio, A. 2017. The use of nanoparticles and nanoformulations in agriculture. Journal of Nnosci. and Nanotech. 17: 8699–8730.

Pomastowski, P., Król-Górniak, A., Railean-Plugaru, V. and Buszewski, B. 2020. Zinc oxide nanocomposites—Extracellular synthesis, physicochemical characterization and antibacterial potential. Materials 13: 4347.

Prakash, T., Jayaprakash, R., Espro, C., Neri, G. and Kumar, E.R. 2014. Effect of Sn doping on microstructural and optical properties of ZnO nanoparticles synthesized by microwave irradiation method. Journal of Materials Sci. 49: 1776–1784.

Preethi, P.S., Narenkumar, J., Prakash, A.A., Abilaji, S., Prakash, C., Rajasekar, A., Nanthini, A.U. and Valli, G. 2019. Myco-synthesis of zinc oxide nanoparticles as potent anti-corrosion of copper in cooling towers. J. of Cluster Sci. 30: 1583–1590.

Pulit-Prociak, J., Chwastowski, J., Kucharski, A. and Banach, M. 2016. Functionalization of textiles with silver and zinc oxide nanoparticles. Appl. Surface Science 385: 543–553.

Quester, K., Avalos-Borja, M. and Castro-Longoria, E. 2013. Biosynthesis and microscopic study of metallic nanoparticles. Micron. 54: 1–27.

Rajamanickam, U., Mylsamy, P., Viswanathan, S. and Muthusamy, P. 2012. Biosynthesis of zinc nanoparticles using actinomycetes for antibacterial food packaging. International Conference on Nutrition and Food Sciences IPCBEE. 2012.

Raliya, R. and Tarafdar, J.C. 2013. ZnO nanoparticle biosynthesis and its effect on phosphorous-mobilizing enzyme secretion and gum contents in Clusterbean (Cyamopsis tetragonoloba L.). Agricultural Res. 2: 48–57.

Ramos-Soriano, J., Reina, J.J., Illescas, B.M., De La Cruz, N., Rodriguez-Perez, L., Lasala, F., Rojo, J., Delgado, R. and Martin, N. 2019. Synthesis of highly efficient multivalent disaccharide/[60] fullerene nanoballs for emergent viruses. J. of the American Chemical Soc. 141: 15403–15412.

Rashmi, S.H., Raizada, A., Madhu, G.M., Kittur, A.A., Suresh, R. and Sudhina, H.K. 2015. Influence of zinc oxide nanoparticles on structural and electrical properties of polyvinyl alcohol films. Plastics, Rubber and Composites 44: 33–39.

Ren, X., Chen, D., Meng, X., Tang, F., Hou, X., Han, D. and Zhang, L. 2009. Zinc oxide nanoparticles/ glucose oxidase photoelectrochemical system for the fabrication of biosensor. J. Colloid and Interface Sci. 334: 183–187.

Riddin, T.L., Gericke, M. and Whiteley, C.G. 2006. Analysis of the inter-and extracellular formation of platinum nanoparticles by *Fusarium oxysporum* f. sp. *lycopersici* using response surface methodology. Nanotech. 17: 3482.

Riggs, R., Karpov, A., Schambony, S. and Best, W. 2012. Modified ZnO Nanoparticles. Google Patents.

Roohani, N., Hurrell, R., Kelishadi, R. and Schulin, R. 2013. Zinc and its importance for human health: An integrative review. J. of Res. in Medical Sciences: The Official Journal of Isfahan University of Medical Sciences 18: 144.

Ryter, S.W., Kim, H.P., Hoetzel, A., Park, J.W., Nakahira, K., Wang, X. and Choi, A.M. 2007. Mechanisms of cell death in oxidative stress. Antioxidants & Redox Signaling 9: 49–89.

Sabir, S., Arshad, M. and Chaudhari, S.K. 2014. Zinc oxide nanoparticles for revolutionizing agriculture: Synthesis and applications. The Scientific World J. 925494.

Sadhukhan, P., Kundu, M., Chatterjee, S., Ghosh, N., Manna, P., Das, J. and Sil, P.C. 2019. Targeted delivery of quercetin via pH-responsive zinc oxide nanoparticles for breast cancer therapy. Materials Sci. and Engineering: C 100: 129–140.

Sahai, A. and Goswami, N. 2014. Probing the dominance of interstitial oxygen defects in ZnO nanoparticles through structural and optical characterizations. Ceramics Internat. 40: 14569–14578.

Sainz, V., Conniot, J., Matos, A.I., Peres, C., Zupanðið, E., Moura, L., Silva, L.C., Florindo, H.F. and Gaspar, R.S. 2015. Regulatory aspects on nanomedicines. Biochem. and Biophysical Res. Communications 468: 504–510.

Salgot, M. and Folch, M. 2018. Wastewater treatment and water reuse. Curr. Opinion in Environmental Sci. & Health 2: 64–74.

Saloga, P.E. and Thünemann, A.F. 2019. Microwave-assisted synthesis of ultrasmall zinc oxide nanoparticles. Langmuir 35: 12469–12482.

Saravanan, A., Kumar, P.S., Karishma, S., Vo, D.V., Jeevanantham, S., Yaashikaa, P.R. and George, C.S. 2021. A review on biosynthesis of metal nanoparticles and its environmental applications. Chemosphere 264: 128580.

Sarkar, J., Ghosh, M., Mukherjee, A., Chattopadhyay, D. and Acharya, K. 2014. Biosynthesis and safety evaluation of ZnO nanoparticles. Bioprocess and Biosystems Engineering 37: 165–171.

Schrauben, J.N., Hayoun, R., Valdez, C.N., Braten, M., Fridley, L. and Mayer, J.M. 2012. Titanium and zinc oxide nanoparticles are proton-coupled electron transfer agents. Science 336: 1298–1301.

Shanmugam, N.R., Muthukumar, S. and Prasad, S. 2017. A review on ZnO-based electrical biosensors for cardiac biomarker detection. Future Sci. OA 3: FSO196.

Sharma, H., Kumar, K., Choudhary, C., Mishra, P.K. and Vaidya, B. 2016. Development and characterization of metal oxide nanoparticles for the delivery of anticancer drug. Artificial Cells, Nanomed. and Biotech. 44: 672–679.

Sharma, P., Mehta, M., Dhanjal, Kaur, D.S., Gupta, G., Singh, Thangavelue, L., Rajeshkumar, S., Tambuwala, M., Bakshi, H.A., Chellappang, H.A., Dua, K. and Satija, S. 2019. Emerging trends in the novel drug delivery approaches for the treatment of lung cancer. Chemico-biological Interactions 309: 108720.

Shirsath, N.R. and Goswami, A.K. 2019. Nanocarriers based novel drug delivery as effective drug delivery: A review. Curr. Nanomater. 4: 71–83.

Simonelli, G. and Arancibia, E.L. 2015. Effects of size and surface functionalization of zinc oxide (ZnO) particles on interactions with bovine serum albumin (BSA). J. of Mole Liquids 211: 742–746.

Singh, A.K., Pal, P., Gupta, V., Yadav, T.P., Gupta, V. and Singh, S.P. 2018. Green synthesis, characterization and antimicrobial activity of zinc oxide quantum dots using *Eclipta alba*. Materials Chemistry and Physics 203: 40–48.

Singh, N.B., Amist, N., Yadav, K., Singh, D., Pandey, J.K. and Singh, S.C. 2013. Zinc oxide nanoparticles as fertilizer for the germination, growth and metabolism of vegetable crops. Journal of Nanoengi. and Nanomanufacturing 3: 353–364.

Singh, R.P., Handa, R. and Manchanda, G. 2020. Nanoparticles in sustainable agriculture: An emerging opportunity. J. of Controlled Release 329: 1234–48.

Soares, S., Sousa, J., Pais, A. and Vitorino, C. 2018. Nanomedicine: Principles, properties, and regulatory issues. Frontiers in Chem. 6: 360.

Sumanth, B., Lakshmeesha, T.R., Ansari, M.A., Alzohairy, M.A., Udayashankar, A.C., Shobha, B., Niranjana, S.R., Srinivas, C. and Almatroudi, A. 2020. Mycogenic synthesis of extracellular zinc oxide nanoparticles from *Xylaria acuta* and its nanoantibiotic potential. Inter. J. of Nanomedicine 15: 8519–8536.

Suresh, A.K., Pelletier, D.A., Wang, W., Broich, M.L., Moon, J.W., Gu, B., Allison, D.P., Joy, D.C., Phelps, T.J. and Doktycz, M.J. 2011. Biofabrication of discrete spherical gold nanoparticles using the metal-reducing bacterium Shewanella oneidensis. Acta Biomaterialia 7: 2148–2152.

Sutradhar, P. and Saha, M. 2016. Green synthesis of zinc oxide nanoparticles using tomato (*Lycopersicon esculentum*) extract and its photovoltaic application. J. of Experiment Nanoscience 11: 314–327.

Tade, R.S., Nangare, S.N. and Patil, P.O. 2020. Agro-industrial waste-mediated green synthesis of silver nanoparticles and evaluation of its antibacterial activity. Nano. Biomed. Eng. 12: 57–66.

Tartaj, P., del Puerto Morales, M., Veintemillas-Verdaguer, S., González-Carreño, T. and Serna, C.J. 2003. The preparation of magnetic nanoparticles for applications in biomedicine. J. of Physics D: Applied Physics 36: R182.

Umrani, R.D. and Paknikar, K.M. 2014. Zinc oxide nanoparticles show antidiabetic activity in streptozotocin-induced Type 1 and 2 diabetic rats. Nanomedicine 9: 89–104.

VanDelden, J. 2018. Nano-enhanced wound dressing. Google Patents.

Velmurugan, P., Shim, J., You, Y., Choi, S., Kamala-Kannan, S., Lee, K.-J., Kim, H.J. and Oh, B.T. 2010. Removal of zinc by live, dead, and dried biomass of Fusarium spp. isolated from the abandoned-metal mine in South Korea and its perspective of producing nanocrystals. J. of Hazardous Materials 182: 317–324.

Venkatesan, G., Vijayaraghavan, R., Chakravarthula, S.N. and Sathiyan, G. 2019. Fluorescent zinc oxide nanoparticles of *Boswellia ovalifoliolata* for selective detection of picric acid. Frontier Resear. Today 2: 2002.

Vera, J.M.D., Rodriguez, N.G., Marcos, M.A.M., Rodriguez-Acosta, F.C., Rodriguez, D.R. and Martin, M.O. 2019. Probiotic bacteria comprising metals, metal nanoparticles and uses thereof. Google Patents.

Verma, N., Kumar, N., Upadhyay, L.S.B., Sahu, R. and Dutt, A. 2017. Fabrication and characterization of cysteine-functionalized zinc oxide nanoparticles for enzyme immobilization. Analyt. Lett. 50: 1839–1850.

Vidya, C., Hiremath, S., Chandraprabha, M., Antonyraj, M.L., Gopal, I.V., Jain, A. and Bansal, K. 2013. Green synthesis of ZnO nanoparticles by *Calotropis gigantea*. Int. J. Curr. Eng. Technol. 1: 118–120.

Wang, L., Lou, Z., Jiang, K. and Shen, G. 2019. Bio-multifunctional smart wearable sensors for medical devices. Adv. Intelligent Systems 1: 1900040.

Wang, P., Jiang, L. and Han, R. 2020. Biosynthesis of Zinc oxide nanoparticles and their application for antimicrobial treatment of burn wound infections. Materials Research Express 7: 095010.

Xing, Y., Li, W., Wang, Q., Li, X., Xu, Q., Guo, X., Bi, X., Liu, X., Shui, Y., Lin, H. and Yang, H. 2019. Antimicrobial nanoparticles incorporated in edible coatings and films for the preservation of fruits and vegetables. Molecules 24: 1695.

Yadav, A., Prasad, V., Kathe, A.A., Raj, S., Yadav, D., Sundaramoorthy, C. and Vigneshwaran, N. 2006. Functional finishing in cotton fabrics using zinc oxide nanoparticles. Bull. Mater. Sci. 29: 641–645.

Yuan, Q. and Hein, S.M.R. 2010. New generation of chitosan-encapsulated ZnO quantum dots loaded with drug: Synthesis, characterization and *in vitro* drug delivery response. Acta Biomaterialia 6: 2732–2739.

Yusefi-Tanha, E., Fallah, S., Rostamnejadi, A. and Pokhrel, LR. 2020. Zinc oxide nanoparticles (ZnONPs) as a novel nanofertilizer: Influence on seed yield and antioxidant defense system in soil grown soybean (Glycine max cv. Kowsar). Science of the Total Environment 738: 140240.

Zamani, M., Rostami, M., Aghajanzadeh, M., Manjini, H.K., Rostamizadeh, K. and danafar, H. 2018. Mesoporous titanium dioxide@ zinc oxide–graphene oxide nanocarriers for colon-specific drug delivery. Journal of Mater. Sci. 53: 1634–1645.

Zhang, L., Pornpattananagkul, D., HU, C.-M. and Huang, C.-M. 2010. Development of nanoparticles for antimicrobial drug delivery. Curr. Med. Chem. 17: 585–594.

Zhang, Y., Yang, Y., Zhao, J., Tan, R., Wang, W., Cui, P. and Song, W. 2011. Optical and electrical properties of aluminum-doped zinc oxide nanoparticles. J. Mater. Sci. 46: 774–780.

Zhou, Y., Bai, Y., Liu, H., Jiang, X., Tong, T., Fang, L., Wang, D., Ke, Q., Liang, J. and Xiao, S. 2018. Tellurium/Bovine serum albumin nanocomposites inducing the formation of stress granules in a protein kinase R-dependent manner. ACS Appl. Mat. & Interfaces 10: 25241–25251.

Zhu, P., Weng, Z., Li, X., Liu, X., Wu, S., Yeung, K., Wang, X., Cui, Z., Yang, X. and Chu, P.K. 2016. Biomedical applications of functionalized ZnO nanomaterials: From biosensors to bioimaging. Adv. Mat. Interfaces 3: 1500494.

14

Myconanotechnology for Antimicrobial Textiles

Magda A. El-Bendary,[1,]* Shimaa R. Hamed,[2]
Samiha M. Abo El-Ola[3] and Mousa A. Allam[4]

Introduction

Nanotechnology is a multidisciplinary and interdisciplinary science that involves the design, synthesis, characterization, and application of particles, materials and devices with at least one dimension in the nanometer scale (Cao 2009). The prefix "nano" is derived from the Greek word "nano" which means "dwarf" and donates one billionth of a meter or 10^{-9} m, which is about the width of 6 carbon atoms or 10 water molecules (Ayele et al. 2019). Nanotechnology has dynamically developed as a very important field of modern research that covers a wide range of applications in many cutting-edge technological areas of science such as physics, chemistry, biology, material, medicine, engineering, technology, environment, energy, etc. (Chaudhari et al. 2012).

Bionanotechnology deals with the synthesis of environmental ecofriendly nanomaterials and their applications (Sobha et al. 2010). Myconanotechnology is the interface of mycology and nanotechnology for developing ecofriendly technologies for the synthesis of various nanoparticles/nanomaterials using fungi (Rai et al. 2009, Adebayo et al. 2021).

[1] Microbial Chemistry Department, Biotechnology Research Institute, National Research Centre, Giza, Egypt.
[2] Microbial Biotechnology Department, Biotechnology Research Institute, National Research Centre, Giza, Egypt.
[3] Protein and Manmade Fibre Department, Textile Research and Technology Institute, National Research Centre, Giza, Egypt.
[4] Spectroscopy Department, Physics Research Institute, National Research Centre, Giza, Egypt.
* Corresponding author: tasnim41@yahoo.com

Nanoparticles are particles in the range of approximately one to several hundred nanometers in at least one dimension. These nanoparticles (i) have different shapes (ii) have unique physico-chemical properties different from the bulk materials (iii) have the large surface area to volume ratio so they have unique magnetic, thermal, electrical, optical, catalytic, chemical, physical, and biological properties compared to their bulk material counterparts depending on their size, shape, charges, and composition, which enhance their proposed applications (Li et al. 2021). These properties led to their application in different fields such as antimicrobial, magnetic devices, microelectronic devices, photo-catalysts, electro-catalysts, anticorrosive coatings, powder metallurgy, and biomedicals. The biotechnological applications of nanoparticles have increased day by day due to its biocompatibility and bioactivity. These applications include effective drug delivery, tumor targeting, bio-absorption, diagnostic biological probes, fabricating biological sensors, display instruments, diagnosis, detecting environmental toxic substances, and therapeutic applications (Menon et al. 2017, Salem and Fouda 2020).

The market analysis report in 2020 by Grand View Research Inc. evaluated the global nanomaterials market size at $8.5 billion in 2019 and is expected to grow at a compound annual growth rate (CAGR) of 13.1% from 2020 to 2027 (Market Analysis Report 2020) (https://www.grandviewresearch.com/industry-analysis/nanotechnology-and-nanomaterials-market).

Nowadays, there is a new revolution in the textile industry with new technologies that could add special functions and properties to the fabrics. The field of textiles is not restricted to garments and fashion but has applications in a wide range of other areas. The industry has a variety of technologies for physically engineering fiber materials into structures, combining different materials to form hybrids, composites, and blends. The research findings have applications in various fields, including medical, physical fitness, sports, intelligent materials, engineering, energy storage, etc. Smart textiles can observe and respond to environmental stimuli such as mechanical, thermal, magnetic, chemical, electrical, optical, and physiological stimuli (Awan et al. 2020).

The textile industry is one of the first industries that applied nanotechnology to offer innovative products, improve the textile processes, develop new products with novel properties and reduce the production cost. The nanoscale modification of textile substrates allows for enhanced fabric performance or incorporating new high-value-added capabilities while keeping aesthetic qualities (e.g., appearance, feel, and breathability), lightness, flexibility, and comfort. Nanomaterials modification improves the washing resistance and mechanical properties as well. The desirable characteristics imparted to textile fibers using nanotechnology include antimicrobial, self-cleaning surfaces, water repellence, wrinkle resistance, soil resistance, stain resistance, flame retardation, electrical conductivity, dyeability, antiodor, fire resistance, UV-protection, abrasion resistance, shrink resistance, etc., as shown in Figure 1 (Saleem and Zaidi 2020, Mehravani et al. 2021, Repon et al. 2021).

There are two ways to incorporate functional and innovative nanomaterials into textiles in the textile industry. At the finishing/coating stage on a ready-made fabric (top-down strategy), through impregnation/dyeing, spray coating, or printing

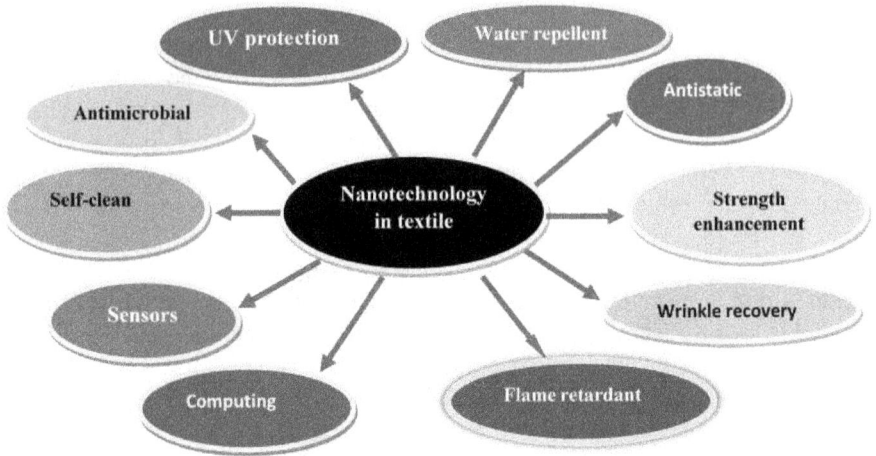

Figure 1. A graphical representation of the various applications of nanotechnology-based textile materials.

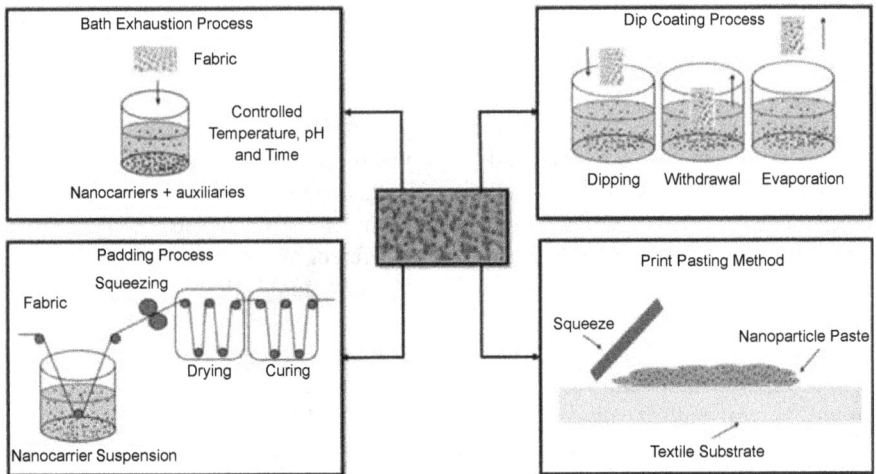

Figure 2. Some methods to functionalize textile materials with nanoparticles (Mehravani et al. 2021).

technologies Figure 2. During fiber manufacturing (bottom-up approach), the fiber can be manufactured from the ground up through spinning processes.

The healthcare-associated infections pose a problem in hospital environments. The pathogenic microorganisms such as viruses and bacteria derived from the patients through skin, vomiting, blood, feces, urine, and others adhere to the hospital textiles (Rajski et al. 2019). Current autoclave laundry and disinfection processes are not sufficient to prevent infection transmission (Sehulster 2015). It is therefore important to impart antimicrobial effect on textile materials to protect the health of the wearer. A number of antimicrobial agents have been used in textile applications. Nanoparticles have a high surface area ratio and a high-energy surface, have a better affinity for the fabric, due to their nanometric size and can penetrate deeper into the fibers (Wong et al. 2006). Recently, increased awareness about hygiene and health led to the development of the antimicrobial feature for all clothes, medical textiles and household products. The antimicrobial finished textile can find wide application in the health and hygiene textile sector, filtration devices, and wound dressings (Gokarnshan et al. 2017).

Various metal oxide (copper oxide, zinc oxide, titanium dioxide) and metal (silica, titanium, gold, zinc, copper, silver) nanoparticles are receiving considerable research attention as prospective antimicrobial agents.

Synthesis of Nanoparticles

A number of physical and chemical approaches are available for the synthesis of nanoparticles (Li et al. 2011). The products of conventional chemical and physical synthesis methods contain toxic chemicals adsorbed on the surface of nanoparticles that may have adverse effects in different applications. In addition, they are complicated, costly, have more energy consumption, and produce hazardous toxic wastes that are harmful, not only to the environment but also to human health (Srivastava and Bhargava 2022). In contrast, the green synthesis of nanoparticles is an interesting method because it is clean, non/less toxic, economic, requires lower energy, commercially viable, simpler processing, environment friendly and can be scaled up for large scale synthesis (Deepak et al. 2011, Dobias 2013, Elamawi et al. 2018).

It was reported that biological systems like fungi, bacteria, actinomycetes, yeast, algae, plants, and enzymes have been used to achieve the green synthesis of nanoparticles (Shedbalkar et al. 2014, Menon et al. 2017). Microorganisms are regarded as potent green nanofactories. They have been used for the synthesis of nanoparticles since they are ease to handle, grow in a low-cost medium, maintain the safety levels, have the potential to adsorb the metal ions and reduce them into nanoparticles by the enzymes and biomolecules produced by microbial metabolic processes (Li et al. 2011, Menon et al. 2017).

Accordingly, the use of biological methods for the synthesis and production of engineered nanomaterials in both scientific research and industrial application has achieved tremendous amount of interest (DeSimone 2002).

Mycofabrication of Nanoparticles and Nanomaterials

Over the last decades, fungi have been widely used for the production of nanoparticles. Recently, nanotechnology research has been focused to develop eco-friendly methods to attain metallic nanoparticles with antimicrobial properties. Fungi are considered as ideal nano-factories for the fabrication of nanoparticles due to (i) rapid growth rate (ii) simple nutrient requirements (iii) form large amounts of biomass (iv) easy handling of biomass/cultures (v) possess a great tendency to grow even in hard conditions with the ability to secrete a huge quantity of specific enzymes and proteins that are assumed to have enhanced reducing and stabilizing capacities (vi) can produce nanoparticles intracellularly or extracellularly (vii) produce biocompatible and non/less toxic nanoparticles (viii) present a cleaner alternative to produce new materials with a wide range of potential applications in biomedicine and industry (Prasad et al. 2016, Guilger-Casagrande and de Lima 2019). In addition, mycosynthesis of nanoparticles is an energy-saving, eco-friendly process and cost-effective (Madakka et al. 2018).

Several fungal species have demonstrated excellent bioreducing candidates to synthesize metallic nanoparticles. Mycofabrication of nanoparticles can be done extracellularly or intracellularly. Extracellular synthesis can produce uniform nanoparticles, less toxicity, easy preparation protocols, and a wide range of applications according to their physico-chemical characteristics (Sarkar et al. 2011). In addition, they support downstream processing, are feasible for different kinds of nanoparticles' production on large scale, and can be directly used in various applications (Sobha et al. 2010), particularly, antimicrobial finishing of textiles (Sharma et al. 2009).

Li et al. (2021) reported that fungi can synthesize many nanoparticles and nanominerals including transition metals (Ag, Au, Pt, Ti, Zr, Mn, Cu, Fe, Co, Ni), post-transition metals (Zn, Al, Cd, Bi), alkaline earth metals (Mg, Sr, Ba), metalloids (Te, Si), non-metals (H, C, P, O, N, S, Se), lanthanides (La, Ce) and actinide (U).

Physicochemical properties and biological activities of mycosynthesized nanoparticles such as shape, surface morphology, size, particle composition, agglomeration, dissolution rate, particle reactivity, biocompatibility, antimicrobial potency, antioxidant and anticancer activities determine the application type of the synthesized nanoparticles (Abd-Elsalam 2022).

Guilger-Casagrande and de Lima (2019) mentioned that fungi are considered as important bionanofactories for the synthesis of nanoparticles such as silver, gold, platinum, titanium, CdS, etc. Myconanofactories have synthesized nanoparticles of desired dimensions, shapes, and monodispersity. Fungi are characterized by the production of large amounts of proteins, enzymes, polysaccharides, and other macromolecules, which are responsible for large-scale nanoparticles fabrication. These fungal biomolecules are mostly secreted extracellularly, which is preferred for easy downstream processing and biomass handling (Bourzama et al. 2021, Adebayo et al. 2021).

Due to their metal tolerance and metal accumulation potential, fungi are of great interest in the development of metallic nanoparticles synthesis (Sastry et al. 2003). Also, fungi can be easily scaled up in solid substrate fermentation techniques, which is an important issue in nanoparticle production. In comparison to other microorganisms, fungi are excellent protein producers, therefore, a higher yield of nanoparticles can be synthesized (Mughal et al, 2021).

Different fungal species are able to fabricate nanoparticles as shown in Table 1 such as *Fusarium, Alternaria, Verticillium, Phanerochaete, Pleurotus, Coriolus, Schizophyllum, Humicola, Metarhizium, Phoma, Phytophthora, Trichoderma, Aspergillus, Neurospora, Scedosporium, Penicillium, Guignardia, Macrophomina, Beauveria, Cladosporium, Ganoderma, Cryphonectria, Bipolaris, Cordyceps, Volvariella, Amylomyces, Botrytis, Hormoconis, Aureobasidium, Colletotrichum, Cylindrocladium, Sclerotium, Hypoxea, Stereum, Saccharomyces, Pichia* and *Yarrowia* (Shedbalkar et al. 2014, Balakumaran et al. 2016, Elamawi

Table 1. Examples of mycosynthesized nanoparticles (Castro-Longoria et al. 2017, Adebayo et al. 2021, Li et al. 2021, El-Bendary et al. 2021).

Fungal Strain	NPs	Size (nm)	Activity against Pathogenic Bacteria
Alternaria alternata	Ag	~ 25–30	*B. subtilis, E. coli, P. aeruginosa, S. aureus*
Alternaria alternata	Fe	5.4–12.1	*B. subtilis, E. coli, P. aeruginosa, E. aureus*
Alternaria solani	Ag	5–20	*E. coli, E. faecalis, S. pyogenes*
Amylomyces rouxii KSU-09	Ag	5–27	*B. subtilis, Citrobacter* sp., *E. coli, P. aeruginosa, S. aureus, S. dysenteriae* type I
Aspergillus clavatus	Ag	10–25	*E. coli, P. fluorescens*
Aspergillus niger BSC-1	FeO	10–30	ND
Aspergillus clavatus	Ag	20–30	Multidrug resistant strains of *S. aureus, S. epidermidis*
Aspergillus flavus	Ag	5–30	*A. baumannii, Bacillus* sp., *E. coli, E. faecalis, K. pneumoniae, M. luteus, P. aeruginosa, S. aureus*
Aspergillus flavus	TiO$_2$	62–74	*B. subtilis, E. coli, K. pneumoniae, P. aeruginosa, S. aureus*
Aspergillus flavus	Au	39.5	ND
Aspergillus fumigatus	ZnO	1.2–6.8	ND
Aspergillus niger	Ag	1–20	*E. coli, P. aeruginosa, S. aureus*
	Ag	3–30	*Bacillus* sp., *E. coli, Staphylococcus* sp.
	Ag	5–35	*K. pneumoniae, L. monocytogenes, M. luteus, P. aeruginosa, S. aureus*
	Ag	40–60	*Bacillus* sp., *E. coli, P. aeruginosa, S. aureus*
Aspergillus nidulans	CoO	20	ND
Aspergillus terreus	Ag	1–20	*E. coli, P. aeruginosa, S. aureus*
	Ag-Au	10–50 8–20	*B. subtilis, E. coli, E. faecalis, K. pneumoniae, P. aeruginosa, S. aureus*
	Ag	5–30	*S. aureus*
	Ag	5–50	*E. faecalis, S. enterica, S. aureus, S. pyogenes*
Aspergillus tubingensis	Ag	35 ± 10	*M. luteus, P. aeruginosa, S. aureus*
Aspergillus versicolor	Ag	3–40	*K. pneumoniae, P. aeruginosa, S. aureus, S. pneumoniae*
Aspergillus versicolor mycelia	Hg	1.8–20.5	ND
Aspergillus caespitosus	Ag	22–57	*Bacillus cereus, Staphylococcus aureus, Escherichia coli, Pseudomonas aeruginosa, Candida albicans*
Bionectria ochroleuca	Ag	35 ± 10	*M. luteus, P. aeruginosa, S. aureus*
Bipolaris nodulosa	Ag	10–60	*B. subtilis, B. cereus, E. coli, M. luteus, P. vulgaris, P. aeruginosa*
Beauveria bassiana	Ag	20–34	*E. coli, S. aureus*
Bjerkandera sp. R1	Ag	70–90	ND

Table 1 contd. ...

...Table 1 contd.

Fungal Strain	NPs	Size (nm)	Activity against Pathogenic Bacteria
Cordyceps militaris	Ag	15	*A. punctata, B. subtilis, E. coli, P. aeruginosa, S. aureus, V. alginolyticus, V. anguillarum, V. parahaemolyticus*
Cryphonectria sp.	Ag	30–70	*E. coli, S. typhi, S. aureus*
Colletotrichum sp.	Au	20–40	ND
Fusarium sp.	Ag	5–50	*E. faecalis, S. enterica, S. aureus, S. pyogenes*
Fusarium acuminatum	Ag	5–40	*E. coli, S. typhi, S. aureus, S. epidermidis*
Fusarium oxysporum	Ag	50	*E. coli, K. pneumoniae, P. aeruginosa, S. typhi, S. aureus*
	Ag	10–45	*S. typhi, S. aureus*
	Fe	20–40	*Bacillus, E. coli, Klebsiella* sp., *Proteus* sp., *Pseudomonas* sp., *Staphylococcus* sp., *Streptococcus* sp., *Vibrio* sp.
Fusarium sp.	Zn	100–200	ND
Fusarium oxysporum	Pt	4–5	ND
Fusarium oxyporum	Cd	9–15	ND
Fusarium oxyporum	CdS	5–20	ND
Fusarium oxyporum	ZnS	42	
Cladosporium oxysporum	Au		ND
Neurospora intermedia	Ag	19–24	*E. coli*
Macrophomina Phaseolina	Ag	8–25	*S. typhi, S. aureus*
Monascus purpureus	Ag	1–7	*E. coli, P. aeruginosa, S. typhimurium, S. aureus, S. epidermidis, S. pyogenes*
Mariannaea sp. HJ	Se		ND
Nigrospora sphaerica	Ag	20–70	*E. coli, P. mirabilis, P. aeruginosa, S. typhi, S. aureus*
Nigrospora oryzae	Ag	30–90	*B. cereus, B. subtilis, E. coli, P. aeruginosa, P. vulgaris, M. luteus*
Paecilomyces lilacinus	Ag	5–50	*E. faecalis, S. enterica, S. aureus, S. pyogenes*
Penicillium (K1 and K10)	Ag	1–100	*B. cereus, E. coli, P. aeruginosa, S. aureus, S. marcescens*
Penicillium sp.	Ag	25	Multidrug resistant strains of *E. coli, S. aureus*
	Ag	20–200	*K. pneumoniae, P. mirabilis, S. dysenteriae, S. aureus*
Penicillium decumbens	Ag	30–60	*E. coli, P. vulgaris, S. aureus, V. cholerae*
Penicillium diversum	Ag	5–45	*E. coli, S. paratyphi, S. typhi, V. cholerae*
Penicillium funiculosum	Ag	5–20	*E. coli, E. faecalis, S. pyogenes*
Penicillium janthinellum	Ag	8–14	*B. subtilis, E. coli, Enterobacter, Enterococcus, K. pneumoniae, Micrococcus, S. typhi, S. aureus, Streptococcus, V. cholera*

Table 1 contd. ...

...Table 1 contd.

Fungal Strain	NPs	Size (nm)	Activity against Pathogenic Bacteria
Pestalotia sp.	Ag	10–40	*S. aureus, S. typhi*
Pleurotus sajor-caju	Ag	30.5	*K. pneumoniae, S. aureus*
	Ag	5–50	*E. coli, P. aeruginosa, S. aureus*
Phoma glomerata	Ag	60–80	*E. coli, P. aeruginosa, S. aureus*
Phomopsis sp.	Ag	10–16	*B. subtilis, E. coli, Enterobacter, Enterococcus, K. pneumoniae, Micrococcus, S. typhi, S. aureus, Streptococcus, V. cholerae*
Rhizopus sp.	Ag	10–30	Multidrug resistant strains of *E. coli*
Rhizopus oryzae	All	10	*B. subtilis, E. coli, P. aeruginosa, Salmonella* sp., *S. aureus*
Rhizopus stolonifera	Ag	5–50	Multidrug resistant strains of *P. aeruginosa*
Schizophyllum commune	Ag	51–93	*B. subtilis, E. coli, K. pneumoniae, P. fluorescens*
Schizophyllum radiatum	Ag	10–40	*B. subtilis, E. coli, S. aureus, P. vulgaris*
Stachybotrys chartarum	Ag	65–108	*B. subtilis, E. coli*
Trichoderma viride	Ag	5–40	*E. coli, M. luteus, S. typhi, S. aureus*
Trichoderma viride	MgO	45–95	ND
Trichosporon beigelii	Ag	50–100	*E. coli*
Tricholoma harzianum	Ag	51 ± 10	*K. pneumoniae, S. aureus*
Tricholoma crissum	Ag	5–50	Multidrug resistant strains of *E. coli*
Volvariella volvacea	Au	20–150	ND

ND: not determined

et al. 2018, Siddiqi et al. 2018, Guilger-Casagrande and de Lima 2019, Mikhailov and Mikhailova 2019, Panchangam and Upputuri 2019, Singh 2019, Salem and Fouda 2020, Adebayo et al. 2021, Li et al. 2021, Rai et al. 2021).

It has been reported that myconanoparticles can be used as nano-fertilizers, nano-fungicides, nano-bactericides, nano-virucides, nano-pesticides, and nano-insecticides by several authors (Prasad et al. 2014, 2017, Bhattacharyya et al. 2016, Avila-Quezada et al. 2022).

Mycosynthesized Nanoparticles Characterization

Nanoparticles/nanomaterials have demonstrated unique physico-chemical properties unlike those of their bulk state. Properties and activities of nanoparticles are dependent on their physico-chemical characteristics provided by the size, shape, surface area, charges, homogeneity, surface composition, and other features. Nanoparticles characteristics are influenced by the synthesis conditions such as the reaction temperature, medium pH, stirring rate, precursors to reducing agents ratio, duration of the reaction, and use of stabilizing agents (Toma et al. 2010).

Physico-chemical characterization of fungal fabricated nanoparticles is an important stage that should be carefully considered before nanoparticles' application. The common techniques used for nanoparticles' characterization includes UV-visible spectrophotometry, Fourier transformation infrared spectroscopy (FTIR), electron microscopy including transmission, high-resolution, scanning, and field emission scanning (TEM, HR-TEM, SEM, FE-SEM), energy-dispersive spectroscopy (EDX), dynamic light scattering (DLS) and X-ray diffraction (XRD) (Pourzahedi and Eckelman 2015, Alheety et al. 2019).

- UV–visible spectrophotometer analysis is important for detecting the surface plasmon resonance (SPR) peaks of nanoparticles through monitoring absorption spectra between 200 nm to 900 nm (Madakka et al. 2018).

- Transmission electron microscopy (TEM) was used for the imaging and mapping of nanoparticles to elucidate the size, aggregation, and morphological shapes of nanoparticls. Under HRTEM, the lattice fringes of nanoparticles can be observed to confirm its crystalline nature through the selected area electron diffraction (SAED) analysis (Nadaf and Kanase 2016).

- Energy-dispersive X-ray spectroscopy (EDX) is a technique used to study the elemental composition and purity of green synthesized nanoparticles (Menon et al. 2017).

- Dynamic light scattering spectroscopy (DLS) technique is used to detect the distribution and size of nanoparticles in colloidal solution and detect agglomerations of nanoparticles. In addition, it detects Zeta potential for the determination of surface charge of nanoparticles, which is responsible for nanoparticles stability (Salem and Fouda 2020).

- Fourier transform infrared spectroscopy (FTIR) measurements detect functional groups, which are responsible for reducing, capping, and stabilizing the synthesized nanoparticles (Elavazhagan et al. 2011).

- X-ray diffraction (XRD) detects the crystallographic shape of nanoparticles and crystalline particle size (Bennur et al. 2016).

Mycosynthesis Mechanism of Nanoparticles

In nature, microorganisms can survive in environments containing high levels of metals by evolving mechanisms to cope with them. These mechanisms may involve altering the chemical nature of the toxic metals so that they no longer cause toxicity, resulting in the formation of metal nanoparticles. Thus, nanoparticles formation is the byproduct of microbial resistance mechanism against a specific metal and this feature can be used as an alternative way for nanoparticles production (Pantidos and Horsfall 2014).

It is known that fungi can produce many enzymes and metabolites for their toxic metal ions survival. These enzymes and metabolites are assumed to reduce the metal ions to the non-toxic nanoparticles and stabilize those (Prasad et al. 2016). It was reported hypothesis that large quantities of extracellular proteins and fungal biomolecules such as polysaccharides, amino acids, peptides, vitamins, enzymes, flavonoids, alkaloids, steroids, quinones, saponins, tannins, carboxylic acids, and others play an important role in the reduction of metal ions to form and stabilize the produced nanoparticles (Salunke et al. 2016, Adebayo et al. 2021, Rai et al. 2021).

Fungi use their enzymes, proteinaceous molecules, or cell membrane-bound molecules to reduce the poisonous ions to nanoparticles and precipitate them intracellularly or extracellularly as shown in Figure 3.

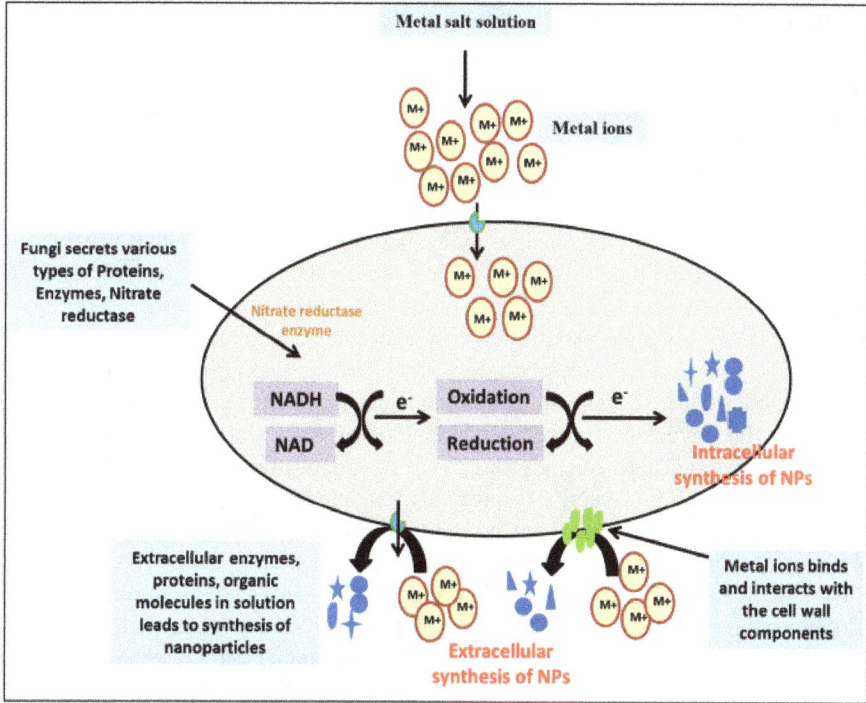

Figure 3. Mechanisms of nanoparticles biosynthesis. Microorganisms synthesize nanoparticles by two mechanisms: Intracellular biosynthesis, which may be mediated by proteins, enzymes and cofactors; and extracellular biosynthesis, which may be mediated by secreted biomolecules and functional groups on the cell wall (Khandel and Shahi 2018).

There are two hypothesized theories about the biosynthesis mechanisms of fungal nanoparticles/nanomaterials. One is an enzymatic catalytic mechanism, and the other is a non-enzymatic catalytic mechanism. In the enzymatic catalytic mechanism, an enzyme acts as an electron transporter to transfer the electrons from reducing substances to the metal ions. However, a possible non-enzymatic reduction mechanism may depend on quinine derivatives of naphthoquinones and anthraquinones produced by fungal cells, which can also act as redox centers and do not require enzymes to reduce metal ions (Liu et al. 2021).

Many types of researches related the fungal nanoparticles biosynthesis to reducing enzymes (as NADH-dependent or NADPH-dependent reductases, nitrate and nitrite reductases, sulfate and sulfite reductases, glutathione reductase (FAD-dependent), cytochrome b5 reductase), and non-enzyme proteins and peptides (as phytochelatins, metallothionein). The formation of nanoparticles involves the nucleation of nanoparticles followed by its stabilization step. The biomolecules produced by fungi are required for the stabilization of synthesized nanoparticles and prevent their agglomeration and aggregation (Srivastava and Bhargava 2022, Li et al. 2021). Proteins can act as capping agents as well as crystal growth regulators to minimize nanoparticle aggregation, regulate their size and shape as reported by Liu et al. 2021. It was reported that electrostatic attractions between free amine groups of fungal proteins and synthesized nanoparticles resulted in their stabilization (Park et al. 2016).

Li et al. (2021) reported that the extracellular and intracellular filtrates of *Verticillium* sp. and *Fusarium oxysporum* are capable of biosynthesis of AgNPs and AuNPs. These fungi can produce large amounts of extracellular proteins and secondary metabolites that act as reductase and stabilizing agents for the produced nanoparticles.

It was established that the low molecular weight cationic fungal proteins of *Fusarium oxysporum* (about 24 and 28 KDa) were responsible for the hydrolysis of anionic hexafluorozirconate precursor to form zirconia (ZrO_2) nanoparticles (Bansal et al. 2011). It has been established that L-cysteine, L-glycine, and glutathione produced by fungi can efficiently bind with metal ions (e.g., Pb^{2+}, Cd^{2+}, Hg^{2+}, As^{3+}, and As^{5+}) and reduce them to form their nanoparticles (Choi and Lee 2020). It was found that the mycelial surface-bound proteins of *Rhizopus oryzae* and *Coriolus versicolor* are associated with the reduction of silver and gold ions and stabilization of the resulting AgNPs and AuNPs, respectively (Das et al. 2012). In the study of Zhang et al. 2011, they found that three proteins (plasma membrane ATPase, 3-glucan-binding protein, and glyceraldehyde-3-phosphatedehydrogenase) produced by *Fusarium oxysporum* were involved in the synthesis of AuNPs. In addition, the triosephosphate isomerase enzyme produced by *Neurospora crassa* plays an important role in the formation and stabilization of copper carbonate nanoparticles (Liu et al. 2021).

Also, it has been documented that fungal cell uses their antioxidative systems such as catalases, methionine sulfoxide reductase, superoxide dismutase, transferases, glutathione, and glutathione peroxidases, thioredoxins and peroxiredoxins to detoxify metal ions and precipitate their nanometals (Jha and Prasad 2016, Gahlawat and Choudhury 2019).

NADH nitrate reductase was reported as an important enzyme in the biosynthesis of AgNPs using *Fusarium oxysporum* extracellular filtrate (Ahmad et al. 2003). While, sulfate reductase is the main enzyme for mycosynthesis of ZnS, CdS, PbS, and MoS_2 nanoparticles using the extracellular filtrate of *Fusarium oxysporum* (Ahmad et al. 2002). In addition, purified NADPH-dependent nitrate reductase and phytochelatin produced by *Rhizopus* stolonifer were successfully used to synthesize AgNPs as reported by Binupriya et al. (2010). Purified hydrogenase enzyme of *Fusarium oxysporum* was efficiently used for reduction of H_2PtCl_6 and $PtCl_2$ into platinum nanoparticles (PtNPs) as reported by Govender et al. 2009. El-Bendary et al. (2018) and (2021) claimed that mycosynthesis of AuNPs and AgNPs by *Aspergillus flavus* and *Aspergillus caespitosus*, respectively are related to the nitrate reductase produced by these organisms.

Furthermore, it was reported that a number of extracellular fungal biomolecules produced under metal ions stressed to mitigate their toxic effect then precipitated them as nanoparticles. For example, the extracellular mucilage materials produced by *Curvularia lunata* in the presence of Cu^{2+}, Pb^{2+}, and Zn^{2+} and pullulans and glomalin glycoprotein produced by *Aureobasidium pullulans* under the stress of Ni^{2+}, Cd^{2+}, and Cu^{2+} (Cornejo et al. 2008).

Myconanoparticles as Antimicrobial Agents

Nanosized materials have unique physical, chemical and biological properties compared with their counterpart materials. The antimicrobial properties of nanoparticles are a contemporary trend of scientists' interest due to the development of multidrug-resistant pathogenic microorganisms and the risk of opportunistic fungal infections (aspergillosis, candidiasis, zygomycosis, cryptococcosis) due to the indiscriminate use of broad-spectrum antibiotics (Rajski et al. 2019). Recently, the development of nano-antibiotics are considered a novel cornerstone to solving many problems concerning the antimicrobial pathogenic microorganisms. Nowadays, nano-antibiotics offer solutions of many problems associated with multidrug microbial resistance. Myconanotechnology is a promising research field to explore active molecules used for nano-antibiotics fabrication. For example, mycosynthesized AgNPs have shown excellent antimicrobial activity as well as a potential synergistic antimicrobial effect when combined with traditional antibiotics (Birla et al. 2009, Fayaz et al. 2010, Dar et al. 2013, Aziz et al. 2016). Therefore, some fungal fabricated AgNPs could be integrated with some products as liquid solutions, creams, cosmetics, dressings, surgical materials, surgical gowns, etc., to prevent or treat microorganisms.

A large number of studies reported the antimicrobial capacity of mycosynthesized nanoparticles to inhibit pathogenic bacteria, fungi, viruses, and others (Prasad 2016). Antimicrobial efficiency of nanoparticles is dependent on the shape, size, monodispersity, charges and chemical composition of the particles.

Due to the antimicrobial activities of AgNPs, there are many studies concentrated on their biosynthesis using different fungi. They showed successful inhibition of microorganisms that cause infectious diseases in humans, especially multidrug-resistant microorganisms (Kim et al. 2007). In addition, they showed antifungal and antiviral activities.

Durán (2007) found that *Fusarium oxysporum* AgNPs were extracellularly synthesized and they showed an antagonistic effect against *Staphylococcus aureus*. Gajbhiye et al. (2009) fabricated AgNPs using *Alternaria alternata* and exhibited antifungal properties against *Candida albicans*, *Trichoderma* sp., *Fusarium semitectum*, *Phoma herbarum* and *Phoma glomerata*. Also, they enhanced the antifungal activities of the antifungal drug (fluconazole) when combined with *Alternaria alternata* AgNPs against *Candida albicans*, *Trichoderma* sp. and *Phoma glomerata*. Another study reported that AgNPs synthesized using biomass water extract of *Amylomyces rouxii* were highly antimicrobial against *E. coli*, *Staphylococcus aureus*, *Pseudomonas aeruginosa*, *Bacillus subtilis*, *Citrobacter* sp.,

Shigella dysenteriae type I, *Fusarium oxysporum* and *Candida albicans* (Musarrat et al. 2010). *Bipolaris nodulosa* was reported for the biosynthesis of AgNPs, which present antimicrobial effects against *Bacillus cereus, Bacillus subtilis, E. coli, Proteus vulgaris, Pseudomonas aeruginosa,* and *Micrococcus luteus* (Saha et al. 2010). Also, *Aspergillus terreus* AgNPs showed antimicrobial activity against *E. coli, Pseudomonas aeruginosa,* and *Staphylococcus aureus* (Li et al. 2012). Likewise, fungal-mediated AgNPs using *Rhizopus stolonifer* were highly antimicrobial against two multidrug-resistant strains of *Pseudomonas aeruginosa* (Rathod and Ranganath 2011). Qian et al. (2013) observed the antimicrobial effect of *Epicoccum nigrum* AgNPs against *Candida albicans, Sporothrix schenckii, Fusarium solani, Aspergillus flavus, Cryptococcus neoformans* and *Aspergillus fumigatus.* Ishida et al. (2014) synthesized AgNPs from *Fusarium Oxysporum* and they found that they exhibited antifungal activities against *Candida* and *Cryptococcus.* AgNPs synthesized by *Guignardia mangiferae* were reported as potential antifungal activities against *Colletotrichum* sp., *Rhizoctonia solani,* and *Curvularia lunata* (Balakumaran et al. 2015). Extracellular biosynthesis of AgNPs using isolated *Fusarium solani* showed good surface sterilization for wheat, barley, and maize seeds contaminated with phytopathogenic fungi as reported by Abd El-Aziz et al. 2015. The inhibitory effect of *Aspergillus versicolor* AgNPs against *Sclerotinia sclerotiorum* and *Botrytis cinerea* (plant pathogenic fungi) was observed by Elgorban et al. 2016. *Penicillium decumbens* AgNPs displayed antibacterial activities against tested multidrug-resistant pathogens (*E. coli, Staphylococcus aureus, Vibrio cholerae* and *Proteus vulgaris*) which isolated from urine, stool, and blood (Majeed et al. 2016). Xue et al. (2016) isolated *Arthroderma fulvum* from soil and used it for the biosynthesis of AgNPs which exhibited fungicidal properties against *Candida albicans, Candida krusei, Candida parapsilosis, Candida tropicalis, Aspergillus flavus, Aspergillus fumigatus, Aspergillus terreus, Fusarium solani, Fusarium oxysporum* and *Fusarium moniliforme.* Also, they stated that these nanoparticles were potential as an antibacterial agent. Ottoni et al. (2017) revealed that AgNPs biosynthesized using *Rhizopus arrhizus, Aspergillus niger* and *Trichoderma gamsii* were active against *Staphylococcus aureus, Escherichia coli,* and *Pseudomonas aeruginosa.* Halkai et al. (2018) used *Fusarium semitectum* for the biogenic synthesis of AgNPs. Mycosynthesized AgNPs were found to be highly antibacterial against *Enterococcus faecalis.* Recently, El-Bendary et al. (2021) biosynthesized AgNPs using both extracellular and intracellular filtrates of *Aspergillus caespitosus,* which were characterized by different techniques. These nanoparticles showed high antimicrobial activities against different Gram-positive bacteria (*Bacillus cereus* and *Staphylococcus aureus*), Gram-negative bacteria (*Escherichia coli* and *Pseudomonas aeruginosa,* and the yeast *Candida albicans.* The authors claimed that *Aspergillus caespitosus* could be used as a cost-effective candidate for the biosynthesis of AgNPs using both extracellular and intracellular fungal filtrates. Wang et al. 2021 investigated the mycosynthesis of AgNPs using *Aspergillus sydowili* and their antifungal activities. They found that *Aspergillus sydowili* AgNPs have antifungal activities against *Candida albicans, Candida glabrata, Candida parapsilosis, Candida tropicalis, Fusarium solani, Fusarium moniliforme, Fusarium oxysporum,*

Aspergillus flavus, *Aspergillus fumigatus*, *Aspergillus terreus*, *Sporothrix schenckii*, and *Cryptococcus neoformans*.

Besides fungal mediated AgNPs synthesis, very scarce studies of mycosynthesized AuNPs have demonstrated antimicrobial capacity. AuNPs from *Rhizopus oryzae* showed remarkable antimicrobial activities against six different pathogens. The maximum growth inhibition was recorded against *E. coli*, *Pseudomonas aeruginosa*, *Bacillus subtilis*, *Staphylococcus aureus*, *Saccharomyces cerevisiae*, and the opportunistic fungus *Candida albicans* as reported by Das et al. 2009. Also, Thakker et al. (2013) synthesized AuNPs using *Fusarium oxysporum*, which efficiently inhibited the growth of *Pseudomonas* sp. It was assumed that AuNPs biosynthesized using different fungal species are alternative antimicrobial agents that are biocompatible and non/less toxic (Prasad 2016).

Other metallic myconanoprticles have also been reported as nano-antibiotics. For instance, TiO_2NPs fabricated using *Aspergillus flavus* were reported as a novel antimicrobial nanomaterial against *Bacillus subtilis*, *E. coli*, *Klebsiella pneumoniae*, *Pseudomonas aeruginosa*, and *Staphylococcus aureus* (Rajakumar et al. 2012). Also, iron nanoparticles (FeNPs) prepared using *Alternaria alternata* were reported to have antibacterial activities against *E. coli*, *Staphylococcus aureus*, *Bacillus subtilis*, and *Pseudomonas aeruginosa* (Mohamed et al. 2015). Gupta and Chundawat (2019) fabricated PtNPs using *Fusarium oxysporum* and they have a bactericidal effect against *E. coli*, which was relatively better than that of ampicillin. El-Sayed and El-Sayed (2020) synthesized AgNPs, CuNPs, and ZnONPs from *Fusarium solani* to combat multidrug-resistant pathogens. They found that AgNPs displayed a significant effect against *Pseudomonas aeruginosa* while ZnONPs were the most effective against *Staphylococcus aureus* and *Fusarium oxysporum*.

It was also reported that the use of nanoparticles provides an interesting opportunity for novel antiviral therapies. Since metals may attack a broad range of targets in the virus there is a lower possibility to develop resistance as compared to conventional antivirals (Galdiero et al. 2011). Accordingly, AgNPs have become one of the strongest candidates as antiviral agents. They have proven to be active against several types of viruses including human immunodeficiency virus, hepatitis B virus, herpes simplex virus, respiratory syncytial virus, monkeypox virus, influenza virus and Tacaribe virus (Ge et al. 2014, Murugan et al. 2015, Lysenko et al. 2018). Gaikwad et al. (2013) synthesized AgNPs using *Fusarium oxysporum* and they reduced the replication of HSV-1, HSV-2, and HPIV-3 in cell cultures.

Antimicrobial Mechanism of Mycosynthesized Nanoparticles

The actual antimicrobial mechanisms of nanoparticles are not clearly known. The antibacterial mechanisms for the majority of nanoparticles include inhibition of cell wall and membrane functions, disruption of energy transduction, photocatalysis, production of toxic reactive oxygen species (ROS) and inactivation of cellular enzyme and DNA (Shaikh et al. 2019). Hypothesized antimicrobial mechanisms by some authors are as follows and illustrated in Figure 4.

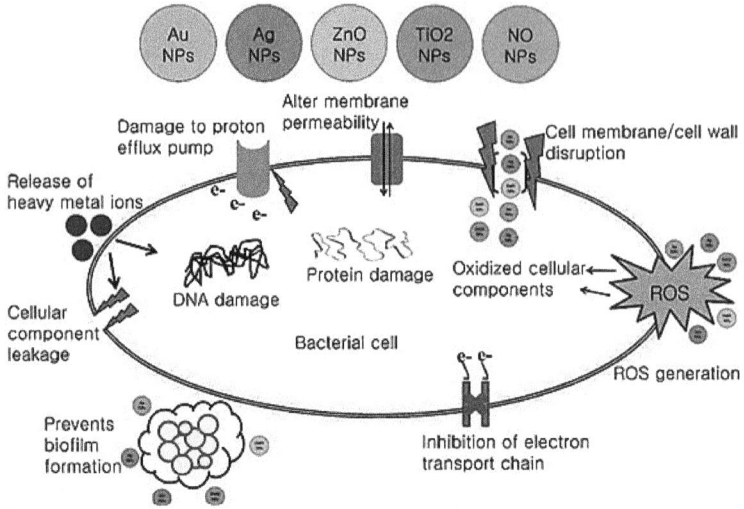

Figure 4. Schematic representations of the antibacterial mechanisms of various nanoparticles (Shaikh et al. 2019).

1. large relative surface area of nanoparticles enhance their attachment to the cell wall biomolecules as phosphorus, sulfur, or nitrogen through electrostatic attraction, Vander Waals forces, receptor-ligand and hydrophobic interactions causing pits formation of the cell wall through which the internal cell contents are effluxed and subsequently cause cell death.

2. Nanoparticles can cross microbial membranes causing changes in membrane shape and function.

3. Nanoparticles penetrate inside the microbial cells and react with sulfur and phosphorus-containing biomolecules like DNA and proteins causing their denaturation and losing their functions.

4. Nanoparticles generate ROS, which triggers the oxidative stress of the cell causing cellular toxicity. ROS is detrimental to bacterial cell where they alter the bacterial cells permeability and cause damage to cellular membranes.

5. Metal ions are released from nanoparticles in aqueous medium and subsequently absorbed through cell membranes and react with the functional groups of proteins and nucleic acids leading to vital effects on bacterial physiological processes.

6. Nanoparticles can disrupt the cellular signal transduction pathways.

Salleh et al. (2020) and Rai et al. (2021) proposed all the above-mentioned mechanisms for the antimicrobial action of AgNPs.

Also, several studies have reported the activities of biosynthesized nanoparticles as antifungal agents against multicellular and unicellular fungi. The proposed antifungal activity of nanoparticles (Figure 5) are as follows (Cruz-Luna et al. 2021):

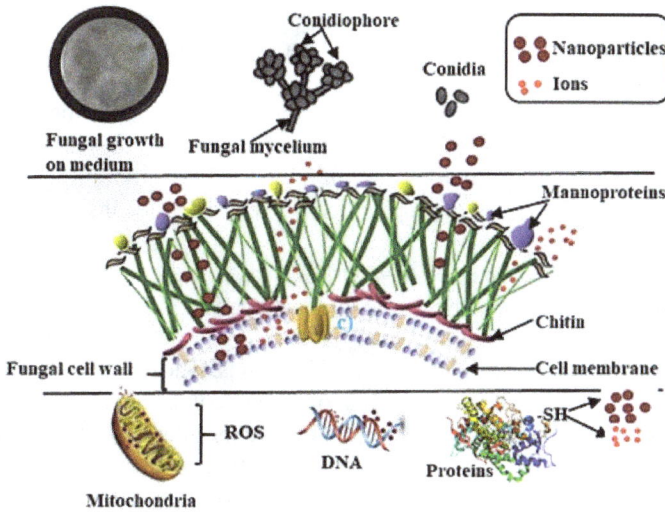

Figure 5. The possible antifungal mechanisms of metal nanoparticles (Cruz-Luna et al. 2021).

1. Ions are released from the nanoparticles and bind to fungal cellular and membrane-active groups losing their permeability.

2. The nanoparticles inhibit the germination and development of the conidia.

3. Nanoparticles release ions that disrupt the electron transport, and destroy protein and DNA.

4. Nanoparticles interfere with protein oxidative electron transport.

5. Nanoparticles generate ROS causing damage to proteins, membranes, and DNA causing cell death.

It was reported that disrupting the structure of the cell membrane by destructing the membrane integrity, thereby the inhibition of the budding process has been attributed to be responsible for the antifungal action of AgNPs against *Candida albicans* species (Srikar et al. 2016).

It was reported that the toxicity of AgNPs is dependent on the size, morphology, concentration, pH of the medium, and exposure time to pathogens. The smaller size is related to the greater surface area of nanoparticles and their agglomeration around the cell wall inhibits the cell division of microbes (Siddiqi et al. 2018).

The suggested mechanism for the antibacterial activity of ZnO is based mainly on the catalysis of formation of the reactive oxygen species (ROS) from water and oxygen that disrupt the integrity of the bacterial membrane, although additional mechanisms have also been suggested. Since the catalysis of radical formation occurs on the particle surface, particles with a larger surface area demonstrate stronger antimicrobial activity (Rajendran et al. 2010).

It was reported that the use of nanoparticles could be extended to the development of antivirals that act by interfering with a viral infection, particularly during attachment and entry (Salleh et al. 2020).

There are a few types of research on the effects of AgNPs on viruses. However, the details of the interaction are limited. The complexity of virus structures can contribute to the limited knowledge of nanoparticles' mechanisms towards the viruses. There are two ways the AgNPs interacts with the pathogenic virus: (1) the AgNPs bind to the outer coat of the virus thus inhibiting the attachment of the virus towards cell receptors and (2) the AgNPs bind to the DNA or the RNA of the virus thus inhibiting the replication or propagation of the virus inside the host cells (Salleh et al. 2020).

With the recent outbreak of new viral diseases such as COVID-19 that has reached millions of cases worldwide, the application of AgNPs could be implemented as a treatment. There is an opinion letter on using AgNPs as the antiviral therapy to treat COVID-19 patients with minimum side effects. The hypothesis was that the AgNPs will bind to the spike glycoprotein of the virus thus inhibiting the binding of the virus towards the cells and the release of silver ions decrease the environmental pH of respiratory epithelium (where the COVID-19 virus usually reside) becoming more acidic and hence hostile towards the virus (Salleh et al. 2020).

Application of Mycosynthesized Nanoparticles for Preparation of Antimicrobial Textiles

Textiles are carriers of microorganisms such as pathogenic bacteria, odour generating bacteria, and fungi (Shojaei et al. 2015). Bacteria are odour-causing, interact with fibers in several phases including the initial adherence, subsequent growth, and damage to the fibers. The attachment of bacteria to fabrics is dependent upon the type of bacteria and the physicochemical characteristics of the fabric substrate. Natural and synthetic fibers vary greatly in their response to microbial growth (Landage and Wasif 2012). The spreading of microorganisms from fabric surfaces to the human skin is a major health concern. Sterile fabrics are one of the common goals of researchers; therefore, fabrics can be treated to avoid microbial infections (Deshmukh et al. 2019). Moreover, the spread of disease and antibiotic-resistant microorganisms make many people focused on protecting themselves against harmful pathogens. It soon became more important for antimicrobially finished textiles to protect the wearer from microorganisms than it was to simply protect the garment from fiber degradation. The need for antimicrobial textiles increased with the rise in resistant strains of microorganisms (Rajendran et al. 2010). A number of traditional antimicrobial agents such as metal salts, quaternary ammonium salts, and triclosan have been used in textile antimicrobial preparation (Gokarneshan et al. 2012).

Antibacterial finishes of fabrics inhibit the bacterial growth, odour-causing germs and reduce the threat of infection to wounds. Such finishes are used in clothing that encounters the skin, skin linings, hospital linens, and contract carpeting. The antibacterial textiles are desirable for medical, technical and hygienic purposes including products such as bandages, sutures, surgical gloves and clothing (Raza et al. 2019).

Metal and metal oxide antimicrobial nanoparticles immersed in textiles provide them with antimicrobial properties. The development of advanced antimicrobial textiles through immobilization of metal nanoparticles has grown rapidly in the past decades. Mehravani et al. (2021) reported that impregnation of nanoparticles in textiles has been a promising strategy to achieve multifunctional textiles especially with antimicrobial properties. AuNPs were used for the functionalization of fabrics, yarns, and fibers to accept their antimicrobial properties for health and hygiene products.

Metal oxide nanoparticles of TiO_2, Al_2O_3, ZnO, and MgO showed UV absorption, photocatalytic ability, electrical conductivity, and photo-oxidizing capacity against biological species and chemicals. Many researchers focused on coating the textile fibers with metal oxides nanoparticles to have some properties such as antimicrobial, self-decontaminating and UV shielding for both military protection gears and civilian health products (Joshi and Bhattacharyya 2011, Qian and Hinestroza 2004, Coyle et al. 2007). As an example, nylon fibers treated with ZnO nanoparticles can provide UV blocking function, antimicrobial activity, and reduce static electricity on nylon fibers. Treatment of textile fibers with TiO_2 or MgO nanoparticles can provide them antimicrobial function (Harholdt 2003).

Durán (2007) efficiently prepared AgNPs and AuNPs using the extracellular filtrate of *Fusarium oxysporum*. They proposed that these synthesized particles could be incorporated into different textile fabrics for antimicrobial finish. In their study, they incorporated AgNPs into the cotton fabric, which exhibited high antibacterial activity against *Staphylococcus aureus*. Namasivayam and Avimanyu (2011) found that cotton fabrics coated with AgNPs synthesized by extracellular filtrate of *Lecanicillium lecanii* displayed an inhibition activity against *Staphylococcus aureus* and *E. coli*. Also, Paul et al. (2015) treated the cotton fabrics with AgNPs of rot fungus *Ganoderma lucidium* and reported that the treated fabrics exhibited antibacterial activity against different Gram-positive and Gram-negative pathogenic bacteria. They claimed that this technology could be used for preparing sterile dressing material that could be used for wounds and infections treatment. Abou Elmaaty et al. (2018) stated that textile functionalization with noble metal AgNPs, and AuNPs has increased in recent years. AgNPs are preferable as compared with traditional antimicrobial agents such as metal salts, quaternary ammonium compounds and triclosan (Wagener et al. 2016). The advantage of AgNPs compared to bulk metal or salts is the slow and regulated release of silver from nanoparticles, thereby causing long-lasting protection against microorganisms (Rajarathinam and Kalaichelvan 2013).

Krishna et al. (2018) reported that natural fibers such as linen or cotton are more susceptible to microbial attack compared to synthetic fibers. They fabricated AgNPs using two white-rot fungi (*Trametes ljubarskyi* and *Ganoderma enigmaticum*) filtrates. These biosynthesized AgNPs were immobilized on cotton fabrics and tested for antibacterial activity. They exhibited high antibacterial activity against *Pseudomonas aeruginosa*, *Staphylococcus aureus*, *Klebsiella pneumoniae* and *Micrococcus luteus*. They concluded that these cotton fabrics treated with nanoparticles could be used as an alternative source to pharmaceutical band-aids in treating wounds. The cotton fabric treatment with AgNPs of *Alternaria alternata* showed high antibacterial activities against *E. coli* and *Staphylococcus aureus* as investigated by Ibrahim and Hassan 2016. In the study of Balakumaran et al. (2016), they prepared AgNPs using *Aspergillus terreus* and studied the antimicrobial activity of cotton textiles finished with these nanoparticles. They found that the cotton fabrics coated with AgNPs exhibited broad-spectrum antibacterial activity against all tested pathogens. Also, these fabrics have excellent antibacterial activity even after 15 wash cycles. Ballottin et al. (2017) found that cotton fibers impregnated with mycosynthesized AgNPs using *Fusarium oxysporum* showed strong antimicrobial activities against *Candida parapsilosis* and *Xanthomonas axonopodis* pv. citri (Xac) even after repeated mechanical washing cycles up to 10 times. They proposed using the treated textile in agriculture clothing and the medical environment to avoid microbial spreading. Fouda et al. (2017) biosynthesized AgNPs using isolated *Aspergillus niger* and coated the cotton fabrics with these nanoparticles. AgNPs treated cotton textile showed high antibacterial activities against *Pseudomonas aeruginosa, Escherichia coli, Staphylococcus aureus* and *Bacillus subtilis* (88–95% growth reduction). Mohmed et al. (2017) synthesized AgNPs using biomass filtrate of Egyptian isolate *Fusarium keratoplasticum* A1-3 and used them in finishing the cotton fabric. These fabrics exhibited 86%–93% growth reduction of *Staphylococcus aureus, Bacillus subtilis, Pseudomonas aeruginosa*, and *Escherichia coli*. In another study, Shaheen and Abd El Aty (2018) used the biomass filtrates of five different isolated endophytic fungi for synthesis of AgNPs on cotton fabrics followed *in-situ* and *ex-situ* methods. Fungal AgNPs were efficiently biosynthesized on cotton fabrics (*in-situ* or *ex-situ*) which acquired antimicrobial activities against different pathogenic bacteria and fungi (*Staphylococcus aureus, E. coli, Candida albicans* and *Aspergillus niger*). They related the antimicrobial activities of AgNPs-cotton fabrics to AgNPs leaching out from pores of the cotton fibrils. In the study of Othman et al. (2019) they mycosynthesized AgNPs using *Aspergillus fumigatus* DSM819 filtrate and studied their application as antimicrobial finishing agent in polyester/cotton blend (PET/C 50/50) fabrics. AgNPs treated polyester/cotton blend fabric displayed antimicrobial activity against *E. coli, Bacillus mycoides*, and *Candida albicans*. The impregnation of AgNPs synthesized by extracellular filtrate of the epiphytic fungus *Bionectria ochroleuca* on cotton and polyester fabrics showed a potent antimicrobial effect against *Staphylococcus aureus, Escherichia coli, Candida albicans, Candida glabrata* and *Candida parapsilosis* as reported by Rodrigues et al. (2019). Recently, El-Bendary et al. (2021) fabricated AgNPs using the extracellular and intracellular filtrates of the isolated strain of *Aspergillus caespitosus*. These nanoparticles were applied to different textile fabrics (cotton, viscose, polyester and wool/polyester blend) for antimicrobial textiles finishing. They reported that different textile samples treated with these nanoparticles exhibited antimicrobial activities against Gram-positive bacteria (*Bacillus cereus* and *Staphylococcus aureus*), Gram-negative bacteria (*E. coli* and *Pseudomonas aeruginosa*) and the yeast, *Candida albicans*. They also found that most of the AgNPs treated fabrics preserved their antimicrobial characteristics even after five washing cycles. Also, Shaheen et al. (2021) used biomass filtrate of *Aspergillus terreus* for mycosynthesis of CuONPs and treated the cotton fabric with these nanoparticles. These fabrics revealed their antibacterial activities against different pathogenic bacteria.

Nanomaterials from the Textile Industry to Environmental Hazards

Nano-finished textiles may be a substantial source of nanomaterials discharged into aquatic systems and soils. The assessment of environmental risks is highly reliant on product life cycles and the worldwide production of nanomaterials. The binding of nanomaterials to the textile determines its stability and influences on the fabric during its life cycle (manufacture, use, disposal/recycling).

The product's life cycle and design govern the environmental and health exposure circumstances. For example, engineered nanomaterials mistakenly discharged from geotextiles may end up in soils, whereas engineered nanomaterials involuntarily expelled from T-shirts may come into direct contact with humans, eventually entering wastewater. As a result, advocating regulation based on the behavior and impacts of nanotextiles necessitates complete product characterization based on particle stability, surface morphology, shape, porosity, size, chemical composition, and tendency to aggregation and accumulation from the nanomaterials of the textile industry to the environmental risks (Saleem and Zaidi 2020).

Conclusion and Future Perspectives

Nanoparticles and nanominerals have shown great potential in different fields, including medicine, environmental biotechnology, electronics and chemical engineering. However, it is still a challenge to develop sustainable methods for the synthesis of these nanomaterials. Compared with conventional physico-chemical methods, biological systems may provide a sustainable, resource efficient and economical method of synthesis. Mycosynthesised nanoparticles/nanomaterials are an alternative to conventional physical and chemical methods. Fungi are versatile and produce nanoparticles in a wide range of culture conditions of diverse temperature, pH, fungal biomass, and metal concentration leading to nanoparticles production that has different physico-chemical characteristics. Additionally, fungi produce large amounts of proteins and biomolecules, which act as reducing, stabilizing, and capping agents. Mycosynthesis of nanoparticles is considered to be cost-effective, and economically viable.

Many mycosynthesized metal and metal oxide nanoparticles showed excellent antimicrobial activities against Gram-positive and Gram-negative bacteria, filamentous fungi and yeasts. Therefore, antimicrobial mycosynthesized nanoparticles are encouraged to use for antimicrobial textile finishing.

Despite the benefits of using mycosynthesized nanoparticles, there are some challenges that face their large-scale applications. There is a need to understand the bioreduction mechanism for nanoparticles synthesis inside and outside the fungal cells. A clear understanding of the role of enzymes, proteins, and other fungal biomolecules in the biosynthesis of nanoparticles will lead to the optimization of mycosynthesis processes. In addition, purification techniques were needed after harvesting or downstream processing of mycosynthesized nanoparticles.

Furthermore, scaling up laboratory processes of fungal biosynthesized nanoparticles to the industrial and the economic large scales have been considered as the important challenge. In addition, the preservation of the nanoparticles with the desired characteristics face major challenge.

Safety and risk assessment of mycosynthesized nanoparticles are not clearly understood. The disposal of metal nanoparticles in the environment has life-threatening consequences. Thus, ecofriendly nanoparticle mycosynthesis should be accomplished based on product purity and safer disposal. There is an urgent need for government bodies of all developed countries to set strict regulations for environment and health protection agencies to prevent overexposure and accumulation of nanoparticles in the field.

Nano finishes have been applied to fabrics for the attainment of desirable functional properties. Of these, the antimicrobial property has been one of the most prominent and promising. Such types of antimicrobial finish can find wide applications in the health and hygiene textile sectors. However, the probable toxicity of biosynthesized nanoparticles and nanomaterials should be considered. In addition, nanoparticles enter the environment through water, soil, and air during numerous human activities. Therefore, the toxicity and environmental impact studies of these nanoparticles would give valuable information.

References

Abd El-Aziz, A.R.M., Al-Othman, M.R., Mahmoud, M.A. and Metwaly, H.A. 2015. Biosynthesis of silver nanoparticles using *Fusarium solani* and its impact on grain borne fungi. Dig. J. Nanomater. Biostruc. 10: 655–662.

Abd-Elsalam, K.A. 2022. Nanobiotechnology for Plant Protection: Green Synthesis of Silver Nanomaterials, Elsevier, Netherlands.

Abou Elmaaty, T.A., El-Nagare, K.H., Raouf, S., Abdelfattah, K.H., El-Kadi, S. and Abdelazizd, E. 2018. One-step green approach for functional printing and finishing of textiles using silver and gold NPs. RSC Advances 8: 25546–25557. https://doi.org/10.1039/c8ra02573h.

Adebayo, E.A., Azeez, M.A., Alao, M.B., Oke, M.A. and Aina, D.A. 2021. Mushroom nanobiotechnology: Concepts, developments and potentials. pp. 257–286. *In*: Lateef, A., Gueguim-Kana, E.B., Dasgupta, N. and Ranjan, S. (eds.). Microbial Nanobiotechnology—Principles and Application. Springer Nature Singapore Pte. Ltd.

Ahmad, A., Mukherjee, P., Mandal, D., Senapati, S., Khan, M.I., Kumar, R. and Sastry, M. 2002. Enzyme mediated extracellular synthesis of CdS nanoparticles by the fungus, *Fusarium oxysporum*. J. Am. Chem. Soc. 124: 12108e12109.

Ahmad, A., Mukherjee, P., Senapati, S., Mandal, D., Khan, M.I. and Kumar, R. 2003. Extracellular biosynthesis of silver nanoparticles using the fungus *Fusarium oxysporum*. Colloids Surf. B Biointerfaces 28: 313–318.

Alheety, M.A., Al-Jibori, S.A., Ali, A.H., Mahmood, A.R., Akbaş, H., Karadağ, A., Orhan, U. and Ahmed, M.H. 2019. Ag(I)-benzisothiazolinone complex: Synthesis, characterization, H$_2$ storage ability, nano transformation to different Ag nanostructures and Ag nanoflakes antimicrobial activity. Mater. Res. Express 6(12). https://doi.org/10.1088/2053-1591/ab5ab4.

Avila-Quezada, G.D., Golinska, P. and Rai, M. 2022. Engineered nanomaterials in plant diseases: Can we combat phytopathogens? Appl. Microbiol. Biotech. 106(1): 117–129. https://doi.org/10.1007/s00253-021-11725-w.

Awan, T.I., Tehseen, A. and Bashir, A. 2020. Introduction to nanomaterials. Chapter 3–25. Chemistry of Nanomaterials Fundamentals and Applications. 1st edn., Elsevier Science Publishing Co Inc.

Ayele, A., Mujmdar, R.S., Addisu, T. and Woinue, O. 2019. Green synthesis of silver nanoparticles for various biomedical and agro industrial application. J. Nanosci. Technol. 5: 694–698. https://doi.org/10.30799/jnst.233.19050211.

Aziz, N., Pandey, R., Barman, I. and Prasad, R. 2016. Leveraging the attributes of *Mucor hiemalis*-derived silver nanoparticles for a synergistic broad-spectrum antimicrobial platform. Front. Microbiol. 7: 1984. https://doi.org/10.3389/fmicb.2016.01984.

Balakumaran, M.D., Ramachandran, R., Balashanmugam, P., Mukeshkumar, D.J. and Kalaichelvan, P.T. 2016. Mycosynthesis of silver and gold nanoparticles: Optimization, characterization and antimicrobial activity against human pathogens. Microbiol. Res. 182: 8–20. Doi: 10.1016/j.micres.2015.09.009.

Balakumaran, M.D., Ramachandran, R. and Kalaichelvan, P.T. 2015. Exploitation of endophytic fungus, *Guignardia mangiferae* for extracellular synthesis of silver nanoparticles and their *in vitro* biological activities. Microbiol. Res. 178: 9–17. http://dx.doi.org/10.1016/j.micres.2015.05.009.

Ballottin, D., Fulaz, S., Cabrini, F., Tsukamoto, J., Oswaldo, N.D., Alves, L. and Tasic, I. 2017. Antimicrobial textiles: Biogenic silver nanoparticles against *Candida* and *Xanthomonas*. Mater. Sci. Eng. C: Mater. Biol. Appl. 75(1): 582–589.

Bansal, V., Ramanathan, R. and Bhargava, S.K. 2011. Fungus-mediated biological approaches towards 'Green' synthesis of oxide nanomaterials. Aust. J. Chem. 64: 279–293.

Bennur, T., Khan, Z., Kshirsagar, R., Javdekar, V. and Zinjarde, S. 2016. Biogenic gold nanoparticles from the *Actinomycete Gordoniaamarae*: Application in rapid sensing of copper ions. Sens. Actuators B: Chem. 233: 684–690.

Bhattacharyya, A., Duraisamy, P., Govindarajan, M., Buhroo, A.A. and Prasad, R. 2016. Nano biofungicides: Emerging trend in insect pest control. pp. 307–319. *In*: Prasad, R. (ed.). Advances and Applications through Fungal Nanobiotechnology. Springer International Publishing, Switzerland.

Binupriya, A.R., Sathishkumar, M. and Yun, S.I. 2010. Biocrystallization of silver and gold ions by inactive cell filtrate of *Rhizopus stolonifer*. Colloids Surf. B Biointerfaces 79: 531–534.

Birla, S.S., Tiwari, V.V., Gade, A.K., Ingle, A.P., Yadav, A.P. and Rai, M.K. 2009. Fabrication of silver nanoparticles by *Phoma glomerata* and its combined effect against *Escherichia coli, Pseudomonas aeruginosa* and *Staphylococcus aureus*. Lett. Appl. Microbiol. 48: 173–179.

Bourzama, G., Ouled-Haddar, H., Marrouche, M. and Aliouat, A. 2021. Iron uptake by fungi isolated from arcelor Mittal-Annaba—in the northeast of Algeria. Braz. J. Poult. Sci. 23(01). https://doi.org/10.1590/1806-9061-2020-1321.

Cao, G.Z. 2009. Nanostructures and Nanomaterials: Synthesis, Properties and Applications. Imperical College Press. London.

Castro-Longoria, E., Garibo-Ruiz, D. and Martínez-Castro, S. 2017. Myconanotechnology to treat infectious diseases: A perspective. Chapter 12 of Fungal Nanotechnology Applications in Agriculture, Industry, and Medicine. Springer, ISBN 978-3-319-68424-6 (eBook).

Chaudhari, P.R., Masurkar, S.A., Shidore, V.B. and Kamble, S.P. 2012. Antimicrobial activity of extracellular synthesized silver nanoparticles using Lactobacillus species obtained from VIZYLAG Capsule. J. Appl. Pharm. Sci. 02: 25–29.

Choi, Y. and Lee, S.Y. 2020. Biosynthesis of inorganic nanomaterials using microbial cells and bacteriophages. Nat. Rev. Chem. 4: 638–656.

Cornejo, P., Meier, S., Borie, G., Rillig, M.C. and Borie, F. 2008. Glomalin-related soil protein in a Mediterranean ecosystem affected by a copper smelter and its contribution to Cu and Zn sequestration. Sci. Total Environ. 406: 154–160.

Coyle, S., Wu, Y., Lau, K.T., Rossi, D.D., Wallace, G. and Diamond, D. 2007. Smart nanotextiles: A review of materials and applications. MRS Bull. 32: 434–442.

Cruz-Luna, A.R., Cruz-Martínez, H., Vásquez-López, A. and Medina, D. 2021. Metal nanoparticles as novel antifungal agents for sustainable agriculture: Current advances and future directions. J. Fungi 7: 1033. https://doi.org/10.3390/jof7121033.

Dar, M.A., Ingle, A. and Rai, M. 2013. Enhanced antimicrobial activity of silver nanoparticles synthesized by *Cryphonectria* sp. evaluated singly and in combination with antibiotics. Nanomedicine 9: 105–110.

Das, S.K., Das, A.R. and Guha, A.K. 2009. Gold nanoparticles: Microbial synthesis and application in water hygiene management. Langmuir 25: 8192–8199.

Das, S.K., Dickinson, C., Laffir, F., Brougham, D.F. and Marsili, E. 2012. Synthesis, characterization and catalytic activity of gold nanoparticles biosynthesized with *Rhizopus oryzae* protein extract. Green Chem. 14: 1322–1344.

Deepak, V., Kalishwaralal, K., Pandian, S.R.K. and Gurunathan, S. 2011. An insight into the bacterial biogenesis of silver nanoparticles, industrial production and scale-up. pp. 17–35. *In*: Rai, M. and Duran, N. (eds.). Metal Nanoparticles in Microbiology. Berlin: Springer-Verlag.

Deshmukh, S.P., Patil, S.M., Mullani, S.B. and Delekar, S.D. 2019. Silver nanoparticles as an effective disinfectant: A review. Mater. Sci. Eng. C 97: 954–965.

DeSimone, J.M. 2002. Practical approaches to green solvents. Science 297: 799–803.

Dobias, J. 2013. Nanoparticles and Microorganisms: from Synthesis to Toxicity. PhD thesis, Ecole Polythechnique Fédérale de Lausanne (EPFL), Doctoral Studies in the Environmental Program EDEN, Lausanne University, Swiss.

Durán, N. 2007. Antibacterial effect of silver nanoparticles produced by fungal process on textile fabrics and their effluent treatment. J. Biomed. Nanotechnol. 3(2): 203–208.

Elamawi, R.M., Al-Harbi, R.E. and Hendi, A.A. 2018. Biosynthesis and characterization of silver nanoparticles using *Trichoderma longibrachiatum* and their effect on phytopathogenic fungi. Egypt. J. Biol. Control 28: 28. https://doi.org/10.1186/s41938-018-0028-1.

Elavazhagan, T., Arunachalam, K.D. and Edule, M. 2011. Leaf extract mediated green synthesis of silver and gold nanoparticle. Int. J. Nanomed. 6: 1265–1278.

El-Bendary, M.A., Maysa, E.M., Hamed, S.R., Abo El-Ola, S.M., Khalil, S.K.H., Mounier, M.M., Roshdy, A.M. and Allam, M.A. 2021. Mycosynthesis of silver nanoparticles using *Aspergillus caespitosus*: Characterization, antimicrobial activities, cytotoxicity, and their performance as an antimicrobial agent for textile materials. Appl. Organomet. Chem. 35: e6338. https://doi.org/10.1002/aoc.6338.

El-Bendary, M.A., Moharam, M.E., Hamed, S.R., Khalil, S.K.H., Elkomy, G.M. and Mounier, M.M. 2018. Myco-synthesis of gold nanoparticles using *Aspergillus flavus*: Characterization, optimization and cytotoxic activity. Curr. Trends Microbiol. 12: 67–79.

Elgorban, A.M., Aref, S.M., Seham, S.M., Elhindi, K.M., Bahkali, A.H., Sayed, S.R. and Manal, M.A. 2016. Extracellular synthesis of silver nanoparticles using *Aspergillus versicolor* and evaluation of their activity on plant pathogenic fungi. Mycosphere 7: 844–852.

El Sayed, M.T. and El-Sayed, A.S.A. 2020. Biocidal activity of metal nanoparticles synthesized by *Fusarium solani* against multidrug-resistant bacteria and mycotoxigenic fungi. J. Microbiol. Biotechnol. 30: 226–236. https://doi.org/10.4014/jmb.1906.06070.

Fayaz, A.M., Balaji, K., Girilal, M., Yadav, R., Kalaichelvan, P.T. and Venketesan, R. 2010. Biogenic synthesis of silver nanoparticles and their synergistic effect with antibiotics: A study against gram-positive and gram-negative bacteria. Nanomedicine 6: 103–109.

Fouda, A., Mohamed, A.A., Elgamal, M.S., El-Din, H.S., Salem, S.S. and Shaheen, T.I. 2017. Facile approach towards medical textiles via myco-synthesis of silver nanoparticles. Der. Pharma Chemica 9: 11–18.

Gahlawat, G. and Choudhury, A.R. 2019. A review on the biosynthesis of metal and metal salt nanoparticles by microbes. RSC Adv. 9(23): 12944–12967.

Gaikwad, S.C., Birla, S.S., Ingle, A.P., Gade, A.K., Marcato, P.D., Rai, M.K. and Durán, N. 2013. Screening of different *Fusarium* species to select potential species for the synthesis of silver nanoparticles. J. Braz. Chem. Soc. 24: 1974–1982. https://doi.org/10.5935/0103-5053.20130247.

Gajbhiye, M., Kesharwani, J., Ingle, A., Gade, A. and Rai, M. 2009. Fungus-mediated synthesis of silver nanoparticles and their activity against pathogenic fungi in combination with fluconazole. Nanomed. Nanotechnol. Biol. Med. 5: 382–386.

Galdiero, S., Falanga, A., Vitiello, M., Cantisani, M., Marra, V. and Galdiero, M. 2011. Silver nanoparticles as potential antiviral agents. Molecules 4(16): 8894–8918.

Ge, L., Li, Q., Wang, M., Ouyang, J., Li, X. and Xing, M.M. 2014. Nanosilver particles in medical applications: Synthesis, performance and toxicity. Int. J. Nanomed. 9: 2399–2407.

Gokarneshan, N., Gopalakrishnan, P.P. and Jeyanthi, B. 2012. Influence of Nanofinishes on the Antimicrobial Properties of Fabrics. ISRN Nanomater. 2012, Article ID 193836. https://doi.org/10.5402/2012/193836.

Gokarnshan, N., Nagarajan, V.B. and Viswanath, S.R. 2017. Developments in antimicrobial textiles-some insights on current research trends. Biomed. J. Sci. Tech. Res. 1: 230–234. https://doi.org/ 10.26717/BJSTR.2017.01.000160.

Govender, Y., Riddin, T., Gericke, M. and Whiteley, C.G. 2009. Bioreduction of platinum salts into nanoparticles: A mechanistic perspective. Biotechnol. Lett. 31: 95–100.

Guilger-Casagrande, M. and de Lima, R. 2019. Synthesis of silver nanoparticles mediated by fungi: A review. Front. Bioeng. Biotechnol. 22. https://doi.org/10.3389/fbioe.2019.00287.

Gupta, K. and Chundawat, T.S. 2019. Bio-inspired synthesis of platinum nanoparticles from fungus *Fusarium oxysporum*: Its characteristics, potential antimicrobial, antioxidant and photocatalytic activities. Mater. Res. Express 6: 1–10. Doi: 10.1088/2053-1591/ab4219.

Halkai, K.R., Mudda, J.A., Shivanna, V., Rathod, V. and Halkai, R. 2018. Antibacterial efficacy of biosynthesized silver nanoparticles against *Enterococcus faecalis* biofilm: An *in vitro* study. Contemp. Clin. Dent. 9: 237–241.

Harholdt, K. 2003. Carbon fiber, past and future. Ind. Fabric Prod. Rev. 88(4): 14–28.

Ibrahim, H.M.M. and Hassan, M.S. 2016. Characterization and antimicrobial properties of cotton fabric loaded with green synthesized silver nanoparticles. Carbohydr. Polym. 151: 841–850.

Ishida, K., Cipriano, T.F., Rocha, G.M., Weissmüller, G., Gomes, F., Miranda, K., Rozental, S., Cruz, M. and de Janeiro, R. 2014. Silver nanoparticle production by the fungus *Fusarium oxysporum*: Nanoparticle characterisation and analysis of antifungal activity against pathogenic yeasts. Mem. Inst. Oswaldo Cruz. 109: 220–228. https://doi.org/10.1590/0074-0276130269.

Jha, A.K. and Prasad, K. 2016. Understanding mechanism of fungus mediated nanosynthesis: A molecular approach. pp. 1–22. *In*: Prasad, R. (ed.). Advances and Applications through Fungal Nanobiotechnology. Springer, Cham.

Joshi, M. and Bhattacharyya, A. 2011. Nanotechnology—A new route to high performance functional textiles. Text. Prog. 43(3): 155–233.

Khandel, P. and Shahi, S.K. 2018. Mycogenic nanoparticles and their bio-prospective applications: Current status and future challenges. J. Nanostructure Chem. 8: 369–391. https://doi.org/10.1007/s40097-018-0285-2.

Kim, J.S., Kuk, E., Yu, K.N., Kim, J.H., Park, S.J., Lee, H.J., Kim, S.H. and Park, Y.K. 2007. Antimicrobial effects of silver nanoparticles. Nanomed. Nanotechnol. 3: 95–101.

Krishna, G., Swetha, G. and Charya, M.A.S. 2018. Bio Fabrication of silver nanoparticles using white rot fungi and their antibacterial efficacy. Glob. J. Nanomed. 4(2): GJO.MS.ID.555635. https://doi.org/10.19080/GJN.2018.04.555635.

Landage, S.M. and Wasif, A.I. 2012. Nanosilver—An effective antimicrobial agent for finishing of textiles. Int. J. Eng. Sci. Emerg. Technol. 4: 66–78.

Li, G., He, D., Qian, Y., Guan, B., Gao, S., Cui, Y., Yokoyama, K. and Wang, L. 2012. Fungus-mediated green synthesis of silver nanoparticles using *Aspergillus terreus*. Int. J. Mol. Sci. 13(1): 466–476.

Li, P., Wang, Y., Huang, H., Ma, S., Yang, H. and Zhen-liang, X. 2021. High efficient reduction of 4-nitrophenol and dye by filtration through Ag NPs coated PAN-Si catalytic membrane. Chemosphere 263: 127995.

Li, X., Xu, H., Chen, Z.S. and Chen, G. 2011. Biosynthesis of nanoparticles by microorganisms and their applications. J. Nanomater. Vol. 2011, Article ID 270974. https://doi.org/10.1155/2011/270974.

Liu, X., Chen, J.L., Yang, W.Y., Qian, Y.C., Pan, J.Y., Zhu, C.N., Liu, L., Ou, W.B., Zhao, H.X. and Zhang, D.P. 2021. Biosynthesis of silver nanoparticles with antimicrobial and anticancer properties using two novel yeasts. Sci. Rep. 11: 15795. | https://doi.org/10.1038/s41598-021-95262-6.

Lysenko, V., Lozovski, V., Lokshyn, M., Gomeniuk, Y.V., Dorovskih, A., Rusinchuk, N., Pankivska, Y., Povnitsa, O., Zagorodnya, S., Tertykh, V. and Bolbukh, Y. (2018). Nanoparticles as antiviral agents against adenoviruses. Adv. Nat. Sci.: Nanosci. Nanotechnol. 9(2): 025021. Doi: 10.1088/2043-6254/aac42a.

Madakka, M., Jayaraju, N. and Rajesh, N. 2018. Mycosynthesis of silver nanoparticles and their characterization. Methods 5: 20–29. https://doi.org/10.1016/j.mex.2017.12.003.

Majeed, S., bin Abdullah, M.S., Dash, G.K., Ansari, M.T. and Nanda, A. 2016. Biochemical synthesis of silver nanoparticles using filamentous fungi *Penicillium decumbens* (MTCC-2494) and its efficacy against A-549 lung cancer cell line. Chin. J. Nat. Med. 14: 615–620.

Market analysis report. 2020. Nanomaterials market size, share & trends analysis report by product (carbon nanotubes, titanium dioxide), by application (medical, electronics, paints & coatings), by region, and segment forecasts, 2020–2027, Report ID: GVR-4-68038-565-6.

Mehravani, B. Ana, Isabel Ribeiro, A.I. and Zille, A. 2021. Gold nanoparticles synthesis and antimicrobial effect on fibrous materials. Nanomaterials 11(5): 1067. https://doi.org/10.3390/nano11051067.

Menon, S., Rajeshkumar, S. and Venkat Kumar, S. 2017. A review on biogenic synthesis of gold nanoparticles, characterization, and its applications. Resource-Efficient Technologies 3: 516–527.

Mikhailov, O.V. and Mikhailova, E.O. 2019. Elemental silver nanoparticles: Biosynthesis and bio applications. Mater. (Basel, Switzerland) 12(19): 3177. https://doi.org/10.3390/ma12193177.

Mohamed, Y.M., Azzam, A.M., Amin, B.H. and Safwat, N.A. 2015. Mycosynthesis of iron nanoparticles by *Alternaria alternata* and its antibacterial activity. Afr. J. Biotechnol. 14: 1234–1241.

Mohmed, A.A., Fouda, A., Elgamal, M.S., Hassan, S.E.D., Shaheen, T.I. and Salem, S.S. 2017. Enhancing of cotton fabric antibacterial properties by silver nanoparticles synthesized by New Egyptian strain Fusarium Keratoplasticum A1-3. Egypt. J. Chem. 60: 63–71.

Mughal, B., Zaidi, S.Z.J., Zhang, X. and Hassan, S.U. 2021. Biogenic nanoparticles: Synthesis, characterisation and applications. Appl. Sci. 11(6): 2598. https://doi.org/10.3390/app11062598.

Murugan, K., Aruna, P., Panneerselvam, C., Madhiyazhagan, P.M., Paulpandi, M., Subramaniam, J., Rajaganesh, R., Wei, H., Alsalhi, M.S., Devanesan, S., Nicoletti, M., Syuhei, B., Canale, A. and Benelli, G. 2015. Fighting arboviral diseases: Low toxicity on mammalian cells, dengue growth inhibition (*in vitro*), and mosquitocidal activity of *Centroceras clavulatum*—synthesized silver nanoparticles. Parasitol. Res. 015: 4783–4786.

Musarrat, J., Dwivedi, S., Singh, B.R., Al-Khedhairy, A.A., Azam, A. and Naqvi, A. 2010. Production of antimicrobial silver nanoparticles in water extracts of the fungus *Amylomyces rouxii* strain KSU-09. Bioresour. Technol. 101: 8772–8776.

Nadaf, N.Y. and Kanase, S.S. 2016. Biosynthesis of gold nanoparticles by *Bacillus marisflavi* and its potential in catalytic dye degradation. Arab. J. Chem. 12: 4806–4814.

Namasivayam, S.K.R. and Avimanyui. 2011. Silver nanoparticle synthesis from *Lecanicillium lecanii* and evolutionary treatment on cotton fabrics by measuring their improved antibacterial activity with antibiotics against *Staphylococcus aureus* (ATCC 29213) and *E. coli* (ATCC 25922) strains. Int. J. Pharm. Pharm. Sci. 1(4): 190–195.

Othman, A.M., Elsayed, M.A., Al-Balakocy, N.G., Mohamed, M., Hassan, M.M. and Elshafei, A.M. 2019. Biosynthesis and characterization of silver nanoparticles induced by fungal proteins and its application in different biological activities. J. Genet. Eng. Biotechnol. 17(1): 8. https://doi.org/10.1186/s43141-019-0008-1.

Ottoni, C.A., Simoes, M.F., Fernandes, S., dos Santos, J.M., da Silva, E.S., de Souza, R.F.B. and Maiorano, A.E. 2017. Screening of filamentous fungi for antimicrobial silver nanoparticles synthesis. AMB Express 7: 31.

Panchangam, R.L. and Upputuri, R.T.P. 2019. *In vitro* biological activities of silver nanoparticles synthesized from *Scedosporium* sp. isolated from soil. Braz. J. Pharm. Sci. 55: e00254. http://dx.doi.org/10.1590/s2175-97902019000200254.

Pantidos, N. and Horsfall, L.E. 2014. Biological synthesis of metallic nanoparticles by bacteria, fungi and plants. J. Nanomed. Nanotechnol. 5: 5–14.

Park, T.J., Lee, K.G. and Lee, S.Y. 2016. Advances in microbial biosynthesis of metal nanoparticles. Appl. Microbiol. Biotechnol. 100: 521–534.

Paul, S., Singh, A.R. and Sasikumar, C.S. 2015. Green synthesis of bio-silver nanoparticles by *Parmelia perlata, Ganoderma lucidum* and *Phellinus igniarius* & their fields of application. Ind. J. Res. Pharm. Biotechnol. 3(2): 100–110.

Pourzahedi, L. and Eckelman, M.J. 2015. Comparative life cycle assessment of silver nanoparticle synthesis routes. Environ. Sci. Nano. 2: 361–369.

Prasad, R. 2016. Advances and Applications through Fungal Nanobiotechnology. Springer, International Publishing, Cham.

Prasad, R., Kumar, V. and Prasad, K.S. 2014. Nanotechnology in sustainable agriculture: Present concerns and future aspects. Afr. J. Biotechnol. 13(6): 705–713.

Prasad, R., Pandey, R. and Barman, I. 2016. Engineering tailored nanoparticles with microbes: Quo vadis. WIREs Nanomed. Nanobiotechnol. 8: 316–330. https://doi.org/10.1002/wnan.1363.

Prasad, R., Bhattacharyya, A. and Nguyen, Q.D. 2017. Nanotechnology in sustainable agriculture: Recent developments, challenges, and perspectives. Front. Microbiol. 20; 8: 1014. https://doi.org/10.3389/fmicb.2017.01014.

Qian, L. and Hinestroza, J.P. 2004. Application of nanotechnology for high performance textiles. J. Text. Appar. Technol. Manag. 4(1): 1–7.

Qian, Y., Yu, H., He, D., Yang, H., Wang, W., Wan, X. and Wang, L. 2013. Biosynthesis of silver nanoparticles by the endophytic fungus *Epicoccum nigrum* and their activity against pathogenic fungi. Bioprocess Biosyst. Eng. 36(11): 1613–1619. https://doi.org/10.1007/s00449-013-0937-z.

Rai, M., Yadav, A., Bridge, P. and Gade, A. 2009. Myconanotechnology: A new and emerging science. pp. 258–267. *In*: Rai, M. and Bridge, P. (eds.). Applied Mycology, Chapter 14, CAB International.

Rai, M., Bonde, S., Golinska, P., Trzcińska-Wencel, J., Gade, A., Abd-Elsalam, K.A., Shende, S., Gaikwad, S. and Ingle, A.P. 2021. *Fusarium* as a novel fungus for the synthesis of nanoparticles: Mechanism and applications. J. Fungi. 15; 7(2): 139. https://doi.org/10.3390/jof7020139.

Rajakumar, G., Rahuman, A.A., Roopan, S.M., Khanna, V.G., Elango, G., Kamaraj, C., Abduz Zahir, A. and Velayutham, K. 2012. Fungus-mediated biosynthesis and characterization of TiO_2 nanoparticles and their activity against pathogenic bacteria. Spectrochim. Acta A Mol. Biomol. Spectrosc. 91: 23–29.

Rajarathinam, M. and Kalaichelvan, P.T. 2013. Biogenic nanosilver as a potential antibacterial and antifungal additive to commercially available dish wash and hand wash for an enhanced antibacterial and antifungal activity against selected pathogenic strains. Int. Res. J. Pharm. 4: 68–75. https://doi.org/10.7897/2230-8407.04715.

Rajendran, R., Balakumar, C., Ahammed, H., Jayakumar, S., Vaideki, K. and Rajesh, E.M. 2010. Use of zinc oxide nano particles for production of antimicrobial textiles. Int. J. Eng. Sci. Technol. 2(1): 202–208.

Rajski, L., Juda, M., Los, A., Witun, E. and Malm, A. 2019. Medical textiles with silver/nanosilver and their potential application for the prevention and control of healthcare-associated infections—mini-review. Curr. Issues Pharm. Med. Sci. 32: 104–107. https://doi.org/10.2478/cipms-2019-0020.

Rathod, V. and Ranganath, E. 2011. Synthesis of monodispersed silver nanoparticles by *Rhizopus stolonifer* and its antibacterial activity against MDR strains of *Pseudomonas aeruginosa* from burnt patients. Int. J. Environ. Sci. 1(7): 1830–1840.

Raza, Z.A., Bilal, U., Noreen, U., Munim, S.A., Riaz, S., Abdullah, M.U. and Abid, S. 2019. Chitosan mediated formation and impregnation of silver nanoparticles on viscose fabric in single bath for antibacterial performance. Fibers Polym. 20: 1360–1367.

Repon, M.R., Islam, T., Sadia, H.T., Mikucioniene, D., Hossain, S., Kibria, G. and Kaseem, M. 2021. Development of antimicrobial cotton fabric impregnating Ag nanoparticles utilizing contemporary practice Md. Reazuddin. Coatings 11: 1413. https://doi.org/10.3390/coatings11111413.

Rodrigues, A.G., Oliveira Gonçalves, P.J.R., Ottoni, C.A., Ruiz, R.C., Morgano, M.A., Araújo, W.L., Melo, I.S. and De Souza, A.O. 2019. Functional textiles impregnated with biogenic silver nanoparticles from *Bionectria ochroleuca* and its antimicrobial activity. Biomed. Microdevices 21(3): 56. https://doi.org/10.1007/s10544-019-0410-0.

Saha, S., Sarkar, J., Chattopadhyay, D., Patra, S., Chakraborty, A. and Acharya, K 2010. Production of silver nanoparticles by a phytopathogenic fungus *Bipolaris nodulosa* and its antimicrobial activity. Dig. J. Nanomater. Biostruc. 5(4): 887–895.

Saleem, H. and Zaidi, S.J. 2020. Sustainable use of nanomaterials in textiles and their environmental impact. Materials 13: 5134. https://doi.org/10.3390/ma1322513.

Salem, S.S. and Fouda, A. 2020. Green synthesis of metallic nanoparticles and their prospective biotechnological applications: An overview. Biol. Trace Elem. Res. 199: 344. https://doi.org/10.1007/s12011-020-02138-3.

Salleh, A., Naomi, R., Utami, N.D., Mohammad, A.W., Mahmoudi, E., Mustafa, N. and Fauzi, M.B. (2020). The potential of silver nanoparticles for antiviral and antibacterial applications: A mechanism of action. Nanomaterials 10: 1566. https://doi.org/10.3390/nano10081566.

Salunke, B.K., Sawant, S.S., Lee, S-I. and Kim, B.S. 2016. Microorganisms as efficient biosystem for the synthesis of metal nanoparticles: Current scenario and future possibilities. World J. Microbiol. Biotechnol. 32: 88. https://doi.org/10.1007/s11274-016-2044-1.

Sarkar, J., Chattopadhyay, D., Patra, S. and ET, A.L. 2011. *Alternaria alternata* mediated synthesis of protein capped silver nanoparticles and their genotoxic activity. Dig. J. Nanomat. Biostruct. 6(2): 563–573.

Sastry, M., Ahmad, A., Islam Khan, M. and Kumar, R. 2003. Biosynthesis of metal nanoparticles using fungi and actinomycetes. Curr. Sci. 85(2): 162–170.

Sehulster, L.M. 2015. Healthcare laundry and textiles in the united states: Review and commentary on contemporary infection prevention issues. Infect. Control Hosp. Epidemiol. 36(09): 1073–1088.

Shaheen, T.I. and Abd El Aty, A.A. 2018. *In-situ* green myco-synthesis of silver nanoparticles onto cotton fabrics for broad spectrum antimicrobial activity. Int. J. Biol. Macromol. 118(Pt B): 2121–2130. https://doi.org/10.1016/j.ijbiomac.2018.07.062.

Shaheen, T.I., Fouda, A. and Salem, S.S. 2021. Integration of cotton fabrics with biosynthesized CuO nanoparticles for bactericidal activity in the terms of their cytotoxicity assessment. Ind. Eng. Chem. Res. 60(4): 1553–1563. https://doi.org/10.1021/acs.iecr.0c04880.

Shaikh, S., Nazam, N., Rizvi, S.M.D., Ahmad, K., Baig, M.H., Lee, E.J. and Choi, I. 2019. Mechanistic insights into the antimicrobial actions of metallic nanoparticles and their implications for multidrug resistance. Int. J. Mol. Sci. 20: 2468. https://doi.org/10.3390/ijms20102468.

Sharma, V.K., Yngard, R.A. and Lin Y. 2009. Silver nanoparticles: Green synthesis and their antimicrobial activities. Adv. Colloid Interface Sci. 145: 83–96.

Shedbalkar, U., Singh, R., Wadhwani, S., Gaidhani, S. and Chopade, B.A. 2014. Microbial synthesis of gold nanoparticles: Current status and future prospects. Adv. Colloid Interface Sci. 209: 40. https://doi.org/10.1016/j.cis.2013.12.011.

Shojaei, K.M., Farrahi, A., Farrahi, H. and Farrahi, A. 2015. The stabilization of nanosilver on polyester filament for a machine-made carpet. Mater. Technol. 49: 461–464.

Siddiqi, K.S., Husen, A. and Rao, R.A.K. 2018. A review on biosynthesis of silver nanoparticles and their biocidal properties. J. Nanobiotechnol. 16(1): 14. https://doi.org/10.1186/s12951-018-0334-5.

Singh, I. 2019. Biosynthesis of silver nanoparticle from fungi, algae and bacteria. Eur. J. Biol. Res. 9: 45. https://doi.org/10.5281/ zenodo.2617168.

Sobha, K., Surendranath, K. and Meena, V. 2010. Emerging trends in nanobiotechnology. Biotechnol. Mol. Biol. Rev. 5: 001–012.

Srikar, S.K., Giri, D.D., Pal, D.B., Mishra, P.K. and Upadhyay, S.N. 2016. Green synthesis of silver nanoparticles: A review. Green Sustain. Chem. 6: 34–56. https://doi.org/10.4236/gsc.2016.61004.

Srivastava, S. and Bhargava, A. 2022. Green Nanoparticles: The Future of Nanobiotechnology, Springer Nature Singapore Pte Ltd, pp. 101–126. https://doi.org/10.1007/978-981-16-7106-7_6.

Thakker, J.N., Dalwadi, P. and Dhandhukia, P.C. 2013. Biosynthesis of gold nanoparticles using *Fusarium oxysporum* f. Sp. *cubense* JT1, a plant pathogenic fungus. Biotechnol. 2013: 1–5.

Toma, H.E., Zamarion, V.M., Toma, S.H. and Araki, K. 2010. The coordination chemistry at gold nanoparticles. J. Braz. Chem. Soc. 21: 1158–1176.

Wagener, S., Dommershausen, N., Jungnickel, H., Laux, P., Mitrano, D., Nowack, B., Schneider, G. and Luch, A. 2016. Textile functionalization and its effects on the release of silver nanoparticles into artificial sweat. Environ. Sci. Technol. 50: 5927–5934.

Wang, D., Xue, B., Wang, L., Zhang, Y., Liu, L. and Zhou, Y. 2021. Fungus-mediated green synthesis of nano-silver using *Aspergillus sydowii* and its antifungal/antiproliferative activities. Sci. Rep. 11: 10356. https://doi.org/10.1038/s41598-021-89854-5.

Wong, Y.W.H., Yuen, C.W.M., Leung, M.Y.S., Ku, S.K.A. and Lam, H.L.I. 2006. Selected applications of nanotechnology in textiles. Autex Res. J. 6(1). http://www.autexrj.org/No1-2006/0191.pdf 1.

Xue, B., He, D., Gao, S., Wang, D., Yokoyama, K. and Wang, L. 2016. Biosynthesis of silver nanoparticles by the fungus *Arthroderma fulvum* and its antifungal activity against genera of *Candida*, *Aspergillus* and *Fusarium*. Int. J. Nanomed. 11: 1899–1906.

Zhang, X., He, X., Wang, K. and Yang, X. 2011. Different active biomolecules involved in biosynthesis of gold nanoparticles by three fungus species. J. Biomed. Nanotechnol. 7: 245–254.

Index

Biography of Editors

Dr. Mahendra Rai is a Senior Professor and UGC-Basic Science Research Faculty at the Department of Biotechnology, Sant Gadge Baba Amravati University, Maharashtra, India. Presently, he is a Visiting Professor at the Department of Microbiology, Nicolaus Copernicus University, Poland. He was Visiting Scientist at University of Geneva; Debrecen University, Hungary; University of Campinas, Brazil; Nicolaus Copernicus University, Poland, VSB Technical University of Ostrava, Czech Republic; National University of Rosario, Argentina; and Federal University of Piaui, Teresina, Brazil. Dr. Rai has published more than 425 research papers in national and international journals with 70 h-index. In addition, he has edited/authored more than 69 books and 6 patents. The main focus of his research is Myconanotechnology and its applications in medicine and sustainable agriculture.

Dr. Patrycja Golińska, currently works at the Department of Microbiology at Nicolaus Copernicus University (NCU), Poland as an Associate Professor. She received a fellowship of the Marshal of Kuyavian-Pomeranian Voivodeship for the best doctoral students in 2006 and from the NCU in 2011 to complete a one-year postdoctoral internship at the School of Biology, Newcastle University, UK. She has 14 years of teaching and 18 years of research experience. Dr. Golińska's scientific contribution encompasses 44 original and 14 review articles, 13 book chapters and edition of one book. She received a grant under the Horizon Europe programme of European Commission. The main focus of Dr. Golińska's research is the biosynthesis of metal nanoparticles, mainly by actinobacteria and fungi, and their use to combat bacterial and fungal pathogens, the study of actinobacterial diversity in extreme biomes and the use of actinobacteria as biocontrol agent of fungal phytopathogens.

For Product Safety Concerns and Information please contact our EU
representative GPSR@taylorandfrancis.com
Taylor & Francis Verlag GmbH, Kaufingerstraße 24, 80331 München, Germany